Studies in Computational Intelligence

Volume 1054

The series "Studies in Computational Intelligence" (SCI) publishes new developments and advances in the various areas of computational intelligence—quickly and with a high quality. The intent is to cover the theory, applications, and design methods of computational intelligence, as embedded in the fields of engineering, computer science, physics and life sciences, as well as the methodologies behind them. The series contains monographs, lecture notes and edited volumes in computational intelligence spanning the areas of neural networks, connectionist systems, genetic algorithms, evolutionary computation, artificial intelligence, cellular automata, self-organizing systems, soft computing, fuzzy systems, and hybrid intelligent systems. Of particular value to both the contributors and the readership are the short publication timeframe and the world-wide distribution, which enable both wide and rapid dissemination of research output.

This series also publishes Open Access books. A recent example is the book Swan, Nivel, Kant, Hedges, Atkinson, Steunebrink: The Road to General Intelligence https://link.springer.com/book/10.1007/978-3-031-08020-3

Indexed by SCOPUS, DBLP, WTI Frankfurt eG, zbMATH, SCImago.

All books published in the series are submitted for consideration in Web of Science.

Anupam Biswas · Can B. Kalayci ·
Seyedali Mirjalili
Editors

Advances in Swarm Intelligence

Variations and Adaptations for Optimization Problems

Springer

Editors
Anupam Biswas
Department of Computer Science
and Engineering
National Institute Of Technology Silchar
Cachar, Assam, India

Seyedali Mirjalili
Centre for Artificial Intelligence Research
and Optimisation
Torrens University Australia
Brisbane, QLD, Australia

University Research and Innovation Center
Obuda University
Budapest, Hungary

Can B. Kalayci
Department of Industrial Engineering
Pamukkale University
Pamukkale, Turkey

ISSN 1860-949X ISSN 1860-9503 (electronic)
Studies in Computational Intelligence
ISBN 978-3-031-09834-5 ISBN 978-3-031-09835-2 (eBook)
https://doi.org/10.1007/978-3-031-09835-2

This Springer imprint is published by the registered company Springer Nature Switzerland AG
The registered company address is: Gewerbestrasse 11, 6330 Cham, Switzerland

Preface

Swarm Intelligence (SI) has grown significantly, both from the perspective of algorithmic development and applications covering almost all disciplines in both science and technology. Generally speaking, the algorithms that come under SI domain are typically population-based meta-heuristics techniques and are mostly inspired by nature. This book strives to cover all the major SI techniques, including particle swarm optimization, ant colony optimization, firefly algorithm, dragonfly algorithm, cuckoo search, and many more. Different variants of such SI techniques have been developed depending on applications and often hybridized with other techniques to improve performance. This also includes some of the variants and hybridized versions of popularly used SI techniques.

On the application front, SI techniques are widely used in optimization problems in different domains, including but not limited to Electrical and Power Systems, Electronics and Communication Engineering, Machine Learning, Deep Learning, Social Network Analysis, Pattern Recognition, Speech Processing, Image Processing, Bioinformatics, Health Informatics, Manufacturing and Operation Research, just to name a few. This book comprises some of the major applications of SI techniques in different domains as mentioned above. The book focuses on the adaptive nature of SI techniques from the context of applications. For specific problems, how representation of population is done, how objective function is designed, what kind of changes are done in SI technique itself, how to deal with constraints, how to manage conflicting objectives, all these issues are addressed in the book from the view point of applications.

This book emphasizes the studies of existing SI techniques, their variants and applications. The book also contains reviews of new developments in SI techniques and hybridizations. Algorithm-specific studies covering basic introduction and analysis of key components of these algorithms, such as convergence, balance of solution accuracy, computational costs, tuning, and control of parameters. Application-specific studies incorporating the ways of designing objective functions, solution representation, and constraint handling. The book also includes studies on application domain-specific adaptations in the SI techniques.

The book is organized into four parts. Part I covers state of the art in SI, which includes reviews on SI techniques, various optimization problems, and performance analysis of SI techniques. Part II covers applications of SI techniques in various engineering problems. Part III comprises applications of SI techniques in different machine learning problems such as clustering and prediction. The last part includes several other applications of SI techniques.

We cordially thank all the authors for their valuable contributions. We also thank the reviewers for their input and valuable suggestions.

Finally, we thank all the stakeholders who have contributed directly or indirectly to making this book a success.

Silchar, India — Anupam Biswas
Pamukkale, Turkey — Can B. Kalayci
Brisbane, Australia — Seyedali Mirjalili
March 2022

Contents

State-of-the-Art

A Brief Tutorial on Optimization Problems, Optimization Algorithms, Meta-Heuristics, and Swarm Intelligence 3
Seyedali Mirjalili, Can B. Kalayci, and Anupam Biswas

Introductory Review of Swarm Intelligence Techniques 15
Thounaojam Chinglemba, Soujanyo Biswas, Debashish Malakar, Vivek Meena, Debojyoti Sarkar, and Anupam Biswas

Swarm Intelligence for Deep Learning: Concepts, Challenges and Recent Trends 37
Vandana Bharti, Bhaskar Biswas, and Kaushal Kumar Shukla

Advances on Particle Swarm Optimization in Solving Discrete Optimization Problems 59
M. A. H. Akhand, Md. Masudur Rahman, and Nazmul Siddique

Performance Analysis of Hybrid Memory Based Dragonfly Algorithm in Engineering Problems 89
Sanjoy Debnath, Ravi Singh Kurmvanshi, and Wasim Arif

Engineering Problems

Optimum Design and Tuning Applications in Structural Engineering via Swarm Intelligence 109
Gebrail Bekdaş, Sinan Melih Nigdeli, and Aylin Ece Kayabekir

Bee Colony Optimization with Applications in Transportation Engineering 135
Dušan Teodorović, Miloš Nikolić, Milica Šelmić, and Ivana Jovanović

Application of Swarm Based Approaches for Elastic Modulus Prediction of Recycled Aggregate Concrete 153
Harish Narayana and Prashanth Janardhan

Grey Wolf Optimizer, Whale Optimization Algorithm, and Moth Flame Optimization for Optimizing Photonics Crystals 169
Seyed Mohammad Mirjalili, Seyedeh Zahra Mirjalili, Nima Khodadadi, Vaclav Snasel, and Seyedali Mirjalili

Intelligent and Reliable Cognitive 5G Networks Using Whale Optimization Techniques 181
Jasiya Bashir, Javaid Ahmad Sheikh, and Zahid A. Bhat

Machine Learning

Automatic Data Clustering Using Farmland Fertility Metaheuristic Algorithm 199
Farhad Soleimanian Gharehchopogh and Human Shayanfar

A Comprehensive Review of the Firefly Algorithms for Data Clustering 217
MKA Ariyaratne and TGI Fernando

A Hybrid African Vulture Optimization Algorithm and Harmony Search: Algorithm and Application in Clustering 241
Farhad Soleimanian Gharehchopogh, Benyamin Abdollahzadeh, Nima Khodadadi, and Seyedali Mirjalili

Estimation Models for Optimum Design of Structural Engineering Problems via Swarm-Intelligence Based Algorithms and Artificial Neural Networks 255
Melda Yücel, Sinan Melih Nigdeli, and Gebrail Bekdaş

A Novel Codebook Generation by Lévy Flight Based Firefly Algorithm 269
Ilker Kilic

Novel Chaotic Best Firefly Algorithm: COVID-19 Fake News Detection Application 285
Miodrag Zivkovic, Aleksandar Petrovic, K. Venkatachalam, Ivana Strumberger, Hothefa Shaker Jassim, and Nebojsa Bacanin

Other Applications

Artificial Bee Colony and Genetic Algorithms for Parameters Estimation of Weibull Distribution 309
Muhammet Burak Kılıç

Graph Structure Optimization for Agent Control Problems Using ACO 327
Mohamad Roshanzamir, Mahdi Roshanzamir, Navid Hoseini Izadi, and Maziar Palhang

A Bumble Bees Mating Optimization Algorithm for the Discrete and Dynamic Berth Allocation Problem 347
Eleftherios Tsakirakis, Magdalene Marinaki, and Yannis Marinakis

Applying the Population-Based Ant Colony Optimization to the Dynamic Vehicle Routing Problem 369
Michalis Mavrovouniotis, Georgios Ellinas, Iaê S. Bonilha, Felipe M. Müller, and Marios Polycarpou

An Improved Cuckoo Search Algorithm for the Capacitated Green Vehicle Routing Problem 385
Kenan Karagul and Yusuf Sahin

Multi-Objective Artificial Hummingbird Algorithm 407
Nima Khodadadi, Seyed Mohammad Mirjalili, Weiguo Zhao, Zhenxing Zhang, Liying Wang, and Seyedali Mirjalili

State-of-the-Art

A Brief Tutorial on Optimization Problems, Optimization Algorithms, Meta-Heuristics, and Swarm Intelligence

Seyedali Mirjalili, Can B. Kalayci, and Anupam Biswas

Abstract This chapter provides preliminaries and essential definitions in optimization, meta-heuristics, and swarm intelligence. It starts with different components of optimization problems, formulations, and categories. Conventional and recent optimization algorithms to optimize such problems are then discussed. The chapter is finished by focusing on meta-heuristics and swarm intelligence algorithms as the emerging and most widely used optimization algorithms lately in both science and industry.

1 Optimization Problems

The pace of changes in emerging technologies has led to ever-increasing complexity in the problem that we solve. Among these, optimization problems can be widely found in different fields. Scientists and practitioners around the world constantly try to design and build models, processes, and procedures that are more efficient, less costly, and more reliable. Despite the differences in optimization problems, they have the common components:

S. Mirjalili (✉)
Center for Artificial Intelligence Research and Optimization, Torrens University, Brisbane, QLD, Australia
e-mail: ali.mirjalili@gmail.com

Yonsei Frontier Lab, Yonsei University, Seoul, South Korea

University Research and Innovation Center, Obuda University, 1034 Budapest, Hungary

C. B. Kalayci
Department of Industrial Engineering, Pamukkale University, Denizli, Turkey
e-mail: cbkalayci@pau.edu.tr

A. Biswas
Department of Computer Science and Engineering, National Institute of Technology Silchar, Assam, India
e-mail: anupam@cse.nits.ac.in

A. Biswas et al. (eds.), *Advances in Swarm Intelligence*, Studies in Computational Intelligence 1054, https://doi.org/10.1007/978-3-031-09835-2_1

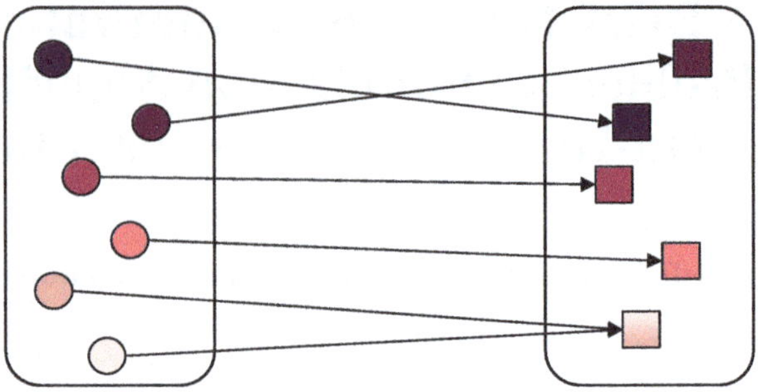

Fig. 1 An objective function assigns to each member of set X exactly one member in the set of objectives

- Decision variables: x_i
- Objective functions: $f(x)$
- Constraints: $g(x)$, $h(x)$

Decision variables are unknown to the problem, which serve as the inputs of the objective functions. Some examples of decision variables are several units, sequence of tasks, length, etc. Decision variables (x_i) can be considered as independent parameters of optimization problems. A dependent parameter is calculated using an objective function (f). In other words, an objective function takes decision variables as the input and return objective values as the output as shown in Fig. 1. It can be seen that an objective function assigns to each member of set X exactly one member in a set of objectives.

Due to the special rules of functions, an objective function work for every possible input value for the decision variables. Each solution is related to exactly one (one-to-many is not allowed but many-to-one is allowed).

Constraints are presented and applied to the decision variables using functions too. When solving an optimization problem, the goal is to find the best values for the decision variables to minimize or maximize the objectives while not violating the constraints. An optimization problem can be formulated with decision variables, objectives, and constraints. Note that in the following paragraphs, a maximization problem is considered, but the same concepts are applied to minimization problems too.

$$\text{Max:} f_i(x) \qquad \text{for i} = 1, 2, \ldots, \text{I} \tag{1}$$

$$\text{Subject to:} g_j(x) \geq c_j \qquad \text{for j} = 1, 2, \ldots, \text{J} \tag{2}$$

$$h_k(x) = d_k \qquad \text{for k} = 1, 2, \ldots, \text{K} \tag{3}$$

where f_i shows the *ith* objective function, g_j indicates the *jth* inequality constraint, h_k is the *kth* equality constraint, c_j is a constraint for *jth* in equality constraint, d_k is a constant for *kth* equality constraint, I shows the number of objectives J is the number of inequality constraints, and K is the number of equality constraints.

Problem formulation is an important step for any optimization problem, but the above generic formulation shows that the three components of decision variables, objectives, and constraints are common between all of them. To better understand this process, we can take a look at an example:

$$\text{Max: } f(x) = -x^2 - 10 \sin \sin(1.5x) \tag{4}$$

$$\text{Subject to: } g(x) = x + 2 \geq 0 \tag{5}$$

$$\text{Where: } -5 \leq x \leq 5 \tag{6}$$

The shape of this function is shown in Fig. 2. The maximum is when x is approximately 1. There is also a local maximum when $x \cong 2.8$.

The same principle is applied to optimization with more than one variable. The following equation and Fig. 3 define and visualize a maximization problem with two variables.

$$\text{Max: } f(x_1, x_2) = -x_1^2 - 10 \sin \sin(1.5x_1) - x_2^2 - 10 \sin \sin(1.5x_2) \tag{7}$$

$$\text{Where: } -5 \leq x_1, x_2 \leq 5 \tag{8}$$

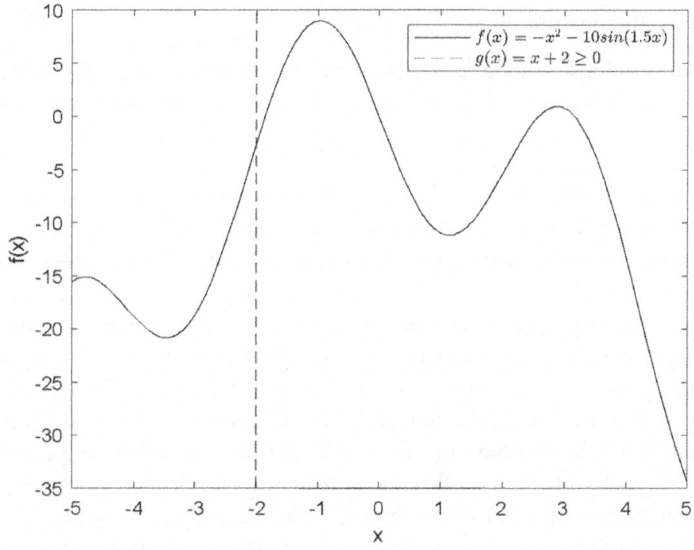

Fig. 2 The shape of $f(x) = -x^2 - 10 \sin \sin(1.5x)$ as the objective function and $g(x) = x+2 \geq 0$ as the constraint

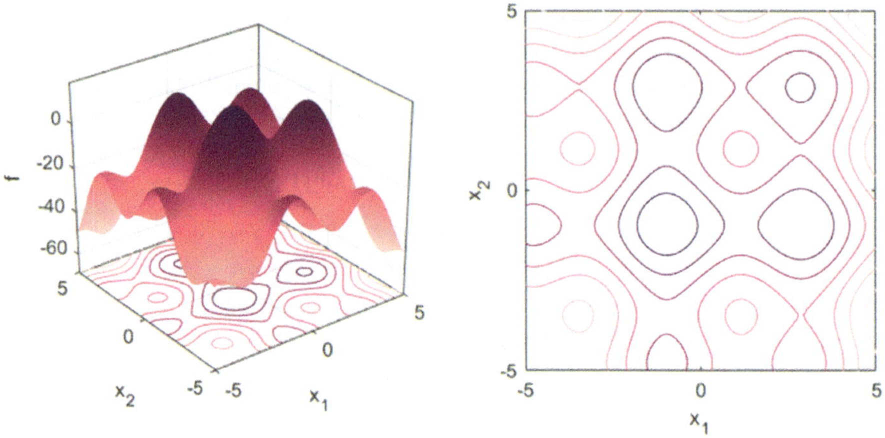

Fig. 3 The shape of $f(x_1, x_2) = -x_1^2 - 10\sin\sin(1.5x_1) - x_2^2 - 10\sin\sin(1.5x_2)$

Despite the similarity of most optimization problems, they can be classified into two categories based on similarity in their formulation or their characteristic.

The following list provides a classification of optimization problems based on formulation:

- Decision variables: continuous, discrete, binary, and mixed
- Objectives: single, multi, many
- Constraint: unconstrained and constrained

Decision variables can be assigned with continuous or discrete values. When having continuous decision variables, the search space is also continuous and infinite. On the other hand, discrete optimization problems have variables of discrete nature that lead to having a finite search space. Most real-world optimization problems have mixed decision variables. It was discussed also above that optimization problems are also divided into two classes unconstrained or constrained. Even a single constraint changes the nature of the optimization problems and required different strategies to solve them.

In terms of the objective functions, optimization problems may have one or more than one objective. A single-objective optimization problem has one objective, so there is usually one global optimum. In problems with multiple objectives, however, there is more than one objective and multiple solutions can be found representing the best tradeoffs between the objectives. The category of many-objective refers to problems with many objectives. What is considered large is probably very arbitrary depending on the complexity of the problems. The name, many-objective, was coined by researchers in the area of evolutionary computation to highlight the importance of an ever-increasing number of objectives and the complexity of addressing them all simultaneously.

Optimization problems can also be categorized based on the characteristics of their objective function into, linear, non-linear, unimodal, multi-modal, convex, non-convex, continuous, discontinuous, static, and dynamic.

2 Optimization Algorithms

Identifying the key components of optimization problems and understanding their characteristics will help us in finding or designing an efficient optimization algorithm. The No Free Lunch theorem logically proves and clearly states that there is no optimization algorithm to solve all optimization problems. After covering different optimization problems, in this sub-section, it is time to discuss optimization algorithms.

There are different classifications for optimization algorithms. One of the most popular ways to classify them is based on the differentiability of the problem into:

- Differentiable: algorithms that require derivative information
- Non-differentiable: algorithms that do not require derivative information.

2.1 Derivative Dependent Algorithms

In the first class, first-order or second-order derivatives are used for continuous optimization problems with one variable to quantify the amount of the changes in the objective function or its derivative, respectively. If a problem has multiple variables (multivariate), these are called gradient and hessian matrices. An algorithm in this class relies on such derivative information to iteratively converge towards an optimum.

One of the most popular and well-regarded algorithms in this category is gradient ascent (descent for minimization problems). In this algorithm, the optimization process starts from a given point in the search space. The gradient ascent then calculates the gradient for those initial solutions which gives a vector indicating the rate of change in the function across all variables. To calculate this, a partial derivative is used. So, the new solution is generated as follows:

$$x_{t+1} = x_t + \alpha \nabla f(x_t) \tag{9}$$

where α is the step size, t shows the current iteration, $\nabla f(x_t)$ is the gradient of $f(x_t)$.

The results of this algorithm on the following problem are shown in Fig. 4.

$$\text{Max: } f(x_1, x_2) = -x_1^2 - 10\sin(1.5x_1) - x_2^2 - 10\sin\sin(1.5x_2) \tag{10}$$

$$\text{Where: } -5 \le x_1, x_2 \le 5 \tag{11}$$

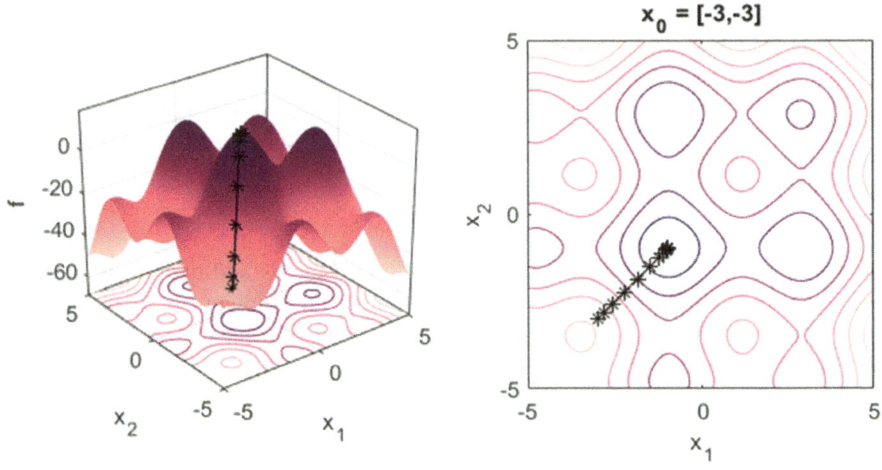

Fig. 4 Gradient ascent finds the global maximum starting from $[-3, -3]$ ($\alpha = 0.02$)

It can be seen in this figure that the gradient ascent algorithm converges to an optimum, which is the global optimum for this problem. Despite the merits of this algorithm, there are two main issues. The first one is the fact that this algorithm is a local search. This means that if the starting point is not on the slope towards the global optimum, it will be trapped in a local optimum. This can be seen in Fig. 5.

Another challenge when using this algorithm is the need to fine-tune the step size. Large step size values lead to unstable results as seen in Fig. 6. In this experiment,

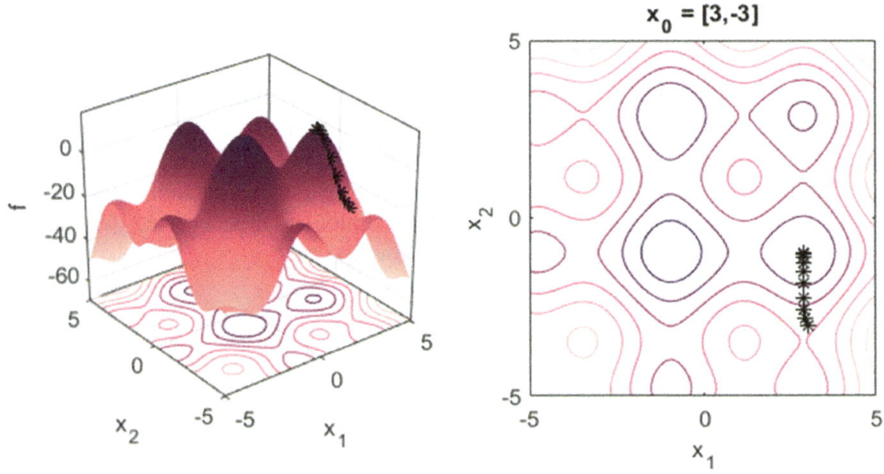

Fig. 5 Gradient ascent gets trapped in a local maximum when starting from $[3, -3]$ ($\alpha = 0.02$)

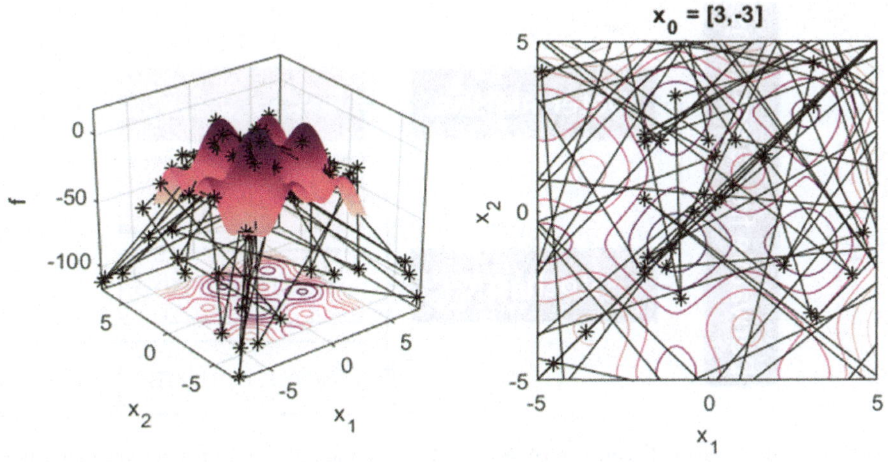

Fig. 6 Gradient ascent shows unstable results when the step size is not tuned property ($\alpha = 0.5$). Note that the algorithm steps after 100 iterations without hitting other stopping conditions (threshold)

the step size was set to 0.2, and it can be seen it even led to overshooting the solution outside the search space ($[-5, 5]$).

The last limitation of a derivative-based algorithm is the need to calculate the derivative of the objective function(s). If the derivative cannot be calculated. Also, most engineering problems have objectives with no analytical descriptions due to the use of simulators. Therefore, such algorithms become inefficient. This has been the reason for the popularity of heuristics, meta-heuristics, and hyper-heuristics lately.

2.2 *Derivative Free Algorithms*

As the name implies, algorithms in this class are not dependent on calculating the derivative of the objective function. This means they consider the problem as the black box. A classification of such algorithms can be found in Fig. 7 that is done based on their randomization behavior.

If an algorithm does not show any random behavior and provides the same results given the same initial solution, it is considered deterministic. Popular algorithms in this class are pattern (direct) search methods [1]. In such algorithms, for any given solution a set of new solutions is created using a pattern metaphor (e.g. rectangle: left, right, up, and down) and the best one is selected. The distance of the generated solutions from the existing solution is reduced if a not better solution is found. This process is repeated until the accuracy of the solution is acceptable or the algorithm reaches a certain maximum number of iterations.

The lack of derivative is a challenge because it always gives accurate value to the challenges in the objective function (or its derivative in case of second-order)

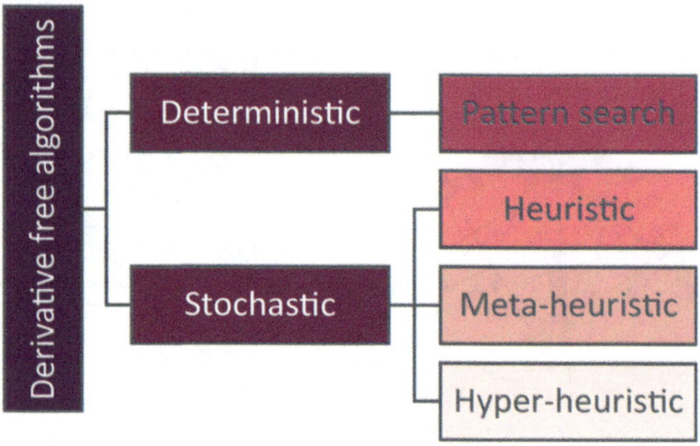

Fig. 7 Classification of derivative-free optimization algorithms based on their randomization behavior

and assists an algorithm to create or choose potentially better solutions compared to an existing one. Heuristic algorithms compensate for this by considering a heuristic function. They are considered stochastic algorithms (see Fig. 7) as they provide a different solution in each run.

During the optimization process, a heuristic function can help an algorithm rule out potentially poor solutions or decisions that will lead to poor solutions. A heuristic function can be used completely instead of the cost function or can be considered in conjunction with it. In robot path planning, for instance, an algorithm can consider the length of a straight line from the robot to the goal, but the movements are still limited to the environment and obstacles. However, it still gives the robot to choose a better move from the set of possible moves. The benefit of such an algorithm is the speed, but the accuracy is compromised. Another limitation is that the heuristic function is usually problem-specific. This means it is hard to re-use the heuristic function for other optimization problems.

Meta-heuristic alleviates this limitation but only relies on the objective function. This means that deciding how to create new solutions and choose the best ones will be done under the objective value of the solutions. They typically use intelligence mechanisms to increase the likelihood of finding better solutions.

One of the limitations of meta-heuristics is the many search operators and controlling parameters to be tuned to ensure maximum accuracy on different optimization problems. Hyper-heuristics include mechanisms to automate this process by adaptively tuning the parameters of meta-heuristics, switching mechanisms between operators, and assassin algorithms to deploy proper search mechanisms based on the improvement of solutions throughout the optimization.

3 Meta-Heuristics

As discussed above, meta-heuristics are stochastic optimization techniques that estimate the global optimum when the solution set of a problem is too large to be explored entirely. Such algorithms leverage intelligence mechanisms to generate a set of random solutions and improve them iteratively until a certain accuracy is achieved. These algorithms are divided into different categories, of which the most popular one is based on the inspiration as follows (see also Fig. 8).

- Evolution: evolutionary phenomena are the source of inspiration in evolution-based meta-heuristics.
- Swarm Intelligence: collective intelligence of nature creates are used in swarm-based meta-heuristics to update solutions during the optimization process
- Physical rules: several meta-heuristics mimic physical phenomena in nature in reaching a stable state to optimize problems
- Other inspiration: there are several meta-heuristics too that mimic phenomena other than evolution, swarm, of physics. For example, the source of inspiration might be society advancement, political decisions, human collective decision making, etc.

The literature shows that meta-heuristics have been widely used in both science and industry. There are several reasons for this, of which one mentioned above, which was due to the derivative-free mechanism. The other reason is the high tendency of such algorithms in avoiding locally optimal solutions. It was observed in Fig. 5 how a gradient ascent algorithm can be easily trapped in locally optimal solutions.

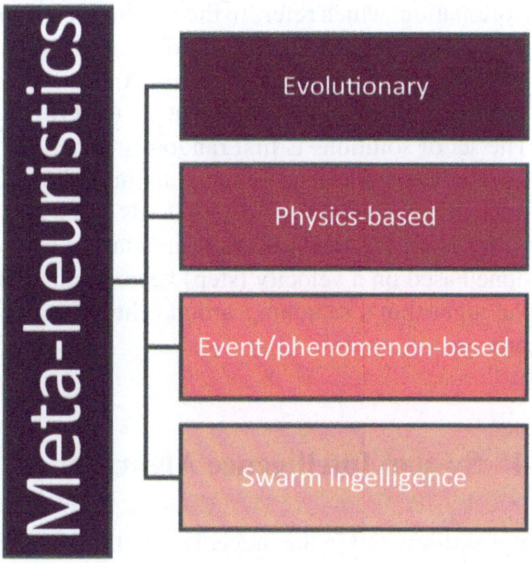

Fig. 8 A classification of meta-heuristics based on the inspiration

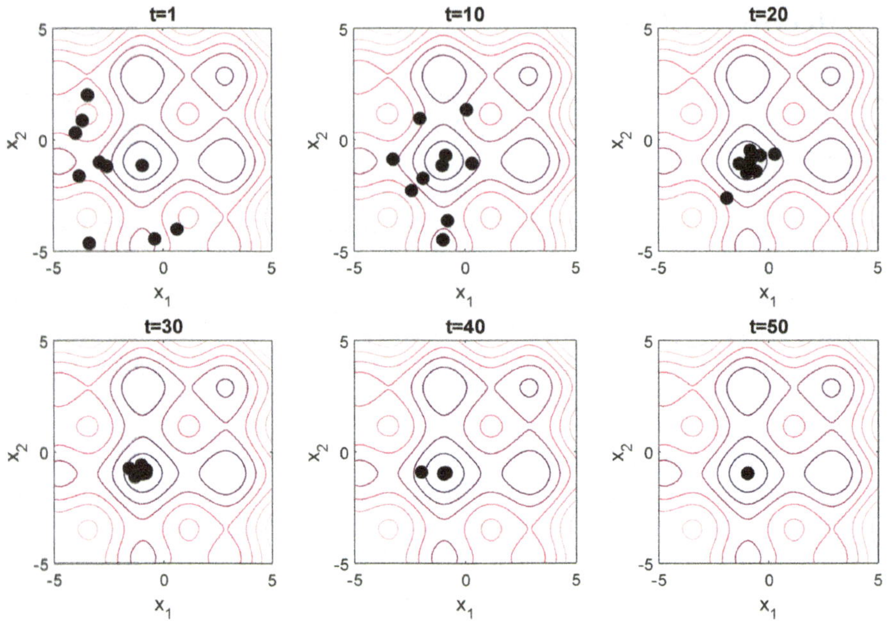

Fig. 9 The movement of 10 particles in PSO over 50 iterations on $f(x_1, x_2) = -x_1^2 - 10\sin\sin(1.5x_1) - x_2^2 - 10\sin\sin(1.5x_2)$

Meta-heuristics leverage intensive random mechanisms to initially explore large portions of the search space (also known as global search). They then focus on exploitation, which refers to the targeted search around the promising solutions found during the exploration (also known as Local search). This increases the chance of avoiding locally optimal solutions. An example is given in Fig. 9. It is an experiment using Particle Swarm Optimization (PSO) [2] with 10 particles and 50 iterations. The set of solutions is first randomly generated around the search space. This set is then moved toward the global maximum. PSO follows the principles of foraging in a flock of birds, school of fish, or herds. Each particle considers its best solution found so far as well as the best solution obtained by the whole swarm. Position updating is done based on a velocity (step) vector using personal and swarm best. This ensures the algorithm's searching around the promising solutions during the optimization process.

4 Swarm Intelligence Algorithms

It was discussed in the preceding section that Swarm Intelligence algorithms belong to one of the biggest sub-classes of meta-heuristics. Algorithms in this category are typically nature-inspired and mimic the collective (swarm) intelligence in nature that leads to problem-solving. Such collaborative efforts are achieved by efficient information sharing and decision making, which are the foundations of swarm intelligence algorithms.

For instance, stigmergy is the process of communication and planning through the manipulation of environments. This means an individual in a swarm tends to change the environment that triggers actions by others without direct contact with others. For instance, ants deposit pheromone on a surface to indicate whether they have found a source of food or not. A certain type of pheromone is used for this, which indicates to others to follow the path and help with bringing the food to the nest. An example of swarm intelligence in nature can be found in a flock of birds, a school of fish, a herd of sheep, a colony of termites, a colony of bacteria, etc.

One question that one might ask is why swarm intelligence is good to solve optimization problems as they have been largely used in Computer Science and Engineering for this purpose. This is mainly because the complex behavior by individuals in a swarm is usually an attempt to solve problems in an optimized manner. For instance, birds navigate in an environment to minimize energy consumption given the complicated nature of airflow in different areas. Schools of fish show swarming maneuvers to better avoid predators and find food in oceans. Therefore, when there is a complexity in natural swarms, it is for the reason of optimization. Therefore, if we mimic that complexity, we can develop optimization algorithms capable of solving optimization problems.

One of the early attempts to create artificial swarms was made by Craig Reynold [3]. He created a system to simulate flocking behaviors in computers using three simple principles of separation (keeping a distance from neighboring individuals to avoid collision), alignment (aligning the flying direction with those in the neighborhood), and cohesion (tendency to the center of a neighborhood to "stay connected" with the swarm). Other mechanisms such as predator avoidance and gravitating towards food sources were also introduced later to improve the model.

The most successful early swarm intelligence algorithms are Ant Colony Optimization (ACO) [4] and PSO [2]. The ACO algorithm is limited to combinatorial optimization that mimics the pathfinding ants in nature. This algorithm finds a reasonably short path on a graph using artificial ants that are evaluated using a cost matrix and communicate using a pheromone matrix. Using these two principles this algorithm can solve a wide range of combinatorial optimization problems. It has been extended to solve continuous problems as well.

The PSO algorithm is indeed a meta-heuristic capable of estimating the global optima for optimization problems with continuous variables. This algorithm has two vectors: position and velocity. The position of particles, which represent solutions to an optimization problem, is updated by adding the current position vector to the velocity vector. The velocity vector is the summation of three components: current velocity, cognitive, and social. The first component allows PSOs to maintain their flying direction and speed to some extent. The cognitive component mimics the individual thinking of birds in a flock by saving the best solution that each has obtained so far and tendency to fly towards them. The last component provides the social intelligence of birds by maintaining the best solution obtained by the swarm and a tendency towards it. With these mechanisms, PSO finds promising solutions during the optimization problems and searches around them until the satisfaction of an end condition.

After the success of these two algorithms, many researchers and practitioners started to improve their performance, hybridize, and develop different variables, and proposed new swarm intelligence algorithms to solve a wide range of optimization problems in both science and industry. Some of the most recent and well-regarded ones are Artificial Bee Colony (ABC) [5], Firefly Algorithm [6], Dragonfly Algorithm (DA) [7], Grasshopper Optimization Algorithm (GOA) [8], etc.

This book emphasizes the studies of existing Swarm Intelligence techniques, their variants, and their applications. The book also contains reviews of new developments in SI techniques and hybridizations. Algorithm-specific studies cover the basic introduction and analysis of key components of these algorithms, such as convergence, the balance of solution accuracy, computational costs, tuning, and control of parameters. Application-specific studies incorporate the ways of designing objective functions, solution representation, and constraint handling. The book also includes studies on application domain-specific adaptations in the Swarm Intelligence techniques.

The book is organized into four sections. The first section covers the state-of-the-art in SI, which includes reviews on SI techniques, various optimization problems, and performance analysis of SI techniques. The second section covers applications of SI techniques in various engineering problems. The third section comprises applications of SI techniques in different machine learning problems such as clustering and prediction. The last section includes several other applications of SI techniques.

References

1. Hooke, R., Jeeves, T.A.: Direct search solution of numerical and statistical problems. J. ACM (JACM) **8**(2), 212–229 (1961)
2. Kennedy, J., Eberhart, R.: Particle swarm optimization. In: Proceedings of ICNN'95-International Conference on Neural Networks, vol. 4, pp. 1942–1948. IEEE (1995)
3. Reynolds, C.W.: Flocks, herds and schools: a distributed behavioral model. In: Proceedings of the 14th Annual Conference on Computer Graphics and Interactive Techniques, pp. 25–34 (1987)
4. Dorigo, M., Birattari, M., Stutzle, T.: Ant colony optimization. IEEE Comput. Intell. Mag. **1**(4), 28–39 (2006)
5. Karaboga, D., Basturk, B.: A powerful and efficient algorithm for numerical function optimization: artificial bee colony (ABC) algorithm. J. Global Optim. **39**(3), 459–471 (2007)
6. Yang, X.-S., Slowik, A.: Firefly algorithm. In: Swarm Intelligence Algorithms, pp. 163–174. CRC Press (2020)
7. Mirjalili, S.: Dragonfly algorithm: a new meta-heuristic optimization technique for solving single-objective, discrete, and multi-objective problems. Neural Comput. Appl. **27**(4), 1053–1073 (2016)
8. Meraihi, Y., Gabis, A.B., Mirjalili, S., Ramdane-Cherif, A.: Grasshopper optimization algorithm: theory, variants, and applications. IEEE Access **9**, 50001–50024 (2021)

Introductory Review of Swarm Intelligence Techniques

Thounaojam Chinglemba, Soujanyo Biswas, Debashish Malakar, Vivek Meena, Debojyoti Sarkar, and Anupam Biswas

Abstract With the rapid upliftment of technology, there has emerged a dire need to 'fine-tune' or 'optimize' certain processes, software, models or structures, with utmost accuracy and efficiency. Optimization algorithms are preferred over other methods of optimization through experimentation or simulation, for their generic problem-solving abilities and promising efficacy with the least human intervention. In recent times, the inducement of natural phenomena into algorithm design has immensely triggered the efficiency of optimization process for even complex multi-dimensional, non-continuous, non-differentiable and noisy problem search spaces. This chapter deals with the Swarm intelligence (SI) based algorithms or Swarm Optimization Algorithms, which are a subset of the greater Nature Inspired Optimization Algorithms (NIOAs). Swarm intelligence involves the collective study of individuals and their mutual interactions leading to intelligent behavior of the swarm. The chapter presents various population-based SI algorithms, their fundamental structures along with their mathematical models.

Keywords Optimization problems · Optimization algorithms · Nature inspired optimization algorithm · Swarm intelligence

T. Chinglemba · S. Biswas · D. Malakar · V. Meena · D. Sarkar · A. Biswas (✉)
Department of Computer Science and Engineering, National Institute of Technology Silchar, Silchar 788010, Assam, India
e-mail: anupam@cse.nits.ac.in

T. Chinglemba
e-mail: thounaojam_pg@cse.nits.ac.in

S. Biswas
e-mail: soujanyo_ug@cse.nits.ac.in

V. Meena
e-mail: vivekmeena_ug@cse.nits.ac.in

D. Sarkar
e-mail: debojyoti_rs@cse.nits.ac.in

A. Biswas et al. (eds.), *Advances in Swarm Intelligence*, Studies in Computational Intelligence 1054, https://doi.org/10.1007/978-3-031-09835-2_2

1 Introduction

Optimization is the simple action to make the best utilization of resources. Mathematically, it is finding the most useful solutions out of all possible solutions for a particular problem. There are two types of optimization problems leaning on whether the variables are *continuous* or *discrete*. Comparatively, continuous optimization problems tend to be easier to solve than discrete optimization. A problem with continuous variables is known as continuous optimization. It aims at achieving maximum efficiency to reduce the cost or increase the overall performance. Variables in continuous optimization are permitted to take on any value within a range of values. In discrete optimization, some or all the variables in a problem are required to belong to a discrete set. Improvements in algorithms coupled with advancements in computing technology have dramatically increased the size and complexity of discrete optimization problems that can be solved efficiently. With the introduction of Swarm Intelligence (SI), conventional methods are less frequently utilized. Still, we need better SI techniques. Optimizing solutions comes with some challenges, for instance, it is time consuming, needs to be updated continuously and can consider only a small range of problems and the solutions obtained may be far from being optimal.

Let us take a small example to visualize the importance of SI over conventional techniques. Suppose, there are four explorers searching for a treasure which is located in a mountain region with several ridges and the treasure is located under one such ridge. The other ridges have small treasures which are not worth the effort. The treasure can be visualized as the optimal solution and the other small regions as the local solutions to the problem. Clearly, the four explorers do not want the small treasures and are definitely going to search for the bigger treasure. They start off with a conventional search for the treasure with a conventional technique known as the gradient descent. They randomly land on a place in the region and go downwards on the nearest slope to get to the nearest ridge and dig it up. What are the chances that they will find the treasure on their first try? It's obviously quite low. In fact, in the real world, finding optimal solutions with this approach is really time consuming if the data to search in reaches a sufficiently large value. Next, they try a SI approach. One explorer each lands in different locations of the region, and they start exploring locally. But this time, they have phones to converse with each other. They communicate with each other and search areas which are optimally closer to the treasure. Finally, after some 'exploration' they find the actual treasure they are searching for. It is to be noted that since they could converse with each other, they can easily skip areas that have been searched by the others. Also, the area of search increases if we consider the whole unit, and also they are searching locally individually. Simply put, the unit *explores* the regions, and individually, they *exploit* a region to find the treasure. It is a lot quicker and less time consuming than the conventional mechanisms to reach the same solution. Mathematically speaking, optimal solutions refer to the minimums of the curve represented by the problem. There can be problems which have only one global minimum, i.e., there is only one slope and any point on the curve can reach that point if we follow the downward slope of the curve. However, most problems in the real world have multiple minimums, which means, there are local minimums effective to only a particular area in the curve and global minimums which are the optimal solution to the problem (or curve).

Swarm intelligence can solve complex and challenging problems due to its dynamic properties, wireless communications, device mobility and information exchange. Some examples of swarm are that of bees, ants, birds, etc. They are smarter together than alone and this can be seen in nature too, that creatures when working together as unified system can outperform the majority of individual members when they have to make a decision and solve problems. Swarm Intelligence is based on clustering individually or adding to existing clustering techniques. Besides the applications to the already present conventional optimization problems, SI is useful in various other problems like communication, medical data-set classification, tracking moving objects, and prediction. To put it simply, SI can be applied to a variety of fields in fundamental research, engineering and sciences. However, to think that an algorithm developed matches the criteria for solving every problem out there in the real world is quite naive. Techniques developed using SI are more problem based, and several changes are made already in present algorithms to reach the solutions of the specific problems. Thus, developing an algorithm without a clear objective (, or problem) is not possible. SI techniques are especially useful to find solutions of non-deterministic polynomial-time problems, which would take exponential time to solve if we take a direct approach to solve the problem. Thus, swarm-based techniques are used to find approximate solutions which are relatively good for the problem.

Rest of the chapter is organized as follows. Section 2 presents a generic framework for SI techniques and briefed primary components of SI techniques. Section 3 shows how different SI techniques evolved with time. Section 4 briefly explains working principles of prominent SI techniques. Section 5 highlights various application domains of SI techniques. Lastly, Sect. 6 concludes highlighting different nuances, shortcomings, and applicability of SI techniques.

2 Generic Framework of Swarm Intelligence Techniques

When we need to find a solution to an optimization problem, we first observe the attributes, constraints, whether it will have a single objective or multiple objectives etc. With traditional optimization, we try to find the solution by traditional methods such as brute force, simulation or apply conventional optimization techniques which is time consuming, requires huge computation cost and requires frequent human intervention. To understand how we approach an optimization problem conventionally, a flow chart is shown in Fig. 1.

As conventional techniques work best for finding solutions to simple continuous, linear optimization problems, real world optimization are often non-linear and discontinuous in nature. When solving a real world optimization problem complication such as a large number of local solution and constraints, discrete variables, multiple objectives, deceptive search space etc. needs to be addressed. When we plug in

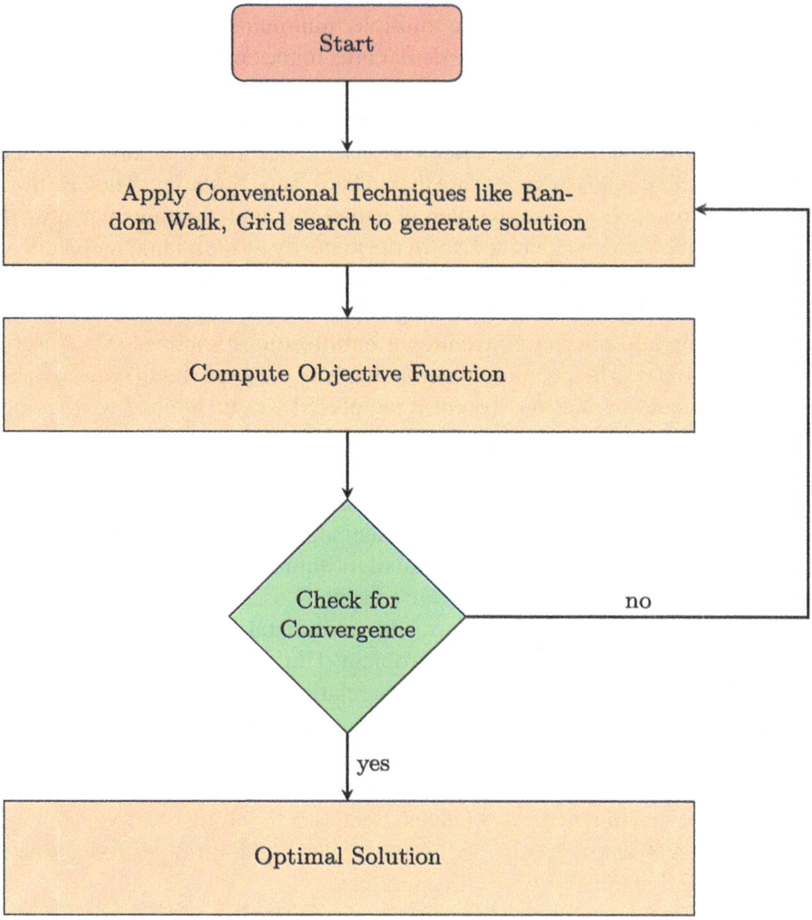

Fig. 1 Generic Flowchart of optimization using Traditional Techniques

these conventional techniques in the real world problem it leads to problems such as non-optimal solutions, being stuck on a local solution, low convergence rate etc. A need to develop an intelligent system to efficiently solve these real world problems have become a necessity. To develop these intelligent systems over the years, researchers and scholars have taken inspiration from various phenomena of nature as these phenomena always tend to self-organize even in the most complex environments of nature. Swarm Intelligence techniques are a subset of these Nature Based algorithms which is based on the swarming behavior of insects, birds, fishes, animals etc. to develop a collective intelligence, self-organized way of solving complex, discontinuous, nonlinear systems efficiently with higher convergence and less time consumption as compared to conventional techniques. In Swarm Based algorithms members of the swarm interact with each other and the environment to develop a collective intelligence.

Over the years as more and more Swarm Intelligence techniques are developed and used, let us look at some important aspects of these swarm based algorithms:

- **Exploration**: One of the most important aspects of every swarm intelligence techniques is its exploration also known as diversification. Exploration traverses the search landscape to search for new solutions which are different from the current solution. This helps us in finding solution which is better than the current solution and helps us in diversifying our solution. It also helps us to escape from a local solution to find new and better solutions. But we should be careful as too much exploration will lead to slow convergence, which is not desired at any swarm intelligence techniques.
- **Exploitation**: While exploration traverses the whole landscape, exploitation, also known as intensity, focuses on a local search area, which will lead to finding better solution in the local search space. This will lead to high convergence in our algorithms but too much exploitation will lead to being stuck in a local solution only. So, a fine balance is needed between the exploration and exploitation [1] of every swarm intelligence techniques and based on the optimization problem we can fine tune the exploration and exploitation to our requirement.
- **Convergence**: Another important aspect of swarm based algorithms is its high convergence rate. Convergence rate can be defined as the speed in which our algorithm converges to a solution over some iterations beyond which a repeating sequence is generated. While some algorithms use defined global best to converge faster [2] to a solution while other algorithms use their own exploration and exploitation [3] to converge to a solution.
- **Randomization**: In most swarm intelligence techniques, randomization parameters are used for better exploration in the search landscape to find alternate solutions which might be better than the current solutions.

As we try to understand these swarm based algorithms and how they work collectively as a member of a swarm to achieve collective intelligence, we look at a generic framework of how these swarm based techniques operates in Fig. 2.

3 Evolution of Swarm Intelligence Techniques

As swarm based optimization techniques provide self-organized, collective, intelligent and faster convergence in solving complex, discontinuous, nonlinear systems, researchers and scholar are looking to develop novel swarm based optimization techniques that are inspired by biological [4], physical [5] and chemical phenomena [6] to solve various optimization problems efficiently and effectively compared to traditional techniques. As of now there are more than 140 optimization techniques based on natural phenomenon to solve various optimization problems in field of Sci-

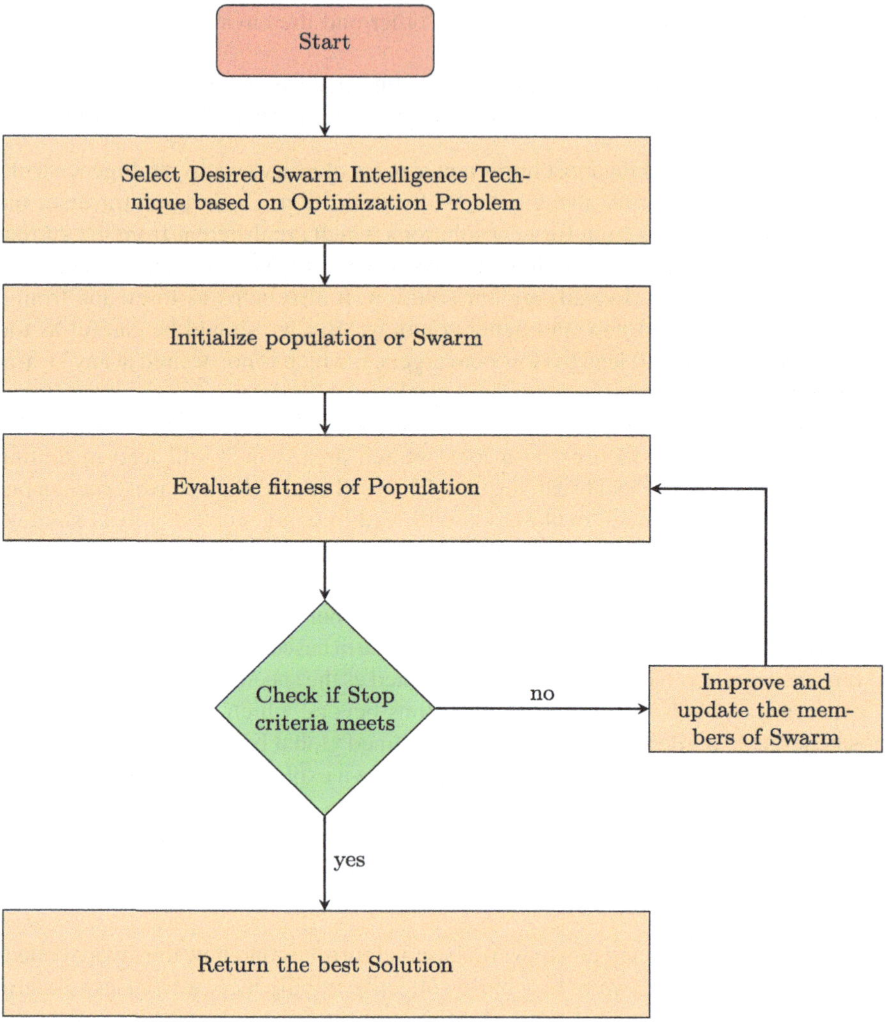

Fig. 2 Generic Flowchart of Swarm Optimization Techniques

ence, Medicine, Artificial Intelligence, Engineering etc. Some of the popular swarm intelligence algorithms that have been developed over the years are listed in Table 1.

As we can see from Table 1, there are various novel nature based optimization techniques to choose from. While some algorithms focus on exploration and exploitation, some focus on faster convergence and parameter tuning etc.,. So, before choosing an optimization technique, we need to understand what type of optimization problem we are dealing with, the constraints associated with the optimization problem, the attributes and the mechanism of the algorithm.

Table 1 Some of the popular bio-inspired meta heuristic algorithm inspired by swarm intelligence

Year	Algorithm proposed	Inspiration
2021	Flamingo Search Algorithm [7]	It is based on migratory and foraging behaviour of flamingos
2021	Horse herd Optimization Algorithm [8]	It implements what horses do at different ages using six important features: grazing, hierarchy, sociability, imitation, defense mechanism and roam
2020	Chimp Optimization Algorithm [9]	It is inspired from the individual intelligence and sexual behaviours of chimps, when they find a group to be in
2020	Black Widow Optimization Algorithm [10]	It is based especially on the interesting sibling cannibalism behaviour offered by Black Widow spiders
2020	Sparrow Search Algorithm [11]	This algorithm is based on intelligent techniques used by sparrows to search for food depending on the situation they are in
2020	Rat Swarm Optimizer [12]	It is inspired by the chasing and attacking behaviour of rats
2019	The Sailfish Optimizer [13]	This algorithm uses sailfish population while searching, and sardines population for diversifying the search space, based on a group of hunting sailfish
2018	Meerkat Clan Algorithm [14]	It is based on Meerkat with their exceptional intelligence, tactical organizational skills, and remarkable directional cleverness in its traversal of the desert when searching for food
2018	Grasshopper Optimization Algorithm [15]	It is inspired by the foraging and swarming behaviour of grasshoppers
2017	Salp Swarm Algorithm [16]	It is inspired by the swarming behaviour of salps in oceans
2017	Camel Herds Algorithm [17]	This algorithm is based on camels, and how they have a leader for each herd and how they search for food and water depending on humidity of neighbouring places
2017	Duck Pack Algorithm [18]	It is based on the foraging behaviours of ducks depending on imprinting behaviour and food orientation
2016	Dragonfly Algorithm [19]	It is based on the static and dynamic behaviour of dragon flies
2016	Sperm Whale Algorithm [20]	It is based on the sperm whale's lifestyle
2016	Dolphin Swarm algorithm [21]	It is based on the biological characteristics and living habits such as echolocation, information exchanges, cooperation, and division of labor of Dolphins

(continued)

Table 1 (continued)

Year	Algorithm proposed	Inspiration
2016	Crow Search Algorithm [22]	It is based on how crows search for food, and hide their food from other crows and remember their hiding places
2015	Ant Lion Optimizer [23]	This algorithm mimics the hunting nature of ant-lions in nature
2015	Elephant Herding Optimization [24]	It is based on the herding behaviour of elephants, different group elephants living under a matriarch
2015	Moth-flame Optimization algorithm [25]	It is based on the navigation method of moths in nature called transverse orientation
2014	Grey Wolf Optimizer [26]	It mimics the living hierarchy and hunting behaviour of grey wolves in nature
2014	Pigeon Optimization algorithm [27]	It is based on the swarming behaviour of passenger pigeons
2014	Spider Monkey Optimization Algorithm [28]	It is inspired by the Fission-Fusion social structure of spider monkeys during foraging
2013	Spider Optimization [29]	It is based on the cooperative characteristics of social spiders
2012	Bacterial Colony Optimization [30]	It is based on the life cycle of a bacteria named, E. Coli
2012	Zombie Survival Optimization [31]	It is based on the foraging behaviour of zombies, when they find a hypothetical airborne antidote which cures their ailments
2010	Bat Algorithm [32]	It is based on the echolocation of micro-bats in nature, with varying pulse rates of emission and loudness
2010	Termite Colony Optimization	It is based on the intelligent behaviour of termites
2010	Fireworks Algorithm [33]	In this method, two types of explosion processes are performed, and the diversity of them are kept based on fireworks
2009	Cuckoo Search [34]	It is inspired by the behaviour of cuckoos to lay eggs in the nests of birds of other species
2009	Gravitational Search Algorithm [35]	It is based on Newton's laws and laws of gravity to search for solutions
2009	Glowworm Swarm Optimization [36]	It simulates the behavior of lighting worms or glow worms
2008	Fast Bacterial Swarming Algorithm [37]	This algorithm combines the foraging behaviour of E. Coli from Bacteria Colony Algorithm and flocking mechanism of birds from Particle Swarm Algorithm
2007	Firefly Algorithm [3]	It is inspired from the fireflies in nature

(continued)

Table 1 (continued)

Year	Algorithm proposed	Inspiration
2006	Cat Swarm Optimization [38]	It is based on the behaviour of cats, and includes two sub-processes—tracing mode and seeking mode
2005	Artificial Bee Colony [39]	This algorithm simulates how honey bee colonies work, including the employed, workers and scouts
2004	Honey Bee Algorithm [40]	It is based on how honey bees forage for food in nature
1995	Particle Swarm Optimization [2]	It is based on how flocks of birds move in the world and search for food, both individually and considering the entire group

As the nature of optimization problem is different with one another, it has become a necessity to either modify an existing SI algorithm, hybridize with another algorithm or to develop a novel SI based algorithm to obtain the best optimal solution of different optimization problem as it is impossible for a single algorithm to get the best solution for every optimization problem as the nature of the problems are complex in nature.

4 Prominent Swarm Intelligence Algorithms

Swarm intelligence techniques in comparison with traditional algorithms provide a novel way to address complex problems and processes efficiently and effectively. These algorithms are based on the swarming behaviours of various insects, birds, animals found in nature where they work collectively to achieve a collective intelligence. While some algorithms work towards achieving faster convergence, some focuses on the exploration and exploitation of optimization problem. In this section, we will briefly discuss some swarm intelligence algorithms and analyze them.

4.1 Particle Swarm Optimization

Particle Swarm Optimization was introduced by Kennedy and Eberthart in 1995 [2], by observing the social foraging behavior of animals such as flock of birds or school of fishes. In flocking of birds, whether it is flocking towards a destination or searching for food, the goal is achieved by cooperating with all the birds in the flock. Each bird learns from its current experience and updates it with the other birds to achieve its goal. The birds follow a leader which is closer to the best solution and any changes made by the leader is followed by all the birds in the flock hence achieving a collective, intelligent and a self-organized behavior as a swarm. Observing this, Kennedy and

Eberthart summarized the Particle Swarm Optimization into two equations, one for the position of every particle in the swarm and other equations for the velocity of the particles in the swarm.

In Particle Swarm Optimization we initialize a group of particle randomly and then search for the best optimal solution over many iterations. Each particle in a swarm maintains three things its personal best solution also known as pbest, its global best solution also known as gbest and its current direction. To search for the best optimal solution each particle in the swarm needs to update the velocity equation and the position equation over many iterations. The velocity and position of a particle, V_i and X_i, respectively can be updated over the iterations as follows:

$$V_i(t+1) = V_i(t) + C_1 R_1 \left[X_{pb}(t) - X_i(t)\right] + C_2 R_2 \left[X_{gb}(t) - X_i(t)\right] \tag{1}$$

$$X_i(t+1) = X_i(t) + V_i(t+1) \tag{2}$$

where R_1, R_2 are two random vector in the range (0,1) and C_1, C_2 are cognitive acceleration coefficient and social acceleration coefficient respectively. In Eq. (1), $V_i(t)$ denotes the current direction of the particle, $X_{pb}(t) - X_i(t)$ describes the cognitive component of the particle where, $X_{pb}(t)$ denotes the position vector of a particle pbest and $X_{gb}(t) - X_i(t)$ describes the social component of the particle where, $X_{gb}(t)$ denotes the particle gbest. The new velocity and position of a particle, $V_i(t+1)$ and $X_i(t+1)$ respectively can be calculated over many iterations to get our desired solution.

Over the years researchers have been proposing new methods to the PSO algorithm due to the simplicity of the algorithm, its low computation cost and effectiveness as compared with traditional optimization algorithms, many algorithms has been proposed which improve the PSO algorithm further. Some of the proposed modification are in the form of hybridization with other nature based optimization algorithms like Genetic Algorithm, ACO [42–44] etc., modification of the PSO algorithms like fuzzy PSO, QPSO [45] etc. and even extension of PSO algorithms to fields like discrete [46] and binary optimization. Some of the modification done to the PSO algorithm over the years are given in Table 2.

4.2 Firefly Algorithm

Inspired by the flashing pattern and behavior of fireflies at night, Yang proposed the Firefly algorithm (FA) [3] in 2008. The flashing light in fireflies serves two purposes; one is to attract the mating partners and the other is to warn of predators. Yang formalized the Firefly algorithm based on the flashing lights and how each firefly is attracted to one another based on the flashing lights. To formalize the FA algorithm Yang idealized some characteristics of the fireflies as:

Table 2 Some modified algorithms based on the PSO algorithm

Year published	Algorithm name	References
2018	PSO-CATV	Time-varying Cognitive avoidance PSO [47]
2013	PSOCA	Cognitive avoidance PSO [48]
2013	MOPSO	Multiple objective PSO [49]
2013	PSO-RANDIW	Random weighted PSO [50]
2013	PSO-GA	Hybridization of PSO with GA [43]
2013	PSO-SA	Hybridization of PSO with SA [44]
2013	QPSO	Quantum-behaved PSO [45]
2012	DPSO	Discrete PSO [46]
2011	PSO-ACO	Hybridization of PSO with ACO [42]
2011	CPSO	Chaotic PSO [51]

1. The fireflies are considered as unisex i.e. fireflies are only attracted to one another based on the flashing pattern not on the gender of the firefly.
2. The attractiveness of the firefly is directly proportional to the light intensity i.e. a firefly will move towards a brighter one and if there is no light intensity, it will perform a random walk.
3. The light Intensity of the fireflies are determined from the objective function.

With these assumptions, Yang proposed the updated position of a firefly say i to a brighter firefly j at iteration $t+1$ is given as:

$$x_i^{t+1} = x_i^t + \beta_0 e^{-\gamma r_{ij}^2}\left(x_j^t - x_i^t\right) + \alpha\epsilon_i^t, \tag{3}$$

In the above equation, x_i^t represents the initial position of the firefly i, $\beta_0 e^{-\gamma r_{ij}^2}\left(x_j^t - x_i^t\right)$ represents the attractiveness of the firefly x_i to another firefly x_j, β_0 represents the brightness at r $= 0$ and $\alpha\epsilon_i^t$ represents the randomization term with α representing a randomization parameter in which ϵ_i^t is a vector of random numbers derived from a Gaussian, uniform or others distribution.

As the FA algorithm is based on the attraction and attractiveness decreases with the distance between fireflies, the whole population can be divided into sub groups which in turns provides an efficiently and faster way of searching for the global best and local best in the search landscape. Other important aspect of FA algorithm is that can be reduced to other meta-heuristic algorithms such as SA, DE, APSO thereby having the characteristics of all these algorithms. Over the years FA algorithm has either been modified or hybridize to obtain better results, low computation cost etc. A detailed survey on firefly algorithm and its variants are done by Fister et al. [52].

4.3 Bacteria Colony Algorithm

The Bacteria Colony Optimization is based on how bacteria move about in the environment. The major inspiration comes from a certain bacteria called *E. Coli*. This method of optimization follows how bacteria move with their flagellum, how only the ones better at searching survive and with the change of environment, the weaker ones die, and how new offsprings are made from the stronger ones and the cycle continues as bacteria search for more optimal conditions to survive longer.

There are four major characteristics of bacteria which make up the bacteria colony optimization.

- **Chemotaxis**: This model very uniquely reproduces how bacteria, especially *E. Coli* move in the environment by the movement of their flagellum, or whip-like structures present in the bacteria. Thus, a better individual means that it has better chances to survive and reproduce.
- **Elimination, reproduction and migration**: We said that natural selection offers the survival of only the bacteria that can search better for the nutrition around. The weaker ones will be eliminated, and the stronger ones reproduce to give better offspring and this cycle will repeat. Chemotaxis, elimination and reproduction is followed by a process called migration. After the nutrition of a place has depleted, the bacteria have to migrate to a newer place with abundant nutrition to continue the former three processes.
- **Communication**: In the bacteria colony model, we find individuals communicating with its neighbours, random individuals in the group, and also with the whole group. This makes the search for better nutrition more efficient.

The bacteria optimization model can be formulated combining three sub-models:

- In the Chemotaxis and Communication model, bacteria run and tumble and communicate with each other. The position of the bacteria can be formulated using the formula:

$$\text{Position}_i(T) = \text{Position}_i(T-1) + R_i * (Ru_{Infor}) + R\Delta(i), \tag{4}$$

$$\text{Position}_i(T) = \text{Position}_i(T-1) + R_i * (Tumb_{Infor}) + R\Delta(i) \tag{5}$$

- The Elimination checks that only the bacteria that has an energy level greater than a given energy level can survive. Mathematically, it can be determined as:

$$\text{if } L_i > L_{\text{given}}, \text{ and } i \in \text{healthy, then } i \in \text{Candidate}_{repr}, \tag{6}$$

$$\text{if } L_i < L_{\text{given}}, andi \in \text{healthy, then } i \in \text{Candidate}_{eli}, \tag{7}$$

$$\text{if } i \in \text{unhealthy, then } i \in \text{Candidate}_{eli}. \tag{8}$$

- Migration model tells the bacteria to move to a newer location with better nutrition, using the formula:

$$\text{Position}_i(T) = rand * (ub - lb) + lb \tag{9}$$

4.4 Crow Search Algorithm

With recent advances in the field of swarm intelligence, the Crow search algorithm, as its name suggests, is a novel method which mimics how crows act in the environment. It is majorly inspired from the fact that crows are intelligent and how they are very efficient in hiding their food in different places. They can remember the location of the food they have hidden for months. Also, interestingly, it has also been observed that crows also keep track of the hiding places of other crows so that they can steal food. To counter this, whenever a crow finds that another crow is following it, they fly away far from the hiding place to trick the other crow and save its food. The crow search algorithm, very cleverly mimics these behaviours of crows to create a very efficient algorithm.

While implementing this algorithm, two matrices, one for the crows, and the other for the memory of the hiding places of the crows are kept. These keep track of the location of the crow and also the location of the hiding place in the search space.

$$\begin{bmatrix} x_{11}^t & x_{12}^t & \cdots & x_{1d}^t \\ x_{21}^t & x_{22}^t & \cdots & x_{2d}^t \\ \vdots & \vdots & \vdots & \vdots \\ x_{N1}^t & x_{N2}^t & \cdots & x_{Nd}^t \end{bmatrix}$$

During the movement of crow i towards crow j, two things can happen - If crow j is unable to discover that crow i was following it, then crow i updates its position as follows:

$$x_i(t+1) = x_i(t) + r_i * fl_i(t) * (m_j(t) - x_i(t)) \tag{10}$$

Where r_i is a random number, and *fl* decides if the search will be local or global. If $fl > 1$ then the crow i moves far away from crow j, and vice versa. If crow j is able to detect crow i following it, then it moves away from its hiding place. The new position of crow j is now a random position on the search space.

This whole equation can be summarised into two parts based on a value which is called *Awareness Probability*. If it is high, then the search happens to be global, and if it is lowered then the search is local. This happens because if a crow can search for another crow better then it can track long distances better and search for more optimal food.

4.5 Grey Wolf Optimization

Grey wolf optimizer (GWO) [26] was proposed by Mirjali in 2014 after taking inspiration from hunting patterns of grey wolves in nature. He also observed the hierarchy of leadership among grey wolves in a pack. Grey wolves live and hunt for prey together in a group. First, the pack tracks the prey and chases it, after which it encircles the prey and starts attacking until the prey stops moving. Taking inspiration from this he proposed the GWO which offers high convergence speed, simple and greater precision as compared to other nature based optimization algorithms. In GWO Mirjali categorize the wolves in the pack into four groups namely the alpha ( ), beta ( ), delta ( ), and lastly the omega ( ). The alpha ( ) is considered as the individual which has the highest authority followed by the beta ( ) and the delta ( ). The rest of the grey wolves are collectively considered as the omega ( ) of the pack.

- **Encircling prey**: In GWO the grey wolves first track and surrounds the prey. This is the encircling part of the algorithm. The mathematical equations representing this is given below:

$$D = \left|\vec{C}.\vec{X}_p(t) - \vec{X}(t)\right| \tag{11}$$

$$\vec{X}(t+1) = \vec{X}_p(t) - \vec{A}.\vec{D} \tag{12}$$

  where $\vec{A}$ and $\vec{C}$ are coefficient vectors, t denotes the current iteration, $\vec{X}_p$ is the position vector of prey, and $\vec{X}$ denotes the position vector of a grey wolf.
  The vectors A and C are calculated as follows:

$$\vec{A} = 2\vec{a}.\vec{r_1} - \vec{a} \tag{13}$$

$$\vec{C} = 2.\vec{r_2} \tag{14}$$

  where components of $\vec{a}$ are linearly decreased from 2 to 0 over the course of iterations and $\vec{r_1}$ and $\vec{r_2}$ are random vectors in [0, 1].

- **Hunting**: After the encircling phase is done, the hunting phase of the algorithm is started. The alpha of the pack leads the pack in hunting for prey followed by the beta and the delta. In GWO to simulate these conditions, it assumes that the alpha has better knowledge about the prey location. The mathematical equations regarding the hunting phase are given below:

$$\vec{D}_\alpha = \left|\vec{C}_1.\vec{X}_\alpha - \vec{X}\right|, \vec{D}_\beta = \left|\vec{C}_2.\vec{X}_\beta - \vec{X}\right|, \vec{D}_\delta = \left|\vec{C}_3.\vec{X}_\gamma - \vec{X}\right| \tag{15}$$

$$\vec{X}_1 = \vec{X}_\alpha - \vec{A}_1.(\vec{D}\alpha), \vec{X}_2 = \vec{X}_\beta - \vec{A}_2.(\vec{D}_\beta), \vec{X}_3 = \vec{X}_\delta - \vec{A}_3.(\vec{D}_\delta) \tag{16}$$

$$\vec{X}(t+1) = \frac{\vec{X}_1 + \vec{X}_2 + \vec{X}_3}{3} \tag{17}$$

After the hunting phase, the grey wolves start attacking the prey. To briefly explain the process of how GWO works, we first start by creating a random population of grey wolves/ individuals. Over many iterations, the individuals based on their hierarchy start to approximate the prey's expected location. The individuals in the population interact with each other and shared information regarding the prey location till the optimal solution is found.

4.6 Sperm Whale Algorithm

This Sperm Whale Algorithm [20] as the name suggests, makes use of how sperm whales interact in nature. This algorithm is based on how sperm whales use their incredibly good hearing to spot food and enemies, and how they use sound to communicate with one another. The main features of this algorithm can be stated as follows:

- Sperm whales come across two opposite poles in their cycle of breathing and feeding—the land for breathing and the sea for food. This feature has been used in the algorithm as two poles of answers, the Best and the Worst answers.
- The sperm whales travel in groups of 6–9 with males and females in the same group. When it's the mating season, the males fight among themselves and the strongest male gets to mate with the females in the group. This has been used in the algorithm to choose better children for the next generations. Obviously, the candidates with higher Objective score wins the fight, and thus, the solution converges towards the optimal value with every generation.

The best and worst individuals in the population of the whales in each gathering are the X_{best} and the X_{worst},individually, showing the following relationships:

$$X_{\text{center}} = X_{\text{worst}} + c \times X_{\text{best}} \tag{18}$$

$$X_{\text{reflex}} = 2X_{\text{center}} - X_{\text{worst}} \tag{19}$$

X_{reflex} is situated outside the inquiry space, c should diminish thus:

$$c = r \times c_i \tag{20}$$

where c_i is the beginning focus factor and r is the constriction coefficient that is less than 1.

5 Applications of SI Techniques

The applications of Swarm Intelligence are limitless to be honest. It has put its name in various fields of research and other real-world usage. One of the biggest advancements in recent years has obviously been machine learning or deep learning and SI techniques have been extensively used to achieve greater results in the field. SI techniques have also shown development in Networking improving the cost of Wireless Sensor Networks (WSNs) [53] and also topological issues, energy issues, energy issues, connectivity and coverage issues and localization issues. SI techniques have been very useful in speech processing [54] as well and have found significant reduction in background noise in audios for recordings. This is important because audio clips with background noises and echo is unintelligent speech and SI techniques have improved a lot in this field. Techniques like PSO, ABC were seen to give better signal to noise ratio.

Techniques like ACO are useful in image processing specifically in feature extraction from images [55]. It has been seen that these techniques give a better perceptual graph while doing feature extraction from images. These techniques also have proved detrimental in image segmentation and also been very useful in bioinformatics [56] by the use of automatic cluster detection and has been extensively used in computer vision for mammography in cancer risk detection and breast cancer detection. Swarm intelligence has also been very useful in data mining in healthcare for better results in finding previous cases of diseases such as cancer, heart diseases, tumors and other health related problems. Swarm Intelligence has also been useful in logistics and transport, in efficient routing of cargo from different one destination to another, especially ACO, with its pheromones have very efficiently given desired outputs from the source to destination nodes. Similarly, for telecommunication as well, ACO has optimized the routing of different telecom networks and the efficient management of users in different networks has been handled. Beehive Algorithm are also useful in segmenting tasks for basic factory operations like transport from one portion to another, packaging etc. Different SI techniques have been useful for mass recruiting for any company because they use pheromones to attract and automatically cluster in more advantageous spots.

Furthermore, SI techniques have been used for automated machine learning by creating Deep Neural Networks (DNNs) [57], which usually require severe expertise to be designed manually, social network analysis for detecting communities [58–61]. They have also shown better results in Resource Allocation; better processing of resources and managing assets in a strategic way. Swarm Intelligence also probably has countless other feats in several other fields like structural engineering [62]. To describe them all would be a very lengthy task to continue, which we will discuss in later chapters, in more detail. To summarize in one sentence, Swarm Intelligence has definitely found its way to every field of study in this world.

6 Discussions and Conclusion

In this chapter, we have presented the various population based swarm intelligence techniques that have been developed so far including particle swarm intelligence, firefly algorithm, bacteria colony algorithm, crow search algorithm, grey wolf optimization, sperm whale algorithm and their fundamental structures along with their mathematical models. SI based search technique and working of a few of these algorithms are highlighted. A brief explanation of applications of swarm intelligence techniques is done. Some major deductions from this chapter are listed below:

1. The major component that differentiates SI optimization algorithms from traditional optimization algorithms is the stochastic nature. Adaptive nature [63] of the algorithms provide a tremendous potential to solve large scale, multi dimensional, complex problems more efficiently.
2. Exploration and exploitation are the fundamental operations that in a SI search. The success rate of finding optimal solutions depend on the balance between exploration and exploitation.
3. As the nature of the optimization problem is different with one another, we need to choose which SI techniques to use based on various factors like working mechanism of the SI techniques, its parameters tuning capabilities, setting of parameters etc.
4. While one or some SI algorithms may perform well on some optimization problems, it might not perform well on other optimization problems. So, according to our requirements we may need to modify parts of SI techniques, hybridize with other algorithms so that we can get the desired solution.
5. Prior analysis of usage of SI techniques is crucial for applications. Though, statistical measures are available, but mostly lacks direct comparison [64]. Alternative techniques like visual analysis, which considers direct comparison of solutions would be useful [65, 66].

As different swarm intelligence based algorithms provide a way to overcome different optimization problems efficiently and effectively as compared to traditional techniques, Swarm Intelligence techniques are becoming more popular to solve various complex optimization problems. It has several advantages over traditional algorithms in terms having stochastic nature, efficient search of the search space using exploration and exploitation and faster convergence.

Acknowledgements This work is supported by the Science and Engineering Board (SERB), Department of Science and Technology (DST) of the Government of India under Grant No. EEQ/2019/000657.

References

1. Blum, C., Roli, A.: Metaheuristics in combinatorial optimization: overview and conceptual comparison. ACM Comput. Surv. (CSUR) **35**(3), 268–308 (2003)
2. Kennedy, J., Eberhart, R.: Particle swarm optimization. In: Proceedings of ICNN'95-international Conference on Neural Networks, vol. 4, pp. 942–1948. IEEE (1995)
3. Yang, X.-S.: Firefly algorithm, levy flights and global optimization. In: Research and Development in Intelligent Systems XXVI, pp. 209–218. Springer, Berlin (2010)
4. Binitha, S., Sathya, S.S. et al.: A survey of bio inspired optimization algorithms. Int. J. Soft Comput. Eng. **2**(2), 137–151 (2012)
5. Biswas, A., Mishra, K.K., Tiwari, S., Misra, A.K.: Physics-inspired optimization algorithms: a survey. J. Optim. (2013)
6. Houssein, E.H., Younan, M., Hassanien, A.E.: Nature-inspired algorithms: a comprehensive review. Hybrid Comput. Intell. 1–25 (2019)
7. Zhiheng, W., Jianhua, L.: Flamingo search algorithm: A new swarm intelligence optimization algorithm. IEEE Access **9**(1), 88564–88582 (2021)
8. MiarNaeimi, F., Azizyan, G., Rashki, M.: Horse herd optimization algorithm: a nature-inspired algorithm for high-dimensional optimization problems. Knowl. Based Syst. **213**, 106711 (2021)
9. Khishe, M., Mosavi, M.R.: Chimp optimization algorithm. Exp. Syst. Appl. **149**, 113338 (2020)
10. Hayyolalam, V., Kazem, A.A.P.: Black widow optimization algorithm: a novel meta-heuristic approach for solving engineering optimization problems. Eng. Appl. Artif. Intell. **87**, 103249 (2020)
11. Xue, J., Shen, B.: A novel swarm intelligence optimization approach: sparrow search algorithm. Syst. Sci. Control Eng. **8**(1), 22–34 (2020)
12. Dhiman, G., Garg, M., Nagar, A., Kumar, V., Dehghani, M.: A novel algorithm for global optimization: Rat swarm optimizer. J. Ambient. Intell. Hum. Comput. 1–26 (2020)
13. Shadravan, S., Naji, H.R., Bardsiri, V.K.: The sailfish optimizer: a novel nature-inspired metaheuristic algorithm for solving constrained engineering optimization problems. Eng. Appl. Artif. Intell. **80**, 20–34 (2019)
14. Al-Obaidi, A.T.S., Abdullah, H.S., Ahmed, Z.O.: Meerkat clan algorithm: a new swarm intelligence algorithm. Indonesian J. Electri. Eng. Comput. Sci. **10**(1), 354–360 (2018)
15. Mirjalili, S.Z., Mirjalili, S., Saremi, S., Faris, H., Aljarah, I.: Grasshopper optimization algorithm for multi-objective optimization problems. Appl. Intell. **48**(4), 805–820 (2018)
16. Mirjalili, S., Gandomi, A.H., Mirjalili, S.Z., Saremi, S., Faris, H., Mirjalili, S.M.: Salp swarm algorithm: a bio-inspired optimizer for engineering design problems. Adv. Eng. Softw. **114**, 163–191 (2017)
17. Al-Obaidi, A.T.S., Abdullah, H.S., et al.: Camel herds algorithm: a new swarm intelligent algorithm to solve optimization problems. Int. J. Perceptive Cogn. Comput. **3**(1) (2017)
18. Wang, W., Wu, S., Lu, K., et al.: Duck pack algorithm-a new swarm intelligence algorithm for route planning based on imprinting behavior. In: 2017 29th Chinese Control And Decision Conference (CCDC), pp. 2392–2396. IEEE (2017)
19. Mirjalili, S.: Dragonfly algorithm: a new meta-heuristic optimization technique for solving single-objective, discrete, and multi-objective problems. Neural Comput. Appl. **27**(4), 1053–1073 (2016)
20. Ebrahimi, A., Khamehchi, E.: Sperm whale algorithm: an effective metaheuristic algorithm for production optimization problems. J. Natural Gas Sci. Eng. **29**, 211–222 (2016)

21. Tian-Qi, W., Yao, M., Yang, J.-H.: Dolphin swarm algorithm. Front. Inf. Technol. Electron. Eng. **17**(8), 717–729 (2016)
22. Askarzadeh, A.: A novel metaheuristic method for solving constrained engineering optimization problems: crow search algorithm. Comput. Struct. **169**, 1–12 (2016)
23. Mirjalili, S.: The ant lion optimizer. Adv. Eng. softw. **83**, 80–98 (2015)
24. Wang, G.-G., Deb, S., dos S Coelho, L.: Elephant herding optimization. In: 2015 3rd International Symposium on Computational and Business Intelligence (ISCBI), pp. 1–5. IEEE (2015)
25. Mirjalili, S.: Moth-flame optimization algorithm: A novel nature-inspired heuristic paradigm. Knowl. Based Syst. **89**, 228–249 (2015)
26. Mirjalili, S., Mirjalili, S.M., Lewis, A.: Grey wolf optimizer. Adv. Eng. softw. **69**, 46–61 (2014)
27. Goel, S.: Pigeon optimization algorithm: a novel approach for solving optimization problems. In: 2014 International Conference on Data Mining and Intelligent Computing (ICDMIC), pp. 1–5. IEEE (2014)
28. Bansal, J.C., Sharma, H., Jadon, S.S., Clerc, M.: Spider monkey optimization algorithm for numerical optimization. Memetic Comput. **6**(1), 31–47 (2014)
29. Cuevas, E., Cienfuegos, M., Zaldívar, D., Pérez-Cisneros, M.: A swarm optimization algorithm inspired in the behavior of the social-spider. Exp. Syst. Appl. **40**(16), 6374–6384 (2013)
30. Niu, B., Wang, H.: Bacterial colony optimization. Discret. Dyn. Nat. Soc. (2012)
31. Nguyen, H.T., Bhanu, B.: Zombie survival optimization: a swarm intelligence algorithm inspired by zombie foraging. In: Proceedings of the 21st International Conference on Pattern Recognition (ICPR2012), pp. 987–990. IEEE (2012)
32. Yang, X.-S.: A new metaheuristic bat-inspired algorithm. In: Nature Inspired Cooperative Strategies for Optimization (NICSO 2010), pp. 65–74. Springer, Berlin (2010)
33. Tan, Y., Zhu, Y.: Fireworks algorithm for optimization. In: International Conference in Swarm Intelligence, pp. 355–364. Springer, Berlin (2010)
34. Yang, X.-S., Deb, S.: Cuckoo search via lévy flights. In: 2009 World Congress on Nature and Biologically Inspired Computing (NaBIC), pp. 210–214. IEEE (2009)
35. Rashedi, E., Nezamabadi-Pour, H., Saryazdi, S.: Gsa: a gravitational search algorithm. Inf. Sci. **179**(13), 2232–2248 (2009)
36. Krishnanand, K.N., Ghose, D.: Glowworm swarm optimisation: a new method for optimising multi-modal functions. Int. J. Comput. Intell. Stud. **1**(1), 93–119 (2009)
37. Chu, Y., Mi, H., Liao, H., Ji, Z., Wu, Q.H.: A fast bacterial swarming algorithm for high-dimensional function optimization. In: 2008 IEEE Congress on Evolutionary Computation (IEEE World Congress on Computational Intelligence), pp. 3135–3140. IEEE (2008)
38. Chu, S.-C. Tsai, P.-W., Pan, J.-S.: Cat swarm optimization. In: Pacific Rim International Conference on Artificial Intelligence, pp. 854–858. Springer, Berlin (2006)
39. Hancer, E., Ozturk, C., Karaboga, D.: Artificial bee colony based image clustering method. In: 2012 IEEE Congress on Evolutionary Computation, pp. 1–5. IEEE (2012)
40. Wedde, H.F., Farooq, M., Zhang, Y.: Beehive: an efficient fault-tolerant routing algorithm inspired by honey bee behavior. In: International Workshop on Ant Colony Optimization and Swarm Intelligence, pp. 83–94. Springer, Berlin (2004)
41. Dorigo, M., Birattari, M., Stutzle, T.: Ant colony optimization. IEEE Comput. Intell. Mag. **1**(4), 28–39 (2006)
42. Sait, S.M., Sheikh, A.T., El-Maleh, A.H.: Cell assignment in hybrid cmos/nanodevices architecture using a pso/sa hybrid algorithm. J. Appl. Res. Technol. **11**(5), 653–664 (2013)
43. Kuo, R.J., Hong, C.W.: Integration of genetic algorithm and particle swarm optimization for investment portfolio optimization. Appl. Math. Inf. Sci. **7**(6), 2397 (2013)
44. Chen, S.-M., Chien, C.-Y.: Solving the traveling salesman problem based on the genetic simulated annealing ant colony system with particle swarm optimization techniques. Exp. Syst. Appl. **38**(12), 14439–14450 (2011)
45. Jau, Y.-M., Kuo-Lan, S., Chia-Ju, W., Jeng, J.-T.: Modified quantum-behaved particle swarm optimization for parameters estimation of generalized nonlinear multi-regressions model based on choquet integral with outliers. Appl. Math. Comput. **221**, 282–295 (2013)

46. Chen, M., Ludwig, S.A.: Discrete particle swarm optimization with local search strategy for rule classification. In: 2012 Fourth World Congress on Nature and Biologically Inspired Computing (NaBIC), pp. 162–167. IEEE (2012)
47. Biswas, A., Biswas, B., Kumar, A., Mishra, K.K.: Particle swarm optimisation with time varying cognitive avoidance component. Int. J. Comput. Sci. Eng. **16**(1), 27–41 (2018)
48. Biswas, A., Kumar, A., Mishra, K.K.: Particle swarm optimization with cognitive avoidance component. In: 2013 International Conference on Advances in Computing, Communications and Informatics (ICACCI), pp. 149–154. IEEE (2013)
49. Qiu, C., Wang, C., Zuo, X.: A novel multi-objective particle swarm optimization with k-means based global best selection strategy. Int. J. Comput. Intell. Syst. **6**(5), 822–835 (2013)
50. Biswas, A., Lakra, A.V., Kumar, S., Singh, A.: An improved random inertia weighted particle swarm optimization. In: 2013 International Symposium on Computational and Business Intelligence, pp. 96–99. IEEE (2013)
51. Chuang, L.-Y., Tsai, S.-W., Yang, C.-H.: Chaotic catfish particle swarm optimization for solving global numerical optimization problems. Appl. Math. Comput. **217**(16), 6900–6916 (2011)
52. Fister, I., Fister Jr, I., Yang, X.-S., Brest, J.: A comprehensive review of firefly algorithms. Swarm Evol. Comput. **13**, 34–46 (2013)
53. Singh, O., Rishiwal, V., Chaudhry, R., Yadav, M.: Multi-objective optimization in wsn: opportunities and challenges. Wirel. Pers. Commun. **121**(1), 127–152 (2021)
54. Wright, S.J., Kanevsky, D., Deng, L., He, X., Heigold, G., Li, H.: Optimization algorithms and applications for speech and language processing. IEEE Trans. Audio, Speech, Lang. Process. **21**(11), 2231–2243 (2013)
55. Jino Ramson, S.R., Lova Raju, K., Vishnu, S., Anagnostopoulos, T.: Nature inspired optimization techniques for image processing-a short review. In: Nature Inspired Optimization Techniques for Image Processing Applications, pp. 113–145 (2019)
56. Handl, J., Kell, D.B., Knowles, J.: Multiobjective optimization in bioinformatics and computational biology. IEEE/ACM Trans. Comput. Biol. Bioinform. **4**(2), 279–292 (2007)
57. Darwish, A., Hassanien, A.E., Das, S.: A survey of swarm and evolutionary computing approaches for deep learning. Artif. Intell. Rev. **53**(3), 1767–1812 (2020)
58. Biswas, A.: Community detection in social networks using agglomerative and evalutionary techniques. Ph.D. thesis (2016)
59. Biswas,A., Gupta, P., Modi, M., Biswas, B.: Community detection in multiple featured social network using swarm intelligence. In: International Conference on Communication and Computing (ICC 2014). Bangalore (2014)
60. Biswas, A., Gupta, P., Modi, M., Biswas, B.: An empirical study of some particle swarm optimizer variants for community detection. In: Advances in Intelligent Informatics, pp. 511–520. Springer, Berlin (2015)
61. Garg, A., Biswas, A., Biswas, B.: Evolutionary computation techniques for community detection in social network analysis. In: Advanced Methods for Complex Network Analysis, pp. 266–284. IGI Global (2016)
62. Parpinelli, R.S., Teodoro, F.R., Lopes, H.S.: A comparison of swarm intelligence algorithms for structural engineering optimization. Int. J. Numer. Methods Eng. **91**(6), 666–684 (2012)
63. Biswas, A., Biswas, B.: Swarm intelligence techniques and their adaptive nature with applications. In: Complex System Modelling and Control Through Intelligent Soft Computations, pp. 253–273. Springer, Berlin (2015)
64. Biswas, A., Biswas, B.: Regression line shifting mechanism for analyzing evolutionary optimization algorithms. Soft Comput. **21**(21), 6237–6252 (2017)
65. Biswas, A., Biswas, B.: Visual analysis of evolutionary optimization algorithms. In: 2014 2nd International Symposium on Computational and Business Intelligence, pp. 81–84. IEEE (2014)
66. Biswas, A., Biswas, B.: Analyzing evolutionary optimization and community detection algorithms using regression line dominance. Inf. Sci. **396**, 185–201 (2017)
67. Revathi, J., Eswaramurthy, V.P., Padmavathi, P.: Bacterial colony optimization for data clustering. In: 2019 IEEE International Conference on Electrical, Computer and Communication Technologies (ICECCT), pp. 1–4 (2019)

68. Niu, B., Wang, H.: Bacterial colony optimization: principles and foundations. In: Emerging Intelligent Computing Technology and Applications, pp. 501–506 (2012)
69. Hussien, A.G., Amin, M., Wang, M., Liang, G., Alsanad, A., Gumaei, A., Chen, H.: Crow search algorithm: theory, recent advances, and applications. IEEE Access **8**, 173548–173565 (2020)
70. Zolghadr-Asli, B., Bozorg-Haddad, O., Chu, X.: Crow Search Algorithm (CSA), pp. 143–149. Springer Singapore, Singapore (2018)
71. Niu, P., Niu, S., Liu, N., Chang, L.: The defect of the grey wolf optimization algorithm and its verification method. Knowl. Based Syst. **171**, 37–43 (2019)
72. Mirjalili, S., Mirjalili, S., Lewis, A.: Grey wolf optimizer. Adv. Eng. Softw. **69**, 46–61, 03 (2014)
73. Ebrahimi, A., Khamehchi, E.: Sperm whale algorithm: an effective metaheuristic algorithm for production optimization problems. J. Nat. Gas Sci. Eng. **29**, 211–222 (2016)
74. Yang, J., Qu, L., Shen, Y., Shi, Y., Cheng, S., Zhao, J., Shen, X.: Swarm intelligence in data science: applications, opportunities and challenges. In: Tan, Y., Shi, Y., Tuba, M. (eds.) Advances in Swarm Intelligence, pp. 3–14. Springer International Publishing, Cham (2020)
75. Bhatnagar, V., Balochian, S., Yan, J., Zhang, Y., Agarwal, P.: Swarm intelligence and its applications. Sci. World J. (2013)
76. Ganesan, R., Sarobin, M.V.R.: Swarm intelligence in wireless sensor networks: a survey. Int. J. Pure Appl. Math. **101** (2015)
77. Devi, K.U., Sarma, D. and Laishram, R.: Swarm intelligence based computing techniques in speech enhancement. In: 2015 International Conference on Green Computing and Internet of Things (ICGCIoT), pp. 1199–1203 (2015)
78. Zhuang, X., Mastorakis, N.: Image processing with the artificial swarm intelligence. WSEAS Trans. Comput. **4**, 333–341, 04 (2005)
79. Kumar, S., Datta, D., Singh, S.: Swarm Intelligence for Biometric Feature Optimization, pp. 147–181. 01 (2015)
80. Khan, I.R., Alam, M., Khan, A.H.: Swarm intelligence in manets: a survey. Int. J. Emerg. Res. Manag. Technol. **5**, 141–150, 05 (2016)
81. Nayar, N., Ahuja, S., Jain, S.: Swarm Intelligence and Data Mining: A Review Of Literature and Applications in Healthcare, pp. 1–7, 06 (2019)
82. Anghinolfi, D., Boccalatte, A., Grosso, A., Paolucci, M., Passadore, A., Vecchiola, C.: A Swarm Intelligence Method Applied to Manufacturing Scheduling, pp. 65–70, 01 (2007)

Swarm Intelligence for Deep Learning: Concepts, Challenges and Recent Trends

Vandana Bharti, Bhaskar Biswas, and Kaushal Kumar Shukla

Abstract Machine learning and deep learning have undoubtedly contributed to tremendous achievements in Artificial Intelligence (AI) in recent years, and more are likely to follow. They have demonstrated extraordinary superiority in various real-world applications like computer vision, medical diagnostic systems, agriculture, robotics, and many more. It enables automating the computer-aided system and drastically reducing the human workload where correct prediction with accurate precision is needed. On the other side, as technology advances, a vast amount of data is generated, raising the problem complexity and computational challenges of real-world applications. Furthermore, machine learning, deep learning, and the majority of real-world applications have complex optimization problems within themselves that must be adequately addressed for better and more accurate analysis. Nonetheless, we believe that swarm intelligence-based approaches to deep learning have traditionally been understudied and may ultimately deliver similar advances in AI capabilities - either building on those provided by deep learning or offering whole new ones. Swarm intelligence approaches are frequently employed to solve a wide range of optimization issues. Nowadays, swarm intelligence-based methods are attracting a lot of attention from the research communities of different domains because previous research in complex optimization has shown that behavioral patterns and phenomena observed in nature have the ability to facilitate the foundation for many optimization algorithms and solve problems efficiently. Swarm intelligence, machine learning, and deep learning, on the other hand, each has its own set of advantages and disadvantages. Recently, research communities have discovered an interest in integrating these concepts in order to overcome the limitations of each domain and give rise to a new paradigm known as evolutionary machine learning or evolution-

V. Bharti (✉) · B. Biswas (✉) · K. K. Shukla
Department of Computer Science and Engineering, Indian Institute of Technology (BHU), Varanasi 221005, U.P., India
e-mail: vandanabharti.rs.cse17@iitbhu.ac.in

B. Biswas
e-mail: bhaskar.cse@iitbhu.ac.in

K. K. Shukla
e-mail: kkshukla.cse@iitbhu.ac.in

A. Biswas et al. (eds.), *Advances in Swarm Intelligence*, Studies in Computational Intelligence 1054, https://doi.org/10.1007/978-3-031-09835-2_3

ary deep learning. In the case of machine learning and deep learning, the "curse of dimensionality," non-convex optimization, automatic parameter optimization, and optimal architecture are just a few of the issues that can be efficiently addressed with swarm intelligence, whereas in the case of swarm intelligence, slow convergence, local optima stagnation, and extensive computation cost can be addressed with the machine learning and deep learning community. Therefore, a robust and self-efficient model can be developed by integrating these concepts to solve the complex problem associated with real-world applications. This hybrid approach benefits the majority of research domains. Thus, this chapter will primarily present the ideas, challenges, and recent trends of an integrative approach of swarm intelligence with deep learning, which is currently in high demand for addressing industrial problems.

Keywords Swarm intelligence · Deep learning · Neuroevolution · Hyperparameter optimization

1 Introduction

A fundamental paradigm in the field of knowledge is the translation of scientific theories that describe natural occurrences into Computational Intelligence (CI) algorithms. Over thousands of years, nature has flourished through various stages of evolution, becoming a rich source of natural phenomena that lead to intelligent species, animals, and so on. These species, such as bird flocks, squirrels, fish, and other animals, have exceptional behavior, adaptability, and intelligence to carry out daily tasks necessary for survival. They are self-learners as well as group learners, which allows them to perform tasks such as food search, migration, and survival from attackers in an efficient manner. These traits are commonly referred to as Swarm Intelligence (SI), and they are used to design mathematical rules for solving complex optimization problems by balancing exploration and exploitation of the search space of specific problems. Researchers are intrigued by natural phenomena and social behavior that have the potential to solve the real-life optimization problem of these biological agents.

Moreover, special attention has been given to addressing complex modeling, simulation, and other complex optimization problem associated with Artificial Intelligence (AI) systems by utilizing the inherent advantages of the optimizer based on these natural phenomena. As far as the optimization problems are concerned, they can be convex optimization, non-convex optimization, unimodal or multimodal, single objective or multiobjective, and large-scale optimization problems. There is a large range of efficient optimization algorithms accessible for problems formulated as convex optimization [71]. However, many problems that are associated with primary application fields are non-convex. In precise, these problems are usually noisy, stochastic, and have dynamically shifting constraints.

SI approaches are widely utilized in this context since they have been demonstrated to be extremely effective for such complex problem areas. Over the last two decades,

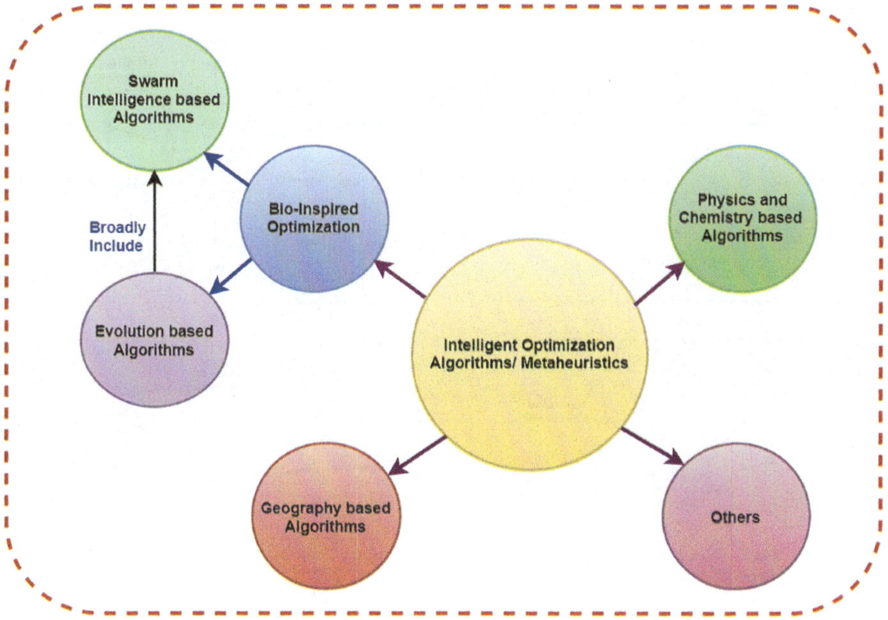

Fig. 1 Overview of Intelligent algorithms

plenty of promising SI algorithms has emerged [20]. Among them, few most promising are Particle Swarm Optimization (PSO), Grey wolf optimization (GWO), Harris Hawks Optimizer (HHO), Whale Optimization Algorithm (WOA), GDWCN-PSO, Harmony Search (HS), Ant colony Optimization (ACO), and Corona Virus Optimization algorithm [12, 68]. At this stage, it is necessary to have a comprehensive understanding of the concepts, ideas, applications, challenges, and solutions related to the deployment of SI algorithms. An overview of intelligent algorithms is presented in Fig. 1, and different subcategories of computational intelligent algorithms are presented in Fig. 2.

On the other hand, the advent of deep learning has made significant advances and demonstrated phenomenal performance in a broad array of applications, including adaptive testing, bioinformatics, computer vision, object detection, fault detection, face recognition, anomaly detection, stock market analysis, image synthesis, smart cities, and several others. Nowadays, it is one of the popular choices in the healthcare industry for analyzing crucial diseases like cancer diagnosis using complex images [52, 82]. In addition to this, deep learning is also a suitable methodological choice for the researchers in modelling secure medical data feature classification [54] and recently its application on COVID-19 prediction has been shown in [53], where a secure model for federated learning utilizing the concepts of model inversion and knowledge distillation in a hostile environment was proposed. Deep learning has led to significant breakthroughs in a wide variety of domains, which is due to deep learning's strong automatic representation capabilities. In [52], authors presented

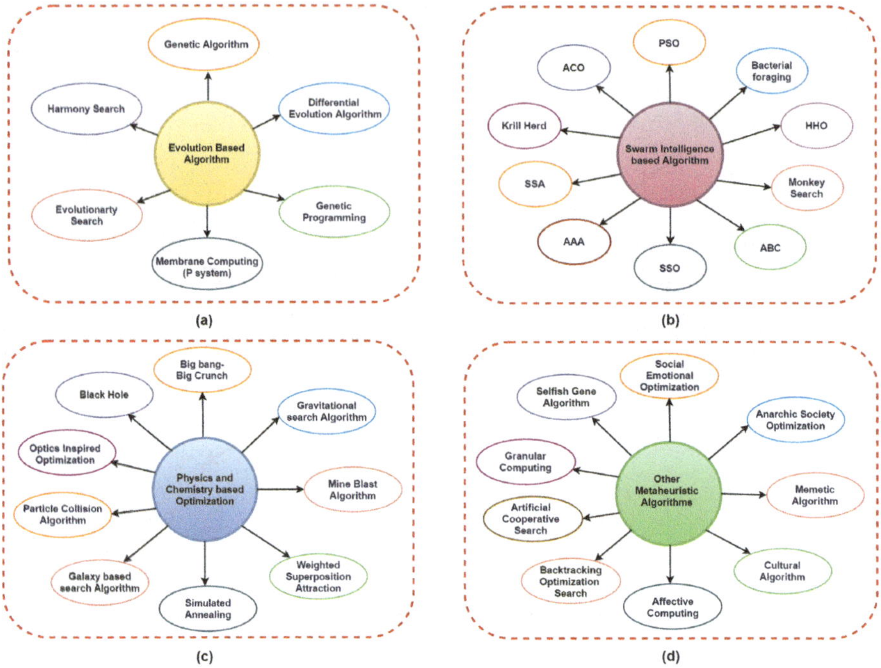

Fig. 2 **a** Computational intelligence based on EA **b** Swarm intelligence based metaheuristics **c** CI inspired from physics **d** Others

a novel deep feature extractor for cancer prediction with extensive experimental results, which showcased the potential of representation capabilities of deep model. Deep learning also adheres to a variety of techniques and topographies that can be used to a number of challenging problems [88]. The approach learns illustrative and differential features in a highly layered manner. In precise, deep learning prioritizes high-level data representation learning over task-specific learning [56]. Furthermore, deep learning can be utilized to analyze and mine intriguing Big Data models [15]. As a result, it is widely acknowledged to be of critical importance in the field of AI. It has been shown to be the best solution for identifying complex architecture in high-dimensional data using the backpropagation algorithm.

Despite its broad success, network architecture design has been shown to be critical to data feature representation and overall performance. In order to achieve a meaningful feature representation of data, the researchers created a number of complex network architectures. However, the design of the network architecture is strongly dependent on the researchers' past knowledge and expertise. Due to the constraints of inherent human knowledge, it is challenging for people to break free from the underlying thinking paradigm and build an optimal model. Besides this, deep learning, like many other areas of research, consists of an optimization problem within itself, which can be single or multiobjective, convex or non-convex, and constrained or unconstrained. As a result, it is needed to reduce human involvement to the greatest extent possible and let the algorithm automatically design the network architecture. For this, we need an efficient and intelligent optimization algorithm, and in such cases, SI algorithms have proven their potential due to inherent capability and derivative-free nature.

SI, machine learning, and deep learning, on the other hand, each has their own set of strengths and shortcomings. Recently, researchers have been motivated to integrate these concepts to overcome their limitations respectively and give rise to a new paradigm of evolutionary deep learning or evolutionary AI. In the following sections, firstly, we give the overview of SI and deep learning, and after that, we primarily focus on deep learning with SI. Furthermore, we reviewed current developments in this field, as well as highlighted future challenges, before concluding the chapter with a brief summary.

2 Swarm Intelligence Algorithms: An Overview

A swarm, in general, is a huge or dense collection of flying insects. Nature is a great witness to several natural phenomena that solve complex optimization problems. Nature has evolved various swarm behaviors in the animal kingdom as a way for different creatures to collaborate and achieve global mutual objectives such as nest-building, foraging, or fighting enemies. Although the objective of the group can be quite complex, it has been seen in animal systems that each individual swarm member follows a set of very basic rules and interacts locally with other individuals and the environment. Usually, a single swarm member cannot find an efficient solution on its own, but under the right conditions, the swarm can find effective solutions. Beni and Wang [9] coined the term SI in 1989, to explain the behavior of a group of cellular robots that may be viewed as a type of intelligent collective behavior. Although animal behavior has always been a primary source of inspiration for SI innovations, this was the first time swarm behaviors were investigated outside of natural sciences. Basically, all such type of natural phenomena is capable of handling complex optimization problems because of their inherent intelligence and are broadly termed as CI or intelligent optimization algorithms. A brief overview of such optimization algorithms is provided in Table 1. However specific examples are presented in Fig. 2b. In Table 1, we have covered the proposals till 2021 while there are other recent proposals which are generally come under hybrid optimizer or some are with new inspirations. Few of them are GDWCNPSO [12], HHO [36] and manta ray foraging algorithm [99].

Table 1 Brief summary of intelligent swarm optimization algorithms

References	Algorithm	Motivation	Year
Kennedy et al. [25]	Particle Swarm Optimization (PSO)	Smart social behavior of bird flock	1995
Hersovici et al. [37]	Shark Search Algorithm	Feeding mechanism and coordinated movement of fish	1998
Nara et al. [70]	Sheep Flock Heredity Model	Natural evolution of sheep flocks	1999
Passino [74]	Bacterial Foraging Algorithm	Foraging strategy of *E. Coli* bacteria	2002
Li [58]	Artificial Fish Swarm Algorithm (AFSA)	Fish swarm's collective intelligence	2003
Martin et al. [61]	Termite Algorithm	Termite colony	2006
Dorigo [13]	Ant Colony Optimization (ACO)	Behaviour of real ant colony	2006
Karaboga and Basturk [48]	Artificial Bee Colony (ABC)	Honey Bee	2007
Mucherino et al. [69]	Monkey Search (MS)	Monkey's food-seeking behavior while climbing trees	2007
He et al. [35]	Group Search Optimizer (GSO)	Foraging behavior in animals	2009
Yang [92]	Firefly Algorithm	Firefly flashing behavior	2009
Yang and Deb [94]	Cuckoo Search	Obligate brood parasitism of cuckoo	2009
Yang [93]	Bat Algorithm	Bat echolocation behavior	2010
Pan [73]	Fruit Fly Optimization algorithm (FFOA)	Fruit fly's fruit-seeking behavior	2012
Gandomi et al. [32]	Krill Herd (KH)	Krill's herd herding behavior in nature	2012
Kaveh et.al [50]	Dolphin Echolocation	Dolphin echolocation ability	2013
Cuevas et al. [19]	Social Spider Optimization Algorithm	Social spider's cooperative behavior	2014
Uymaz et al. [87]	Artificial Algae Algorithm	Microalgae living behaviors	2015
Mirjalili [63]	Ant Lion Optimizer (ALO)	Ant lions hunting mechanism in nature	2015
Mirjalili et al.[64]	Dragonfly Algorithm	Dragonfly swarming behaviors, both static and dynamic	2016

(continued)

Table 1 (continued)

References	Algorithm	Motivation	Year
Abedinia et al. [1]	Shark Smell Optimization(SSO)	Shark's ability to locate prey using its smell sense	2016
Yong et al. [95]	Dolphin Swarm Optimization Algorithm (DSOA)	Mechanism of dolphins to detect, chase, and prey on sardine swarms	2016
Li et al. [57]	Virus Colony Search	Virus infection and diffusion strategies	2016
Mirjalili and Lewis [65]	Whale Optimization Algorithm (WOA)	Humpback whale social behavior	2016
Mirjalili et al. [66]	Multi-verse Optimizer (MVO)	Based on a cosmology theory concept	2016
Askarzadeh [4]	Crow Search Algorithm (CSA)	Crows' intelligent food-hiding behavior	2016
Mirjalili et al. [67]	Salp Swarm Algorithm	Salps' swarming behavior when navigating and foraging in the oceans	2017
Saremi et al. [79]	Grasshopper Optimization Algorithm	Grasshopper swarming behavior	2017
Fausto et al. [29]	Selfish Herd Optimizer (SHO)	Hamilton's selfish theory	2017
Dhiman et al. [22]	Spotted Hyena Optimizer	Spotted hyenas' social behavior	2017
Qi et al. [75]	Butterfly-inspired Algorithm	Butterfly mate-searching mechanism	2017
Jahani et al. [42]	Mouth Brooding Fish Algorithm	Life cycle of mouth brooding fish	2018
Kaur and Arora [49]	Chaotic Whale Optimization	Hybrid approach to improve the efficiency of WOA	2018
Torabi et al.[86]	Improved Raven Roosting Optimization Algorithm	Ravens' social roosting and foraging behavior	2018
Jain et al. [43]	Squirrel Search Algorithm (SSA)	Dynamic foraging behavior	2019
Bharti et al. [12]	Genetic Directed Weighted Complex Network PSO	Dynamic network topology of swarms	2021
Dhiman et al. [23]	Rat Swarm Optimization	Rat chasing and attacking behavior	2021
Salehan et al. [78]	Corona Virus Optimization	Corona virus characteristics and behavior	2021

3 Deep Learning: An Overview

Deep learning is a prominent technique in machine learning and AI, and it has evolved into a core learning method of the revolutionary Industry, Industry 4.0. Deep learning is emanated from the neural network and due to its exceptional learning capabilities from data make it worthy and popular among industries and research communities. However, neural networks were widely acceptable models in machine learning and AI in late 1980. Following that, various innovative models like multi-layer perceptron trained through backpropagation, radial basis function (RBF) networks, etc., were developed. However, after a certain time, researchers were losing their interest in the neural network. Again, in 2006 Hinton et al. [38] proposed a new generation neural network which is basically a rebirth of neural network in the form of deep learning, and now it is accepted as a more successful technique in every field. The rise of deep learning has a long history which was further divided in (i) First generation had time period (1958–1969), that started with single-layer perceptron neural network developed by Rosenblatt [76]. But once the observation made by Minsky [62] in 1969 was presented that single-layer perceptron was incapable of solving problems that are linearly inseparable, resulting in the research on the neural network being halted around 20 years. (ii) Second generation lied in 1986–1998 where Hinton et al. [77] proposed multi-layer perceptron with backpropagation, sigmoid function for nonlinear embedding which is capable of solving nonlinear classification problems. Further, Hornik [40] came with a universal approximation theorem, and after that, LeCun et al. [55] proposed a Convolutional Neural Networks (CNN), but it takes three days for training, and in 1991 gradient vanishing problem was reported. Besides, unclear theoretical and mathematical foundations and trial-error approaches weaken the usability of this model while other statistical learning methods with strong mathematical foundations gain popularity. Finally, the third generation neural network models (2006 to present) change the perception of researchers when a graphical model of the brain is explored and presented in the form of a deep learning model [39]. After that, several inventions like AutoEncoders (AE), deep belief network, AlexNet, VGGNet, and many more are presented. Even to overcome the problem of vanishing gradient, solutions like pre-training then tuning were given. Authors in [34] also proposed a ReLU activation function, which is capable of suppressing the gradient vanishing problem. Due to technological innovations like GPU, deep learning has become more popular in different domains. Deep learning refers to a collection of methods and models that includes, but is not limited to, LSTM, CNN, Generative adversarial network (GAN), Auto Encoders, and many others. The broad classification of an artificial neural network is presented in Fig. 3.

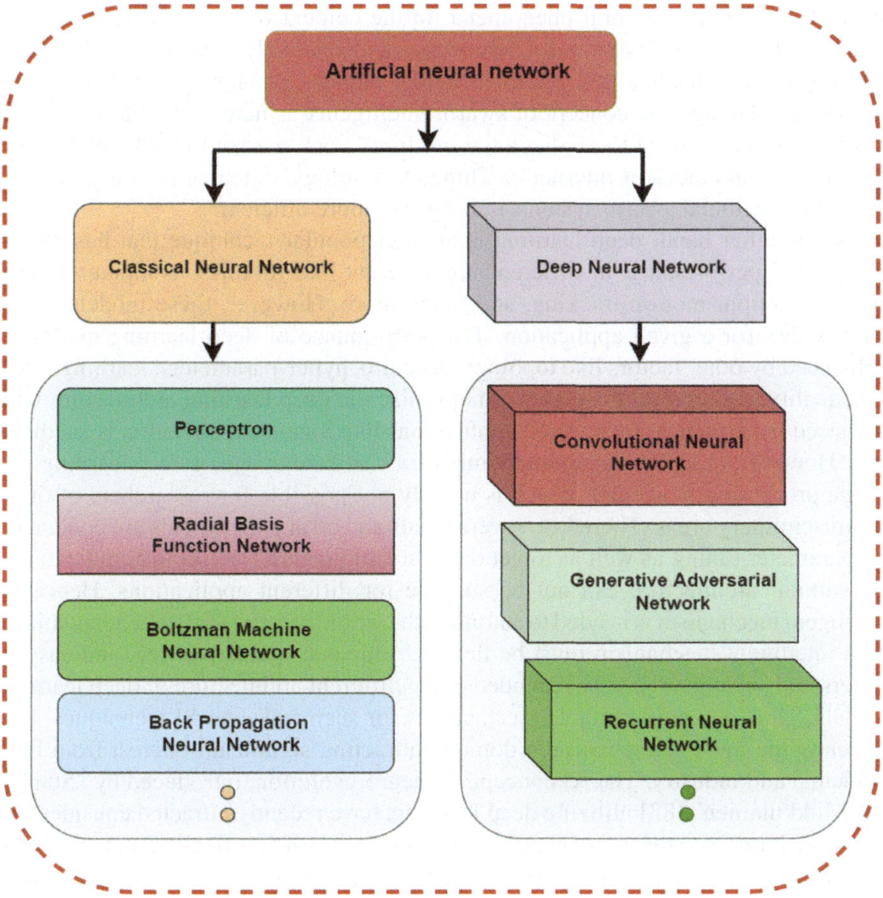

Fig. 3 Broad categorization of neural network

4 Swarm Intelligence with Deep Learning

The mechanisms of biological evolution and natural events performed by organisms serve as inspiration for designing CI. CI has discovered creative and innovative solutions to complex practical problems with algorithmic advancements and increased computing resources. This chapter is mainly dedicated to assessing the applicability of such intelligent techniques for solving large-scale optimization problems. How the strength of such methods like flexibility, derivative-free, complex mapping, and information transition from one to other in an optimal way can be utilized for complex networks and to overcome the limitations of both approaches are discussed. The flaws of existed approaches give a roadmap for future invention to narrow down the gaps of our understanding, mathematical formulations of natural events to the original transition of natural phenomena. In the context of complex deep models, Swarm Artificial Intelligence is a fascinating and relatively new field of study for researchers. To enhance and optimize model choices, it blends global and local knowledge. Though the concept of swarm intelligence is new to the literature, it is increasingly being used to predict anything from stock market movements to sales forecasting. Advances in Internet of Things technology, machine learning, and 5G have made artificial swarm systems quicker and more efficient.

On the other hand, deep learning is another popular technique that has shown remarkable performance in different applications like robotics, computer vision, threat detection, motion tracking, and many more. However, these models act as a black box for a given application. The performance of deep learning models is influenced by other factors like topology structure, hyper-parameters, learning rules, data quality, and optimum weights. To optimize the deep learning architecture with enhanced performance for specific applications like image classification is a tedious task. However, to make it an efficient model, a vast knowledge of deep learning and image processing is needed, which is usually not possible for researchers of other interdisciplinary areas. Therefore, several trials and error experiments are conducted for parameter tuning as well as to get the best architecture for specific applications, and similar architecture can not be suitable for different applications. Hence an intelligent mechanism is needed to automate the whole process for better adaptability. That intelligent mechanism must be flexible, efficient, derivative-free, and easy to understand such that it can be extended to the different architecture of deep learning models. Thus, the most suitable mechanisms for such tasks are SI techniques. It is currently the most active research domain, attracting significant interest from both academia and industry. The AI concepts of neuro-evolution, introduced by "Stanley and Miikkulainen" [83] utilizing deep learning, have recently attracted considerable attention. It refers to the use of simulated evolution to build artificial neural networks, including their learning mechanism, optimal weights, and network structure. Hence, the optimization of deep learning models using CI techniques is a topic of interest and debate in computer science and other fields. We focused primarily on neuroevolution and deep learning parameter optimization using intelligent swarm approaches in this chapter, which are covered in detail in the following sections.

4.1 Neuroevolution

An effective architecture design for deep learning is a crucial point in solving any problem. Nonetheless, it is incredibly challenging to explore and determine the ideal architecture in neural architecture search space that best matches the data of a specific problem. Neural Architecture Search (NAS) or Neuroevolution has recently gained prominence in this area. The major goal of neuroevolution is to search for the optimum neural network design to tackle the problem at hand automatically. In precise, a process of constructing neural architectures automatically for a specific task in order to avoid the tedious effort in designing a task-specific model is known as neuroevolution or NAS. Generally, NAS is a complex combinatorial optimization problem with complex constraints, discrete representation, and bi-level formulation. This optimization contains objective functions that are usually non-convex and non-differential. Numerous search strategies such as random-search, CI techniques, reinforcement learning, bayesian approach, and derivative information-based methods can be employed for exploring the neural networks space. Intelligent computing techniques are widely utilized for NAS in different applications [2, 31]. In [30], deep neural network has been searched and pruned extra parameter by using evolution search strategy and validated on medical images. Few recent works also show its potential in optimizing scalable architecture [98] and complex cyclic GAN [11] architecture in a unique way. This chapter mainly deals with intelligent computing search strategy so we will elaborate accordingly. Mathematically, neuroevolution can be formulated as optimization technique with possible architecture search space **S** and an architecture N under consideration is shown as below:

$$\begin{cases} argmin_N \ \Psi(N, D^{\text{train}}, D^{\text{Fitness}}) \\ \qquad\qquad s.t \ N \in \mathbf{S} \end{cases}$$

Here $\Psi(.)$ denotes the performance measure of N on dataset D^{Fitness}, after training on D^{train} which is a task-specific training dataset. The basic presentation of neuroevolution in AI is shown in Fig. 4. Researchers are quite fascinated by neuroevolution and hence applied different intelligent techniques such as in [46] authors update the positions of particles using the information of layers in spite of layer's parameters. Further, low convergence and efficiency issues of PSO in large search space of hyperparameters were addressed in [33] by proposing gradient-priority PSO. In another works [14, 28] Ant colony optimization is used for deep architecture search. Besides these, firefly optimization is also used to discover the best architecture of CNN for classification tasks [80]. However, multiobjective SI [45, 89] for NAS have been explored from the last few years, and there is a long way to go in this direction. Recent works and applications of intelligent computing in deep learning, along with challenges and future directions, are presented in [10, 100].

The representation scheme of solutions, also known as the encoding scheme, is an important factor in neuroevolution and model training. The majority of NAS approaches are characterized by their varied solution encoding schemes. The relationship between encoding space, initial space, and search space is shown in Fig. 5. Encoding scheme is further divided into two types: (i) direct encoding and (ii) indirect encoding.

- **Direct encoding scheme**: In NEAT (Neuroevolution of Augmenting Topologies) [83], authors suggested the benefit of direct encoding over the indirect scheme. Based on a linear representation, they proposed a direct encoding scheme in which each model parameter, like connection list, set of neurons, etc., is directly expressed as it is in this method. In this model, parameters are expressed as a vector [6, 91], with each member denoting a different model parameter. It is a phenotype construction of the model, and hence there are sufficient well-known data structures are

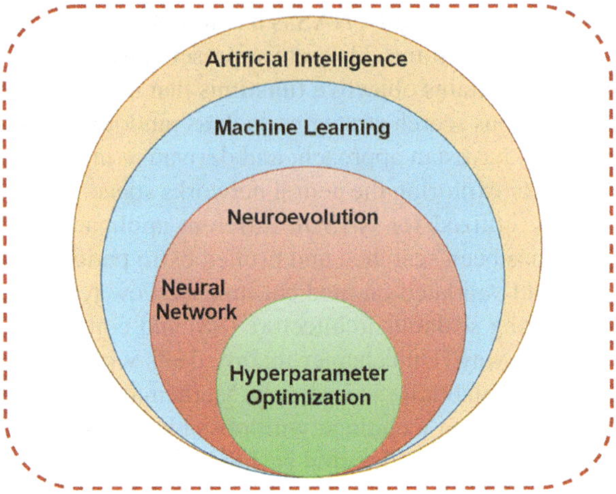

Fig. 4 Conceptual representation of neuroevolution in AI

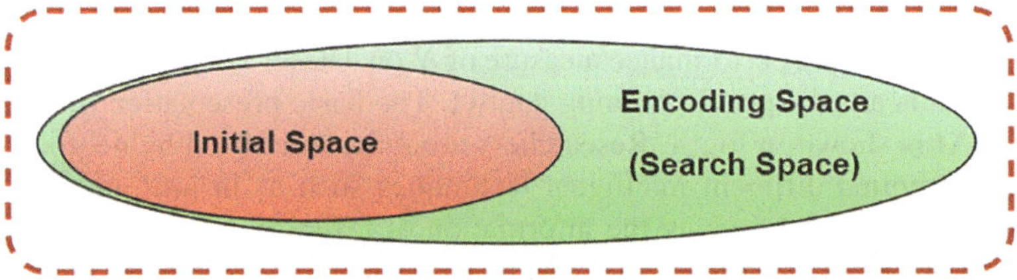

Fig. 5 Pictorial representation of relationship between search space, encoding space and initial space

available for such representation. Also, different variation operators are available that efficiently perform manipulation. Block-chained encoding and graph-based encoding are employed by several networks are direct encoding techniques.

- **Indirect encoding scheme**: An indirect encoding is a genotype that must be converted into a phenotype using a set of rules. This encoding family is divided into two main categories: binary encodings and grammar encodings. In this case, model parameters are encoded using specified mappings. This encoding paradigm is still less explored. Authors [84] represents network weights as a set of coefficients that are used to linearly combine some specified basis vectors to obtain network weights. While in another work [85], mean value and variance of Gaussian distribution are used to encode, and the weight of the deep model is sampled from this distribution.

4.2 Other Popular Deep Frameworks with SI

In this section, other popular deep learning frameworks with intelligent techniques have been discussed.

4.2.1 Deep Convolutional Neural Network

Manually expanding the width and depth of CNN often takes a significant period of time and expertise. Over the years, it has encouraged an increase in the need for NAS. On the other hand, most common NAS techniques only consider predictive performance criteria for optimization, while structural complexity was penalized. One year ago, the authors in [45] present MOPSO/D-Net, where NAS for CNN has been formulated as a multicriteria optimization problem with classification error rate and total number of parameters in a network as two objective functions, which was then minimized using multiobjective PSO based on decomposition. In another recent paper [72], authors presented IntelliSwAS, a swarm-based technique to NAS for improving CNN designs utilizing a particle swarm-based approach. The authors claimed that it is a generic framework that can be utilized for classification or regression. Directed acyclic graph recurrent network is used to evaluate the searched model. Further, multi-level PSO is recently utilized to optimize the architecture of CNN, and its hyperparameters, as well as the performance of the model, are validated on benchmark datasets for classification [81]. Various applications like intrusion detection also utilized this approach by applying adaptive PSO in CNN for better performance [47]. Authors [24] have also applied WOA to CNN for texture detection by optimizing filters and parameters of CNN. In addition to this, another application of distributed denial of service attack detection, social lion optimizer, is utilized with CNN [3]. Authors in [21] provided a new way to overcome the overfitting problem of CNN for image classification tasks by using PSO. PSO was used to identify the appropriate dropout ratio, which acts as a regularization parameter of CNN. The dropout ratio selection was driven by the loss function as a fitness function across the validation set. Experimental results on four benchmark datasets show the comparative analysis of different swarm optimizers like cuckoo search, bat algorithm, and firefly optimizer. They also presented the results with conventional dropout ratio and without dropout in CNN. Junior et al. [46] introduced psoCNN, a novel PSO-based strategy with unique direct encoding and velocity operator for automatic search of CNN for a classification task. It is capable of identifying an efficient CNN model with fast convergence. Further, Biomedical image segmentation for cancer detection also applied different SI approaches for optimizing specific CNN model U-Net. In this, parameters of U-Net are represented in the search space. It was observed that the model trained using swarm techniques gives a significant performance in contrast to the backpropagation deep model with a 25% quicker training time. Further, in [80] authors suggested an approach to allow persons with no expertise to design their own CNN model automatically without prior understanding in this domain. To generate the problem's search space, the recommended approach encoded the skip connections, which reflect the normal CNN block. An improved firefly algorithm utilizing the neighborhood attraction model was applied for exploring the search space of the targeted network and validated on two datasets (CIFAR-10, CIFAR-100) and resulted in better accuracy.

4.2.2 Generative Adversarial Network

GAN is a deep learning model that has proven to be a powerful image generation tool in recent years. It is made up of two parts: the generator (G) and the discriminator (D). The G learns how to generate data that looks similar to genuine training data in order to fool the D, whereas the D learns how to distinguish between data from a real training set and data generated by the G. Following training, the G has a high likelihood of producing data that is useful in a number of real-world applications. It started a new revolution in deep learning and computer vision. This revolution has resulted in some notable advances in research. Hence this section investigates a swarm footprint for GAN's generator training improvement. Generally, the training of the G can be expressed in the form of an optimization problem. Although the fitness function for the G cannot be directly created, its local state may be assessed as a function of D performance. So, the overall training process is basically a min-max optimization problem. Recently, the potential of intelligent computation techniques has also been experimented on this framework.

E-GAN [90] was the first proposal in this direction where they design training of GAN with different evolutionary operators and number of generators (population) are trained. They tried to give a solution for stable training that could alleviate the problem of mode collapse. Further, authors in [17] have introduced a novel model incorporating neuroevolution and co-evolution in the GAN training method. They employed the loss function (fitness function) for the discriminator and the Frchet Inception Distance (FID) for the generator. Further, they experimented on the MNIST dataset and showed that for the generator, the FID score is a good evaluation metric. However, it is crucial to measure competing metrics for discriminators. It is further extended by the author in [16], presenting an enhanced model CO-EGAN based on neuroevolution and co-evolution techniques by incorporating the adversarial feature of GAN components in the training process to construct coevolutionary methods.

Further, EvoGAN [59] used for facial image generation, CG-GAN (Composite generating GAN) [96] for facial composite generation and finally multiobjective EGAN [7] have been proposed. MO-EGAN further improved in [8] by improving the training of a single objective generator by using a multicriteria training process. Another recent work [11] proposed a new training scheme for Cyclic GAN using multiobjective optimization and simulated annealing concept. Different selection schemes and mutation operators have been proposed for cyclic GAN and resulted in the model EMOCGAN, which was further validated for unpaired image translation application. Recently, the authors in [41] presented GANs-PSO that tackle typical GAN problems such as mode collapse. In this work, GANs-PSO used the PSO algorithm with GAN to overcome the mode collapse problem during training for self-collision avoidance of arm robots. While PSO is also applied in [18], where authors designed a GAN by utilizing the concept of neuroevolution for biomedical image chest X-ray of COVID-19 generation. The whole training process, along with architecture search, was achieved using PSO. FID score is used as a fitness value, and a progressive growth approach is utilized to design GAN. The suggested approach achieves a superior FID score for image generation as compared to the baseline approach. Another work [5] utilized fractional Harris Hawk optimization to optimize GAN training for Osteosarcoma detection at an early stage to enhance the survival rate. PSO is also utilized in [97] to optimize the GAN for face generation, and based on quality and diversity evaluation metrics, the position of particles is updated.

4.2.3 Long Short Term Memory Network

LSTM models are recurrent deep networks with a wide range of applications, including time series analysis, speech and voice recognition. Recently, an APSO-based optimization approach has been proposed by [60], addressing the difficulties in the conventional LSTM model, such as poor convergence speed and ease of falling into local extremum in weight correction. The results of this approach reveal that when compared to the GALSTM, ACO-LSTM, and PSO-LSTM models, the average SSE value of the APSO-LSTM model under three data sets (SSD, OLD, and MDTD) is lowered by 9.7–22.6%. When compared to other algorithms, the prediction error of APSO-LSTM is the smallest, implying that prediction performance is greatly improved.

Ant colony optimization was used by ElSaid et al. [26] to evolve an LSTM network for airplane engine vibrations. By using an ant colony optimizer for constructing the LSTM cell structure, the authors optimized the analytical calculations prediction of engine vibration. Furthermore, the same authors [27] used an ant colony optimizer is applied to develop LSTMs cells, and while taking earlier cell output into account, the path between the input and hidden layer connections are produced. Another work [28] presented ant swarm neuroevolution where ant colony is applied for designing optimal RNN topologies. The process chooses from a variety of modern recurrent cell types as well as recurrent connections. PSO, another swarm optimizer, is also explored to search and optimize the vast hyperparameter search space of the LSTM-CNN model for classifying the role of role-based access control system to ensure the database security [51]. Further, in another work [44], CNN-LSTM architecture is automatically designed and optimized by a sine cosine algorithm for solar irradiance forecasting.

5 Challenges and Future of Intelligent Deep Learning

Intelligent approaches have shown their potential in designing and optimizing deep neural network models. It seeks a lot of attention from researchers and industrial people due to its remarkable performance. Although it automates the tedious task in deep learning and reduces the intervention of a human, but it also brings several challenges which need to be addressed in the future for its success.

- Most of the neuroevolution and optimal model training are computationally expensive. These methods usually consider the model performance for evaluation without considering computation cost. Several search mechanisms take multiple GPUs for several days, which is too costly to utilize for real-world applications. Further, lots of efforts have been made to speed up the process by restricting the search space, which again ends up with a solution that may not be best. Therefore, in the future, effort must be made to find the tradeoff between model performance and cost for optimization.
- There must be a systematic method and specific benchmarks for evaluating NAS techniques and the effectiveness of optimizing models because, in some cases, random search or hybrid search mechanisms work. As a result, it is still unclear which approach is appropriate for a specific application that is quite challenging and can be explored in the future.
- Most of the NAS and parameter optimization are evaluated on small-scale datasets like MNIST, CIFAR-10 and CIFAR 100, etc. It is challenging to see the performance on large-scale real-world datasets in terms of performance and computation cost.

Besides these, deep learning with intelligent techniques needs a huge amount of datasets, while in the real world, the availability of huge annotated datasets is unrealistic, so it becomes a big challenge to train a model in such a scenario. It will be a good direction if we can automate the process of data annotation. Moreover, we need to explore other natural phenomena that can be formulated as an efficient optimization algorithm suitable for designing such architecture efficiently. Recently, designing search space and encoding schemes is another interesting area that is still in the exploration stage. Designing an automated model for EDGE-based applications is also unexplored.

6 Conclusion

Deep learning is a popular and powerful framework that gives a new edge to computer vision. Due to its exceptional representation ability, it is successfully accepted by different domains. Although it is quite successful but at the same time, it is like a black box technique. To construct the best model architecture requires lots of expertise in a specific domain, as well as tedious effort for manual tuning using the trial and error method. On the other hand, intelligent computation techniques are flexible, derivative-free, and inspired by natural phenomena with the capability of solving a complex optimization problems. Both methods can be integrated to overcome each other's limitations and enhance their own strength in designing efficient automated models that reduce human intervention. Thus, this chapter briefly reviewed the recent works in deep learning with swarm intelligence and provide insight into ongoing work, recent trends, and challenges in this area. It also pointed out the future directions and subdomains that are still unexplored and need attention from academia and industry researchers.

References

1. Abedinia, O., Amjady, N., Ghasemi, A.: A new metaheuristic algorithm based on shark smell optimization. Complexity **21**(5), 97–116 (2016)
2. Angeline, P.J., Saunders, G.M., Pollack, J.B.: An evolutionary algorithm that constructs recurrent neural networks. IEEE Trans. Neural Netw. **5**(1), 54–65 (1994)
3. Arivudainambi, D., KA, V.K. and Sibi Chakkaravarthy, S.: Lion ids: a meta-heuristics approach to detect ddos attacks against software-defined networks. Neural Comput. Appl. **31**(5), 1491–1501 (2019)
4. Askarzadeh, A.: A novel metaheuristic method for solving constrained engineering optimization problems: crow search algorithm. Comput. Struct. **169**, 1–12 (2016)
5. Badashah, S.J., Basha, S.S., Ahamed, S.R., Subba Rao, S.: Fractional-harris hawks optimization-based generative adversarial network for osteosarcoma detection using renyi entropy-hybrid fusion. Int. J. Intell. Syst. **36**(10), 6007–6031 (2021)
6. Badem, H., Basturk, A., Caliskan, A., Yuksel, M.E.: A new efficient training strategy for deep neural networks by hybridization of artificial bee colony and limited-memory bfgs optimization algorithms. Neurocomputing **266**, 506–526 (2017)
7. Baioletti, M., Coello, C.A.C., Di Bari, G., Poggioni, V.: Multi-objective evolutionary gan. In: Proceedings of the 2020 Genetic and Evolutionary Computation Conference Companion, pp. 1824–1831 (2020)
8. Baioletti, M.., Di Bari, G., Poggioni, V., Coello, C.A.C.: Smart multi-objective evolutionary gan. In: 2021 IEEE Congress on Evolutionary Computation (CEC), pp. 2218–2225. IEEE (2021)
9. Beni, G., Wang, J.: Swarm intelligence in cellular robotic systems. In: Robots and Biological Systems: Towards a New Bionics?, pp. 703–712. Springer, Berlin (1993)
10. Bharti, V., Biswas, B., Shukla, K.K.: Recent trends in nature inspired computation with applications to deep learning. In: 2020 10th International Conference on Cloud Computing, Data Science and Engineering (Confluence), pp. 294–299. IEEE (2020)
11. Bharti, V., Biswas, B., Shukla, K.K.: Emocgan: a novel evolutionary multiobjective cyclic generative adversarial network and its application to unpaired image translation. Neural Comput. Appl. 1–15 (2021a)
12. Bharti, V., Biswas, B., Shukla, K.K.: A novel multiobjective gdwcn-pso algorithm and its application to medical data security. ACM Trans. Internet Technol. (TOIT) **21**(2), 1–28 (2021)
13. Bianchi, L., Dorigo, M.: Ant colony optimization and local search for the probabilistic traveling salesman problem: a case study in stochastic combinatorial optimization (2006)
14. Byla, E., Pang, W.: Deepswarm: Optimising convolutional neural networks using swarm intelligence. In: UK Workshop on Computational Intelligence, pp. 119–130. Springer, Berlin (2019)
15. Chen, X.W., Lin, X.: Big data deep learning: challenges and perspectives. IEEE Access **2**, 514–525 (2014)

16. Costa, V., Lourenço, N., Correia, J., Machado, P.: Coegan: evaluating the coevolution effect in generative adversarial networks. In: Proceedings of the Genetic and Evolutionary Computation Conference, pp. 374–382 (2019a)
17. Costa, V., Lourenço, N., Machado, P.: Coevolution of generative adversarial networks. In: International Conference on the Applications of Evolutionary Computation (Part of EvoStar), pp. 473–487. Springer, Berlin (2019b)
18. Rodríguez-de-la Cruz, J.A., Acosta-Mesa, H.G., Mezura-Montes, E.: Evolution of generative adversarial networks using pso for synthesis of covid-19 chest x-ray images. In: 2021 IEEE Congress on Evolutionary Computation (CEC), pp. 2226–2233. IEEE (2021)
19. Cuevas, E., Cienfuegos, M.: A new algorithm inspired in the behavior of the social-spider for constrained optimization. Expert. Syst. Appl. **41**(2), 412–425 (2014)
20. Darwish, A.: Bio-inspired computing: algorithms review, deep analysis, and the scope of applications. Futur. Comput. Inform. J. **3**(2), 231–246 (2018)
21. De Rosa, G.H., Papa, J.P., Yang, X.S.: Handling dropout probability estimation in convolution neural networks using meta-heuristics. Soft Comput. **22**(18), 6147–6156 (2018)
22. Dhiman, G., Kumar, V.: Spotted hyena optimizer for solving complex and non-linear constrained engineering problems. In: Harmony Search and Nature Inspired Optimization Algorithms, pp. 857–867. Springer, Berlin (2019)
23. Dhiman, G., Garg, M., Nagar, A., Kumar, V., Dehghani, M.: A novel algorithm for global optimization: rat swarm optimizer. J. Ambient. Intell. Hum. Comput. **12**(8), 8457–8482 (2021)
24. Dixit, U., Mishra, A., Shukla, A., Tiwari, R.: Texture classification using convolutional neural network optimized with whale optimization algorithm. SN Appl. Sci. **1**(6), 1–11 (2019)
25. Eberhart, R., Kennedy, J.: Particle swarm optimization. In: Proceeding of IEEE International Conference on Neural Network, pp 1942–1948. Perth, Australia (1995)
26. ElSaid, A., Wild, B., Jamiy, F.E., Higgins, J., Desell, T.: Optimizing lstm rnns using aco to predict turbine engine vibration. In: Proceedings of the Genetic and Evolutionary Computation Conference Companion, pp. 21–22 (2017)
27. ElSaid, A., Jamiy, F.E., Higgins, J., Wild, B., Desell, T.: Using ant colony optimization to optimize long short-term memory recurrent neural networks. In: Proceedings of the Genetic and Evolutionary Computation Conference, pp. 13–20 (2018)
28. ElSaid, A.A., Ororbia, A.G., Desell, T.J.: The ant swarm neuro-evolution procedure for optimizing recurrent networks (2019). arXiv:1909.11849
29. Fausto, F., Cuevas, E., Valdivia, A., González, A.: A global optimization algorithm inspired in the behavior of selfish herds. Biosystems **160**, 39–55 (2017)
30. Fernandes, F.E., Yen, G.G.: Automatic searching and pruning of deep neural networks for medical imaging diagnostic. IEEE Trans. Neural Netw. Learn. Syst. **32**(12), 5664–5674 (2020)
31. Floreano, D., Dürr, P., Mattiussi, C.: Neuroevolution: from architectures to learning. Evol. Intell. **1**(1), 47–62 (2008)
32. Gandomi, A.H., Alavi, A.H.: Krill herd: a new bio-inspired optimization algorithm. Commun. Nonlinear Sci. Numer. Simul. **17**(12), 4831–4845 (2012)
33. Gao, Z., Li, Y., Yang, Y., Wang, X., Dong, N., Chiang, H.D.: A gpso-optimized convolutional neural networks for eeg-based emotion recognition. Neurocomputing **380**, 225–235 (2020)
34. Glorot, X., Bordes, A., Bengio, Y.: Deep sparse rectifier neural networks. In: Proceedings of the Fourteenth International Conference on Artificial Intelligence and Statistics, JMLR Workshop and Conference Proceedings, pp. 315–323 (2011)
35. He, S., Wu, Q.H., Saunders, J.: Group search optimizer: an optimization algorithm inspired by animal searching behavior. IEEE Trans. Evol. Comput. **13**(5), 973–990 (2009)
36. Heidari, A.A., Mirjalili, S., Faris, H., Aljarah, I., Mafarja, M., Chen, H.: Harris hawks optimization: algorithm and applications. Futur. Gener. Comput. Syst. **97**, 849–872 (2019)
37. Hersovici, M., Jacovi, M., Maarek, Y.S., Pelleg, D., Shtalhaim, M., Ur, S.: The shark-search algorithm. an application: tailored web site mapping. Comput. Netw. ISDN Syst. **30**(1–7), 317–326 (1998)
38. Hinton, G.E., Osindero, S., Teh, Y.W.: A fast learning algorithm for deep belief nets. Neural Comput. **18**(7), 1527–1554 (2006)

39. Hinton, G.E., et al.: What kind of graphical model is the brain? IJCAI **5**, 1765–1775 (2005)
40. Hornik, K., Stinchcombe, M., White, H.: Multilayer feedforward networks are universal approximators. Neural Netw. **2**(5), 359–366 (1989)
41. Iklima, Z., Adriansyah, A., Hitimana, S.: Self-collision avoidance of arm robot using generative adversarial network and particles swarm optimization (gan-pso). SINERGI **25**(2) (2021)
42. Jahani, E., Chizari, M.: Tackling global optimization problems with a novel algorithm-mouth brooding fish algorithm. Appl. Soft Comput. **62**, 987–1002 (2018)
43. Jain, M., Singh, V., Rani, A.: A novel nature-inspired algorithm for optimization: squirrel search algorithm. Swarm Evol. Comput. **44**, 148–175 (2019)
44. Jalali, S.M.J., Ahmadian, S., Kavousi-Fard, A., Khosravi, A., Nahavandi, S.: Automated deep cnn-lstm architecture design for solar irradiance forecasting. IEEE Trans. Syst. Man Cybern.: Syst. **52**(1), 54–65 (2021)
45. Jiang, J., Han, F., Ling, Q., Wang, J., Li, T., Han, H.: Efficient network architecture search via multiobjective particle swarm optimization based on decomposition. Neural Netw. **123**, 305–316 (2020)
46. Junior, F.E.F., Yen, G.G.: Particle swarm optimization of deep neural networks architectures for image classification. Swarm Evol. Comput. **49**, 62–74 (2019)
47. Kan, X., Fan, Y., Fang, Z., Cao, L., Xiong, N.N., Yang, D., Li, X.: A novel iot network intrusion detection approach based on adaptive particle swarm optimization convolutional neural network. Inf. Sci. **568**, 147–162 (2021)
48. Karaboga, D., Basturk, B.: Artificial bee colony (abc) optimization algorithm for solving constrained optimization problems. In: International Fuzzy Systems Association World Congress, pp. 789–798. Springer, Berlin (2007)
49. Kaur, G., Arora, S.: Chaotic whale optimization algorithm. J. Comput. Des. Eng. (2018)
50. Kaveh, A., Farhoudi, N.: A new optimization method: dolphin echolocation. Adv. Eng. Softw. **59**, 53–70 (2013)
51. Kim, T.Y., Cho, S.B.: Optimizing cnn-lstm neural networks with pso for anomalous query access control. Neurocomputing **456**, 666–677 (2021)
52. Kumar, A., Singh, S.K., Saxena, S., Lakshmanan, K., Sangaiah, A.K., Chauhan, H., Shrivastava, S., Singh, R.K.: Deep feature learning for histopathological image classification of canine mammary tumors and human breast cancer. Inf. Sci. **508**, 405–421 (2020)
53. Kumar, A., Purohit, V., Bharti, V., Singh, R., Singh, S.K.: Medisecfed: private and secure medical image classification in the presence of malicious clients. IEEE Trans. Indus. Inf. (2021a)
54. Kumar, A., Singh, S.K., Lakshmanan, K., Saxena, S., Shrivastava, S.: A novel cloud-assisted secure deep feature classification framework for cancer histopathology images. ACM Trans. Internet Technol. (TOIT) **21**(2), 1–22 (2021)
55. LeCun, Y., Boser, B., Denker, J.S., Henderson, D., Howard, R.E., Hubbard, W., Jackel, L.D.: Backpropagation applied to handwritten zip code recognition. Neural Comput. **1**(4), 541–551 (1989)
56. LeCun, Y., Bengio, Y., Hinton, G.: Deep learning. Nature **521**(7553), 436–444 (2015)
57. Li, M.D., Zhao, H., Weng, X.W., Han, T.: A novel nature-inspired algorithm for optimization: Virus colony search. Adv. Eng. Softw. **92**, 65–88 (2016)
58. Li, X.: A new intelligent optimization-artificial fish swarm algorithm. Doctor thesis, Zhejiang University of Zhejiang, China (2003)
59. Liu, F., Wang, H., Zhang, J., Fu, Z., Zhou, A., Qi, J., Li, Z.: Evogan: An evolutionary computation assisted gan. Neurocomputing **469**, 81–90 (2022)
60. Lu, A., Yu, L., Tan, L.: Apso-based optimization algorithm of lstm neural network model. In: 2021 IEEE 5th Advanced Information Technology, Electronic and Automation Control Conference (IAEAC), vol. 5, pp. 2194–2200 (2021). https://doi.org/10.1109/IAEAC50856.2021.9390997
61. Martin, R., Stephen, W.: Termite: a swarm intelligent routing algorithm for mobilewireless ad-hoc networks. In: Stigmergic Optimization, pp. 155–184. Springer, Berlin (2006)

62. Minsky, M., Papert, S.A.: Perceptrons, Reissue of the 1988 Expanded Edition with a new foreword by Léon Bottou: An Introduction to Computational Geometry. MIT Press (2017)
63. Mirjalili, S.: The ant lion optimizer. Adv. Eng. Softw. **83**, 80–98 (2015)
64. Mirjalili, S.: Dragonfly algorithm: a new meta-heuristic optimization technique for solving single-objective, discrete, and multi-objective problems. Neural Comput. Appl. **27**(4), 1053–1073 (2016)
65. Mirjalili, S., Lewis, A.: The whale optimization algorithm. Adv. Eng. Softw. **95**, 51–67 (2016)
66. Mirjalili, S., Mirjalili, S.M., Hatamlou, A.: Multi-verse optimizer: a nature-inspired algorithm for global optimization. Neural Comput. Appl. **27**(2), 495–513 (2016)
67. Mirjalili, S., Gandomi, A.H., Mirjalili, S.Z., Saremi, S., Faris, H., Mirjalili, S.M.: Salp swarm algorithm: a bio-inspired optimizer for engineering design problems. Adv. Eng. Softw. **114**, 163–191 (2017)
68. Mirjalili, S., Dong, J.S., Lewis, A.: Nature-Inspired Optimizers: Theories, Literature Reviews and Applications, vol. 811. Springer, Berlin (2019)
69. Mucherino, A., Seref, O.: Monkey search: a novel metaheuristic search for global optimization. AIP Conf. Proc., AIP **953**, 162–173 (2007)
70. Nara, K., Takeyama, T., Kim, H.: A new evolutionary algorithm based on sheep flocks heredity model and its application to scheduling problem. In: 1999 IEEE International Conference on Systems, Man, and Cybernetics, 1999. IEEE SMC'99 Conference Proceedings, vol. 6, pp. 503–508. IEEE (1999)
71. Nesterov, Y.: Smooth convex optimization. In: Lectures on Convex Optimization, pp. 59–137. Springer, Berlin (2018)
72. Nistor, S.C., Czibula, G.: Intelliswas: optimizing deep neural network architectures using a particle swarm-based approach. Exp. Syst. Appl. **187**, 115945 (2022)
73. Pan, W.T.: A new fruit fly optimization algorithm: taking the financial distress model as an example. Knowl. Based Syst. **26**, 69–74 (2012)
74. Passino, K.M.: Biomimicry of bacterial foraging for distributed optimization and control. IEEE Control Syst. **22**(3), 52–67 (2002)
75. Qi, X., Zhu, Y., Zhang, H.: A new meta-heuristic butterfly-inspired algorithm. J. Comput. Sci. **23**, 226–239 (2017)
76. Rosenblatt, F.: The perceptron: a probabilistic model for information storage and organization in the brain. Psychol. Rev. **65**(6), 386 (1958)
77. Rumelhart, D.E., Hinton, G.E., Williams, R.J.: Learning representations by back-propagating errors. Nature **323**(6088), 533–536 (1986)
78. Salehan, A., Deldari, A.: Corona virus optimization (cvo): a novel optimization algorithm inspired from the corona virus pandemic. J. Supercomput. 1–32 (2021)
79. Saremi, S., Mirjalili, S., Lewis, A.: Grasshopper optimisation algorithm: theory and application. Adv. Eng. Softw. **105**, 30–47 (2017)
80. Sharaf, A.I., Radwan, E.S.F.: An automated approach for developing a convolutional neural network using a modified firefly algorithm for image classification. In: Applications of Firefly Algorithm and Its Variants, pp. 99–118. Springer, Berlin (2020)
81. Singh, P., Chaudhury, S., Panigrahi, B.K.: Hybrid mpso-cnn: multi-level particle swarm optimized hyperparameters of convolutional neural network. Swarm Evol. Comput. **63**, 100863 (2021)
82. Singh, R., Bharti, V., Purohit, V., Kumar, A., Singh, A.K., Singh, S.K.: Metamed: Few-shot medical image classification using gradient-based meta-learning. Pattern Recognit. 108111 (2021)
83. Stanley, K.O., Miikkulainen, R.: Evolving neural networks through augmenting topologies. Evol. Comput. **10**(2), 99–127 (2002)
84. Sun, Y., Yen, G.G., Yi, Z.: Evolving unsupervised deep neural networks for learning meaningful representations. IEEE Trans. Evol. Comput. **23**(1), 89–103 (2018)
85. Sun, Y., Xue, B., Zhang, M., Yen, G.G.: Evolving deep convolutional neural networks for image classification. IEEE Trans. Evol. Comput. **24**(2), 394–407 (2019)

86. Torabi, S., Safi-Esfahani, F.: Improved raven roosting optimization algorithm (irro). Swarm Evol. Comput. **40**, 144–154 (2018)
87. Uymaz, S.A., Tezel, G., Yel, E.: Artificial algae algorithm (aaa) for nonlinear global optimization. Appl. Soft Comput. **31**, 153–171 (2015)
88. Wan, S., Ding, S., Chen, C.: Edge computing enabled video segmentation for real-time traffic monitoring in internet of vehicles. Pattern Recognit. **121**, 108146 (2022)
89. Wang, B., Sun, Y., Xue, B., Zhang, M.: Evolving deep neural networks by multi-objective particle swarm optimization for image classification. In: Proceedings of the Genetic and Evolutionary Computation Conference, pp. 490–498 (2019a)
90. Wang, C., Xu, C., Yao, X., Tao, D.: Evolutionary generative adversarial networks. IEEE Trans. Evol. Comput. **23**(6), 921–934 (2019)
91. Wu, M., Su, W., Chen, L., Liu, Z., Cao, W., Hirota, K.: Weight-adapted convolution neural network for facial expression recognition in human-robot interaction. IEEE Trans. Syst. Man Cybern.: Syst. **51**(3), 1473–1484 (2019)
92. Yang, X.S.: Firefly algorithms for multimodal optimization. In: International Symposium on Stochastic Algorithms, pp. 169–178. Springer, Berlin (2009)
93. Yang, X.S.: A new metaheuristic bat-inspired algorithm. In: Nature Inspired Cooperative Strategies for Optimization (NICSO 2010), pp. 65–74. Springer, Berlin (2010)
94. Yang, X.S., Deb, S.: Cuckoo search via lévy flights. In: World Congress on Nature and Biologically Inspired Computing. NaBIC 2009, pp. 210–214. IEEE (2009)
95. Yong, W., Tao, W., Cheng-Zhi, Z., Hua-Juan, H.: A new stochastic optimization approach-dolphin swarm optimization algorithm. Int. J. Comput. Intell. Appl. **15**(02), 1650011 (2016)
96. Zaltron, N., Zurlo, L., Risi, S.: Cg-gan: an interactive evolutionary gan-based approach for facial composite generation. Proc. AAAI Conf. Artif. Intell. **34**, 2544–2551 (2020)
97. Zhang, L., Zhao, L.: High-quality face image generation using particle swarm optimization-based generative adversarial networks. Futur. Gener. Comput. Syst. **122**, 98–104 (2021)
98. Zhang, T., Lei, C., Zhang, Z., Meng, X.B., Chen, C.P.: As-nas: adaptive scalable neural architecture search with reinforced evolutionary algorithm for deep learning. IEEE Trans. Evol. Comput. **25**(5), 830–841 (2021)
99. Zhao, W., Zhang, Z., Wang, L.: Manta ray foraging optimization: an effective bio-inspired optimizer for engineering applications. Eng. Appl. Artif. Intell. **87**, 103300 (2020)
100. Zhou, X., Qin, A., Gong, M., Tan, K.C.: A survey on evolutionary construction of deep neural networks. IEEE Trans. Evol. Comput. **25**(5), 894–912 (2021)

Advances on Particle Swarm Optimization in Solving Discrete Optimization Problems

M. A. H. Akhand, Md. Masudur Rahman, and Nazmul Siddique

Abstract Particle Swarm Optimization (PSO) is a well-known optimization method which optimizes a problem by having a population of candidate solutions, here dubbed particles, and moving these particles around in the search-space according to simple mathematical formulae over the particle's position and velocity. PSO initially proposed for continuous optimization and, to till, different discrete optimizations methods are developed adapting PSO in different time periods. The major concerns to solve discrete optimization task with PSO are the adaptation of particle encoding, velocity measurement and position update. The aim of this study is to demonstrate the evolution of PSO in solving discrete optimizations conceiving different adaptations in its operations. This study explains adaptation of PSO for four different discrete optimization problems: knapsack problem (KP), traveling salesman problem (TSP), vehicle routing problem (VRP), and university course scheduling problem (UCSP). The selected problems are well diverse having different constraints and objectives; KP seems a simplest one and UCSP is the most complex optimization task. The rhythmic presentation of PSO adaptation in solving KP, TSP, VRP and UCSP in this chapter may be a proper demonstration of PSO transformation from its original continuous domain to different discrete domains. The study will be made easy to understand other PSO-based discrete optimization methods as well as will be helpful to solve any new discrete optimization task adopting PSO.

Keywords Particle swarm optimization · Knapsack problem · Traveling salesman problem · Vehicle routing problem · University course scheduling problem

M. A. H. Akhand (✉) · Md. M. Rahman
Department of Computer Science and Engineering, Khulna University of Engineering & Technology, Khulna 9203, Bangladesh
e-mail: akhand@cse.kuet.ac.bd

N. Siddique
School of Computer Engineering and Intelligent Systems, Ulster University, Northern Ireland, UK
e-mail: nh.siddique@ulster.ac.uk

A. Biswas et al. (eds.), *Advances in Swarm Intelligence*, Studies in Computational Intelligence 1054, https://doi.org/10.1007/978-3-031-09835-2_4

1 Introduction

Computational methods inspiring from natural phenomenon have gain much interest in the recent years. The natural phenomenon from microscopic matter to living behavior of insects or large animals are studied to developed distinct computing techniques in last few decades. Genetic algorithm, inspired by DNA based biological systems' fitness improvement, is the pioneer one and widely used to solve many scientific and engineering problems [1, 2]. In the last decade, collective behaviors of various insects (e.g., ant, bee, fish, firefly, glowworm) are investigated and proposed a number of computing methods in the category of swarm intelligence (SI) [3]. Recently, living and hunting behaviors of animals (e.g., cattle, gray wolf, lion) are considered to develop better SI based optimization techniques [4]. Among the developed algorithms, particle swarm optimization (PSO) [5], mimicking behavior of bird flocking or fish schooling, seems the most famous method due to its simplicity as well as performance. PSO was proposed for continuous problems (e.g., function optimization) [5] and it has been proven as an effective as well as well-performed method to solve such optimization tasks. A large number of studies with moderation of PSO's components and adaptation it for different tasks with diverse techniques are performed in the last two decades [5–21].

PSO is also adopted to solve different discrete optimization tasks in last several years. PSO-based methods are found efficient to solve diverse discrete optimization tasks having different constraints and objectives. To solve discrete optimization tasks with PSO, individual tasks require distinct formulations of different components and operations of PSO. Among the formulations, the major issues are the particle encoding and managing interaction with justified techniques. Individual discrete optimization tasks hold to satisfy diverse constraints and hence adaptation of PSO for the tasks are quite different from one another. A number of studies available to solve different discrete optimization tasks including resource optimization [16–21], optimal route generation [8–13, 22–32], and different scheduling [14, 33–35].

The aim of this study is to demonstrate the evolution of PSO in solving discrete optimizations conceiving different modification/adaptations in its operations. This study explains adaptation of PSO for four different discrete optimization problems: knapsack problem (KP), traveling salesman problem (TSP), vehicle routing problem (VRP), and university course scheduling problem (UCSP). The selected problems are well diverse having different constraints and objectives; KP seems a simplest one and UCSP is the most complex optimization task. The rhythmic presentation of PSO adaptation, in solving KP, TSP, CVRP and UCSP, in this chapter may be a proper demonstration of PSO transformation from its original continuous domain to different discrete domains. The study will be made easy to understand other PSO-based discrete optimization methods as well as will be helpful to solve any new discrete optimization task adopting PSO.

The organizations of the rest of the chapter are as follows. For better understanding as well as to make the chapter self-readable, basic PSO and its significant properties are described first in Sect. 2. The main contributions of the chapter are then presented in Sects. 3–6 for PSO adoptions for KP, TSP, VRP and UCSP. Finally, Sect. 7 concludes the chapter with a brief summary as well as several future directions.

2 Particle Swarm Optimization (PSO)

PSO is a population-based optimization technique on metaphor of social behavior of flocks of birds or schools of fishes [5]. It is a simple model of social learning whose emergent behavior has found popularity in solving difficult optimization problems. PSO was proposed for continuous problems and also found efficient to solve such problems (e.g., function optimization) [5]. It firstly generates a random initial population, the population contains numbers of particles, each particle represents a potential solution of system, and each particle is represented by three indexes: position, velocity and fitness. At every step, each particle calculates its new velocity considering its previous best position and the best one among all the particles in the population. Each particle then moves to new position (i.e., search a new point) based on the calculated velocity. These processes of iteration continue until the stopping criterion is reached. PSO has been investigated on various continuous [5–7] and combinatorial optimization tasks [8–15].

PSO optimizes a problem by having a population of candidate solutions, here dubbed particles, and moving these particles around in the search-space according to simple mathematical formulae over the particle's position and velocity. PSO was originally designed and introduced by Eberhart and Kennedy [5]. Each particle's movement is influenced by its local best-known position (P_i) and is also guided toward the best-known positions in the search-space *(G)*, which are updated as better positions are found by other particles. Formally in PSO, each particle changes position $X_i = \{x_{i1}, x_{i2}, x_{i3} \ldots, x_{iD}\}$ based on its calculated velocity $V_i = (v_{i1}, v_{i2}, \ldots, v_{iD})$ that depends on its previous best position, $P_i = (p_{i1}, p_{i2} \ldots p_{iD})$ and the best one among all the particles in the population $G = (g_1, g_2 \ldots g_D)$. In each of iteration, each particle calculates its velocity according to the following formula:

$$V_i^{(t)} = \omega V_i^{(t-1)} + c_1 * r_1\left(P_i^{(t-1)} - X_i^{(t-1)}\right) + c_2 * r_2\left(G^{(t-1)} - X_i^{(t-1)}\right), \quad (1)$$

where, ω is inertia factor, c_1 and c_2 are learning factors, and r_1 and r_2 are vectors of random values between (0,1).

After calculating $V_i^{(t)}$ we can get the new position in next iteration by using the following formula:

$$X_i^{(t)} = X_i^{(t-1)} + V_i^{(t)} \quad (2)$$

Figure 1 shows a representation of the particle position update from time step *t-1* to *t*. In the figure $X_i^{(t)}$ is the updated position of particle $X_i^{(t-1)}$ applying velocity

Fig. 1 Movement of particles in PSO

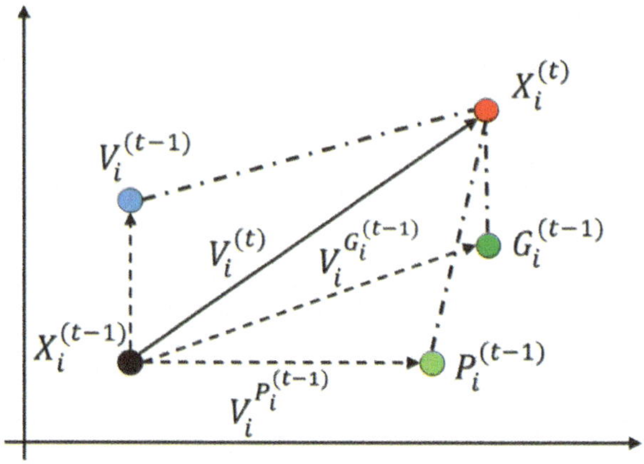

$V_i^{(t)}$ which is measured by following Eq. (1). Algorithm 1 shows the major steps of PSO. In initialization (Step 1), it defines number of particles and assigns a random solution and a random velocity to each of the particle. At this initial stage, previous best solution of each particle (P_i) is considered as the current random tour of it and Global best solution (G). At every iteration step, for each particle, PSO calculates a new velocity value which used to find the next position of particles in Step 2. These processes of iteration continue until reaching stopping condition.

Algorithm 1: Particle Swarm Optimization (PSO)

Step 1: Initialization.
Step 2: Calculate and update the velocity according to Eq. (1).
Step 3: Calculate and update the new position according to Eq. (2).
Step 4: Update P_i if the new solution $X_i^{(t)}$ is superior to P_i .
Step 5: Update Gif there is new solution $X_i^{(t)}$ is superior to G.
Step 6: Loop to **Step 2** until a termination criterion is met.
// Usually, a sufficiently good fitness or a maximum number of iterations.
Step 7: Take the global best solution G as an outcome.

The significant properties of PSO are the mathematical operations in Eqs. (1) and (2). Addition, subtraction, and multiplication are convenient in continuous domain where a particle is represented with numeric values in multidimensional space. Such mathematical operations are not possible or not appropriate for discrete problems and therefore a different formulation of Eqs. (1) and (2) are required to solve a discrete problem conceiving the idea of PSO. Individual discrete optimization tasks hold to satisfy diverse constraints and objective functions; hence; adaptation of PSO for the tasks are quite different from one another. In another words, to solve discrete optimization tasks with PSO, individual tasks require distinct formulations of different components and operations of PSO. Among the formulations, the major issues are the particle encoding, velocity measure and position update which are different for problem to problem. The following sections describes PSO adaptations to solve four popular but diverse discrete optimization tasks: KP, TSP, VRP, and UCSP.

3 Knapsack Problem (KP) and Solution Using PSO

Knapsack problem is a typical problem of combinatorial optimization in operational research, having broad application foreground, such as resource allocation problem, goods shipment problem, project selection problem and so on. It belongs to NP hard complete problem [18]. The problem induces efficient assignment of objects into a knapsack with maximum profit without exceeding the capacity of the knapsack. The methods of solving knapsack problem are accurate method (such as dynamic programming, recursion method, backtracking, branch and bound algorithm and so on), approximation algorithm (such as greedy method, Lagrange method and so on) and intelligent optimization algorithm (such as simulated annealing algorithm, genetic algorithm, ant colony algorithm and so on) [21]. Figure 2 shows a representation of a knapsack problem.

3.1 KP Basics and Constraints

Knapsack is a traditional binary optimization problem, which can be described as follows: given a set of n items and a knapsack; each i th ($i = 1 \dots n$) item has a positive profit p_i and a number of positive resource consumption $r_{i1}, r_{i2}, \dots, r_{im}$ corresponding to m resources; the knapsack has m capacities $C_j (j = 1 \dots m)$ for

Fig. 2 Representation of knapsack problem

each resource; the task is to select an item subset so that the total profit of selected items is maximized while for all resources the total resource consumption does not exceed the knapsack's resource capacity.

Decision Variable. Given that i th item is selected, the decision variable can be denoted as

$$x_i = \begin{cases} 1 & \text{if } i\text{th item is selected} \\ 0 & \text{otherwise} \end{cases} \tag{3}$$

Objective Function. The objective function can be then introduced by

$$\max \sum_{i=1}^{n} p_i * x_i \tag{4}$$

Constraints of KP. The following equations represents the constraints of KP: Equation (5) ensures that the items must not exceed the capacity of the knapsack

$$\sum_{i=1}^{n} r_{ij} * x_i \leq C_j, \ \forall j \in \{1, 2, \ldots, m\} \tag{5}$$

Equation (6) ensures that the decision variable must be a binary, thus no fraction of items is allowed in the knapsack

$$x_i \in 0, 1 \ \forall i \in= \{1, 2, \ldots, n\} \tag{6}$$

Figure 3 shows an example of KP and two possible solutions. The problem table shows the items and their corresponding profits and consumptions. The capacity of the knapsack in the example is 15. The solution is a collection of binary bits that

Problem Table

Item	1	2	3	4	5	6
Profit	5	6	1	2	4	3
Consumption	6	5	3	2	1	4

Knapsack Capacity = 15

Consumption = 6+5+4 = 15 Profit = 5+6+3 = 14

Solution 1	**1**	**1**	**0**	**0**	**0**	**1**

Consumption = 6+5+3+1 = 15 Profit = 5+6+1+4 = 16

Solution 2	**1**	**1**	**1**	**0**	**1**	**0**

Fig. 3 A Sample Knapsack Problem and its Solution

indicate which items are selected. In Solution 1, the selected three items (i.e., Item 1, 2 and 6); consumption and profit of the solution are 15 and 14, respectively. On the other hand, Solution 2 consumption is also 15 with four items (i.e., Item 1, 2, 3 and 5) but achieve profit 16. Clearly, with higher profit Solution 2 is better than Solution 1. The aim of the KP optimization is to find the solution with maximum profit without exceeding the knapsack capacity.

3.2 Solution of KP Using PSO

To solve KP, a variant of PSO, Binary PSO (BPSO) is employed in the exiting studies [18–20]. BPSO was originally developed by Kennedy and Eberhart [36], to solve different combinatorial problems, for example, job-shop scheduling [37] and feature selection [38]. The search space in BPSO is considered as a hypercube in which a particle may be seen to move to nearer and farther corners of the hypercube by flipping various numbers of bits. The moving velocity is defined in terms of changes of probabilities that a bit will be in one state or the other. Thus, a particle moves in a state space restricted to 0 and 1 on each dimension, where each V_i represents the probability of bit X_i taking the value 1. With this definition P_i and X_i are integers in $\{0, 1\}$ and V_i, since it is a probability, must be constrained to the interval [0.0,1.0], and is updated by Eq. (7).

$$V_i^{(t)} = \omega V_i^{(t-1)} + c_1 * r_1\left(P_i^{(t-1)} - X_i^{(t-1)}\right) + c_2 * r_2\left(G^{(t-1)} - X_i^{(t-1)}\right) \quad (7)$$

Rather than adding the velocity to the position to obtain the new position like the continuous PSO, in BPSO, the velocity entry is used to determine the probability that the corresponding position entry takes the value 1, as shown in the position update equation in Eq. (8)

$$X_i^{(t)} = \begin{pmatrix} 1 & if\, rand() \leq s\left(V_i^{(t)}\right) \\ 0 & \text{otherwise} \end{pmatrix} \quad (8)$$

where $s\left(V_i^{(t)}\right) = \frac{1}{1+e^{-v_i^{(t)}}}$. It is known as a transfer function [20].

Algorithm 2 shows the steps of solving Knapsack Problem. To solve KP with BPSO, initially, some random solutions (as particles) are generated in Step 1. Then, the optimal solution can be found by following the rest of the steps the algorithm. The bits in the solution space denote which items are chosen to be in the knapsack. In Step 2, the calculation of fitness sums up the profit and consumption to decide if the solution is eligible to be a best particle. In Steps 3 and 4, $X_i^{(t)}$ decides whether a bit in the solution should be 1 or 0, and velocity $V_i^{(t)}$ acts as a probability to decide the value of $X_i^{(t)}$. Then the global best and particle best are updated in Steps 5 and 6 if the newfound solution is better than the previous one.

Algorithm 2: Binary Particle Swarm Optimization (BPSO) for Knapsack Problem

Step 1: Initialize particles with random solutions
Step 2: Evaluate the particle fitness
Step 3: Calculate and update the velocity according to Eq. (7).
Step 4: Calculate and update the new position according to Eq. (8).
Step 5: Update P_i if the new solution $X_i^{(t)}$ is superior to P_i .
Step 6: Update Gif there is new solution $X_i^{(t)}$ is superior to G.
Step 7: Loop to **Step 2** until a termination criterion is met.
Step 6: Take the global best solution G as an outcome.

4 Traveling Salesman Problem (TSP) and Solution Using PSO

TSP is a well-studied combinatorial optimization problem in which a salesman is required to complete a tour with the minimum distance visiting all the assigned cities exactly for once [39, 40]. It is the most popular combinatorial problem having many real-world applications such as drilling a printed circuit board, computer wiring, order picking problem in warehouses, vehicle routing, X-Ray crystallography [39]. Interest grows in recent years to solve TSP new ways; almost every new approach for solving engineering and optimization problems has been tested on the TSP considering it as a general test bench. Following subsections briefly explains TSP basics and PSO formulation to solve it.

4.1 TSP Basics and Constrains

Suppose graph G is a complete graph, where every pair of distinct vertices is connected by a unique edge. Let the set of vertices be $V = \{v_1, v_2, \ldots, v_n\}$. The cost matrix is given by $C = \left(c_{ij}\right)_{n\times n}$ where the cost of the edge joining node v_i to node v_j, denoted c_{ij}, is given in entry (i, j). In the context of the TSP, the vertices correspond to cities and the edges correspond to the path between those cities. When modeled as a complete graph, paths that do not exist between cities can be modeled as edges of very large cost without loss of generality. Minimizing the sum of the costs for Hamiltonian cycle is equivalent to identifying the shortest path in which each city is visiting only once. Figure 4 shows a simple representation route costs of a TSP with four cities A, B, C and D. The optimal tour, i.e., TSP solution, is A-B-C-D-A with tour cost 97 (= 20 + 30 + 12 + 35).

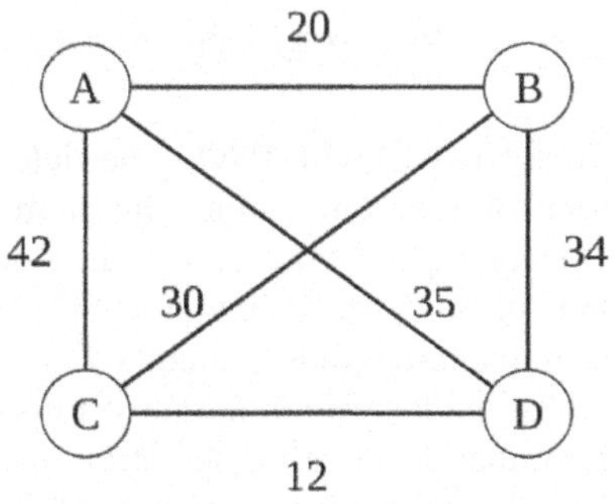

Fig. 4 Graphical representation of a TSP mentioning route costs

Decision Variable. Given a set of n cities enumerated $0, 1, \ldots, n-1$ to be visited with the distance between each pair of cities i and j is given by c_{ij}. Introduce decision variables y_{ij} for each (i, j) such that

$$y_{ij} = \begin{cases} 1 & \text{if city } j \text{ is visited immediately after city } i \\ 0 & \text{otherwise} \end{cases} \tag{9}$$

Objective Function. The objective function is then given by

$$\min \sum_i \sum_j c_{ij} y_{ij} \tag{10}$$

Constraints of TSP. To ensure that the result is a valid tour, several constraints must be added. Equation (11) is the go-to constraint. After visiting a city i, the salesman must visit only one city next:

$$\sum_j y_{ij} = 1, i = 0, 1, \ldots, n-1 \tag{11}$$

Equation (12) is the come-from constraint. When visiting a city, the salesman must have come from only one city:

$$\sum_i y_{ij} = 1, i = 0, 1, \ldots, n-1 \tag{12}$$

Equation (13) is the subtour elimination. It ensures that there is only one path between any two cities, i.e., no subtours exists.

$$\sum_i \sum_j y_{ij} \leq |S| - 1, S \subset V, 2 \leq |S| \leq n-2, \tag{13}$$

where, S is the set of all tours of G.

4.2 Solution of TSP Using PSO

To solve TSP with PSO, a particle represents a tour and velocity is to change the tour towards a new tour. The methods use different techniques and parameters for calculating the velocity and then find new tour for particles. Among the methods, the operators Swap Operator (SO) and Swap Sequence (SS) are used in a number of prominent methods to handle TSP operations [13, 41, 42].

Swap Operator is the most important in solving TSP problem and its operation is similar to mutation operation of genetic algorithm [1, 2]. A Swap Operator (SO) contains a pair of indexes that indicates two cities may swap in a tour. Suppose, a TSP problem has 5 cities and a solution is $S = (1-3-5-2-4)$. Let a Swap Operator is $SO(2,4)$ then new solution with the following *SO* like below $S' = S + SO(2,4) = (1-3-5-2-4) + SO(2,4) = (1-2-5-3-4)$, here ' + ' means to apply *SO* on the solution *S* [22]. A Swap Sequence (SS) is a collection of one or more Swap Operator(s) that might be applied on a solution one after another sequentially. To solve TSP, the velocity of PSO is represented as SS. The Swap Sequence can be defined as:

$$SS_{12} = (SO_1, SO_2, SO_3, \ldots SO_n), \tag{14}$$

where, $SO_1, SO_2, SO_3, \ldots SO_n$ are Swap Operators. Swap Sequence acts on a solution applying all its *SOs* maintaining its sequence and then finally produces the new tour. This can be described by the following formula:

$$S_2 = S_1 + SS_{12} = S_1 + (SO_1, SO_2, SO_3, \ldots, SO_n) \tag{15}$$

The order of *SOs* in a SS_{12} is important because implication of same SOs in different order may give different solutions from the original solution. Considering Eq. (16) SS_{12} may get from solutions S_1 and S_2 in the following formula:

$$SS_{12} = S_2 - S_1 = (SO_1, SO_2, SO_3, \ldots, SO_n) \tag{16}$$

here, '–' means need to apply *SOs* of SS_{12} on solution S_1 to get S_2. As an example, if $S_1 = (1-2-3-4-5)$ and $S_2 = (2-3-1-5-4)$ then $SS_{12} = SO(1,3), SO(2,3), SO(4,5)$.

Moreover, several SSs may be merged into a new SS; the operator defines as merging operation [5]. If $SS_1 = SO(1,2), SO(5,2)$ and $SS_2 = SO(5,3), SO(4,1)$ then new Swap Sequence *SS(new)* merging SS_1 *and* $SS_2$is

$$\begin{aligned} SS(\text{new}) &= SS1SS2 = \{SO(1,2), SO(5,2)\}\{SO(5,3), SO(4,1)\} \\ &= SO(1,2), SO(5,2), SO(5,3), SO(4,1) \end{aligned} \tag{17}$$

To solve TSP with PSO, a particle represents a tour and velocity is the SS is to change the tour towards a new tour. The SOs of the velocity SS of a particle is measured on its previous best tour ($P_i^{(t-1)}$) and the best tour among all the particles ($G^{(t-1)}$) in the population. Swap Sequence PSO (SSPSO) [22] is being the pioneer one to introduce SS as a velocity to transform a tour to a new one applying all its SOs. There have been various modifications of SSPSO are proposed to solve TSP based on SO and SS. Self-Tentative PSO (STPSO) [9] introduces tentative behavior after PSO operation in solving TSP that tries to improve each particle. Enhanced Self-Tentative PSO (ESTPSO) [10] introduces some new features overcoming the limitations of STPSO. Most recent modification is the Velocity Tentative PSO (VTPSO) [13] which conceives a measure called partial search (PS) to apply calculated SS to update particle's position (i.e., TSP tour). Moreover, VTPSO conceives a moderate self-tentative technique to improve its performance.

Equations for velocity calculation and position update in SS based PSO are Eq. (18) and Eq. (19), respectively.

$$V_i^{(t)} = V_i^{(t-1)}\alpha\left(P_i^{(t-1)} - X_i^{(t-1)}\right)\beta\left(G^{(t-1)} - X_i^{(t-1)}\right)\alpha, \beta\epsilon[1, 0] \tag{18}$$

$$X_i^{(t)} = X_i^{(t-1)} + V_i^{(t)}, \tag{19}$$

where α, β are random number between 0 and 1,$\alpha\left(P_i^{(t-1)} - X_i^{(t-1)}\right)$ means all SOs in BSS for $\left(P_i^{(t-1)} - X_i^{(t-1)}\right)$ should be maintained with the probability of α, it is the same for $\beta\left(G^{(t-1)} - X_i^{(t-1)}\right)$. The bigger the value of α (and β the greater the influence of $P_i^{(t-1)}$ (and $G^{(t-1)}$) will be maintained on present velocity calculation selecting more SOs. After velocity SS calculation using Eq. (18), each particle moves to a new tour solution ($X_i^{(t)}$) applying SS on its previous solution ($X_i^{(t-1)}$) using Eq. (19). The PS technique to apply calculated SS to update particle's position (i.e., TSP tour) is main difference of VTPSO from other methods. Since velocity SS is the accumulation of several SOs and each one transforms a tour to a new one, VTPSO measures performance of tours applying SOs one after another, and the final velocity is considered for which it gives better tour. Therefore, the final velocity may be a portion (from the beginning) of calculated velocity SS. It is reported that PS technique may explore better result evaluating intermediate tentative tours without increasing the computational time.

Algorithm 3 shows the steps of the VTPSO which initializes the population with random solutions and tries to improve them at every iteration step. At each iteration step, VTPSO calculates velocity SS (Step 2.a) using Eq. (18) and considers (i) last applied velocity ($v^{(t-1)}$), (ii) previous best solution of the particle (P_i) and (iii) global best solution of the swarm (G). VTPSO calculates fitness of every new position (X_i) of a particle and compares to its previous best P_i. If X_i is found better than P_i then VTPSO applies Self-Tentative (ST) operation on (X_i) owing to improve it furthermore. After ST operation, the method also compares X_i with P_i and G; and updates accordingly if the values are found inferior to X_i (Step 2.d and Step 2.e).

Algorithm 3: Velocity Tentative PSO (VTPSO)

Step 1: Initialization.
Step 2: For each particle $X_i$in the swarm

a. Calculate velocity $V_i^{(t)}$ using Eq. (18)
b. Update $X_i^{(t)}$ using Eq. (19) through Partial Search manner
c. Apply Tentative Operation on $X_i^{(t)}$ if is Superior to P_i
d. Update P_i if the new solution $X_i^{(t)}$ is superior to P_i
e. Update Gif the new solution $X_i^{(t)}$ is superior to$G$

Step 3: Loop to **Step 2** until a termination criterion is met
Step 4: Take the global best solution Gas an outcome

5 Vehicle Routing Problem (VRP) and Solution Using PSO

VRP is a real life constraint satisfaction problem to find minimal travel distances of vehicles to serve customers [43]. Capacitated VRP (CVRP) is the simplest form of VRP considering equal vehicle capacity constraint [44]. In CVRP, all customers have known demands and known locations for the delivery. The delivery for a customer cannot be split. In other words, the demand of a customer must be satisfied via only one visit. Goods are delivered from single depot and the vehicles return to the depot after serving their assigned customers. The delivery or unloading time may or may not be considered. The objective of CVRP is to minimize the total travelling distance for all vehicles. Figure 5 shows a pictorial representation of a CVRP solution having three vehicles.

5.1 CVRP Basics and Constrains

The CVRP can be defined as a directed complete graph $G(V; E)$, where $V = \{v_0, v_1, \ldots, v_n\}$ is the vertex set, $E = \{(v_i, v_j)|v_i, v_j \in V, i \neq j\}$ is the edge set. Normally, v_0 is set as the depot, and $\{v_1, v_2, \ldots, v_n\}$ are N customers. A set of $|K|$ homogenous vehicles with the same capacity Q depart from depot v_0. Each customer v_i has a demand of capacity q_i. Each customer in the network requires to be serviced by one vehicle only once. The travel cost c_{ij} between vertices i and j is represented in proportion to Euclidean distance between them.

Decision Variable. Given the set of N customers and a depot $V_0, V_1, \ldots, V_n$ to be visited using the set of K vehicles with the distance between each pair of customers i and j is given by c_{ij}. Introduce decision variables x_{ij} for each (i, j) such that

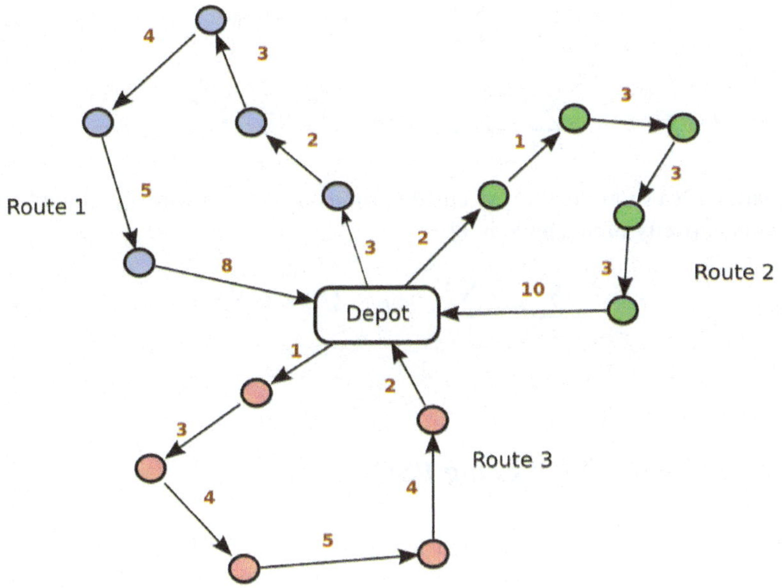

Fig. 5 Representation of a VRP solution having three vehicles

$$x_{ij} = \begin{cases} 1 & \text{if vehicle } k \text{ travels from } v_i \text{ to } v_j \\ 0 & \text{otherwise} \end{cases} \tag{20}$$

Objective Function. The objective function is then given by

$$\min \sum_{k \in K} \sum_{i \in V} \sum_{j \in V} c_{ij} x_{ijk} \tag{21}$$

Constraints of CVRP. All vehicles must start their tour from the depot and return to the depot after end of the tour. Also, there must be at most K vehicles in the fleet. Equation (22) formulates this constraint.

$$\sum_{i \in V} x_{i0k} = \sum_{j \in V} x_{0jk} = 1 (\forall k \in K) \tag{22}$$

Each vehicle must visit a customer once. Equation (23) formulates the constraint.

$$\sum_{j \in V, j \neq i} x_{ijk} = \sum_{j \in V, j \neq i} x_{jik} \leq 1 (\forall i \in V, \forall k \in K) \tag{23}$$

Each customer can be served by only one vehicle. Equation (24) and Eq. (25) formulate this constraint.

$$\sum_{k \in K} \sum_{i \in V, i \neq j} x_{ijk} = 1(\forall j \in V) \tag{24}$$

$$\sum_{k \in K} \sum_{j \in V, j \neq i} x_{ijk} = 1(\forall i \in V) \tag{25}$$

Equation (26) shows that every customer must be served with q_i demand and the maximum capacity of a vehicle is Q.

$$\sum_{i \in V} q_i \sum_{j \in V, j \neq i} x_{ijk} \leq Q(\forall k \in K) \tag{26}$$

5.2 *Solution of CVRP Using PSO*

Various ways have been investigated for solving the CVRP those approaches are broadly categorized as constructive and clustering methods. Constructive methods build a feasible solution gradually while keeping an eye on solution cost, but it may not contain an improvement or optimization phase. Some well-known algorithms of constructive methods are - Clarke and Wright's Savings Algorithm [45–49], Matching Based Algorithm and Multi-route Improvement Heuristics [50]. On the other hand, clustering methods solve problem in two steps and that is why this approach is also called 2-phase or cluster first, route second method. First phase of this approach uses clustering algorithm to generate clusters of customers and second phase uses optimization technique to find optimum routes for each cluster generated in the first step. Cluster First and Route Second (CFRS) is the most studied way to solve CVRP using different clustering and TSP optimization techniques in phase 1 and phase 2, respectively. Figure 6 depicts a generalized block diagram of CFRS method for a sample CVRP problem having three vehicles. PSO is employed for individual vehicles' route optimization in several CFRS CVRP studies. Route generation of individual vehicles is a simple TSP problem considering depot as common point for all the vehicles. Therefore, solution of TSP using PSO explained in Sect. 4.2 is

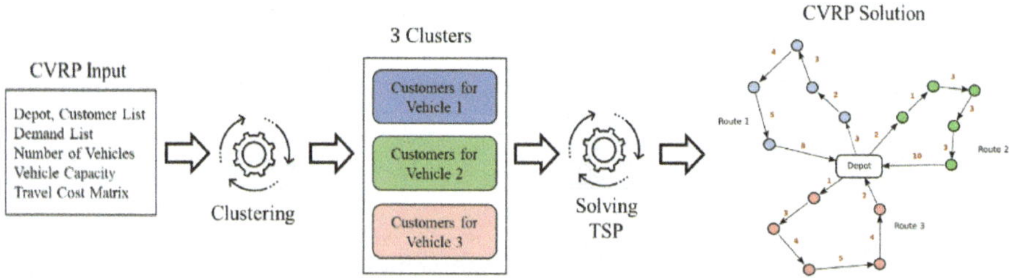

Fig. 6 Block diagram of CFRS method

applicable. Clustering nodes into different vehicles is the main significance in solving CVRP with PSO. Sweep [51, 52] and Fisher and Jaikumar [53] are the two popular clustering algorithms and the following subsections explains these methods briefly.

Clustering Method: Sweep. Among different clustering methods, Sweep [51, 52] is the most popular due to its simplicity. Cluster formation starts from 0° and consequently advance towards 360° to assign all the nodes under different vehicles while maintaining vehicle capacity. This type of sweeping is called *forward sweep*. And in *backward sweep*, clustering direction is clockwise which means though clustering starts from 0°, then it advances algorithm from 360° to 0°. The general formula for calculating polar angle of the customers with respect to depot is,

$$\theta = \left(\frac{y}{x}\right), \tag{27}$$

where

θ Angle of a node (depot/customer).
x, y X, Y co-ordinates of customer.

Figure 7 shows forward Sweep cluster creation starts from 0° and goes in anti-clockwise direction, while Algorithm 4 shows the Sweep steps. The algorithm has two major steps. In the initialization step, each customer's polar angle is calculated and they are sorted according to their increasing order. In the clustering step, customers are added to a cluster by increasing polar angles until a vehicle capacity is exceeded. After that, a new cluster is created in the same way. This process continues until all customers are assigned to vehicles.

Algorithm 4: Sweep

1. ***Initialization***
   a) Compute the polar angle of each customer using Eq. (27).
   b) Sort the customer according to their increasing order of polar angles.

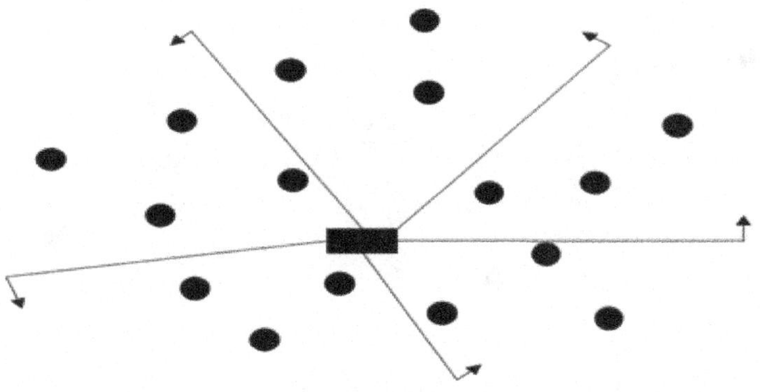

Fig. 7 Sweep clustering

2. ***Clustering***

   a) Set $C = 1$.
   b) Start sweeping all customers by increasing polar angle and assign the customers to the current cluster.
   c) Stop the sweep when adding the next customer would violate the maximum vehicle capacity.
   d) Set $C = C + 1$.
   e) Create a new cluster by resuming the sweep where the last one left off.
   f) Repeat Steps 2–5, until all customers have been included into clusters.

A number of Sweep variants are also investigated in solving CVRP. The Sweep algorithm clusters the nodes solely by polar angle. If the nodes are widely separated but have less angular difference, they may be grouped in the same cluster. This reduces the optimality of the solution cluster. To resolve this problem, an algorithm was proposed named Sweep Nearest algorithm (SN) [54], which combines the classical Sweep and the Nearest Neighbor algorithm. SN first assigns a vehicle to the customer with the smallest polar angle among the remaining customers and then finds the nearest stop to those already assigned and then inserts that customer.

Clustering Method: Fisher and Jaikumar Algorithm. In Fihser and Jaikumar algorithm, the customers are divided into K cones at first where K is the number of total vehicles. Seed customers are selected from the cones, based on some criteria such as customers with maximum demand or most distant customer from origin. Figure 8 shows that the customers ($N = 15$) are divided into four cones ($K = 4$). If the smallest angle is 20° and the largest angle is 340° then each cone is of 80°. From each cone, the farthest node from the origin is chosen as the seed customer. After seed selection, insertion cost is calculated which is the cost of inserting customer within the route going back and forth between seed customer and depot. Then the customers are assigned to vehicles according to the increasing order of insertion cost.

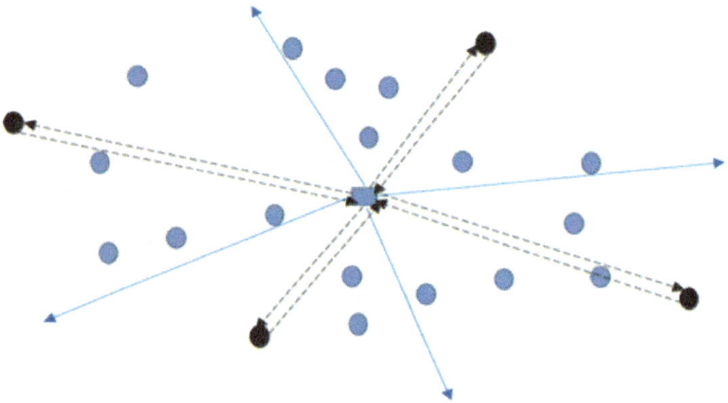

Fig. 8 Seed customer selection in Fisher and Jaikumar

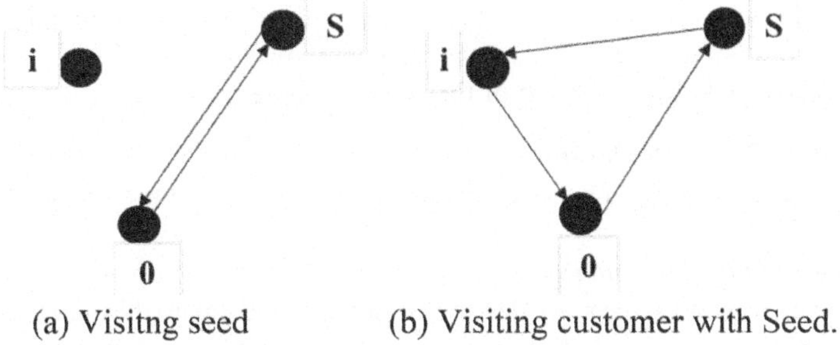

Fig. 9 Demonstration of visiting seed alone and with a customer

Figure 9 demonstrates visiting seed and a customer where 0 denotes the depot, S is seed customer and i is the customer being considered for insertion. Figure 9a shows the back-and-forth route of seed customer and depot. Figure 9b shows the route if customer i is visited in the way of returning from seed customer to depot. From Fig. 9a the travelled distance d_S of visiting the seed customer and returning to depot is shown in Eq. (28).

$$d_S = d_{0S} + d_{S0} = 2d_S \tag{28}$$

And if customer i is visited on the way while visiting seed, then the distance $d_{i,}$

$$d_i = d_S + d_{Si} + d_{i0} \tag{29}$$

Substituting Eq. (28) from Eq. (29), insertion cost of customer i is

$$C_{ik} = d_i - d_s = d_{si} + d_{io} - d_s \tag{30}$$

The procedure of Fisher and Jaikumar algorithm is shown in Algorithm 5. The algorithm works in three major steps: Step 1 selects some customers as 'seed customer'. It is done by obtaining a number of conical regions (same as number of vehicles) by dividing the total angular plane with the number of vehicles; then, setting the maximum demand or farthest customer as seed. Each seed customers are assigned to a vehicle as well. Step 2 calculates insertion cost of each customer with respect to each seed. And finally, Step 3 assigns the customers to the vehicles according their increasing order of insertion cost. The vehicle capacity acts as a constraint since it cannot be exceeded.

Algorithm 5: Fisher and Jaikumar

1. ***Seed Selection***

   a) Divide the total angular planes by the number of vehicle and obtain cones equal to the number of vehicles.

b) Choose a customer from each cone with maximum demand or farthest from the origin and make that seed customer.
c) Assign a vehicle to each of the seed customer.

2. ***Insertion Cost Calculation***
Calculate insertion cost of each customer with respect to each seed using Eq. (30).
3. ***Assignment of Customers***
The customers are assigned to the vehicles according to their increasing order of insertion cost while maintaining vehicle capacity constraint.

6 University Course Scheduling Problem (UCSP) and Solution Using PSO

Scheduling problems evolve optimally assigning limited resources to tasks over time. Different scheduling problems include Job Scheduling [55], Job Shop Scheduling [56], Tournament Scheduling [57], different Educational scheduling [14, 34, 58, 59]. Educational scheduling focuses on course scheduling and exam scheduling of institutions. Among the different educational scheduling, course scheduling of universities is much more complex having a large number of constraints. And almost every university has to solve the general problem of UCSP in regular basis [60]. Hence, University Course Scheduling Problem (UCSP) has become a challenging optimization task and solving UCSP is an ultimate test/achievement for an/any optimization method.

6.1 UCSP Basics and Constraints

UCSPs are to assign days and hours to courses, so that the constraints imposed by the students as well as instructors are satisfied. For UCSP, the constraints are different for different institutions. Whatsoever, it is always a difficult task due to limited resources. In UCSP, each instructor should teach one course section at a time and 'each course section should be taught by an instructor'- and other likelihood constraints that can't be violated are termed as hard constraints. Soft constraints evolve instructors' and students' preferences about days and time periods and leisure periods, etc.; these are less important than hard constraints and can be disobeyed. Tasks of UCSP also include lectures allocation into suitable rooms considering room facility and enrolled students for the lectures [61, 62].

Mathematically, the UCSP is defined as a triple $\langle E, T, C\rangle$, where $E = \{c_i, s_j, i_k\}$ contains three sets: set of classes c_i, set of students s_j, and set of instructors i_k; $T = \{t_1, \ldots, t_n\}$ is a set of time slots and $C = \{C_1, \ldots, C_n\}$ is the set of hard and soft constraints. The task is to assign E_i to the time slot T_i satisfying constraints C_i where $C_i \in C$.

The challenges of the UCSP are the constraints, for example, instructors' dispositions, educational policies of the school, availability of teaching staffs and other physical resources. In UCSP, each instructor can teach one class at a timeslot and students can just go to one class at any given time. Other similar kinds of constraints are treated as hard constraints that must be satisfied. Common soft constraints of the UCSP are instructors' preferences for favored days and timeslots, expected to be satisfied to the extent possible. In the UCSP the main issue is to handle room allocation for lectures considering the maximum capacity of each room, the number of enrolled students in a course and other related facilities [58, 62].

Decision Variable. In case of UCSP, there is several decision variables, since this problem is different and much more complex than other problems discussed in the previous sections. In UCSP, elements of both E and T (i.e., $E = \{c_i, s_j, i_k\}$ and $T = \{t_1, \ldots, t_n\}$ act as decision variables. In particle encoding (Sect. 6.2), the issue of decision variable will be discussed.

Objective Function. The objective of UCSP to satisfy the preferences of the instructors and students while maintaining the constraints. The preferences vary from institution to institution, therefore objective function depends on the specific UCSP for a particular institute. In a later section, the objective function will be discussed through fitness calculation.

Constraints of UCSP. Both hard and soft constraints may vary from institution to institution based on their resources and facilities. Any resource modification or update (including capacity alteration in resources) requires rescheduling of classes, which is very common at the beginning of a term. Constraints that are considered in UCSP are summarized as follows.

Hard Constraints:

1. Each instructor can only teach one course section at a time.
2. Each course section can only be taught by an instructor.
3. Students can only attend one course section at a time.
4. Each classroom can't be used more than one course section at a time period.
5. Certain time periods can't be scheduled for non-academic activities, such as lunch time and sport.
6. Some course sections have to be scheduled in a particular room, such as computer lab.
7. The subjects must match the predefined lab subjects.
8. Two instructors can't take class at the same time.

Soft Constraints:

1. Instructors and students (classes) can indicate their preferences along with their preferred days and time periods in $d \times p$ time-slots across d days and p periods.
2. Instructors can choose to maximize the number of teaching free time periods; i.e., instructors can specify time periods when they prefer not to lecture.

3. A course section within multiple teaching hours has to be scheduled at the consecutive time periods and cannot be separated into more than a day of different time periods if required; i.e., instructors can specify their preferred lecturing format for a specific course section (for example, if a course section requires three teaching periods, the formats of 3–0, 2–1 and 1–1-1 represent three consecutive periods, two consecutive periods with a further single period and three single periods, respectively).
4. Minimizing students' movement between rooms.
5. One instructor for one subject.
6. The no. of students of every class must be permitted considering the lab capacity

6.2 Solution of UCSP Using PSO

To solve UCSP, there have been developments of various modified PSO algorithms. Among these, Modified Hybrid PSO (MHPSO) [33], Hybrid PSO with Local Search [35], and PSO with Selective Search (PSOSS) [34] are worth mentioning. PSOSS method works with a population of particles in which individual particle contains a feasible solution, calculates velocity of each individual particle using swap sequence and updates each particle with the computed velocity through selective search and forceful swap operation with repair mechanism. The particle encoding, swap operator and swap sequence, velocity computation, forceful swap operation with repair mechanism, selective search, fitness calculation and other operations of PSOSS are described in the following subsections.

Particle Encoding. The UCSP in consideration consists of u number of instructors, q number of courses, s number of rooms, and h number of timeslots. Each yearly student intakes are grouped together in a batch and the total number of batches is o. Each batch may be further subdivided into k number of subgroups. The PSOSS uses z number of particles to solve the UCSP. Sets of instructors, courses, rooms, timeslots, batches, subgroups, and particles are represented as follows:

- Set of Instructors, $I = \{e_i | i = 1, \ldots, u\}$
- Set of Courses, $C = \{c_i | i = 1, \ldots, q\}$
- Set of Rooms, $R = \{r_i | i = 1, \ldots, s\}$
- Set of Timeslots, $T = \{t_i | i = 1, \ldots, h\}$
- Set of Batches, $B = \{b_i | i = 1, \ldots, o\}$
- Set of Subgroups, $J = \{v_i | i = 1, \ldots, k\}$
- Set of Particles, $P = \{p_i | i = 1, \ldots, z\}$

Each instructor object contains information about assigned courses and preference for conducting a class in a particular timeslot. Each course object contains information about the number of timeslots required for a class, number of classes per week, course type (theory/laboratory) and number of students per class. Each room object contains information about allowed courses and seat capacity. Each timeslot represents a teaching period and the total number of timeslots (h) is 45 (9 periods in each

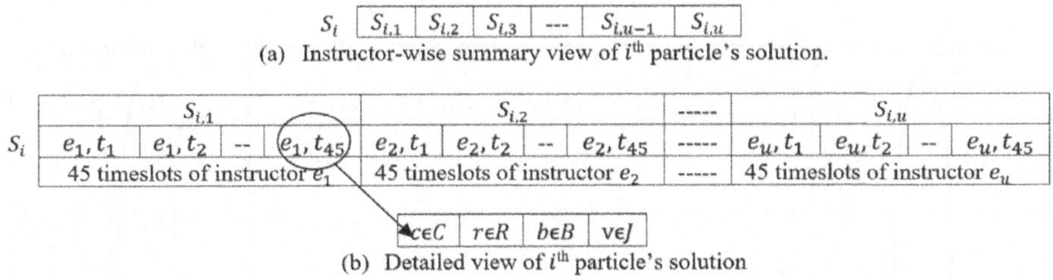

Fig. 10 Particle representation in PSOSS

day for 5 working days of a week). Each batch object contains subgroups and information about number of assigned students to that batch. Each subgroup object contains information about the number of assigned students. Each particle object contains a feasible solution to the UCSP i.e., the complete schedule for instructors, batches, and rooms. Solution of ith particle, S_i is represented by instructor-wise solutions in a one-dimensional matrix as shown in Fig. 10. Figure 10a is the instructor-wise summary view of particle's solution where $S_{i,1}, S_{i,2}, \ldots, S_{i,u}$ denote the 1st, 2nd,..., uth instructor's solution, respectively. Figure 10b shows the detailed particle view of 45 timeslots for each instructor and each timeslot identifies one of the nine teaching periods in a day for five working days in a week. The total number of timeslots is $45 \times u$ for u instructors; timeslots $e_1, t_1 - e_1, t_{45}$ for the first instructor, $e_2, t_1 - e_2, t_{45}$ for the second instructor, and so on. Each timeslot comprises assigned course, room, batch, and subgroup information as shown in Fig. 10b. Figure 11 represents the mapping of timeslots of a particle in days and periods. As an example, timeslot e_1, t_{44} represents 5th working day's 8th period of the first instructor.

Population Initialization. PSOSS method starts with an initial population where each particle is assigned a random solution. A random solution is generated by randomly assigning timeslots to the assigned courses of all instructors. Timeslots are allocated by maintaining all the hard constraints and it is checked that courses requiring multiple consecutive slots do not include any break periods. Weighted random distribution is used for assigning rooms to courses so that, a room having the highest load has the lowest probability of getting selected.

Timeslot		Sun	Mon	Tue	Wed	Thu	------	Sun	Mon	Tue	Wed	Thu
	1	e_1, t_1	e_1, t_{10}	e_1, t_{19}	e_1, t_{28}	e_1, t_{37}	------	e_u, t_1	------	------	------	------
	2	e_1, t_2	e_1, t_{11}	e_1, t_{20}	e_1, t_{29}	e_1, t_{38}	------	e_u, t_2	------	------	------	------
	3	e_1, t_3	e_1, t_{12}	e_1, t_{21}	e_1, t_{30}	e_1, t_{39}	------	e_u, t_3	------	------	------	------
	4	e_1, t_4	e_1, t_{13}	e_1, t_{22}	e_1, t_{31}	e_1, t_{40}	------	e_u, t_4	------	------	------	------
	5	e_1, t_5	e_1, t_{14}	e_1, t_{23}	e_1, t_{32}	e_1, t_{41}	------	------	------	------	------	------
	6	e_1, t_6	e_1, t_{15}	e_1, t_{24}	e_1, t_{33}	e_1, t_{42}	------	------	------	------	------	------
	7	e_1, t_7	e_1, t_{16}	e_1, t_{25}	e_1, t_{34}	e_1, t_{43}	------	------	------	------	------	------
	8	e_1, t_8	e_1, t_{17}	e_1, t_{26}	e_1, t_{35}	e_1, t_{44}	------	------	------	------	------	e_u, t_{44}
	9	e_1, t_9	e_1, t_{18}	e_1, t_{27}	e_1, t_{36}	e_1, t_{45}	------	------	------	------	------	e_u, t_{45}

Fig. 11 Mapping of timeslots in days and periods

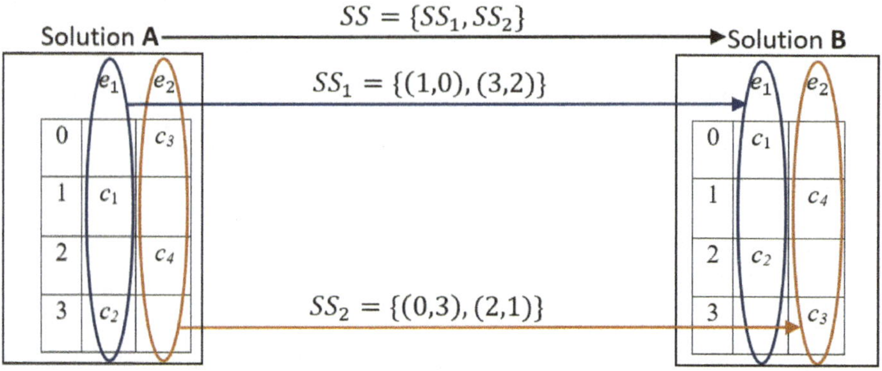

Fig. 12 Instructor wise Swap Sequences (SSs) of complete SS

Swap Operator (SO) and Swap Sequence (SS). A Swap Operator (SO) denotes the index of items to be swapped in a list [4, 22, 63] and a Swap Sequence (SS) is the collection of several SOs. The basics and formulations are SO and SS are similar to TSP operations and are same as discussed in Sect. 4.2

Velocity Computation using SO and SS. SO and SS have been used in the PSOSS method for velocity calculation. The SS to convert one particle's solution to another particle's solution is a collection of swap sequences which is measured in an instructor-to-instructor basis. Consider a UCSP consisting of two instructors e_1 and e_2 each having two courses $\{c_1, c_2\}$ and $\{c_3, c_4\}$, respectively. Figure 12 shows two different solutions A and B for the UCSP in consideration. In solution A, instructor e_1 has a course c_1 in timeslot 1 and another course c_2 in timeslot 3 whereas in solution B, course c_1 is in timeslot 0 and c_2 is in timeslot 2. So, the required SS for converting the schedule of e_1 in solution A to schedule of e_1 in solution B is $SS_1 = \{(1, 0), (3, 2)\}$. Similarly, $SS_2 = \{(0, 3), (2, 1)\}$. So, the complete SS for converting solution A to solution B is $SS = \{SS_{i_1}, SS_{i_2}\} = \{\{(1, 0), (3, 2)\}, \{(0, 3), (2, 1)\}\}$.

In Velocity Computation, swap sequence SS is treated as velocity to update a particle's solution at each iteration which is calculated using Eq. (31)

$$SS = \alpha(SGB - SC) + \beta(SPB - SC)\gamma SSPA, \tag{31}$$

where,SGB is the global best solution of the swarm, SC is the current solution of the particle, SPB is the personal best solution of the particle, $SSPA$ is the previously velocity applied, is merge operation and $\{\alpha, \beta, \gamma\}$ are selection probabilities for selecting a bunch of SOs from the corresponding SSs. $SGB - SC$ represents instructor-wise SSs to reach SGB from SC and $SPB - SC$ is the instructor-wise SSs to reach SPB from SC. If $SSGB = SGB - SC$ and $SSPB = SPB - SC$ then, Eq. (31) can be rewritten as:

$$SS = \alpha SSGB + \beta SSPB\gamma SSPA \tag{32}$$

After selection of SOs with $\{\alpha, \beta, \gamma\}$, SS becomes

$$SS = SSSGB + SSSPBSSSPA = SSSGB + SSM, \tag{33}$$

where, $SSSGB, SSSPB, SSSPA$ are the selected SS from $SSGB, SSPB$ and$SSPA$, respectively and SSM is the swap sequence resulting from merging $SSSPB$ with$SSSPA$.

As $SSSGB$ and SSM may contain redundant SOs, redundant swaps are removed from them and they become $SSSGBM$ and $SSMM$, respectively. Finally, SS becomes:

$$SS = SSSGBM + SSMM \tag{34}$$

This final velocity *SS* is then applied using forceful swap operation with repair mechanism and selective search as described in following sections.

Forceful Swap Operation with Repair Mechanism. Forceful Swap Operation with Repair Mechanism is an added feature of PSOSS. UCSP is highly constrained in nature and most of the constraints are interrelated. Consequently, if a class needs to be shifted to a new timeslot, then all the involved members such as instructor, students, and room need to be free in that timeslot. As a result, most of the selected swaps can't be applied because of violation of constraints. Therefore, a portion of $SSSGBM$ is forcefully applied to current solution *SC* to ensure that *SC* moves a little towards SGB. Forcefully applying an SO can cause conflicts. Therefore, a repair mechanism is involved in forceful SO operation to make sure that no invalid solution results in that process. The repair mechanism works by randomly moving conflicting courses to non-conflicting positions. The repair mechanism is illustrated in Fig. 13. Consider a UCSP instance consisting of two instructors e_1 and e_2. Instructor e_1 has course c_1 of batch b_1, subgroup v_1 in room r_1 and course c_2 of batch b_3, subgroup v_2 in room r_3 at timeslots 5 and 8, respectively. Instructor e_2 has course c_4 of batch b_2, subgroup v_1 in room r_2 and course c_3 of batch b_3, subgroup v_2 in room r_5 at timeslots 3 and 5, respectively. Now, if an SO(5, 8) is applied forcefully on the solution of e_1 then, course c_2 of e_1 comes at timeslot 5 which results in a conflict with e_2 because subgroup v_2 of batch b_3 is already engaged in a class with e_2 (shown with circle in Figs. 3 and 4). This conflict is resolved by moving the conflicting course of e_2 to a randomly chosen non-conflicting timeslot 7.

Selective Search. Selective search is one of the most important features of PSOSS. In PSOSS, velocity *SS* is applied using selective search mechanism. In selective search, each solution generated by applying an SO of *SS* is considered as an intermediate solution and the sequence of SOs generating the best intermediate solution is considered as the final velocity which becomes the previously applied velocity for the next iteration.

Suppose , $SS = \{SO_1, SO_2, SO_3, \ldots, SO_n\}$ then the selective search can be written as

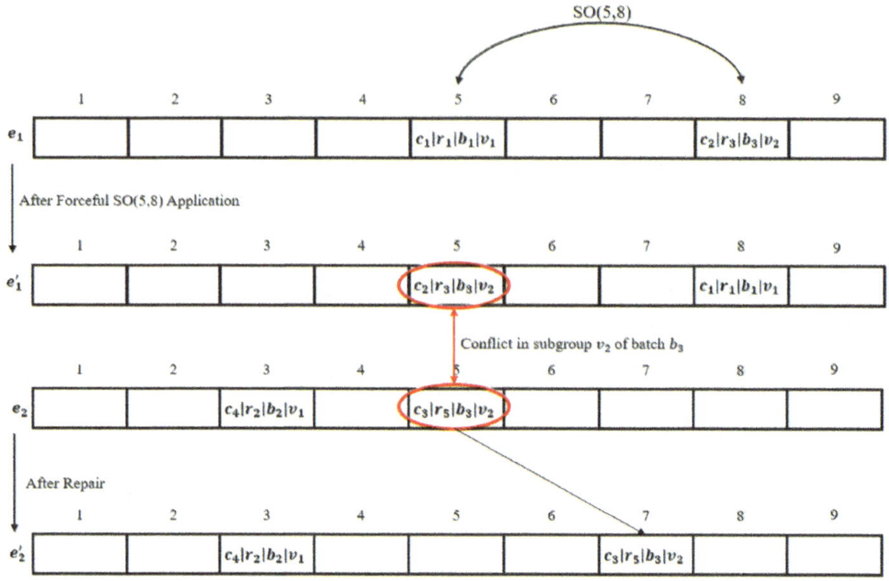

Fig. 13 Illustration of the repair mechanism

$$
\begin{aligned}
SC^1 &= SC + SO_1 \\
SC^2 &= SC^1 + SO_2 \\
&\vdots \\
SC^n &= SC^{n-1} + SO_n
\end{aligned}
$$

In the above cases, $SC^1, SC^2, \ldots, SC^n$ are the intermediate solutions and the intermediate solution having the highest fitness becomes the final solution SC in selective search as defined by the following equation:

$$SC = \{SC^f\}, f = 1, 2, \ldots, n \tag{35}$$

Finally, the velocity is $SS = \{SO_1, SO_2, SO_3, \ldots, SO_f\}, 1 \le f \le n$.

The ultimate solution SC in selective search is the intermediate solution possessing the highest fitness value. Thus, the selective search technique explores the opportunity of keeping better solution from the intermediate solutions.

Fitness Calculation. Each instructor's preference for conducting a class in a particular timeslot is represented by an integer value as shown in Fig. 14. A higher value corresponds to a higher preference of an instructor to conduct the class in that particular timeslot. Whereas, a negative value shows the instructor's non-preference. The fitness of ith particle's solution S_i is calculated by considering fitness of each of the instructor's solution which belongs to that particle's solution using the following equation:

Timeslot		Sun	Mon	Tue	Wed	Thu	....	Sun	Mon	Tue	Wed	Thu
	1	0	0	0	-1	5	....	0	5	4	-1	5
	2	1	3	4	-1	5	....	2	5	4	2	5
	3	1	3	4	-1	5	....	2	5	4	2	5
	4	1	3	4	-1	5	....	2	5	4	2	5
	5	3	5	-1	3	4	....	1	3	-1	1	-1
	6	3	5	-1	3	4	....	1	3	-1	1	-3
	7	3	5	-1	3	-1	....	1	2	-1	1	-5
	8	0	0	-1	3	-1	....	0	2	-1	1	-1
	9	0	0	-1	3	-1	....	0	2	-1	1	-1
		{–		e_1		–}	....	{–		e_u		–}

Fig. 14 Sample preference values for instructors

$$F(S_i) = \sum\nolimits_{j=0}^{u} F\left(S_{i,j}\right), \tag{36}$$

where, $F(S_i)$ is the fitness of the ith particle's solution, and $F(S_{i,j})$ is the fitness of the ith particle's jth instructor's solution. Now, fitness of each instructor's solution is calculated by considering quality and violation of the instructor's solution using the following equation:

$$F\left(S_{i,j}\right) = Q\left(S_{i,j}\right) - V\left(S_{i,j}\right), \tag{37}$$

where, $Q(S_{i,j})$ is the quality of the instructors' solution, and $V(S_{i,j})$ is the violation of the instructor's solution. Preference values of corresponding positions where courses are assigned to an instructor are summed up to calculate the quality of an instructor's solution. Violation of the instructors' solution is calculated using the following equation:

$$V\left(S_{i,j}\right) = \sum_{a=1}^{tc} 2^{l_a}, \tag{38}$$

where, tc is the total number of blocks of consecutive classes in an instructor's solution and l_a is the number of classes in ath block. The exponential in Eq. (38) is used to mimic the human nature. If the number of consecutive classes increases, then the dissatisfaction of an instructor increases rapidly. For example, three consecutive classes are much more difficult (practically almost impossible) to manage for an instructor than two consecutive classes.

PSOSS Algorithm for Solving UCSP. The PSOSS method for solving UCSP is shown in Algorithm 6.

7 Conclusions

Particle Swarm Optimization (PSO) is a well-known optimization technique; and particle encoding, formulation of velocity and solution update with velocity are the key elements to solve a particular optimization task with PSO. PSO was proposed to solve problems in continuous domain where mathematical operations for velocity measure and its implication to update particles' position are straight forward. Due to straight-forward mechanism and simplicity in operation, a number of PSO-based discrete optimization methods are investigated in different time periods. Particle encoding, velocity formulation and solution update are significantly different in individual problem domains.

This study presented a comprehensive view of PSO evolution for four popular discrete optimization tasks: knapsack problem (KP), traveling salesman problem (TSP), capacitated vehicle routing problem (CVRP), and university course scheduling problem (UCSP). KP is the simplest one where a string of 0 and 1 s represent particle and individual bits are updated using binary version of PSO. In solving TSP, Swap Sequence based operation are demonstrated to update routes represented by the individual particles. In solving CVRP, PSO is considered to generate optimal routes of individual vehicles. Finally, UCSP which holds a large number of constraints is considered to understand how PSO operations are evolved for such a complex optimization task.

Algorithm 6: Solution Representation Procedures

1. ***Initialization***

   a) Create z numbers of particles and add them to the list of particles P
   b) Calculate fitness of each particle as described in Section 0
   c) Record the global best particle

2. ***Computation and application of velocity***

   a) For each particles, calculate velocity using Eq. (31), Eq. (32), Eq. (33) and Eq. (34)
   b) Apply f percent of swaps from the minimized SS acquired from the global best SS using forceful swap operation with repair mechanism and selective search.
   c) Apply the remaining swaps using selective search.
   d) Apply the minimized swaps acquired by merging global and personal best SSs.
   e) Update the current solution, personal best and previously applied SS.

3. ***Update the global best solution***
4. ***Repeat Step 2 and Step 3 if not met stopping criteria***
5. ***Return global best solution as result***

At present PSO is established as a generic method, attempts can be taken to solve more range of problems by introducing appropriate adaptation of its elements and operations. The rhythmic presentation of carefully selected PSO-based methods in this chapter will be helpful to realize PSO adaptation from its original continuous domain to different discrete domains. Most importantly, this study might be a guideline to solve any new discrete optimization task adopting PSO.

Although a number of PSO-based methods are available for each of the discussed problem, the presented methods are carefully chosen to understand PSO evolution progressively from simple to complex one. Remaining PSO-based methods for discrete optimization might be studied to realize diversification of PSO adaptations. Self-Adaptive Check and repair operator based PSO [17], PSO with Time Varying acceleration [64], and Set-based PSO [65] considered different mechanisms to solve KP with PSO. For TSP, different modification on PSO [66, 67], and hybridization with different methods [68, 69] are the considerable PSO-based methods. For CVRP, Attractors-based PSO [29], Discrete PSO for CVRP [28] solved the customer assignment to vehicles and route generation together. Finally, Hybrid PSO for Timetable Scheduling [35] and Modified Hybrid PSO for UCSP [33] might be studied for different strategies of PSO adaption in solving UCSP.

References

1. Goldberg, D.E.: Genetic Algorithms in Search, Optimization and Machine Learning. Addison-Wesley Longman Publishing Co., Inc. (1989)
2. Whitley, D.: A genetic algorithm tutorial. Stat. Comput. **4** (1994). https://doi.org/10.1007/BF00175354
3. Brownlee, J.: Clever Algorithms: Nature-inspired Programming Recipe (2011)s
4. Akhand, M.A.H., Shill, P.C., Hossain, M.F., et al.: Producer-Scrounger method to solve traveling salesman problem. Int. J. Intell. Syst. Appl. **7**, 29–36 (2015). https://doi.org/10.5815/ijisa.2015.03.04
5. Eberhart, R., Kennedy, J.: A new optimizer using particle swarm theory. In: MHS'95. Proceedings of the Sixth International Symposium on Micro Machine and Human Science, pp. 39–43. IEEE (1995)
6. Liang, J.J., Qin, A.K., Suganthan, P.N., Baskar, S.: Comprehensive learning particle swarm optimizer for global optimization of multimodal functions. IEEE Trans. Evol. Comput. **10**, 281–295 (2006). https://doi.org/10.1109/TEVC.2005.857610
7. Liao, Y.-F., Yau, D.-H., Chen, C.-L.: Evolutionary algorithm to traveling salesman problems. Comput. Math. Appl. **64**, 788–797 (2012). https://doi.org/10.1016/j.camwa.2011.12.018
8. Xiong, W., Jiang-Wei, Z., Hon-Lin, Z.: Enhanced self-tentative particle swarm optimization algorithm for TSP. J. North China Electr. Power Univ. **36**, 69–74 (2009)
9. Zhang, J., Si, W.: Improved Enhanced Self-Tentative PSO algorithm for TSP. In: 2010 Sixth International Conference on Natural Computation, pp. 2638–2641. IEEE (2010)
10. Shi, X.H., Liang, Y.C., Lee, H.P., et al.: Particle swarm optimization-based algorithms for TSP and generalized TSP. Inf. Process. Lett. **103**, 169–176 (2007). https://doi.org/10.1016/j.ipl.2007.03.010
11. Fan, H.: Discrete particle swarm optimization for TSP based on neighborhood. J. Comput. Inf. Syst. **6**, 3407–3414 (2010)
12. Goldbarg, E.F.G., Goldbarg, M.C., de Souza, G.R.: Particle swarm optimization algorithm for traveling salesman problem. In: Traveling Salesman Problem, pp. 75–96. InTech (2008)

13. Akhand, M.A.H., Akter, S., Rashid, M.A., Yaakob, S.B.: Velocity tentative PSO: an optimal velocity implementation based particle swarm optimization to solve traveling salesman problem. IAENG Int. J. Comput. Sci. **42**, 221–232 (2015)
14. Montero, E., Altamirano, L.: A PSO algorithm to solve a Real Course+Exam Timetabling Problem. In: International Conference on Swarm Intelligence, p. 2 (2011)
15. Yan, X., Zhang, C., Luo, W., et al.: Solve traveling salesman problem using particle swarm optimization algorithm. Int. J. Comput. Sci. Issues **9**, 264–271 (2012)
16. Liu, Z.: An analysis of particle swarm optimization of multi-objective knapsack problem. In: 2020 9th International Conference on Industrial Technology and Management (ICITM), pp. 302–306, IEEE (2010)
17. Chih, M.: Self-adaptive check and repair operator-based particle swarm optimization for the multidimensional knapsack problem. Appl. Soft Comput. **26**, 378–389 (2015). https://doi.org/10.1016/j.asoc.2014.10.030
18. Liang, Y., Liu, L., Wang, D., Wu, R.: Optimizing particle swarm optimization to solve knapsack problem. Commun. Comput. Inf. Sci. **105** CCIS, 437–443 (2010). https://doi.org/10.1007/978-3-642-16336-4_58
19. Li, Y., He, Y., Li, H., et al.: A Binary Particle Swarm Optimization for Solving the Bounded Knapsack Problem, pp. 50–60 (2019)
20. Nguyen, B.H., Xue, B., Andreae, P., Zhang, M.: A new binary particle swarm optimization approach: momentum and dynamic balance between exploration and exploitation. IEEE Trans. Cybern. **51**, 589–603 (2021). https://doi.org/10.1109/TCYB.2019.2944141
21. Hembecker, F., Lopes, H.S., Godoy, W.: Particle Swarm Optimization for the Multidimensional Knapsack Problem, pp. 358–365 (2007)
22. Wang, K.-P., Huang, L., Zhou, C.-G., Pang, W.: Particle swarm optimization for traveling salesman problem. In: Proceedings of the 2003 International Conference on Machine Learning and Cybernetics (IEEE Cat. No.03EX693), pp. 1583–1585. IEEE (2003)
23. Akhand, M.A.H., Sultatana, T., Shuvo, M.I.R., Al-Mahmud, A.-M.: Constructive and clustering methods to solve capacitated vehicle routing problem. Orient. J. Comput. Sci. Technol. **10**, 549–562 (2017). https://doi.org/10.13005/ojcst/10.03.02
24. Akhand, M.A., Zahrul, J.P., Murase, K.: Capacitated vehicle routing problem solving using adaptive sweep and velocity tentative PSO. Int. J. Adv. Comput. Sci. Appl. **8**, 288–295 (2017). https://doi.org/10.14569/IJACSA.2017.081237
25. Akhand, M.A.H., Peya, Z.J., Sultana, T., Rahman, M.M.H.: Solving capacitated vehicle routing problem using variant sweep and swarm intelligence. J. Appl. Sci. Eng. **20**, 511–524 (2017). https://doi.org/10.6180/jase.2017.20.4.13
26. Peya, Z.J., Akhand, M.A.H., Murase, K.: Capacitated vehicle routing problem solving through adaptive sweep based clustering plus swarm intelligence based route optimization. Orient. J. Comput. Sci. Technol. **11**, 88–102 (2018). https://doi.org/10.13005/ojcst11.02.04
27. Marinakis, Y., Marinaki, M., Migdalas, A.: A multi-adaptive particle swarm optimization for the vehicle routing problem with time windows. Inf. Sci. **481**, 311–329 (2019). https://doi.org/10.1016/j.ins.2018.12.086
28. Mauliddina, A.N., Saifuddin, F.A., Nagari, A.L., et al.: Implementation of discrete particle swarm optimization algorithm in the capacitated vehicle routing problem. Jurnal Sistem dan Manajemen Industri **4**, 117–128 (2020). https://doi.org/10.30656/jsmi.v4i2.2607
29. Halassi Bacar, A.-H., Rawhoudine, S.C.: An attractors-based particle swarm optimization for multiobjective capacitated vehicle routing problem. RAIRO—Oper. Res. **55**, 2599–2614 (2021). https://doi.org/10.1051/ro/2021119
30. Islam, M.A., Gajpal, Y., ElMekkawy, T.Y.: Hybrid particle swarm optimization algorithm for solving the clustered vehicle routing problem. Appl. Soft Comput. **110**, 107655 (2021). https://doi.org/10.1016/j.asoc.2021.107655
31. Chirawichitchai, N.: Emotion classification of Thai text based using term weighting and machine learning techniques. In: 2014 11th International Joint Conference on Computer Science and Software Engineering (JCSSE), pp. 91–96. IEEE (2014)

32. Wisittipanich, W., Phoungthong, K., Srisuwannapa, C., et al.: Performance comparison between particle swarm optimization and differential evolution algorithms for postman delivery routing problem. Appl. Sci. **11**, 2703 (2021). https://doi.org/10.3390/app11062703
33. Ferdoushi, T., Das, P.K., Akhand, M.A.H.: Highly constrained university course scheduling using modified hybrid particle swarm optimization. In: 2013 International Conference on Electrical Information and Communication Technology (EICT), pp. 1–5. IEEE (2014)
34. Imran Hossain, S., Akhand, M.A.H., Shuvo, M.I.R., et al.: Optimization of university course scheduling problem using particle swarm optimization with selective search. Exp. Syst. Appl. **127**, 9–24 (2019). https://doi.org/10.1016/j.eswa.2019.02.026
35. Psarra, E., Apostolou, D.: Timetable scheduling using a hybrid particle swarm optimization with local search approach. In: 2019 10th International Conference on Information, Intelligence, Systems and Applications (IISA), pp. 1–8. IEEE (2019)
36. Kennedy, J., Eberhart, R.C.: Particle swarm optimization. In: Proceedings of ICNN'95—International Conference on Neural Networks, pp. 1942–1948. IEEE (1995)
37. Eddaly, M., Jarboui, B., Siarry, P.: Combinatorial particle swarm optimization for solving blocking flowshop scheduling problem. J. Comput. Des. Eng. **3**, 295–311 (2016). https://doi.org/10.1016/j.jcde.2016.05.001
38. Banka, H., Dara, S.: A Hamming distance based binary particle swarm optimization (HDBPSO) algorithm for high dimensional feature selection, classification and validation. Pattern Recogn. Lett. **52**, 94–100 (2015). https://doi.org/10.1016/j.patrec.2014.10.007
39. Matai, R., Singh, S., Lal, M.: Traveling salesman problem: an overview of applications, formulations, and solution approaches. In: Traveling Salesman Problem, Theory and Applications. InTech (2010)
40. Reza, M., Rahimi, M., Shah-Hosseini, H.: Population-based optimization algorithms for solving the travelling salesman problem. In: Traveling Salesman Problem. InTech (2008)
41. Akhand, M.A.H., Akter, S., Sazzadur Rahman, S., Hafizur Rahman, M.M.: Particle swarm optimization with partial search to solve traveling salesman problem. In: 2012 International Conference on Computer and Communication Engineering, ICCCE, pp. 118–121 (2012). https://doi.org/10.1109/ICCCE.2012.6271164
42. Akhand, M.A.H., Akter, S., Rashid, M.A.: Velocity tentative particle swarm optimization to solve TSP. In: 2013 International Conference on Electrical Information and Communication Technology (EICT), pp. 1–6. IEEE (2014)
43. Dantzig, G.B., Ramser, J.H.: The truck dispatching problem. Manag. Sci. **6**, 80–91 (1959). https://doi.org/10.1287/mnsc.6.1.80
44. Christofides, N., Mingozzi, A., Toth, P.: The vehicle routing problem. In: Combinatorial Optimization. Wiley, New York (1979)
45. Venkatesan, S.R., Logendran, D., Chandramohan, D.: Optimization of capacitated vehicle routing problem using particle swarm optimization. Int. J. Eng. Sci. Technol. (IJEST) 3 (2011)
46. Lysgaard, J.: Clarke and Wright's Savings Algorithm (1997)
47. Poot, A., Kant, G., Wagelmans, A.P.M.: A Savings Based Method for Real-Life Vehicle Routing Problems (1999)
48. Doyuran, T., Çatay, B.: A robust enhancement to the Clarke-Wright savings algorithm. J. Oper. Res. Soc. **62**, 223–231 (2011). https://doi.org/10.1057/jors.2009.176
49. Pichpibul, T., Kawtummachai, R.: An improved Clarke and Wright savings algorithm for the capacitated vehicle routing problem. ScienceAsia **38**, 307 (2012). https://doi.org/10.2306/scienceasia1513-1874.2012.38.307
50. Gillett, B.E., Miller, L.R.: A heuristic algorithm for the vehicle-dispatch problem. Oper. Res. **22**, 340–349 (1974). https://doi.org/10.1287/opre.22.2.340
51. Suthikarnnarunai, N.: A sweep algorithm for the mix fleet vehicle routing problem. In: Proceedings of the International MultiConference of Engineers and Computer Scientists, vol II, pp. 19–21. Hong Kong (2008)
52. Nurcahyo, G.W., Alias, R.A., Shamsuddin, S.M., Sap, M.N.: Sweep algorithm in vehicle routing problem for public transport. Jurnal Antarabangsa (Teknologi Maklumat) **2**, 51–64 (2002)

53. Fisher, M.L., Jaikumar, R.: A generalized assignment heuristic for vehicle routing. Networks **11**, 109–124 (1981). https://doi.org/10.1002/net.3230110205
54. Na, B., Jun, Y., Kim, B.-I.: Some extensions to the sweep algorithm. Int. J. Adv. Manuf. Technol. **56**, 1057–1067 (2011). https://doi.org/10.1007/s00170-011-3240-7
55. Xie, X., Liu, R., Cheng, X., et al.: Trust-driven and PSO-SFLA based job scheduling algorithm on Cloud. Intell. Autom. Soft Comput. **22**, 561–566 (2016). https://doi.org/10.1080/10798587.2016.1152770
56. Liao, J., Lin, C.: Optimization and simulation of job-shop supply chain scheduling in manufacturing enterprises based on particle swarm optimization. Int. J. Simul. Model. **18**, 187–196 (2019). https://doi.org/10.2507/IJSIMM18(1)CO5
57. Schauz, U.: The tournament scheduling problem with absences. Eur. J. Oper. Res. **254**, 746–754 (2016). https://doi.org/10.1016/j.ejor.2016.04.056
58. Shiau, D.-F.: A hybrid particle swarm optimization for a university course scheduling problem with flexible preferences. Expert Syst. Appl. **38**, 235–248 (2011). https://doi.org/10.1016/j.eswa.2010.06.051
59. Tassopoulos, I.X., Beligiannis, G.N.: A hybrid particle swarm optimization based algorithm for high school timetabling problems. Appl. Soft Comput. **12**, 3472–3489 (2012). https://doi.org/10.1016/j.asoc.2012.05.029
60. Chiarandini, M., Birattari, M., Socha, K., Rossi-Doria, O.: An effective hybrid algorithm for university course timetabling. J. Sched. **9**, 403–432 (2006). https://doi.org/10.1007/s10951-006-8495-8
61. Song, X.: Hybrid particle swarm algorithm for job shop scheduling problems. In: Future Manufacturing Systems. Sciyo (2010)
62. Naji Azimi, Z.: Hybrid heuristics for Examination timetabling problem. Appl. Math. Comput. **163**, 705–733 (2005). https://doi.org/10.1016/j.amc.2003.10.061
63. Khan, I., Maiti, M.K.: A swap sequence based artificial bee colony algorithm for traveling salesman problem. Swarm Evol. Comput. (2018). https://doi.org/10.1016/j.swevo.2018.05.006
64. Chih, M., Lin, C.-J., Chern, M.-S., Ou, T.-Y.: Particle swarm optimization with time-varying acceleration coefficients for the multidimensional knapsack problem. Appl. Math. Model. **38**, 1338–1350 (2014). https://doi.org/10.1016/j.apm.2013.08.009
65. Chen, W.-N., Zhang, J., Chung, H.S.H., et al.: A novel set-based particle swarm optimization method for discrete optimization problems. IEEE Trans. Evol. Comput. **14**, 278–300 (2010). https://doi.org/10.1109/TEVC.2009.2030331
66. Yousefikhoshbakht, M.: Solving the traveling salesman problem: a modified metaheuristic algorithm. Complexity **2021**, 1–13 (2021). https://doi.org/10.1155/2021/6668345
67. Blamah, N.V., Oluyinka, A.A., Wajiga, G., Baha, Y.B.: MAPSOFT: a multi-agent based particle swarm optimization framework for travelling salesman problem. J. Intell. Syst. **30**, 413–428 (2020). https://doi.org/10.1515/jisys-2020-0042
68. Wei, B., Xing, Y., Xia, X., Gui, L.: A novel particle swarm optimization with genetic operator and its application to TSP. Int. J. Cogn. Inform. Nat. Intell. **15**, 1–17 (2021). https://doi.org/10.4018/IJCINI.20211001.oa31
69. Cui, Y., Zhong, J., Yang, F., et al.: Multi-subdomain grouping-based particle swarm optimization for the traveling salesman problem. IEEE Access **8**, 227497–227510 (2020). https://doi.org/10.1109/ACCESS.2020.3045765

Performance Analysis of Hybrid Memory Based Dragonfly Algorithm in Engineering Problems

Sanjoy Debnath, Ravi Singh Kurmvanshi, and Wasim Arif

Abstract Hybrid Memory Based Dragonfly Algorithm with Differential Evolution (DADE) is one of the most prominent swarm-based optimization techniques due to its better computational complexity and high convergence rate for achieving optimum results. In DADE, the best solution is memorized and processed with Differential Evolution (DE) for enhanced diversity and balanced exploration and exploitation rate. DADE brings two distinct advantages: First, the superior convergence rate due to continuous update of personal best individual in the search process. Second, better exploration due to the inclusion of global best and global worst individuals in the hybridization process. Comparative simulations have been performed on 24 standard benchmark functions along with the benchmark function of CEC2005 and CEC2017. Comparative analysis of the result demonstrates the competitiveness of the DADE algorithm in terms of optimal cost, computational complexity and convergence characteristics compared to other considered optimization algorithms.

Keywords Optimization · Hybrid swarm based algorithms · Dragonfly algorithm · Comparative analysis

1 Introduction

Optimization techniques are becoming an integrated component in finding an accurate and effective solution to general engineering problems. The utilization of efficient optimization techniques could maximize the efficacy of the system while producing

S. Debnath (✉)
Department of ECE, Vel Tech Rangarajan Dr. Sagunthala R&D Institute of Science and Technology, Chennai, Tamil Nadu, India
e-mail: drsanjoydebnath@veltech.edu

R. S. Kurmvanshi · W. Arif
Department of ECE, National Institute of Technology Silchar, Assam, India
e-mail: ravinit18@gmail.com

W. Arif
e-mail: arif.ece.nits@gmail.com

A. Biswas et al. (eds.), *Advances in Swarm Intelligence*, Studies in Computational Intelligence 1054, https://doi.org/10.1007/978-3-031-09835-2_5

quality output in large amounts. Due to this, the researchers are interested in developing problem-dependent meta-heuristic optimization techniques. Moreover, minimizing the algorithm's simplicity while prioritizing the accuracy of the optimal solutions is also a challenging aspect to be encountered in the development of optimization techniques.

Swarm-based optimization techniques are well known for their simplicity and robustness [1]. By developing PSO, the basic building block of swarm-based optimization techniques was laid down by Eberhart and Kennedy [2]. PSO has become very popular for its simplicity and attracts researchers for solving diverse problems of engineering fields. After that, based on the social behavior of animals, birds, and living things, problem-specific development of optimization techniques has started to achieve the optimal solution. Grey wolf optimization [3], Whale optimization algorithm [4], Moth search optimization [5], ant lion optimizer [6], Dragonfly algorithm (DA) [7], Salp Swarm Algorithm [8], Grasshopper optimization algorithm (GOA) [9], Differential Evolution (DE) [10], Buyer Inspired Metaheuristic Algorithm (BIMA) [11], and Ageist Spider Monkey optimization [12] are some of the well-known meta-heuristic optimization techniques build on the platform of social behavior of living things. Toward the applicability of meta-heuristic algorithm in real-life engineering problems, the heuristic algorithm shows significant results. To solve the 3D UAV deployment problem, a learning component is incorporated in DE to improve the algorithm's computational efficiency [13]. This hybridization is highly essential for the high flight time of UAVs. To address the issue of energy consumption in water distribution system, a binary DA algorithm is designed to solve the water pump scheduling problem with high energy efficiency [14]. Towards the application of binary DA in real-world practical problems, an improved version of the binary DA algorithm with a learning strategy is designed to get enhanced feature selection from the extensive data set of Covid-19 disease [15]. The utilization of DA is also analyzed through developing a hybridized DA and Simulated Annealing algorithm to improve the feature selection in real engineering problems [16].

In spite of the enormous advantage of swarm-based optimization techniques, they are frequently stuck down at the local optima. That deficiency can be coverup by tune-up the balancing between the rate of exploitation and exploration. This balancing can only be achieved by hybridizing the key strategy of two algorithms [17]. In comparison to standard models, a hybrid approach tries to reduce complexity, increase stability, speed up convergence, and provide greater accuracy. Hybrid optimization techniques will significantly improve the parent algorithm's performance matrices. In DADE, the deficiencies of DA have been encountered and rectified for achieving improved solutions in the considered search space [18]. In DADE, local and global best solutions are tracked by the memory update scheme; subsequently, these solutions are incorporated in DE to get better mutation towards improved exploitation of the search space. Additionally, this incorporation brings exploration in the initial stage and exploitation in the later stage of the evaluation, which ensures the reach of global optimum with enhanced accuracy.

On the other hand, evolutionary optimization techniques have become famous for solving diverse engineering problems with high computational efficiency [19].

However, DE has some limitations for its inappropriate selection of tuning parameters like mutation and crossover. To improvise these limitations of DE, an adoptive update of mutation and crossover strategy is proposed in jDE [20]. In 2009, Qin et al. proposed SaDE, which incorporates adaptive update of mutation and crossover parameter in the evolution process based on the learning experience.

Thereafter, a new mutation scheme, "DE/current-to-pbest" is introduced in JADE by Zhang and Sanderson for achieving maximum population diversity [21]. Subsequently, an improved version of DE, namely DMPSADE is proposed to enhance the overall exploration rate of the DE by adopting discrete mutation and different crossover strategies for each variable [22]. However, very less literature is available that deals with improving the computational complexity of the algorithm while ensuring global optimal solution. In DADE, the learning component is adopted to update mutation parameter by tracking local and global best solutions for guaranteed convergence rate and computational efficiency. Thus in this work, we analyze the outcome of DADE with other popular meta-heuristic algorithm to draw the comparative results.

The remaining paper is organized as follows: Sect. 2 comprises of mathematical interpretation of the DADE algorithm. Performance evaluation is drawn in Sect. 3 that comprises results of popular meta-heuristic algorithm on different benchmark functions, followed by the presentation of convergence analysis of the DADE algorithm in Sect. 4. Finally, conclusion of the work is drawn in Sect. 5.

2 Mathematical Interpretation of DADE

In DADE algorithm, the internal memory component is invoked to get the tracked version of *pBest* (x^{pBest}), *gBest* (x_{leader}), and *gWorst* (x_{worst}) solutions. From these tracked and updated values, mutation schemes of DE is supervised for acquiring additional exploitation in the algorithm. The mutation vector for a target vector is formulated in DADE algorithm as given below.

$$v_{i,k} = x_{i,k} + r_1^*\left(x_{i,k}, x_{i,k}^{\text{old}}\right) + F_1^*\left(x_{\text{leader}} - x_{i,k}\right) - F_2^*\left(r_n^* x_{\text{worst}} - x_{i,k}^{pBest}\right) \quad (1)$$

where F_1 and F_2 are mutation factor modified as

$$F_1 = M_F + r_1 M_F \quad (2)$$

$$F_2 = M_F - r_1 M_F \quad (3)$$

$$M_F = 0.8 + (0.8 - 0.2)\ \text{iteration}/\text{Max\_iteration} \quad (4)$$

where x^{old} is the position of the particle in the last generation and r is the random number follows normal distribution.

After the evaluation of mutation vector, the subsequent crossover operation is performed with a CR rate of 0.9.

3 Performance Evaluation

The Performance of DADE algorithm is evaluated on total 59 benchmark function consisting of 24 standard benchmark functions [18], 5 CEC2005 [18], and 30 CEC-2017 [23]. These functions are divided into two Suits of functions based on their property. Suit-I has 29 benchmark functions comprised of 24 standard benchmark functions [18] and 5 CEC2005 functions, whereas Suit-II comprises 30 benchmark functions of CEC2017. The Suit-I has a variety of four types of functions: unimodal, multi-modal, fixed-dimensional, and composite functions. The Suit-II has the significance of real parameter numerical for single objective benchmarks.

3.1 Experimental Results

Experiments are conducted on Matlab® R2015a platform. The performance of DADE is comparatively evaluated on some of the popular optimization techniques, namely ALC-PSO, GWO, jDE, SaDE, JADE, DMPSADE, and GSA. The parameters assumption of the DADE algorithm in the evaluations are 30 search agents, MF as 0.5, CR as 0.9. Algorithms are run 30 times, while each run consists of 1000 iterations to observe the global optimal output. Results are evaluated on the statistical mean (Mn) and standard deviation (Sd) of the achieved optimal fitness value. Tables 1 and 2 given below show the comparative result where bold letters indicates the global optimum result.

3.2 Analysis of DADE

The drawn observations of the comparative analysis of experimental results are listed below:

Table 1 Results analysis of DADE with other meta-heuristic algorithms on Suit-I benchmark functions

Function		ALC-PSO	GWO	jDE	SaDE	JADE	DMPSADE	GSA	DADE
F-1	Mn	1.67E−161	6.59E−28	6.12E + 04	3.73E + 03	6.09E + 02	2.61E + 03	2.53E−16	**1.5163E−80**
	Sd	8.21E−161	6.34E−05	7.64E + 04	3.26E + 03	1.18E + 03	8.09E + 03	9.67E−17	**1.5522E−80**
F-2	Mn	1.161E−90	7.18E−17	2.27E−15	**0.00E + 00**	**0.00E + 00**	1.62E + 07	0.055655	**4.8733E−36**
	Sd	4.146E−90	2.91E−02	7.87E−15	**0.00E + 00**	**0.00E + 00**	7.01E + 07	0.194074	**2.2047E−36**
F-3	Mn	1.792E−11	2.69E−21	4.09E−14	**0.00E + 00**	9.86E−04	5.25E + 03	896.5347	2.7599E−75
	Sd	3.538E−11	2.54E−16	2.60E−14	**0.00E + 00**	5.95E−03	1.88E + 04	318.9559	1.9550E−75
F-4	Mn	4.379E−04	5.61E−07	8.53E + 00	**0.00E + 00**	0.00E + 00	13.6266	7.35487	2.363E−37
	Sd	2.53E−04	1.32E + 00	2.16E + 01	**0.00E + 00**	0.00E + 00	23.0639	1.741452	1.9121E−37
F-5	Mn	**7.613**	2.65E + 01	2.03E + 01	2.03E + 01	2.03E + 01	3.19E + 06	67.54309	25.0306
	Sd	**6.658**	69.90499	3.26E−02	4.03E−02	3.23E−02	2.01E + 07	62.22534	0.0038
F-6	Mn	**0.00E + 00**	8.16E−01	5.31E + 00	1.49E + 01	9.15E + 00	2.37E + 03	2.5E−16	4.5065E−05
	Std	**0.00E + 00**	1.26E−04	4.04E + 00	9.42E−01	2.21E + 00	8.81E + 03	1.74E−16	1.3358E−06
F-7	Mn	4.37E−05	2.22E−02	2.96E−04	**0.00E + 00**	**0.00E + 00**	2.078	0.089441	2.3746E−05
	Sd	4.09E−05	1.00E−01	1.48E−03	**0.00E + 00**	**0.00E + 00**	8.2145	0.04339	0.5999
F-8	Mn	21.028	−6123.1	1.19E−01	**0.00E + 00**	**0.00E + 00**	−1077.96	−2821.07	−7862.903
	Sd	54.103	4.08E + 04	3.30E−01	**0.00E + 00**	**0.00E + 00**	2.43E + 03	493.0375	0.0015
F-9	Mn	2.528E−14	3.12E−01	3.81E + 01	3.58E + 01	2.62E + 01	15.1969	25.96841	**0.00E + 00**
	Sd	1.376E−15	4.74E + 01	5.71E + 00	7.01E + 00	4.96E + 00	69.7802	7.470068	**0.00E + 00**
F-10	Mn	1.148E−14	1.06E−13	3.17E + 00	1.11E + 00	8.16E−03	0.5884	0.062087	**8.881E−16**
	Sd	2.941E−15	7.78E−02	3.18E + 00	2.02E + 00	1.18E−02	3.3505	0.23628	**0.00E + 00**
F-11	Mn	1.221E−2	4.49E−03	2.71E + 03	2.28E + 03	1.67E + 03	22.689	27.70154	**0.00E + 00**

(continued)

Table 1 (continued)

Function		ALC-PSO	GWO	jDE	SaDE	JADE	DMPSADE	GSA	DADE
	Sd	1.577E−2	6.66E−03	2.75E + 02	3.45E + 02	2.13E + 02	71.0846	5.040343	**0.00E + 00**
F-12	Mn	**4.393E−32**	5.34E−02	4.77E−01	4.59E−01	2.67E−01	7.25E + 06	1.799617	0.029723
	Sd	**7.062E−32**	2.07E−02	5.41E−02	5.23E−02	3.57E−02	4.44E + 07	0.95114	6.5612E−07
F-13	Mn	**3.167E−31**	6.54E−01	2.84E−01	3.02E−01	2.20E−01	1.04E + 07	8.899084	3.5179E−05
	Sd	**6.918E−31**	4.47E−03	3.55E−02	3.69E−02	3.25E−02	7.66E + 07	−2821.07	1.8005E−08
F-14	Mn	3.55E−03	4.042493	0.9980	0.9980	0.9980	1.0716	5.859838	**0.998**
	Sd	1.65E−03	4.252799	0.9880	0.9980	1.015E−5	0.3920	3.831299	**0.00E + 00**
F-15	Mn	2.11E + 02	0.000337	0.0013	0.0015	0.0012	0.0011	0.003673	**3.0749E−04**
	Sd	2.01E + 01	0.000625	0.0011	0.0017	2.88E−11	0.0041	0.001647	**0.00E + 00**
F-16	Mn	2.84E + 01	−1.03163	−1.0316	−1.0316	−1.0316	−1.0267	−1.03163	**−1.0316**
	Sd	2.34E + 01	− 1.03163	1.23E–11	1.0316	3.52E–12	0.0399	4.88E–16	**0.00E + 00**
F-17	Mn	2.10E + 01	0.397889	0.3979	0.3979	0.3979	N/A	0.397887	**0.3978**
	Sd	1.23E + 01	0.397887	1.01E–03	0.3979	2.19E–12	N/A	0.0E + 00	**0.00E + 00**
F-18	Mn	2.03E + 02	3.00003	3.0000	3.0000	3.0000	3.2269	3.00	**3.0000**
	Sd	2.81E + 01	3.0000	3.0090	2.9000	2.9901	3.1587	4.17E–15	**1.7308E−14**
F-19	Mn	5.89E + 02	−3.86263	−3.8628	−3.8628	−3.8627	−3.8625	−3.86278	**−3.8628**
	Sd	4.36E−02	−3.86278	1.219E-1	2.8128	1.94E–12	0.0022	2.29E–15	**0.00E + 00**
F-20	Mn	− 3.8628	−3.28654	−3.1938	−3.2936	−4.1132	−3.29	**−3.31778**	−3.2031
	Sd	6.29E−06	−3.25056	3.1938	3.2936	5.96E-10	0.1543	**0.023081**	7.2916E−11

(continued)

Table 1 (continued)

Function		ALC-PSO	GWO	jDE	SaDE	JADE	DMPSADE	GSA	DADE
F-21	Mn	– 3.322	−10.1514	−9.0152	−10.1512	−9.5288	−9.0636	−5.95512	**−10.1532**
	Sd	1.22E−05	−9.14015	9.0153	1.10E−03	1.47402	2.2577	3.737079	**0.00E + 00**
F-22	Mn	– 2.6305	−10.4015	−10.1029	−10.3122	−10.4022	−9.1543	−9.68447	**−10.4029**
	Sd	1.88E-04	−8.58441	1.21E−08	10.3742	5.44E-10	2.4274	2.014088	**0.00E + 00**
F-23	Mn	– 10.4028	−10.5343	−10.5364	−10.5364	−10.5364	−9.376	−10.5364	**−10.5364**
	Sd	0.0028	−8.55899	2.17E−13	10.5364	1.18E−11	2.5127	2.6E−15	**0.00E + 00**
F-24	Mn	100.001	4.38E + 01	3.02E−01	2.68E−01	2.41E−01	2.44E + 01	**6.63E−17**	5.1721E−04
	Sd	2.23E–03	6.98E + 01	4.15E−02	1.40E−01	3.18E−02	77.029	**2.78E−17**	5.363E−04
F-25	Mean	21.732	9.18E + 01	5.36E + 00	4.86E + 00	**3.20E + 00**	2.95E + 01	200.6202	20.2817
	Std	1.25E−01	9.55E + 01	7.43E−01	4.17E−01	**4.55E−01**	59.142	67.72087	1.393E−06
F-26	Mn	214.773	6.14E + 01	1.03E + 01	**1.03E + 01**	9.30E + 00	2.37E + 02	180.00	133.4765
	Sd	1.30E−01	6.86E + 01	3.23E−01	**3.42E−01**	4.61E−01	147.999	91.89366	0.0013
F-27	Mn	315.611	**1.23E + 01**	1.62E + 03	8.55E + 02	1.91E + 04	3.54E + 02	170	288.3820
	Sd	7.41E–02	**1.63E + 02**	1.49E + 03	2.80E + 02	1.08E + 05	103.713	82.32726	7.584E−09
F-28	Mn	4.775	1.02E + 01	1.86E + 01	4.92E + 01	1.14E + 02	2.25E + 01	200	**4.1090**
	Sd	4.22E–02	8.12E + 01	1.04E + 01	2.57E + 01	1.97E + 02	81.639	47.14045	**7.16E + 08**
F-29	Mn	951.661	4.31E + 01	4.97E + 00	5.26E + 00	4.48E + 00	5.35E + 02	142.0906	**4.19E−01**
	Sd	8.41E–04	8.44E + 01	9.61E−01	1.15E + 00	7.56E−01	78.693	88.87141	**1.1135E−10**

Table 2 Comparison results of DADE with other meta-algorithms on CEC2017 (Suit-II) benchmark functions

Benchmark Function of CEC 2017 (Suit-II)									
Function		DMPSADE	GWO	ALC-PSO	jDE	SaDE	GSA	JADE	DADE
F-1	Mn	9.0058E + 9	1.5656e + 08	3.0592e + 03	1.90E + 03	**1.00E + 02**	3.5385e + 10	**1.00E + 02**	6.7256e + 02
	Sd	2.7899E + 9	8.0425e + 04	15.0549	1.89E + 03	**1.00E + 02**	2.0061e + 06	**1.00E + 02**	3.9946e + 02
F-2	Mn	1.6358E + 46	6.1983e + 28	3.1345e + 06	1.63e + 8	2.97E + 02	4.7690e + 46	2.91E + 02	**2.904e + 02**
	Sd	8.6906E + 46	7.8898e + 24	3.8005e + 06	1.6394e + 24	2.53E + 02	6.4146e + 19	2.05E + 02	**1.1004e + 02**
F-3	Mn	1.5151E + 05	5.9234e + 04	1.1447e + 04	1.151e + 04	**3.05E + 02**	8.4340e + 04	3.10E + 02	9.3512e + 03
	Sd	3.0408E + 04	32.8237	1.0048e + 04	1.150e + 04	**3.00E + 02**	3.1717e + 04	3.05E + 02	1.1348e−02
F-4	Mn	2.2070E + 03	541.8132	**477.603**	4.91E + 02	4.79E + 02	1.193E + 04	4.79E + 02	478.0026
	Sd	7.6962E + 03	0.0861	**4.1538**	1.90E + 02	4.70E + 02	0.0450	4.70E + 02	0.0643
F-5	Mn	752.7622	**614.8143**	658.2471	6.98E + 02	5.40E + 02	757.3012	5.42E + 02	615.7125
	Sd	93.1763	**0.0443**	36.2180	5.10E + 02	5.09E + 02	1.8152	5.41E + 02	1.1255e−7
F-6	Mn	624.4011	605.7465	630.6217	600.809	**6.00E + 02**	662.8511	6.10E + 02	683.1382
	Std	29.0885	0.1987	0.7589	0.5102	**6.00E + 02**	0.4039	6.07E + 02	0.1201
F-7	Mn	1.1473E + 03	846.1413	801.1150	943.592	7.83E + 02	1.064E + 03	7.79E + 02	**772.1132**
	Sd	585.9710	0.1247	90.1021	155.284	7.81E + 02	0.9541	7.11E + 02	**1.0591e−02**
F-8	Mn	1.0462E + 03	916.2766	857.4102	1.051E + 03	8.76E + 02	968.8488	8.41E + 02	**836.2609**
	Sd	85.9508	0.0177	0.4092	247.7128	8.52E + 02	0.9892	8.09E + 02	**0.0206**
F-9	Mn	5.0930E + 03	2.8280e + 03	3.1016e + 03	906.9024	9.31E + 02	4.808E + 03	**9.01E + 02**	930.0814
	Sd	7.5802E + 03	2.6833e + 03	3.1097e + 03	0.1865	9.30E + 02	89.4428	**9.02E + 02**	3.0280e−06
F-10	Mn	8.1642E + 03	3.2231e + 03	**3.1050e + 03**	8.007E + 03	3.54E + 03	5.835E + 03	3.13E + 03	5.2258e + 03
	Sd	800.0550	0.5471	**1.0034e + 03**	1.301E + 03	3.48E + 03	535.6572	1.48E + 03	587.1356

(continued)

Table 2 (continued)

Benchmark Function of CEC 2017 (Suit-II)									
Function		DMPSADE	GWO	ALC-PSO	jDE	SaDE	GSA	JADE	DADE
F-11	Mn	4.2681E + 03	1.3668e + 03	1.1513e + 03	1.151E + 03	1.31E + 03	9.925E + 03	1.16E + 03	**1.1100e + 03**
	Sd	4.7707E + 03	0.1086	0.7229	120.0301	1.30E + 03	11.4680	1.14E + 03	**0.0015**
F-12	Mn	8.6411E + 08	9.5738e + 03	2.4537e + 03	1.192E + 04	2.37E + 03	1.022E + 10	2.32E + 03	**1.256e + 03**
	Sd	4.2780E + 08	8.1381e + 05	2.0049e + 03	1.029e + 04	2.36E + 03	7.386E + 07	1.59E + 03	**1.1108**
F-13	Mn	5.3185E + 08	1.3791e + 05	2.1801e + 03	1.906E + 03	1.44E + 03	9.159E + 09	1.54E + 03	**1.3128e + 03**
	Sd	4.4461E + 08	5.2842e + 04	29.1105	219.0120	1.40E + 03	2.804E + 07	1.55E + 03	**5.5621e−5**
F-14	Mn	1.6878E + 06	5.7356e + 05	3.2225e + 04	**1.450E + 03**	2.11E + 03	3.209E + 07	1.51E + 03	2.5567e + 03
	Sd	1.6003E + 06	2.0692e + 03	1.5019e + 04	**17.0150**	2.16E + 03	3.029E + 07	1.46E + 03	1.0201e−05
F-15	Mn	1.1546E + 08	3.8406e + 04	3.5182e + 03	**1.584E + 03**	1.62E + 03	1.682E + 04	1.65E + 03	1.7003e + 03
	Sd	1.1539E + 08	1.5880e + 04	3.0002e + 03	**30.1136**	1.62E + 03	1.562E + 04	1.65E + 03	2.2260e−10
F-16	Mn	3.3677E + 03	2.4384e + 03	2.2300e + 03	**2.010E + 03**	2.24E + 03	7.874E + 03	2.26E + 03	2.2122e + 03
	Sd	1.0510E + 03	6.7336	1.2031	**16.8058**	1.03E + 03	1.466E + 03	1.83E + 03	1.0112e + 05
F-17	Mn	2.6054E + 03	2.0689e + 03	2.2201e + 03	2.001E + 03	1.92E + 03	4.149E + 03	1.92E + 03	**1.9019e + 03**
	Sd	509.0968	5.7602	301.1866	32.910	1.57E + 03	4.1096	1.80E + 03	**36.0754**
F-18	Mn	1.7912E + 07	2.0518e + 06	2.0375e + 03	2.0310e + 03	1.97E + 03	2.642E + 07	1.90E + 03	**1.9000e + 03**
	Sd	1.5425E + 07	7.5117e + 03	2.0117e + 07	100.1564	1.90E + 03	1.0059e + 06	1.90E + 03	**1.0867e−03**
F-19	Mn	2.0423E + 08	2.1077e + 04	2.5126e + 03	**1.9328e + 03**	1.99E + 03	4.455E + 06	1.99E + 03	2.0621e + 03
	Sd	2.0345E + 08	6.4039e + 03	2.0984e + 03	**12.9117**	1.90E + 03	2.2252e + 06	1.96E + 03	2.0127e−01
F-20	Mn	2.8185E + 03	2.4662e + 03	2.2018e + 03	**2.100E + 03**	2.12E + 03	3.271E + 03	2.14E + 03	2.1804e + 03
	Sd	245.4724	0.0973	50.130	**41.4030**	2.10E + 03	535.2114	2.11E + 03	2.0948e + 02
F-21	Mn	2.5150E + 03	2.3593e + 03	2.3180e + 03	2.4022e + 03	2.34E + 03	2.777E + 03	2.34E + 03	**2.3160e + 03**
	Sd	90.9628	0.0242	6.6015	40.4185	2.00E + 03	1.5985	2.12E + 03	**0.0035**

(continued)

Table 2 (continued)

Benchmark Function of CEC 2017 (Suit-II)									
Function		DMPSADE	GWO	ALC-PSO	jDE	SaDE	GSA	JADE	DADE
F-22	Mn	4.1229E + 03	5.5180e + 03	2.3020e + 03	2.314E + 03	2.36E + 03	8.531E + 03	2.36E + 03	**2.300e + 03**
	Sd	2.6224E + 03	0.7051	2.0312	11.0024	2.00E + 03	633.7529	2.52E + 03	**1.7029e−08**
F-23	Mn	2.9397E + 03	2.7216e + 03	3.0010e + 03	2.7604e + 03	2.71E + 03	3.560E + 03	2.70E + 03	**2.6947e + 03**
	Sd	254.1425	0.0524	11.2109	51.4058	2.70E + 03	6.6368	2.31E + 03	**0.0307**
F-24	Mn	3.0977E + 03	3.0276e + 03	3.2105e + 03	3.001e + 03	2.90E + 03	4.079E + 03	**2.86E + 03**	2.9285e + 03
	Sd	299.4759	120.1723	10.8624	42.0652	2.89E + 03	2.860E + 03	**2.41E + 03**	0.00e + 00
F-25	Mean	3.5617E + 03	3.0036e + 03	2.8901e + 03	2.9821e + 03	2.91E + 03	4.613E + 03	2.89E + 03	**2.8877e + 03**
	Std	2.6781E + 03	0.0645	16.2012	10.1298	2.90E + 03	6.0932	2.50E + 03	**1.6278e−11**
F-26	Mn	8.8347E + 03	5.0125e + 03	2.9020e + 03	4.830E + 03	3.71E + 03	1.021E + 04	**3.90E + 03**	4.0633e + 03
	Sd	2.4947E + 03	2.1667	11.0148	113.0310	3.11E + 03	1.006E + 04	**3.89E + 03**	5.1338e−07
F-27	Mn	3.4056E + 03	3.2307e + 03	**3.1241e + 03**	3.207E + 03	3.21E + 03	4.596E + 03	3.22E + 03	3.2037e + 03
	Sd	389.9037	0.6443	**10.7161**	4.3350	2.21E + 03	4.1393	2.00E + 03	4.5977e−08
F-28	Mn	5.9211E + 03	3.3292e + 03	3.2336e + 03	3.198E + 03	3.15E + 03	6.016E + 03	3.13E + 03	**3.1100e + 03**
	Sd	1.8808E + 03	0.0082	162.1665	23.4051	2.00E + 03	0.0146	2.11E + 03	**6.1497e−06**
F-29	Mn	4.7719E + 03	3.6576e + 03	3.7017e + 03	4.101E + 03	3.55E + 03	6.334E + 03	3.57E + 03	**3.5420e + 03**
	Sd	733.7443	0.9921	10.0828	201.1096	2.50E + 03	1.214E + 03	2.00E + 03	**145.1249**
F-30	Mn	1.2750E + 08	8.4763e + 06	7.8015e + 03	4.009E + 04	5.20E + 03	1.278E + 09	**5.19E + 03**	8.7543e + 03
	sd	2.5831E + 08	9.4355e + 03	106.0110	1.1011e + 04	3.20E + 03	1.197E + 05	**3.23E + 03**	342.6424

Suit-I:

- Unimodal functions (F1–F7): These functions are utilized to measure the exploitation potential of the optimization method. Observed that in uni-modal benchmark function, the DADE algorithm performs significantly in two benchmark functions (F1 and F2) out of seven. In function F6, DADE is ranked second among the seven other state-of-the-art heuristic algorithms. Results demonstrate the accuracy of the proposed algorithm. The result demonstrates the DADE algorithm's ability to converge and accuracy to find the global optimal result.
- Multimodal Functions (F8–F13): These functions are particularly important for evaluating the heuristic algorithms' exploration ability. Observed that DADE outperformed other algorithms in three functions (F9, F10, and F11) out of six cases and at least acquires fifth-best place in function F8, second-best in function F12, and F13. The suggested algorithm's competitive exploration capability is demonstrated by the results. In comparison to other methods, DADE provides improved swarm divergence and convergence span management, allowing the swarm to obtain an optimal solution accurately and robustly.
- Fixed-Dimensional Multi-modal Function (F14–F23): These functions are used to demonstrate the capacity of optimization algorithm to reach the specified fitness level. Statistical results depict that DADE is able to converge to global optimal result in functions (F14 to F19 and F21 to F23) and also able to achieve the position of second best in function F20 compared to other considered techniques. The result demonstrates the algorithm's supremacy in detecting the best solution as compared to the other algorithms.
- Composite Functions (F24–F29): These functions are important for assessing the algorithm's capacity to avoid local convergence by balancing local and global search. Observed that DADE outperforms in function F24 whereas, in function F27, DADE is able to secure second-best position. The obtained results demonstrate that the DADE has achieved global convergence without being stuck in local convergence.

Suit-II:

- CEC2017 (F1–F30): Results of the analysis of those functions are presented in Table 2. It is observed that DADE beats other algorithms in thirteen benchmark functions (F2, F7, F8, F11, F12, F13, F17, F21, F22, F23, F25, F28, and F29) out of thirty functions. It's also worth noting that DADE comes in second place in four benchmark functions (F4, F16, F18, and F27) and becomes third-best in four functions (F1, F3, F9, and F24). DADE is able to secure fourth place in F5, F14, F15, F19, F20, F26, and F30 functions among the considered eight meta-heuristic algorithms. The result demonstrate how well the algorithm can converge and how accurately it will find the optimum cost of the problem.

3.3 Statistical Analysis

The statistical importance of the algorithm is measured with the Friedman's test and Wilcoxon Ranksum test [24].

Friedman's test and Wilcoxon Ranksum test are used to establish the statistical significance of the results [24]. The better-performing algorithms are identified based on the lower rankings and corresponding p & z values in these two tests. Considering the Friedman's Test's best result as the benchmark, Wilcoxon's Ranksum test [25] is performed. The summary of these two tests are listed in the Tables 3 and 4 given below.

The results of the Friedman test depicts that the DADE with a mean rank of 2.7069 is able to secure the first position compared to all the other algorithm. It is also observed that with the mean rank of 2.5000, DADE secure the best position in Suit-II with 5% level of significance.

Tables 3b and 4b summarize the Wilcoxon Ranksum test results for Suit-I and Suit-II, respectively, where '+' and '−' represent better and poorer outcomes obtained by the considered algorithm in evaluation. The p- and z-value measure the likelihood of receiving a large-valued test statistic and the importance of the acquired results. Results depict that in both suits, the performance of the algorithm DADE is remarkable as compared to the other techniques.

Table 3 Friedman's and Wilcoxon test summary of Suit—I

(a): Friedman test of function 1–29

Algorithm	Mean rank	Rank
ALC-PSO	5.1379	5
GWO	4.4483	3
jDE	5.6034	6
SaDE	4.6034	4
JADE	4.2414	2
DMPSADE	6.9483	8
GSA	5.6379	7
DADE	2.7069	1

(b): Wilcoxon test of function 1–29

DADE Vs	+	−	p-value
ALC-PSO	43	257	0.00016
GWO	63.5	261.5	0.00086
jDE	66	234	0.00544
SaDE	70	206	0.00736
JADE	69	184	0.00694
DMPSADE	32	403	0.00001
GSA	33	318	0.00018

Table 4 Friedman's test and Wilcoxon test summary of function CEC 2017 (Suit—II)

(a): Friedman test of CEC2017 function

Algorithm	Mean rank	Rank
DMPSADE	7.0333	7
GWO	5.4000	6
ALC-PSO	4.000	4
jDE	4.0333	5
SaDE	2.8333	3
GSA	7.6333	8
JADE	2.5667	2
DADE	2.5000	1

(b): Wilcoxon test of CEC2017 function

DADE Vs	+	−	p-value
DMPSADE	1	464	0.00001
GWO	27	438	0.00001
ALC-PSO	107	358	0.00988
jDE	96	369	0.00496
SaDE	268	197	0.4654
GSA	1	464	0.00001
JADE	288	177	0.25428

4 Convergence Analysis

DADE's convergence performance is evaluated using a set of six test functions, including Sphere (F1), Quadric (F3), Rosenbrock (F5), Ackley (F10), Griewank (F11), and Generalized Penalized Function 2 (F13) in order to evaluate its convergence ability. In comparative analysis the most famous and widely used meta-heuristic algorithms namely ALC-PSO, PSO, DA, SSA, and GOA are used. Figures 1, 2, 3, 4, 5 and 6 shows the representation of the results.

The convergence properties of the sphere function is shown in Fig. 1. It is observed from the figure that the DADE's convergence characteristic is clearly superior to those of the other evaluated algorithms.

It is observed that the algorithm SSA leads the DADE in the initial phase of the iteration as shown in Fig. 2; after 10th iteration, DADE becomes the superior algorithm among the other considered algorithms.

From Fig. 3, it is observed that the GOA and SSA algorithm performs better in the initial phase of iteration up to 8th iteration; after that, the DADE is superior to the other. The superior convergence characteristic of the DADE depict its better exploration capability and ability to avoid the local minima.

Figure 4 shows the convergence characteristics of all the considered algorithms in 30D function of F10. Figure also depicts that the convergence feature of the DADE is remarkably superior to the other considered algorithm. That superior convergence

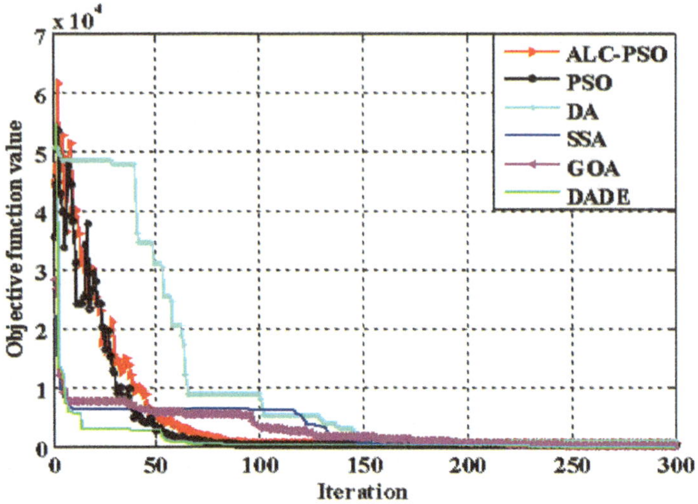

Fig. 1 Convergence characteristics plot of F1 in 30D

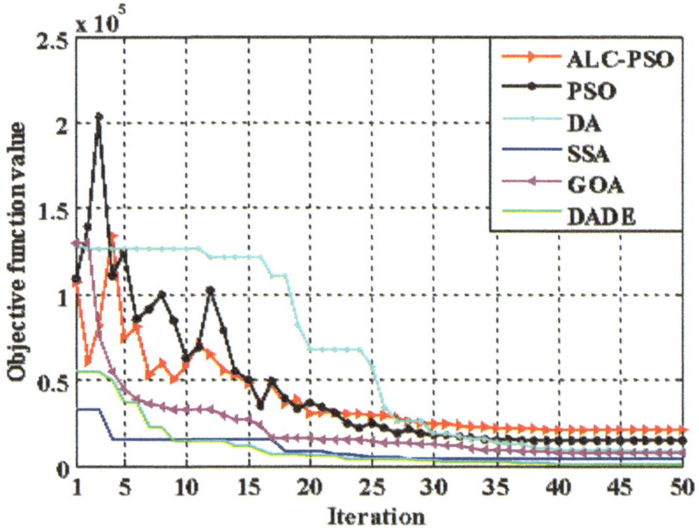

Fig. 2 Convergence characteristics plot of F3 in 30D

also confirms the balancing between the exploration and exploitation rate of the DADE.

Lastly, the convergence performance of algorithms on Griewank (F11) and Generalized Penalized Function 2 (F13) function are represented respectively in Figs. 5 and 6 for 30 dimensions. In comparison to all of the algorithms considered, the

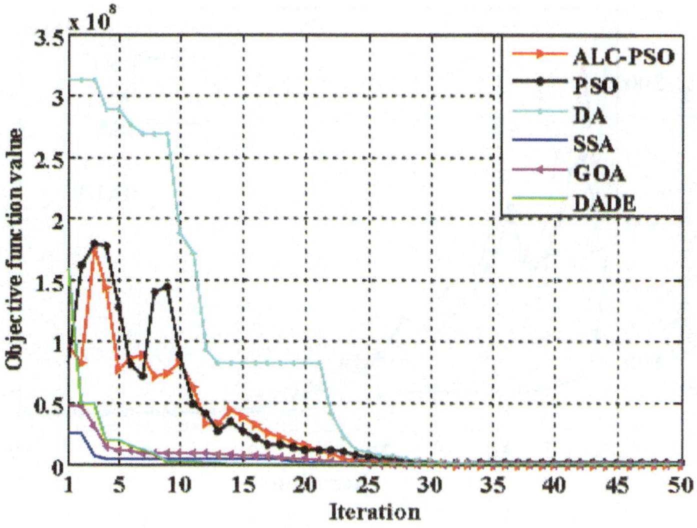

Fig. 3 Convergence characteristics plot of F5 in 30D

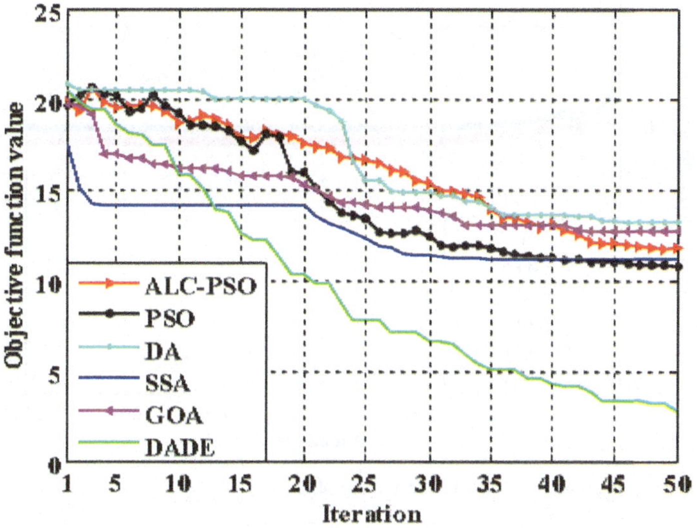

Fig. 4 Convergence characteristics plot of F10 in 30D

DADE provides the fastest convergence in both the 30D functions of Griewank and Generalized Penalized Function 2.

The convergence performance of algorithms is analyzed using both uni-modal (F1, F3, and F5) and multi-modal (F10, F11, and F13) functions. Finding the global solution of a unimodal function is relatively simple due to the occurrence of a single mode.

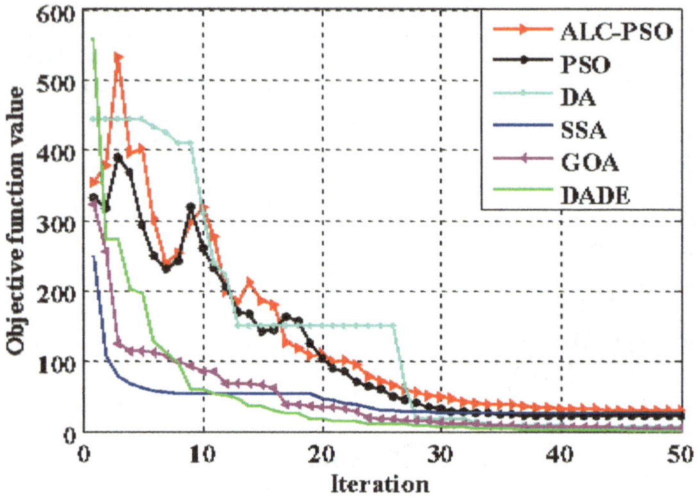

Fig. 5 Convergence characteristics plot of F11 in 30D

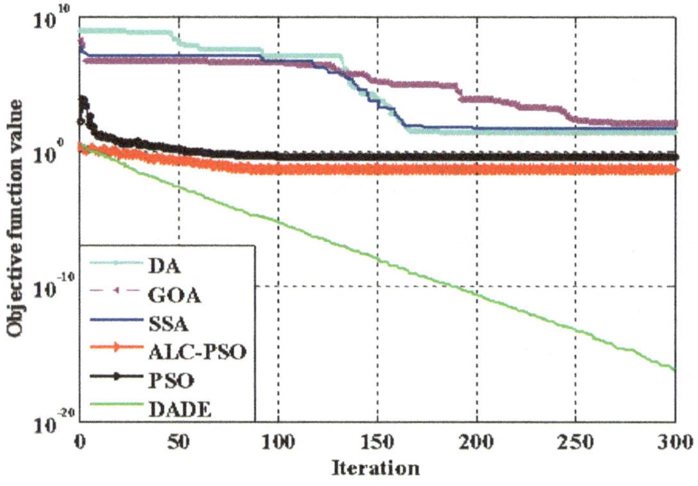

Fig. 6 Convergence characteristics plot of F13 in 30D

On the other hand, evaluation of optimal convergence in a multi-modal function is rather challenging.

The performance of DADE towards finding the global optimal solutions in unimodal function depicts its significant exploration characteristics. Whereas the significant balancing of the exploration and exploitation rate is confirmed by the convergence performance of DADE in multi-modal benchmark functions.

5 Conclusion

Tracking and saving particle bests (pBest) and global best (x_leader) solutions incorporate memory elements in DADE algorithm. These felicitate DADE with high exploration with balancing exploitation when the memory elements are utilized in the further mutation strategy of DE algorithm by hybridization. It is also observed that the inclusion of worst solution along with best solutions in mutation results in fine-tuning of the exploitation capability. The performance of the DADE in uni-modal, multi-modal, and composite benchmark functions shows the ability of DADE to find the global optimal solution efficiently while avoiding the local optimal solutions. DADE's performance on the CEC-2017 (Suit-II) benchmark function demonstrates its capacity to solve complex real-time problems. Generation-wise hybridization and incorporation of DE in DADE felicitates an evolutionary factor to the algorithm for solving critical optimization issues. Analysis of convergence performance is also examined, and the results prove the capability of DADE in escaping local convergence points and declination towards global optimal solutions.

References

1. Dorigo, M., Thomas, S.: Ant Colony Optimization. MIT Press eBooks (2004)
2. Kennedy, J., Eberhart, R.: Particle swarm optimization. In: Proceedings IEEE International Conference on Neural Networks, vol. 4, pp. 41942–1948. Perth, WA (1995). https://doi.org/10.1109/ICNN.1995.488968
3. Mirjalili, S., Mirjalili, S.M., Lewis, A.: Grey wolf optimizer. Adv. Eng. Softw. **69**, 46–61 (2014). https://doi.org/10.1016/j.advengsoft.2013.12.007
4. Mirjalili, S., Lewis, A.: The whale optimization algorithm. Adv. Eng. Softw. **95**, 51–67 (2016). https://doi.org/10.1016/j.advengsoft.2016.01.008
5. Wang, G.G.: Moth search algorithm: a bio-inspired metaheuristic algorithm for global optimization problems. Memetic Comput. 1–14 (2016). https://doi.org/10.1007/s12293-016-0212-3.
6. Mirjalili, S.: The ant lion optimizer. Adv. Eng. Softw. **83**, 80–98 (2015). https://doi.org/10.1016/j.advengsoft.2015.01.010
7. Mirjalili, S.: Dragonfly algorithm: a new meta-heuristic optimization technique for solving single-objective, discrete, and multi-objective problems. Neural Comput. Appl. **27**(4), 1053–1073 (2016). https://doi.org/10.1007/s00521-015-1920-1
8. Mirjalili, S., Gandomi, A.H., Mirjalili, S.Z., Saremi, S., Faris, H., Mirjalili, S.M.: Salp swarm algorithm, Adv. Eng. Softw. **114**, 163–191 (2017). https://doi.org/10.1016/j.advengsoft.2017.07.002
9. Saremi, S., Mirjalili, S., Lewis, A.: Grasshopper optimisation algorithm: theory and application. Adv. Eng. Softw. **105**, 30–47 (2017). https://doi.org/10.1016/j.advengsoft.2017.01.004
10. Storn, R., Price, K.: Differential Evolution–a simple and efficient Heuristic for global optimization over continuous spaces. J. Global Optim. **11**, 341–359 (1997). https://doi.org/10.1023/A:1008202821328
11. Debnath, S., Arif, W., Baishya, S.: Buyer inspired meta-heuristic optimization algorithm. Open Comput. Sci. **10**(1), 194–219 (2020). https://doi.org/10.1515/comp-2020-0101
12. Sharma, A., Sharma, A., Panigrahi, B., Kiran, D., Kumar, R.: Ageist spider monkey optimization algorithm. Swarm Evol. Comput. **28**, 58–77 (2016). https://doi.org/10.1016/j.swevo.2016.01.002

13. Debnath, S., Arif, W., Sen, D., Baishya, S.: Hybrid differential evolution with learning for engineering application. Int. J. Bio-Inspired Comput. Indersci. **19**(1), 29–39 (2021). https://doi.org/10.1504/IJBIC.2022.120744
14. Jafari-Asl, J., Azizyan, G., Monfared, S.A.H., Rashki, M., Andrade-Campos, A.G.: An enhanced binary dragonfly algorithm based on a V-shaped transfer function for optimization of pump scheduling program in water supply systems (case study of Iran). Eng. Failure Anal. **123** (2021). https://doi.org/10.1016/j.engfailanal.2021.105323
15. Too, J., Mirjalili, S.: A hyper learning binary dragonfly algorithm for feature selection: a COVID-19 case study. Knowl. Based Syst. **212** (2021). https://doi.org/10.1016/j.knosys.2020.106553
16. Chantar, H., Tubishat, M., Essgaer, M., Mirjalili, S.: Hybrid binary dragonfly algorithm with simulated annealing for feature selection. SN Comput. Sci. **2**, 295 (2021). https://doi.org/10.1007/s42979-021-00687-5
17. Thangaraj, R., Pant, M., Abraham, A., Bouvry, P.: Particle swarm optimization: hybridization perspectives and experimental illustrations. Appl. Math. Comput. **217**(12), 5208–5226 (2011). https://doi.org/10.1016/j.amc.2010.12.053
18. Debnath, S., Baishya, S., Sen, D., Arif, A.: A hybrid memory-based dragonfly algorithm with differential evolution for engineering application. Eng. Comput. **37**(4), 2775–2802 (2020). https://doi.org/10.1007/s00366-020-00958-4
19. Das, S., Suganthan, P.N.: Differential evolution: a survey of the state-of-the-art. IEEE Trans. Evol. Comput. **15**, 4–31 (2011)
20. Brest, J., Greiner, S., Boskovic, B., Mernik, M., Zumer, V.: Self-adapting control parameters in differential evolution: a comparative study on numerical benchmark problems. IEEE Trans. Evol. Comput. **10**, 646–657 (2006)
21. Zhang, J., Sanderson, A.C.: JADE: adaptive differential evolution with optional external archive. IEEE Trans. Evol. Comput. **13**(5), 945–958 (2009)
22. Fan, Q., Yan, X.: Self-adaptive differential evolution algorithm with discrete mutation control parameters. Exp. Syst. Appl. **44**(3), 1551–1572 (2015)
23. Awad, N.H., Ali, M.Z., Suganthan, P.N., Liang, J.J., Qu. B.Y.: Problem Definitions and Evaluation Criteria for the CEC 2017 Special Session and Competition on Single Objective Real-Parameter Numerical Optimization, Technical Report, Nanyang Technological University, Singapore (2016)
24. Demsar, J.: Statistical comparisons of classifiers over multiple data sets. J. Mach. Learn. Res. **7**, 1–30 (2006)
25. Derrac, J., Garcia, S., Molina, D., Herrera, F.: A practical tutorial on the use of nonparametric statistical tests as a methodology for comparing evolutionary and swarm intelligence algorithms. Swarm Evol. Comput. **1**(1), 3–18 (2011). https://doi.org/10.1016/j.swevo.2011.02.002

Engineering Problems

Optimum Design and Tuning Applications in Structural Engineering via Swarm Intelligence

Gebrail Bekdaş, Sinan Melih Nigdeli, and Aylin Ece Kayabekir

Abstract As all engineering disciplines, structural engineering problems are needed to be optimized and due to the nonlinear behavior of these problems, it is not possible to solve them mathematically, but metaheuristic methods are very successful in iterative optimization by assuming values for the design variables within a desired range of the user. In structural engineering problems, metaheuristic methods including swarm-intelligence-based algorithms are used in two groups of problems. Design optimization is the first group and the design like dimension, amount of material and orientations are optimally found for minimizing objectives related to cost, weight, CO_2 emission and others. In these problems, constraints are found via design codes like steel and reinforced concrete structure design regulations. This group belongs to a design of a structure. The second group includes optimum tuning and it generally covers structural control applications. This group involves the optimum tuning of the additional control system of the structure that can be added to the newly constructed structure for better performance or existing ones to correct the failure or increase the existing performance. The role of engineers is to make the best possible structural design and optimization is important. More especially, tuning optimization is a must to provide acceptable performance. In this chapter, a review of existing studies about the design optimization of structural systems is presented for swarm intelligence-based algorithms. Then, optimum tuning applications are mentioned including the most important studies about tuned mass dampers. Finally, optimization problems are presented for design and tuning optimization. The RC retaining wall optimization was presented for two cases with and without toe projection and the optimization of a toe is 5% effective on reduction of cost. In span length optimization of frame structures, frame models with different stories have similar optimum span lengths.

G. Bekdaş (✉) · S. M. Nigdeli
Department of Civil Engineering, Istanbul University-Cerrahpaşa, 34320, Avcilar Istanbul, Turkey
e-mail: bekdas@iuc.edu.tr

S. M. Nigdeli
e-mail: melihnig@iuc.edu.tr

A. E. Kayabekir
Department of Civil Engineering, Istanbul Gelişim University, 34310 Avcılar, Istanbul, Turkey

A. Biswas et al. (eds.), *Advances in Swarm Intelligence*, Studies in Computational Intelligence 1054, https://doi.org/10.1007/978-3-031-09835-2_6

Active tuned mass dampers are up to 22.08% more effective than passive tuned mass dampers.

Keywords Structural engineering · Optimum design · Optimization · Metaheuristic · Swarm intelligence

1 Introduction

When engineering designs are examined, it will be seen that there is an observation of nature in their foundation. Scientists have put forward various theories over time about why migratory birds fly in a V shape. For example, in one theory, it has been suggested that birds fly in this form to find their way during migration. The main reason is related to energy saving. Air vortices form at the ends of the wings of birds. Birds, who feel that they create a buoyancy force in these vortices formed in the air, move with the most appropriate form to spend less energy on their journeys to long distances during migration.

A similar situation occurs in the wings of airplanes. In the research on the effect of vortices on energy saving in aircraft, it has been observed that energy-saving increases with the shrinkage of the formed air vortices, and for this purpose, small parts called winglets have been added to the aircraft designs. In this way, significant amounts of fuel can be saved, especially on long-haul flights.

As can be understood from the example, it is the fact that a natural event directly or indirectly plays a role in the developments in human history.

Similarly, metaheuristic algorithms inspired by nature are used in engineering to solve optimization problems under certain constraints, depending on the maximum or minimum objective function. Ant Colony Optimization (ACO) [1] is an algorithm that was developed by observing the behavior of ants in searching for the shortest way between nest and food.

Ants communicate through pheromones secreted from their bellies. Within the ant colony, there are worker ants that are constantly circulating outside the colony to provide food. If any of these ants find food, they leave a trace on the road with the help of the scent they secrete when they return to the colony. Thanks to this smell, other ants understand where the food comes from. Each ant that passes over the path to the food re-marks the way between the nest and food with its secretion. In that situation, the amount of secretion substance on the road increases and the road becomes more obvious for the wandering ants. The shorter one of the alternative routes between the colony and the food will accumulate more pheromones than the other routes after a certain time due to its shortness. Since the ants will prefer the route containing more pheromones in the future, the colony generally starts to carry the food using the shortest path.

In addition, the Simulated Annealing Method (SA) used the inspiration of annealing (heating, controlled cooling) process, which is applied to reduce the crystals of metals and reduce the deterioration in the molecular structure [2]. The development of the Genetic Algorithm (GA) was inspired by the evolutionary process (heredity, mutation, crossover, natural selection) [3, 4]. Several algorithms using the behavior of bees have been developed including Honey Bee Algorithm (HBA) [5], Virtual Bee Algorithm (VBA) [6], Honey Bee Mating Algorithm (HBMO) [7] and Artificial Bee Colony (ABC) [8]. As well as Taboo Search (TS) [9] used the human mind or memory analogy. As the most known swarm-intelligence method, Particle Swarm Optimization (PSO) [10] was developed by inspiring the movement of species such as a flock of birds and birds. Another widely used method, the Big Bang-Big Crunch method (BB-BC) [11] is based on mathematical equations that develop based on the big bang theory, which tries to explain the process of the formation of the universe. While Firefly Algorithm (FA) [12] used the specific flashing behavior of fireflies, Bat Algorithm (BA) [13] was developed by utilizing the sound direction-finding character of bats. Krill Herd algorithm [14], which is another algorithm developed in recent years, is an algorithm created by observing the behavior of each krill in time.

Civil engineering deals with many issues such as transportation, infrastructure, superstructure design, water resources and soil stabilization. There are geographical areas with different resources and different natural disaster risks in the world. Due to these differences, material and labor costs may differ in different regions. In civil engineering, structural safety is the most important factor that is needed to be provided in the design. Structural safety standards vary depending on the natural hazards and ground conditions that may be encountered. For this reason, there are many regulations in different countries. Additionally, An important factor is the demands and needs of people. Because of these differences, the optimum design of civil engineering is user-centered and the optimum values may vary from region to region.

2 Review on Design Optimization via Swarm Intelligence and Metaheuristic Algorithm

In this section, optimization applications in structural engineering are discussed. These applications are presented as truss structures, reinforced concrete (RC) elements and frame structures subsections.

2.1 Optimization of Truss Structures

As the most basic structure type, truss structures forming the skeleton of structures have a wide practical use as bridges, towers, roof support systems, etc.

Genetic algorithm (GA) is the oldest and most employed one in optimization applications of truss structures. Koumousis and Georgiou employed GA to handle the plan and size optimization problems of typical steel roofs [15], and Rajan used continuous variables for optimum shape design [16]. Coello and Christiansen used GA as a multi-objective method using the min–max optimization approach in their work [17]. Erbatur et al. used the GA for space and plane truss systems [18]. Krishnamoorthy et al. proposed the application of an object-oriented approach to the genetic algorithm in solving optimization problems of space truss systems [19]. Hasançebi researched and reported the development strategies of optimization methods for truss bridges [20]. Kelesoglu also combined fuzzy formulation and GA and solved the optimization problem of space trusses with multiple objectives [21]. Also, Šešok and Belevičius proposed GA for solving the topology optimization problem of truss systems [22]. Toğan and Daloğlu used GA and they developed the optimization of truss structures based on the initial population strategy [23]. Richardson et al. proposed GA for truss-like structures using a kinematic stability repair-based approach [24]. Li developed a method for truss structures using an improved species-preserving GA [25].

Also, many metaheuristic algorithms are used besides GA. Schutte and Groenwold suggested the use of PSO for the size and topology optimization process of truss systems [26]. Li et al. used the method they developed based on passive ensemble PSO and Harmony search (HS) for the optimization of truss systems [27]. Perez and Behdinan used the PSO to make the optimum design of trusses [28].

ACO is also employed for truss structures [29] and Kaveh and Talatahari proposed a hybrid algorithm that is the combination of PSO, ACO and HS in their study related to the optimization of truss structures [30].

In addition to these algorithms, the teaching–learning-based optimization (TLBO) inspired by the teacher-student relations and knowledge transfer in a classroom, was used to minimize the weights of truss systems depending on the safety coefficient and displacement constraints [31–33]. The weight of the truss systems was minimized by Sonmez [34], by using the ABC together with the compatible penalty equation. Bekdas et al. also suggested using the flower pollination algorithm (FPA) to reduce the weight of two and three-dimensional truss structures [35]. Also; FA [36], Cuckoo search (CS) [37], BA [38] and BB-BC [39–42] were used to obtain the best solution of the truss structures.

Kaveh et al. hybridized swarm intelligence and chaos theory and used them in solving optimum design problems of truss structures [43]. Ho-Huu et al. developed the improved constrained differential evolution, which includes sequential optimization and reliability evaluation in the optimization problems of truss structures [44]. Tort et al. showed that the optimum shape and size of steel truss transfer towers can be obtained by minimizing the weight with the two-stage simulated annealing algorithm they developed following ASCE 10–97 [45]. Yücel et al. predicted optimum values for 3-bar truss structure via generating artificial neural networks (ANNs) model that uses the HS-based optimized solution in machine learning [46]. Bekdaş et al. predicted the optimum design of truss structures including a 10-bar system via using ANNs model [47]. Bekdaş et al. evaluated several metaheuristic algorithms by using Lèvy flight modification in the optimum design of truss structures [48]. Symbiotic organism search [49, 50], chaotic coyote algorithm [51] and heat transfer search [52, 53] algorithm are the recent metaheuristics that are employed in single and multi-objective optimization of truss structures.

Generally, scientific studies are in the form of testing the performance of metaheuristic algorithms on reference sample structures. A significant part of the reference examples is about these structural systems. In this respect, optimization problems of truss structures are among the important examples where the developed theory can be used in practice in a short time.

2.2 Reinforced Concrete Members

It is very difficult to determine the most suitable (economic) cross-section dimensions in terms of cost for RC structures consisting of two different materials that have opposite mechanical properties and unit costs. In addition, because RC structures are highly statically indetermined, the cross-sectional dimensions of any structural element change the internal forces of all structural elements in the system. It is almost impossible to mathematically determine the cross-section dimensions (optimal cross-sections) to which the combination will be made, even by an engineer with design experience. This situation can be shown as one of the main reasons why the concept of optimum design in RC design attracted the attention of researchers with the first applications of RC structures and the development of methods based on various approaches. The concept of optimization has started to be used more frequently, especially with the optimization algorithms, including metaheuristic algorithms developed with the increase in the use of high-speed personal computers.

Metaheuristic algorithms, especially GA, are used in optimization applications of RC elements. While Govindaraj and Ramasamy [54] and Fedghouche and Tiliouine [55] used GA in the optimization of RC beams, Camp et al. used it in the optimization of RC frames [56]. Leps and Sejnoha suggested using metaheuristic algorithms in hybrid form (GA and SA algorithms) for the optimum design of continuous RC beams [57]. Sahab et al. also proposed a hybrid optimization algorithm that was developed using GA and the split form Hook and Jeeves method for the optimization of floor slabs [58].

Another metaheuristic algorithm used in the optimization of RC elements is the HS algorithm. From the HS algorithm, continuous RC beams [59], T-section RC beams [60] and RC columns [61] were used in the design optimization of RC elements. Nigdeli et al. developed a new optimization process based on the HS algorithm and used it for cost optimization of columns loaded with bidirectional moment [62]. In another study, Nigdeli and Bekdaş proposed a random technique for the detailed design of continuous RC beams [63]. Bekdaş et al. proposed the HS optimized ANNs model for estimation of the optimum design RC beams [47]. Yücel et al. generated an ANN model using metaheuristic-based optimum results to predict the optimum design of several RC design optimization problems including T-shaped beams and carbon fiber reinforced polymer (CFRP) retrofitted beams [64].

Reinforced concrete retaining walls are one of the RC elements that are cost-optimized using metaheuristic algorithms. Ceranic et al. used simulated annealing (SA) algorithm by considering material and labor costs together in cost minimization of an RC retaining wall [65]. Yepes et al. also used the SA algorithm to obtain the optimum design values of retaining walls [66]. Ahmadi-Nedushan and Varaee obtained the optimum design values of retaining walls with the method they developed using the PSO algorithm [67]. Kaveh and Abadi suggested using the HS algorithm to obtain optimum design values for RC retaining walls [68]. Ghazavi and Salavati also utilized bacterial foraging algorithm (BFOA) for cost optimization and sensitivity analysis of a cantilever RC retaining wall [69]. Yepes et al. aimed to minimize carbon dioxide emissions and costs. For this purpose, they used the hybrid optimization method they developed using neighbor search and threshold acceptance strategies in the design of cantilever RC retaining walls [70]. BB-BC was used by Camp and Akin in the optimum design of RC retaining walls [71]. Kayabekir et al. employed HS on the optimum design of RC retaining wall as a multi-objective design considering both cost and CO_2 emission minimization [72]. Yücel et al. proposed an adaptive-hybrid HS for the optimization of RC retaining walls [73]. An ANN-based prediction model was developed by Yücel et al. for cantilever-type retaining walls by using machine learning of optimum results found via FPA [74].

2.3 *Frame Structures*

Design engineers perform their analysis on an idealized system, ignoring negligible effects on systems due to constraints arising from software and computer capacity. Such a preference also plays a role in the definition of frame and lattice-type (truss) structures. In the design of frame-type structural systems, in the most general case, they consist of bars, three of which are linear and three of which are angular, a total of six displacements are considered at each node. Articulated joint types of these systems are called truss systems and three linear displacements are taken into account for each node in their design. Looking at these definitions, it is understood that lattice-type systems are actually a sub-type of frame systems and are reduced systems in which the internal forces that do not occur due to the joint state in the system cannot be taken into account.

An engineer determines the cross-section dimensions in the first step of the design and performs safety checks with the analyzes he/she performs accordingly. These selected values differ from engineer to engineer and cause different costly designs. To choose the most suitable design among these differences, different optimization methods have been developed over time. In this process, which started with numerical methods, metaheuristic optimization methods were developed with the invention of fast computers. Some of the optimization applications about frames made are listed below.

Pezeshk et al. employed GA for the optimum design of two-dimensional geometric nonlinear steel frame structures [75]. Li et al. used a non-gradient approach and a method based on evolutionary structure optimization (ESO) in the design optimization of one-way connection-constrained structures [76]. Camp et al. developed an optimization design for steel frame structures using discrete optimization with the ACO [77]. Saka proved that stochastic search methods are powerful compared to mathematical methods for discrete variables of steel frame systems [78].

Perea et al. evaluated multiple heuristics including random walk and the descent local search, and metaheuristic methods including threshold accepting and the simulated annealing in cost optimization of RC frames of bridges [79]. In addition, many methods developed based on metaheuristic algorithms are used in cost optimization of RC frame systems [56, 80–85] and minimizing CO2 emissions [86, 87].

Fesanghary et al. hybridized sequential quadratic programming with HS and it was used in the design optimization of welded beams and steel frames [88]. Hasancebi et al. aimed to increase the performance of the SA algorithm for practically applicable structures and used this algorithm for weight optimization by considering the cross-section dimensions of the 304-member steel braced steel frame and 132-element space frame [89]. Toğan utilized TLBO based discrete optimization technique in the design of plane steel frames [90]. The size and topology optimization of steel roof space frame structures (with hollow square section produced as standard) was applied by Kociecki and Adeli by utilizing a two-GA [91].

Akin and Saka made the design optimization of the RC frame system by using the HS algorithm [59]. Talatahari et al. proposed the combination of the eagle strategy and differential evolution methods and the method was used to minimize the weight of steel frames by using discrete variables [92]. Aydogdu et al. developed ABC with the Lévy flight distribution and used this algorithm on minimizing the weight by selecting the optimum (W-sections) sections of a space steel frame [93]. Saka et al. used various algorithms (ACO, PSO, ABC, FA and CS) to solve the optimization problems of space steel frames and compared the performances of these algorithms [94]. HS was combined with several randomization stages to shorten and improve the performance in the optimum design of RC frames [95, 96]. Kayabekir optimized the span dimensions of frame structures via TLBO and HS [97]. Rakıcı et al. optimized RC frames via Jaya Algorithm (JA) [98].

3 Review on Tuning Optimization of Structural Control Systems Via Swarm Intelligence and Metaheuristic Algorithms

In civil engineering, bridges, towers, buildings, etc. are subjected to vibrations due to effects such as earthquakes, wind and traffic. Various systems are used to reduce vibrations in structures. The systems, which are called control systems in accordance with their intended use, are divided into subgroups as active, semi-active and passive. Structural control is another important area of metaheuristic-based optimization. Optimization is a must for perfect tuning to provide efficiency in vibration control. Metaheuristics have been used in optimum control of various systems [99–103]. In this section, the applications of passive and active tuned mass dampers are briefly reviewed.

3.1 Passive Tuned Mass Dampers (TMD)

Mass dampers, a type of passive control system, are one of the frequently used types of control systems, with application examples including retrofit projects. Examples of these applications are the Berlin television tower with a mass damper used against the wind for which the building photos are given (Fig. 1), the LAX Theme Tower at the Los Angeles airport with a mass damper added as a result of seismic retrofit and the Taipei 101 building with a spherical mass damper.

Studies on the optimization of mass dampers are based on the mass, stiffness and damping parameters of the damper. The studies in question started in the second half of the twentieth century and as a result, various equations were developed by the researchers to find the optimum parameters. However, these equations developed in line with the technical possibilities of the time [104–106] had limited usefulness in the case of multiple vibration modes, random frequency excitations, principal systems with damping, mass dampers with limited motion, and user-defined variables. With the development of metaheuristic algorithms, various studies have been carried out with the methods developed based on these algorithms.

Some of the algorithms used are GA [107–111], PSO [112, 113], HS [114–117], ACO [118], artificial bee optimization [119], BA [120, 121] and JA [122]. In addition, the optimum design of mass dampers to prevent the collision of adjacent buildings was investigated by Nigdeli and Bekdaş [123]. Multiple positioned TMDs for control of structures constrained with axial force capacity were proposed by using FPA in optimization [124]. Yücel et al. developed ANN-based optimum tuning formulations by using optimum values of TMD parameters found via metaheuristics [125].

Fig. 1 Berlin TV tower

3.2 Active Tuned Mass Dampers (TMD)

If a passive control system is combined with a force-producing member like a dynamic actuator that is controlled according to a control algorithm (controller), it will be an active control system and the optimization is more important since both system and controller parameters are needed to be optimized.

Ahlawat and Ramaswamy used ATMD with Fuzzy Logic Control (FLC) to reduce displacement responses, which are important for structural safety, and both structural responses, which are stated to reduce structural accelerations, which are important for structural comfort [126], as Yang et al. sought the optimum parameters of the control system by applying it to the model developed by him [127]. Optimum FLC and ATMD parameters were determined by multi-objective optimization using GA and as a result of the analyzes made under wind loads, it was shown that the optimization method applied was successful in achieving the expected results depending on the objective function.

Pourzeynali et al. suggested using FLC and GA (GFLC), together in the control of tall buildings against earthquake effects with ATMD [111]. During the optimization process using the GA algorithm, ATMD and FLC parameters were sought to ensure that the displacements on the upper floor of the building are minimal. The proposed method was applied to an 11-story shear building and the results were compared with the methods in the literature. Accordingly, it has been observed that the results obtained with the proposed GFLC method are more effective in reducing the maximum displacements compared to the Linear Quadratic Regulator (LQR) control algorithm, whereas the required control force is more than LQR in short. The analysis results showed that GFLC was more effective in reducing the maximum displacements when compared to those obtained with TMD (with optimum parameters obtained by GA) and LQR control methods, but the control force required for the GFLC system was greater than that required for the LQR system.

Ozer et al. proposed the sliding mode-controlled ATMD process to reduce vibrations in a three-layer structure [128]. In the first stage, the optimum SMC parameters were calculated with the help of GA to ensure that the displacements on the third floor are minimal. Due to the high total control energy in the first case, the second stage was performed using GA in the multi-objective optimization process that will ensure that both the displacements on the third floor and the control energy are minimal. It was observed that the ratio of the maximum displacement value on the third floor of the proposed ATMD-controlled structure to the displacement of the uncontrolled structure was 95%. It has been observed that the total energy value consumed by the controller has been reduced to 75%.

PSO and LQR were proposed by Amini et al. to find the optimum active control force in ATMD-controlled structures [129]. In the optimization process, the PSO algorithm is used to minimize the recovery matrix of the LQR control. The superiority of the developed method over the classical LQR control has been proven by testing it on a 10-story structure exposed to near-impact fault motion.

Venanzi et al. used a hybrid control system consisting of a series of ATMDs to reduce the bending and torsion effects of tall buildings exposed to wind [130]. In the design of the hybrid control system, LQR control algorithm and GA were used. In the related study, in the optimization process using GA, firstly, the number and location of ATMDs placed on the upper floor of the building, and secondly, the optimum position of the accelerometer along the height of the building was sought. The method was applied to a 60-story building model with 3 degrees of freedom on each floor, and it was understood from the analysis results that the method was suitable for finding the most effective configuration of ATMD.

FLC with PSO was employed by Shariatmadar and Meshkat Razavi to reduce vibrations in the structure under the influence of earthquakes [131]. The PSO algorithm was used to determine the FLC input parameters, which will ensure that the maximum displacements on the top floor of the building are minimal.

Soleymani and Khodadadi suggested the use of a multi-purpose adaptive genetic-fuzzy controller in the control of high-rise buildings exposed to high earthquake activity and repetitive wind loads with ATMD [132]. The proposed method was applied to a 76-story building model in Melbourne, Australia. First, the optimum TMD design has been made by using a multi-purpose GA and in this process, it is aimed to minimize the base shear force and the displacement between the floors. However, when the optimum TMD design is made by considering the wind effect, it is not sufficient to reduce the earthquake effects. However, to increase the performance of TMD, an active actuator was added to TMD and multi-purpose genetic-fuzzy ATMD was proposed. At this stage, the parameters of the FLC are optimized.

However, better results could not be obtained with this method. In addition, adaptive control based on multi-objective genetic-fuzzy ATMD has been proposed and it has been observed that the proposed multi-objective adaptive genetic-fuzzy control reduces the relative floor displacements and base shear forces under the influence of both earthquake and wind.

Li and Cao proposed a hybrid control system of tuned mass damper (HATMD) to reduce structural vibrations resulting from ground acceleration [133]. The minimization of the dynamic magnification factor (DMF) values was determined as the optimization criterion and obtained with the help of a genetic algorithm (GA).

Heidari et al. proposed a system combining proportional + integral + derivative (PID) type controllers and LQR (linear quadratic regulator) controllers as a hybrid to improve the traditional LQR control algorithm [134]. The developed hybrid control method is used for the 10-story building model under seismic effect and controlled by ATMD. The optimum frequency ratio and the optimum critical damping ratio percentage of ATMD were calculated using the cuckoo search (CS) algorithm, which is one of the metaheuristic algorithms. As a result of the analysis, it was seen that the developed hybrid control method was more successful in reducing the maximum floor displacements and accelerations compared to the classical LQR control method.

Kayabekir et al. optimized PID-controlled ATMDs by using a modified HS. The employed HS algorithm is an adaptive version and uses automatically modified algorithm parameters in each iteration. In the optimization, both ATMD parameters and PID controller parameters are taken as design variables [135]. Then, the optimum parameters are evaluated for robustness against the time delay of the control signal [136]. Kayabekir et al. also proposed a hybrid metaheuristic algorithm for the optimization of the PID-driven ATMDs. In the hybrid algorithm, the specific features of four algorithms are combined. The optimization methodology considers time-delay of the control signal and force capacity of control to propose practical solutions [137].

4 Application Results of Several Problems

4.1 Optimum Design of Retaining Walls

In this section, a problem given in Kayabekir [97] is presented. As seen in Table 1, eight variables were used in the optimum design process. Four of them are geometric variables as seen in Fig. 2. Other variables are related to RC reinforcement layout.

All design constraints that are related to safety verifications (geotechnical constraints) as the first four constraints and structural constraints as the others are shown in Table 2.

In Table 4, M_u represents the bending moment, V_u represents the shear force, A_s symbolizes the area of the reinforcing steel, and S symbolizes the space between two reinforcing bars.

Table 1 Design variables of the retaining wall problem

	Definition	Variable
Design variables related to geometry	Heel length	X_1
	Toe length	X_2
	Wall thickness	X_3
	Base height	X_4
Design variables related to reinforcement	Diameter of reinforcements of wall	X_5
	Spacing of the reinforcements in wall	X_6
	Diameter of reinforcements of base	X_7
	Spacing of the reinforcements of base	X_8

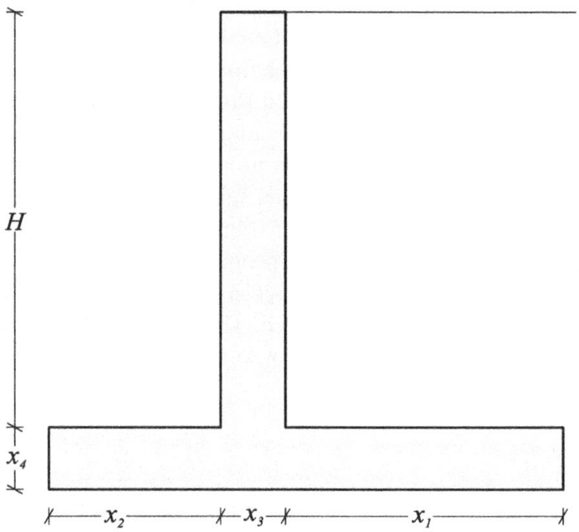

Fig. 2 Geometric design variables of the retaining wall

The objective function is defined as the minimization of the total material cost (concrete and reinforcing steel) and is expressed mathematically by Eq. (1).

$$\min f(X) = C_c V_c + C_s W_s \tag{1}$$

In this equation, C_c is the unit cost of concrete (TL/m^3), C_s is the unit cost of reinforcement (TL/ton), V_c is the unit length concrete volume, W_s is the unit length reinforcing steel weight. The cost of transporting and placing the concrete is included in the unit cost of concrete, and the transportation and labor costs of the reinforcing steel are included in the unit cost of reinforcement.

Table 2 Design constraints of the retaining wall problem

Definition	Constraints
Overturning	$g_1(X): SF_{O,design} \geq SF_O$
Sliding	$g_2(X): SF_{S,design} \geq SF_S$
Maximum bearing capacity	$g_3(X): SF_{B,design} \geq SF_B$
Minimum bearing capacity, q_{min}	$g_4(X): q, \min \geq 0$
Flexural moment capacity, M_d	$g_{5-6}(X): M_d \geq M_u$
Shear force capacity, V_d	$g_{7-8}(X): V_d \geq V_u$
Minimum rebar area, A_{smin}	$g_{9-10}(X): A_s \geq A_{s\min}$
Maximum rebar area, A_{smax}	$g_{11-12}(X): A_s \leq A_{s\max}$
Maximum spacing of rebar, S_{max}	$g_{13-14}(X): S \leq S_{max}$
Minimum spacing of rebar, S_{min}	$g_{15-16}(X): S \geq S_{min}$
Minimum clear cover, c_c	$g_{17}(X): C_c \geq 40\,\text{mm}$

Table 3 Design parameters of the retaining wall problem

Parameter	Value
Material	C25/S420
Bearing capacity of the soil	250 kN/m^2
Friction coefficient	0.6
Internal friction angle of soil	30^0
Unit weight of soil	18 kN/m^3
Wall height	5.6 m
Surcharge load	10 kN/m^2

Two different designs were made, the case without toe (Solution 1) and the case with toe (Solution 2). The parameters used in the design are shown in Table 3. The unit price of concrete is taken as 111 TL/m^3. The unit price of steel is 1400 TL/ton. Analysis results are shown in Tables 4, 5 and 6. The RC design was made by dividing it into three parts from the top of the console and into two parts from the end of the heel. It is clear to see that toe projection is very effective on the reduction of the cost.

Table 4 Optimum results for design variables related with geometry

Variables	Optimum Results (m)	
	Solution 1	Solution 2
X_1	4.20	3.50
X_2	0.00	0.45
X_3	0.35	0.35
X_4	0.35	0.35

Table 5 Optimum results for design variables related with reinforcement

	Solution 1		Solution 2	
	Diameter (mm)/Spacing (mm)	Area (mm^2)	Diameter (mm)/Spacing(mm)	Area (mm^2)
$X_{5(1)}/X_{6(1)}$	$12\phi/190$	595	$12\phi/190$	595
$X_{5(2)}/X_{6(2)}$	$12\phi/90$	1257	$12\phi/90$	1257
$X_{5(3)}/X_{6(3)}$	$28\phi/140$	4398	$28\phi/140$	4398
$X_{7(1)}/X_{8(1)}$	$14\phi/70$	2199	$14\phi/100$	1539
$X_{7(2)}/X_{8(2)}$	$28\phi/110$	5597	$16\phi/40$	5027

Table 6 Optimum costs of the retaining wall problem

	Solution 1	Solution 2
Cost (TL/m)	732.15	696.26

4.2 Span Optimization of Frame Structures

In this section, the optimum span value has been sought by considering the frame structures exposed to a uniformly distributed load over the beam elements as in Fig. 3 [97]. Reducing the maximum (absolute value) normal stress that may occur as a result of internal forces such as bending moment and axial force in structural elements is defined as the objective function. Equation (2) was used to determine the maximum stresses. The locations of the columns were considered as the design variables.

In the frame example in Fig. 3, X_1, X_2 and X_3 show the location of the columns, and q is the distributed load.

$$\sigma_{i,\,\min,\,\max} = \frac{N_i}{A_i} \pm \frac{M_i}{W_i}, \quad i = 1, n \tag{2}$$

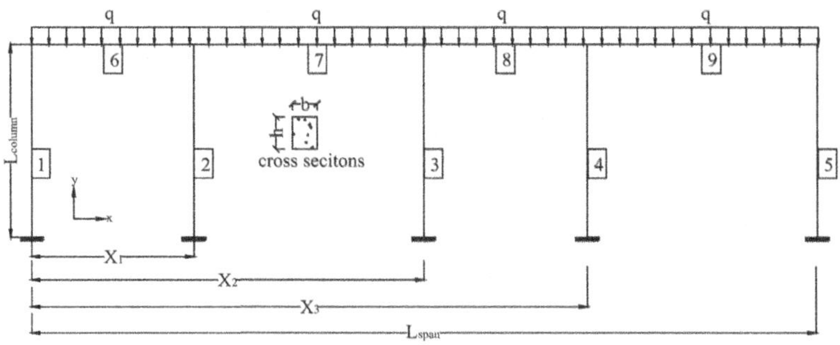

Fig. 3 Frame system subjected to distributed load

In the equation, N_i is the normal force, M_i is the bending moment, A_i is the cross-sectional area, W_i is the section modulus. The Force-Matrix method was used to find the internal forces. The equations used are shown in Eqs. (3)–(8).

Local stiffness matrix [KL]:

$$K_L = \begin{bmatrix} \frac{EA}{l} & 0 & 0 & \frac{EA}{l} & 0 & 0 \\ 0 & \frac{12EI}{l^3} & \frac{6EI}{l^2} & 0 & \frac{12EI}{l^3} & \frac{6EI}{l^2} \\ 0 & \frac{6EI}{l^2} & \frac{4EI}{l} & 0 & \frac{6EI}{l^2} & \frac{2EI}{l} \\ \frac{EA}{l} & 0 & 0 & \frac{EA}{l} & 0 & 0 \\ 0 & \frac{12EI}{l^3} & \frac{6EI}{l^2} & 0 & \frac{12EI}{l^3} & \frac{6EI}{l^2} \\ 0 & \frac{6EI}{l^2} & \frac{2EI}{l} & 0 & \frac{6EI}{l^2} & \frac{4EI}{l} \end{bmatrix} \quad (3)$$

The l in Eq. (3) represents the element length.

Transformation matric [R]:

$$R = \begin{bmatrix} \cos\theta & \sin\theta & 0 & 0 & 0 & 0 \\ -\sin\theta & \cos\theta & 0 & 0 & 0 & 0 \\ 0 & 0 & 1 & 0 & 0 & 0 \\ 0 & 0 & 0 & \cos\theta & \sin\theta & 0 \\ 0 & 0 & 0 & -\sin\theta & \cos\theta & 0 \\ 0 & 0 & 0 & 0 & 0 & 1 \end{bmatrix} \quad (4)$$

In Eq. (4), θ indicates the angle that the structural element makes with the horizontal in the positive direction.

Global stiffness matrix [KG]:

$$[K_G] = [R]^T \cdot [K_L] \cdot [R] \quad (5)$$

Global displacement matrix $[\Delta_G]$:

$$[\Delta_G] = [K_G]^{-1} \cdot [P] \quad (6)$$

[P] in the equation is the force (acting at the nodes) vector. Cross-sectional effects of structural elements are found by retrospective analysis. Using the global displacement vector of the system, the local displacement vector $[\Delta_L]$ (Eq. 7) for each structural element is obtained.

$$[\Delta_L] = [R] \cdot [\Delta_G] \quad (7)$$

By multiplying the local displacement vector with the local stiffness matrix, the cross-sectional effects at both ends of each structural member are obtained $[P_{ic}]$:

$$[P_{ic}] = [K_L] \cdot [\Delta_L] \quad (8)$$

As can be seen in Figs. 4, 5 and 6, optimization applications aiming at the minimization of the maximum normal stresses that may occur in frame systems have been made by considering three different situations. X_1 and X_2 in the figures are design variables and these variables are needed to provide the conditions given as Eqs. (9) and (10).

$$0 < X_1 < X_2 \tag{9}$$

$$X_1 < X_2 < L_{span} \tag{10}$$

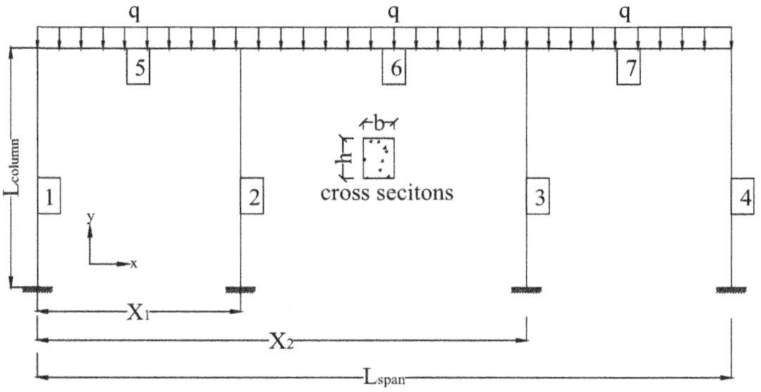

Fig. 4 Mathematical model for Case 1

Fig. 5 Mathematical model for Case 2

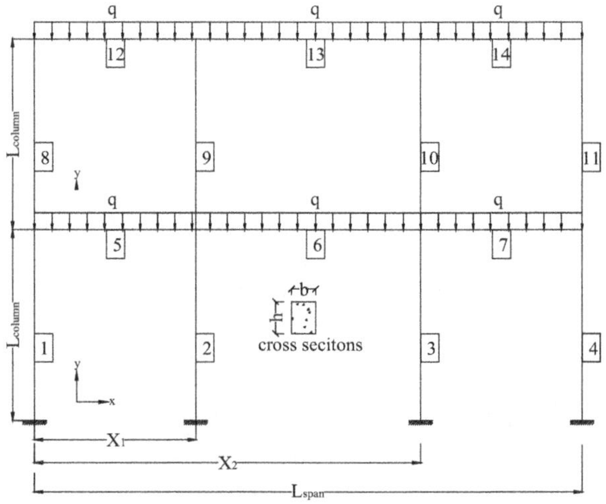

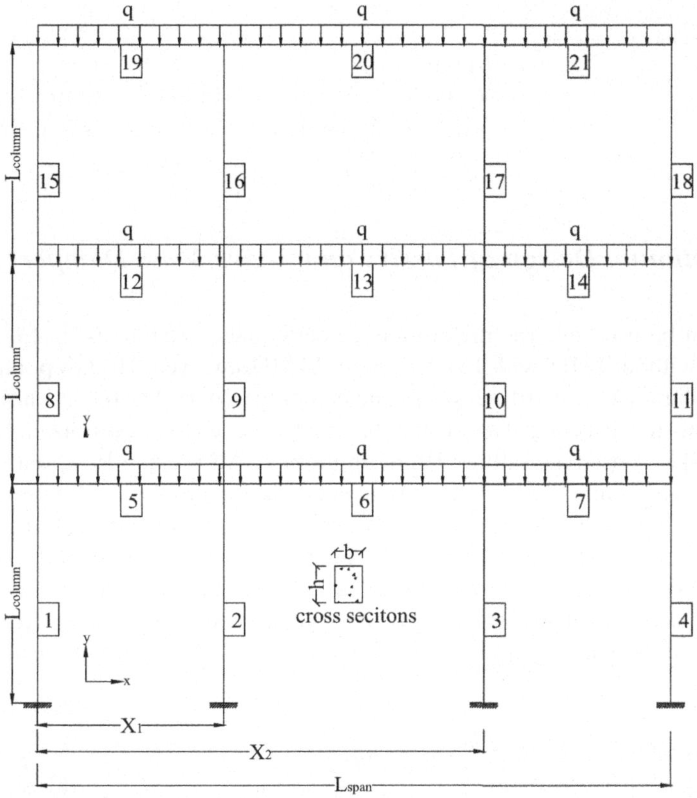

Fig. 6 Mathematical model for Case 3

The design constants taken in the study and the optimum results are given in Tables 7 and 8, respectively. Optimum values found by using the TLBO algorithm and optimum values found by using the HS algorithm in optimization applications carried out considering three different situations are the same.

Table 7 Design constants

Definition	Symbol	Unit	Value
Total length of span	L_{span}	m	15.0
Distributed load	q	kN/m^2	36.0
Elasticity modulus	E	MPa	200,000
Breadth of the elements	b	m	0.4
Height of the elements	h	m	0.4
Length of column	L_{column}	m	3

Table 8 Optimum results of the frame problem

	Case 1	Case 2	Case 3
X_1 (m)	4.85	4.9	4.95
X_2 (m)	10.15	10.10	10.05
$\max\lvert\sigma_{\max}, \sigma_{\min}\rvert$ (kN/m^2)	8015.5	7835.4	7722.3
Critical member	6	12, 14	19, 21

4.3 Optimum Design of Passive and Active Mass Dampers

TMDs can be used in structures to reduce earthquake excitations. In Fig. 7, shear buildings without TMD, with TMD and with ATMD are given. The key point is to take the TMD parameters as stiffness (k_d), and damping (c_d). In ATMD, a control force is produced by using a control algorithm that is also needed to be optimized. Kayabekir et al. [135] proposed a modified HS to optimize ATMDs using PID controllers and found the optimum values of PID parameters that are proportional gain, derivative time and integral time. To minimize the top story displacement (x_N), the optimum parameters are searched for a constrained problem of stroke limit of TMD or ATMD. The constraint (g_1) is given as Eq. (11) must be lower than a user-defined value called st-max. x_d is the displacement of TMD and ATMD respect to ground. According to the results of Kayabekir et al. [135] shown as Fig. 8, for optimum design respect

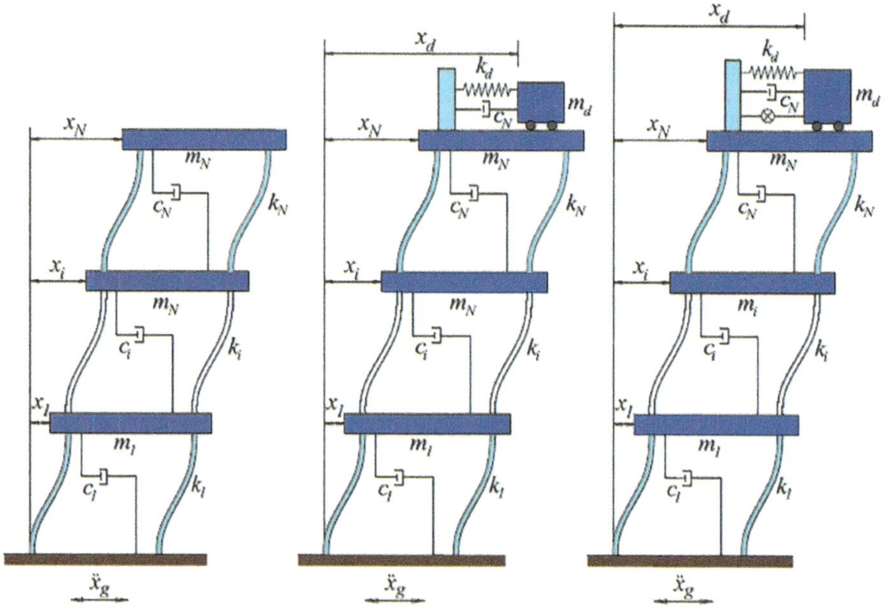

Fig. 7 Uncontrolled, TMD controlled and ATMD controlled N-floor building models, respectively

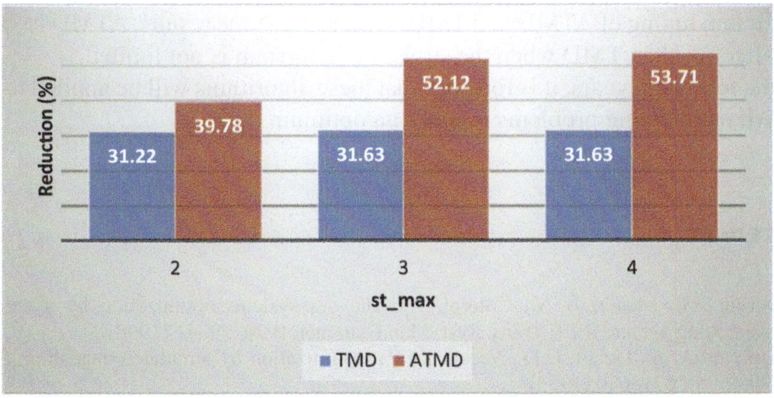

Fig. 8 The reduction performance of TMD and ATMD on top story displacement

to different st-max values, ATMD can be 22.08% more effective than TMD if the stroke capacity is in big value.

$$g_1 = \frac{\max(|x_d - x_N|)_{\text{with control}}}{\max(|x_N|)_{\text{without control}}} < st\_\max \tag{11}$$

5 Conclusion

As seen from the review presented in this paper, swarm-intelligence and metaheuristic methods are the major tools in optimum design in structural engineering. The main application areas of these methods are optimum design and optimum tuning of the control system as presented in this chapter. It must be noted that these algorithms are not only useful for optimization and these methods can be also applied in structural analysis by minimizing the total potential energy of structural systems [138–144].

Another notification of this chapter is related to different algorithms. These algorithms can be always useful for all problems with correct modifications, improvements and hybridization of other ones. This can be seen from the presented applications employing different algorithms. Also, the results of frame optimization presented in this study prove that the same results are exactly found by using two different algorithms.

In this study, three optimization problems are summarized. The first one is the cost optimization of RC retaining walls. This problem was investigated for two cases. If a toe is used and optimized, the cost reduces by 5%. The second problem is a frame structure and optimum span lengths are investigated. Similar length dimensions for spans were found for frames with a different number of stories. The last example is the optimum tuning of ATMD and TMD. According to the results, ATMD is 22.08% more effective than TMD when the stroke of the system is not limited.

In the following years, it is foreseen that these algorithms will be applied to many new civil engineering problems to make an optimum design.

References

1. Dorigo, M., Maniezzo, V., Colorni, A.: The ant system: optimization by a colony of cooperating agents. IEEE Trans. Syst. Man Cybernet. B **26**, 29–41 (1996)
2. Kirkpatrick, S., Gelatt, C.D., Vecchi, M.P.: Optimization by simulated annealing. Science **220**(4598), 671–680 (1983)
3. Holland, J.H.: Adaptation in Natural and Artificial Systems. University of Michigan Press, Ann Arbor, MI (1975)
4. Goldberg, D.E.: Genetic Algorithms in Search, Optimization and Machine Learning. Addison Wesley, Boston, MA (1989)
5. Nakrani, S., Tovey, C.: On honey bees and dynamic server allocation in internet hosting centers. Adapt. Behav. **12**(3–4), 223–240 (2004)
6. Yang, X.S.: Engineering optimizations via nature-inspired virtual Bee algorithms. In: Lecture Notes in Computer Science, vol. 3562, p. 317. Springer, GmbH (2005)
7. Haddad, O.B., Afshar, A., Marino, M.A.: Honey-bees mating optimization (HBMO) algorithm: a new heuristic approach for water resources optimization. Water Resour. Manag. **20**(5), 661–680 (2006)
8. Karaboga, D., Basturk, B.: A powerful and efficient algorithm for numerical function optimization: artificial bee colony (ABC) algorithm. J. Global Optim. **39**(3), 459–471 (2007)
9. Glover, F.: Tabu search—part II. ORSA J. Comput. **2**(1), 4–32 (1990)
10. Kennedy, J., Eberhart, R.C.: Particle swarm optimization. In: Proceedings of IEEE International Conference on Neural Networks No. IV, pp. 1942–1948. Perth Australia (1995)
11. Erol, O.K., Eksin, I.: A new optimization method: big bang–big crunch. Adv. Eng. Softw. **37**(2), 106–111 (2006)
12. Yang, X.S.: Firefly algorithms for multimodal optimization. In: Osamu, W., Thomas, Z. (eds.) Lecture Notes in Computer Sciences, vol. 5792, pp. 169–178. Chapter 14, Springer, London (2009)
13. Yang, X.S.: A new metaheuristic bat-inspired algorithm. In: Nature Inspired Cooperative Strategies for Optimization (NICSO 2010), pp. 65–74. Springer, Berlin, Heidelberg (2010)
14. Gandomi, A.H., Alavi, A.H.: Krill herd: a new bio-inspired optimization algorithm. Commun. Nonlinear Sci. Numer. Simul. **17**(12), 4831–4845 (2012)
15. Koumousis, V.K., Georgiou, P.G.: Genetic algorithms in discrete optimization of steel truss roofs. J. Comput. Civ. Eng. **8**(3), 309–325 (1994)
16. Rajan, S.D.: Sizing, shape, and topology design optimization of trusses using genetic algorithm. J. Struct. Eng. **121**(10), 1480–1487 (1995)
17. Coello, C.A., Christiansen, A.D.: Multiobjective optimization of trusses using genetic algorithms. Comput. Struct. **75**(6), 647–660 (2000)
18. Erbatur, F., Hasançebi, O., Tütüncü, I., Kılıç, H.: Optimal design of planar and space structures with genetic algorithms. Comput. Struct. **75**(2), 209–224 (2000)
19. Krishnamoorthy, C.S., Prasanna Venkatesh, P., Sudarshan, R.: Object-oriented framework for genetic algorithms with application to space truss optimization. J. Comput. Civ. Eng. **16**(1), 66–75 (2002)
20. Hasancebi, O.: Optimization of truss bridges within a specified design domain using evolution strategies. Eng. Optim. **39**(6), 737–756 (2007)
21. Kelesoglu, O.: Fuzzy multiobjective optimization of truss-structures using genetic algorithm. Adv. Eng. Softw. **38**(10), 717–721 (2007)

22. Šešok, D., Belevičius, R.: Global optimization of trusses with a modified genetic algorithm. J. Civ. Eng. Manag. **4**(3), 147–154 (2008)
23. Toğan, V., Daloğlu, A.T.: An improved genetic algorithm with initial population strategy and self-adaptive member grouping. Comput. Struct. **86**(1), 1204–1218 (2008)
24. Richardson, J.N., Adriaenssens, S., Bouillard, P., Coelho, R.F.: Multiobjective topology optimization of truss structures with kinematic stability repair. Struct. Multidiscip. Optim. **46**(4), 513–532 (2012)
25. Li, J.P.: Truss topology optimization using an improved species-conserving genetic algorithm. Eng. Optim. **47**(1), 107–128 (2015)
26. Schutte, J.F., Groenwold, A.A.: Sizing design of truss structures using particle swarms. Struct. Multidiscip. Optim. **25**(4), 261–269 (2003)
27. Li, L.J., Huang, Z.B., Liu, F., Wu, Q.H.: A heuristic particle swarm optimizer for optimization of pin connected structures. Comput. Struct. **85**(7), 340–349 (2007)
28. Perez, R.E., Behdinan, K.: Particle swarm approach for structural design optimization. Comput. Struct. **85**(19), 1579–1588 (2007)
29. Camp, C.V., Bichon, B.J.: Design of space trusses using ant colony optimization. J. Struct. Eng. **130**(5), 741–751 (2004)
30. Kaveh, A., Talatahari, S.: Particle swarm optimizer, ant colony strategy and harmony search scheme hybridized for optimization of truss structures. Comput. Struct. **87**(5), 267–283 (2009)
31. Degertekin, S.O., Hayalioglu, M.S.: Sizing truss structures using teaching-learning-based optimization. Comput. Struct. **119**, 177–188 (2013)
32. Camp, C.V., Farshchin, M.: Design of space trusses using modified teaching–learning based optimization. Eng. Struct. **62–63**, 87–97 (2014)
33. Dede, T., Ayvaz, Y.: Combined size and shape optimization of structures with a new meta-heuristic algorithm. Appl. Soft Comput. **28**, 250–258 (2015)
34. Sonmez, M.: Artificial Bee Colony algorithm for optimization of truss structures. Appl. Soft Comput. **11**(2), 2406–2418 (2011)
35. Bekdaş, G., Nigdeli, S.M., Yang, X.S.: Sizing optimization of truss structures using flower pollination algorithm. Appl. Soft Comput. **37**, 322–331 (2015)
36. Miguel, L.F.F., Lopez, R.H., Miguel, L.F.F.: Multimodal size, shape, and topology optimisation of truss structures using the Firefly algorithm. Adv. Eng. Softw. **56**, 23–37 (2013)
37. Gandomi, A.H., Yang, X.S., Alavi, A.H.: Cuckoo search algorithm: a metaheuristic approach to solve structural optimization problems. Eng. Comput. **29**(1), 17–35 (2013)
38. Talatahariand, S., Kaveh, A.: Improved bat algorithm for optimum design of large-scale truss structures. Iran Univ. Sci. Technol. **5**(2), 241–254 (2015)
39. Camp, C.V.: Design of space trusses using Big Bang-Big Crunch optimization. J. Struct. Eng. **133**(7), 999–1008 (2007)
40. Kaveh, A., Talatahari, S.: Size optimization of space trusses using Big Bang-Big Crunch algorithm. Comput. Struct. **87**(17), 1129–1140 (2009)
41. Kaveh, A., Talatahari, S.: A discrete big bang-big crunch algorithm for optimal design of skeletal structures. Asian J. Civ. Eng. **11**(1), 103–122 (2010)
42. Hasançebi, O., Kazemzadeh Azad, S.: Discrete size optimization of steel trusses using a refined big bang–big crunch algorithm. Eng. Optim. **46**(1), 61–83 (2014)
43. Kaveh, A., Sheikholeslami, R., Talatahari, S., Keshvari-Ilkhichi, M.: Chaotic swarming of particles: a new method for size optimization of truss structures. Adv. Eng. Softw. **67**, 136–147 (2014)
44. Ho-Huu, V., Nguyen-Thoi, T., Le-Anh, L., Nguyen-Trang, T.: An effective reliability-based improved constrained differential evolution for reliability-based design optimization of truss structures. Adv. Eng. Softw. **92**, 48–56 (2016)
45. Tort, C., Şahin, S., Hasançebi, O.: Optimum design of steel lattice transmission line towers using simulated annealing and PLS-TOWER. Comput. Struct. **179**, 75–94 (2017)
46. Yücel, M., Bekdaş, G., Nigdeli, S.M.: Prediction of optimum 3-bar truss model parameters with an ANN model. In: International Conference on Harmony Search Algorithm, pp. 317–324. Springer, Singapore (2020)

47. Bekdaş, G., Yücel, M., Nigdeli, S.M.: Estimation of optimum design of structural systems via machine learning. Front. Struct. Civ. Eng. (2021). https://doi.org/10.1007/s11709-021-0774-0
48. Bekdaş, G., Yucel, M., Nigdeli, S.M.: Evaluation of metaheuristic-based methods for optimization of truss structures via various algorithms and lèvy flight modification. Buildings **11**(2), 49 (2021)
49. Tejani, G.G., Savsani, V.J., Patel, V.K., Mirjalili, S.: Truss optimization with natural frequency bounds using improved symbiotic organisms search. Knowl. Based Syst. **143**, 162–178 (2018)
50. Tejani, G.G., Pholdee, N., Bureerat, S., Prayogo, D.: Multiobjective adaptive symbiotic organisms search for truss optimization problems. Knowl. Based Syst. **161**, 398–414 (2018)
51. Pierezan, J., dos Santos Coelho, L., Mariani, V.C., de Vasconcelos Segundo, E.H., Prayogo, D.: Chaotic coyote algorithm applied to truss optimization problems. Comput. Struct. **242**, 106353 (2021)
52. Kumar, S., Tejani, G.G., Pholdee, N., Bureerat, S.: Multi-objective modified heat transfer search for truss optimization. Eng. Comput. **37**(4), 3439–3454 (2021)
53. Tejani, G.G., Kumar, S., Gandomi, A.H.: Multi-objective heat transfer search algorithm for truss optimization. Eng. Comput. **37**(1), 641–662 (2021)
54. Govindaraj, V., Ramasamy, J.V.: Optimum detailed design of reinforced concrete continuous beams using genetic algorithms. Comput. Struct. **84**, 34–48 (2005)
55. Fedghouche, F., Tiliouine, B.: Minimum cost design of reinforced concrete T-beams at ultimate loads using Eurocode2. Eng. Struct. **42**, 43–50 (2012)
56. Camp, C.V., Pezeshk, S., Hansson, H.: Flexural design of reinforced concrete frames using a genetic algorithm. J. Struct. Eng. **129**(1), 105–115 (2003)
57. Leps, M., Sejnoha, M.: New approach to optimization of reinforced concrete beams. Comput. Struct. **81**(18), 1957–1966 (2003)
58. Sahab, M.G., Ashour, A.F., Toropov, V.V.: Cost optimisation of reinforced concrete flat slab buildings. Eng. Struct. **27**(3), 313–322 (2005)
59. Akin, A., Saka, M.P.: Optimum detailed design of reinforced concrete continuous beams using the harmony search algorithm. In: Proceedings of the Tenth International Conference on Computational Structures Technology, p. 131. Valencia, Civil-Comp Press, Stirlingshire ,UK (2010)
60. Bekdaş, G., Nigdeli, S.M.: Cost optimization of T-shaped reinforced concrete beams under flexural effect according to ACI 318. In: 3rd European Conference of Civil Engineering, pp. 122–126. Paris, France, WSEAS (2012). ISBN: 978–1–61804–137–1
61. Bekdaş, G., Nigdeli, S.M.: Optimization of slender reinforced concrete columns. Proc. Appl. Math. Mech. **14**(1), 183–1884 (2014)
62. Nigdeli, S.M., Bekdas, G., Kim, S., Geem, Z.W.: A novel harmony search based optimization of reinforced concrete biaxially loaded columns. Struct. Eng. Mech. **54**(6), 1097–1109 (2015)
63. Nigdeli, S.M., Bekdaş, G.: Optimum design of RC continuous beams considering unfavourable live-load distributions. KSCE J. Civ. Eng. **21**(4), 1410–1416 (2017)
64. Yücel, M., Nigdeli, S.M., Kayabekir, A.E., Bekdaş, G.: Optimization and artificial neural network models for reinforced concrete members. In: Carbas, S., Toktas, A., Ustun, D. (eds.) Nature-Inspired Metaheuristic Algorithms for Engineering Optimization Applications. Springer Tracts in Nature-Inspired Computing. Springer, Singapore (2021a). https://doi.org/10.1007/978-981-33-6773-9_9
65. Ceranic, B., Freyer, C., Baines, R.W.: An application of simulated annealing to the optimum design reinforced concrete retaining structure. Comput. Struct. **79**(17), 1569–1581 (2001)
66. Yepes, V., Alcala, J., Perea, C., Gonzalez-Vidosa, F.: A parametric study of optimum earth-retaining walls by simulated annealing. Eng. Struct. **30**(3), 821–830 (2008)
67. Ahmadi-Nedushan, B., Varaee, H.: Optimal design of reinforced concrete retaining walls using a swarm intelligence technique. In: The First International Conference on Soft Computing Technology in Civil, Structural and Environmental Engineering, pp. 1–12. UK, Civil-Comp Press, Stirlingshire, Scotland (2009)

68. Kaveh, A., Abadi, A.S.M.: Harmony search based algorithms for the optimum cost design of reinforced concrete cantilever retaining walls. Int. J. Civ. Eng. **9**(1), 1–8 (2011)
69. Ghazavi, M., Salavati, V.: Sensitivity analysis and design of reinforced concrete cantilever retaining walls using bacterial foraging optimization algorithm. In: 3rd International Symposium on Geotechnical Safety and Risk (ISGSR), pp. 307–314. Karlsruhe, München, Germany, Bundesanstalt für Wasserbau (2011)
70. Yepes, V., Gonzalez-Vidosa, F., Alcala, J., Villalba, P.: CO2-optimization design of reinforced concrete retaining walls based on a VNS-threshold acceptance strategy. J. Comput. Civ. Eng. **26**(3), 378–386 (2011)
71. Camp, C.V., Akin, A.: Design of retaining walls using big bang–big crunch optimization. J. Struct. Eng. **138**(3), 438–448 (2012)
72. Kayabekir, A.E., Arama, Z.A., Bekdaş, G., Nigdeli, S.M., Geem, Z.W.: Eco-friendly design of reinforced concrete retaining walls: multi-objective optimization with harmony search applications. Sustainability **12**(15), 6087 (2020)
73. Yücel, M., Kayabekir, A.E., Bekdas, G., Nigdeli, S.M., Kim, S., Geem, Z.W.: Adaptive-hybrid harmony search algorithm for multi-constrained optimum eco-design of reinforced concrete retaining walls. Sustainability **2021**(13), 1639 (2021)
74. Yücel, M., Bekdaş, G., Nigdeli, S.M., Kayabekir, A.E.: An artificial intelligence-based prediction model for optimum design variables of reinforced concrete retaining walls. Int. J. Geomech. **21**(12), 04021244 (2021)
75. Pezeshk, S., Camp, C.V., Chen, D.: Design of nonlinear framed structures using genetic optimization. J. Struct. Eng. **126**(3), 382–388 (2000)
76. Li, W., Li, Q., Steven, G.P., Xie, Y.M.: An evolutionary approach to elastic contact optimization of frame structures. Finite Elem. Anal. Des. **40**(1), 61–81 (2003)
77. Camp, C.V., Bichon, B.J., Stovall, S.P.: Design of steel frames using ant colony optimization. J. Struct. Eng. **131**(3), 369–379 (2005)
78. Saka, M.P.: Optimum design of steel frames using stochastic search techniques based on natural phenomena: a review. Civ. Eng. Comput. Tools Tech. **6**, 105–147 (2007)
79. Perea, C., Alcala, J., Yepes, V., Gonzalez-Vidosa, F., Hospitaler, A.: Design of reinforced concrete bridge frames by heuristic optimization. Adv. Eng. Softw. **39**(8), 676–688 (2008)
80. Rajeev, S., Krishnamoorthy, C.S.: Genetic algorithm-based methodology for design optimization of reinforced concrete frames. Comput. Aided Civ. Infrastruct. Eng. **13**, 63–74 (1998)
81. Lee, C., Ahn, J.: Flexural design of reinforced concrete frames by genetic algorithm. J. Struct. Eng. **129**(6), 762–774 (2003)
82. Govindaraj, V., Ramasamy, J.V.: Optimum detailed design of reinforced concrete frames using genetic algorithms. Eng. Optim. **39**(4), 471–494 (2007)
83. Paya, I., Yepes, V., González-Vidosa, F., Hospitaler, A.: Multiobjective optimization of concrete frames by simulated annealing. Comput. Aided Civ. Infrastruct. Eng. **23**(8), 596–610 (2008)
84. Akin, A., Saka, M.P.: Harmony search algorithm based optimum detailed design of reinforced concrete plane frames subject to ACI 318–05 provisions. Comput. Struct. **147**, 79–95 (2015)
85. Kaveh, A., Sabzi, O.: A comparative study of two meta-heuristic algorithms for optimum design of reinforced concrete frames. Int. J. Civil Eng. **9**(3), 193–206 (2011)
86. Paya-Zaforteza, I., Yepes, V., Hospitaler, A., Gonzalez-Vidosa, F.: CO_2-optimization of reinforced concrete frames by simulated annealing. Eng. Struct. **31**(7), 1501–1508 (2009)
87. Camp, C.V., Huq, F.: CO2 and cost optimization of reinforced concrete frames using a big bang-big crunch algorithm. Eng. Struct. **48**, 363–372 (2013)
88. Fesanghary, M., Mahdavi, M., Minary-Jolandan, M., Alizadeh, Y.: Hybridizing harmony search algorithm with sequential quadratic programming for engineering optimization problems. Comput. Methods Appl. Mech. Eng. **197**(33), 3080–3091 (2008)
89. Hasançebi, O., Çarbaş, S., Saka, M.P.: Improving the performance of simulated annealing in structural optimization. Struct. Multidiscip. Optim. **41**(2), 189–203 (2010)

90. Toğan, V.: Design of planar steel frames using teaching–learning based optimization. Eng. Struct. **34**, 225–232 (2012)
91. Kociecki, M., Adeli, H.: Two-phase genetic algorithm for topology optimization of free-form steel space-frame roof structures with complex curvatures. Eng. Appl. Artif. Intell. **32**, 218–227 (2014)
92. Talatahari, S., Gandomi, A.H., Yang, X.S., Deb, S.: Optimum design of frame structures using the eagle strategy with differential evolution. Eng. Struct. **91**, 16–25 (2015)
93. Aydoğdu, İ, Akın, A., Saka, M.P.: Design optimization of real world steel space frames using artificial bee colony algorithm with Levy flight distribution. Adv. Eng. Softw. **92**, 1–14 (2016)
94. Saka, M.P., Carbas, S., Aydogdu, I., Akin, A.: Use of swarm intelligence in structural steel design optimization. In: Yang, X.S., Bekdaş, G., Nigdeli S.M. (eds.) Metaheuristics and Optimization in Civil Engineering, vol. 7, pp. 43–73. Springer International Publishing, London (2016)
95. Bekdaş, G., Nigdeli, S.M.: Modified harmony search for optimization of reinforced concrete frames. In: International Conference on Harmony Search Algorithm, pp. 213–221. Springer, Singapore (2017)
96. Ulusoy, S., Kayabekir, A.E., Bekdaş, G., Nigdeli, S.M.: Optimum design of reinforced concrete multi-story multi-span frame structures under static loads. Int. J. Eng. Technol **10**(5), 403–407 (2018)
97. Kayabekir, A.E.: Yapı Mühendisliğinde Metasezgisel Algoritmalar ile Optimizasyon Uygulamaları, MSc Thesis, Istanbul University, Istanbul, Turkey (2018)
98. Rakıcı, E., Bekdaş, G., Nigdeli, S.M.: Optimal cost design of single-story reinforced concrete frames using jaya algorithm. In: International Conference on Harmony Search Algorithm, pp. 179–186. Springer, Singapore (2020)
99. Ulusoy, S., Niğdeli, S.M., Bekdaş, G.: Optimization of PID controller parameters for active control of single degree of freedom structures. Challenge **5**(4), 130–140 (2019)
100. Ulusoy, S., Bekdas, G., Nigdeli, S.M.: Active structural control via metaheuristic algorithms considering soil-structure interaction. Struct. Eng. Mech. **75**(2), 175–191 (2020)
101. Ulusoy, S., Nigdeli, S.M., Bekdaş, G.: Novel metaheuristic-based tuning of PID controllers for seismic structures and verification of robustness. J. Build. Eng. **33**, 101647 (2021)
102. Ulusoy, S., Bekdaş, G., Nigdeli, S.M., Kim, S., Geem, Z.W.: Performance of optimum tuned PID controller with different feedback strategies on active-controlled structures. Appl. Sci. **11**(4), 1682 (2021b)
103. Ulusoy, S., Kayabekir, A.E., Nigdeli, S.M., Bekdaş, G.: Metaheuristic-based structural control methods and comparison of applications. In: Nature-Inspired Metaheuristic Algorithms for Engineering Optimization Applications, pp. 251–276. Springer, Singapore (2021c)
104. Den Hartog, J.P.: Mechanical Vibrations. McGraw-Hill, New York, USA (1956)
105. Warburton, G.B.: Optimum absorber parameters for various combinations of response and excitation parameters. Earthq. Eng. Struct. Dyn. **10**(3), 381–401 (1982)
106. Sadek, F., Mohraz, B., Taylor, A.W., Chung, R.M.: A method of estimating the parameters of tuned mass dampers for seismic applications. Earthq. Eng. Struct. Dyn. **26**(6), 617–636 (1997)
107. Hadi, M.N., Arfiadi, Y.: Optimum design of absorber for MDOF structures. J. Struct. Eng. **124**(11), 1272–1280 (1998)
108. Marano, G.C., Greco, R., Chiaia, B.: A comparison between different optimization criteria for tuned mass dampers design. J. Sound Vib. **329**(23), 4880–4890 (2010)
109. Singh, M.P., Singh, S., Moreschi, L.M.: Tuned mass dampers for response control of torsional buildings. Earthq. Eng. Struct. Dyn. **31**(4), 749–769 (2002)
110. Desu, N.B., Deb, S.K., Dutta, A.: Coupled tuned mass dampers for control of coupled vibrations in asymmetric buildings. Struct. Control. Health Monit. **13**(5), 897–916 (2006)
111. Pourzeynali, S., Lavasani, H.H., Modarayi, A.H.: Active control of high rise building structures using fuzzy logic and genetic algorithms. Eng. Struct. **29**(3), 346–357 (2007)
112. Leung, A.Y.T., Zhang, H.: Particle swarm optimization of tuned mass dampers. Eng. Struct. **31**(3), 715–728 (2009)

113. Leung, A.Y., Zhang, H., Cheng, C.C., Lee, Y.Y.: Particle swarm optimization of TMD by non-stationary base excitation during earthquake. Earthq. Eng. Struct. Dynam. **37**(9), 1223–1246 (2008)
114. Bekdaş, G., Nigdeli, S.M.: Estimating optimum parameters of tuned mass dampers using harmony search. Eng. Struct. **33**(9), 2716–2723 (2011)
115. Bekdaş, G., Nigdeli, S.M.: Optimization of tuned mass damper with harmony search. In: Gandomi, A.H., Yang, X.S., Alavi A.H., Talatahari, S. (eds.) Metaheuristic Applications in Structures and Infrastructures, vol. 14, pp. 345–372. Elsevier, Londra (2013a)
116. Bekdaş, G., Nigdeli, S.M.: Mass ratio factor for optimum tuned mass damper strategies. Int. J. Mech. Sci. **71**, 68–84 (2013)
117. Nigdeli, S.M., Bekdas, G.: Optimum tuned mass damper design for preventing brittle fracture of RC buildings. Smart Struct. Syst. **12**(2), 137–155 (2013)
118. Farshidianfar, A., Soheili, S.: Ant colony optimization of tuned mass dampers for earthquake oscillations of high-rise structures including soil–structure interaction. Soil Dyn. Earthq. Eng. **51**, 14–22 (2013)
119. Farshidianfar, A.: ABC optimization of TMD parameters for tall buildings with soil structure interaction. Interact. Multiscale Mech. **6**, 339–356 (2013)
120. Bekdaş, G., Nigdeli, S.M.: Metaheuristic based optimization of tuned mass dampers under earthquake excitation by considering soil-structure interaction. Soil Dyn. Earthq. Eng. **92**, 443–461 (2017)
121. Bekdaş, G., Nigdeli, S.M., Yang, X.S.: A novel bat algorithm based optimum tuning of mass dampers for improving the seismic safety of structures. Eng. Struct. **159**, 89–98 (2018)
122. Bekdaş, G., Kayabekir, A.E., Nigdeli, S.M., Toklu, Y.C.: Tranfer function amplitude minimization for structures with tuned mass dampers considering soil-structure interaction. Soil Dyn. Earthq. Eng. **116**, 552–562 (2019)
123. Nigdeli, S.M., Bekdas, G.: Optimum tuned mass damper approaches for adjacent structures. Earthq. Struct. **7**(6), 1071–1091 (2014)
124. Nigdeli, S.M., Bekdaş, G.: Optimum design of multiple positioned tuned mass dampers for structures constrained with axial force capacity. Struct. Design Tall Spec. Build. **28**(5), e1593 (2019)
125. Yucel, M., Bekdaş, G., Nigdeli, S.M., Sevgen, S.: Estimation of optimum tuned mass damper parameters via machine learning. J. Build. Eng. **26**, 100847 (2019)
126. Ahlawat, A.S., Ramaswamy, A.: Multiobjective optimal fuzzy logic control system for response control of wind-excited tall buildings. J. Eng. Mech. **130**(4), 524–530 (2004)
127. Yang, J.N., Agrawal, A.K., Samali, B., Wu, J.C.: Benchmark problem for response control of wind-excited tall buildings. J. Eng. Mech. **130**(4), 437–446 (2004)
128. Ozer, H.O., Sayin, A., Korkmaz, N., Yagız, N.: Sliding mode control optimized by genetic algorithm for building model. In: 11th Biennial International Conference on Vibration Problems (ICOVP-2013). Lisbon, Portugal (2013)
129. Amini, F., Hazaveh, N.K., Rad, A.A.: Wavelet PSO-based LQR algorithm for optimal structural control using active tuned mass dampers. Comput. Aided Civil Infrastruct. Eng. **28**(7), 542–557 (2013)
130. Venanzi, I., Ubertini, F., Materazzi, A.L.: Optimal design of an array of active tuned mass dampers for wind-exposed high-rise buildings. Struct. Control. Health Monit. **20**(6), 903–917 (2013)
131. Shariatmadar, H., Meshkat Razavi, H.: Seismic control response of structures using an ATMD with fuzzy logic controller and PSO method. Struct. Eng. Mech. **51** (2014)
132. Soleymani, M., Khodadadi, M.: Adaptive fuzzy controller for active tuned mass damper of a benchmark tall building subjected to seismic and wind loads. Struct. Design Tall Spec. Build. **23**(10), 781–800 (2014)
133. Li, C., Cao, B.: Hybrid active tuned mass dampers for structures under the ground acceleration. Struct. Control. Health Monit. **22**(4), 757–777 (2015)
134. Heidari, A.H., Etedali, S., Javaheri-Tafti, M.R.: A hybrid LQR-PID control design for seismic control of buildings equipped with ATMD. Front. Struct. Civ. Eng. **12**(1), 44–57 (2018)

135. Kayabekir, A.E., Bekdaş, G., Nigdeli, S.M., Geem, Z.W.: Optimum design of PID controlled active tuned mass damper via modified harmony search. Appl. Sci. **10**(8), 2976 (2020)
136. Kayabekir, A.E., Nigdeli, S.M., Bekdaş, G.: Robustness of Structures with active tuned mass dampers optimized via modified harmony search for time delay. In: International Conference on Harmony Search Algorithm, pp. 53–60. Springer, Singapore (2020)
137. Kayabekir, A.E., Nigdeli, S.M., Bekdaş, G.: A hybrid metaheuristic method for optimization of active tuned mass dampers. Comput. Aided Civil Infrastruct. Eng. (2021)
138. Kayabekir, A.E., Toklu, Y.C., Bekdaş, G., Nigdeli, S.M., Yücel, M., Geem, Z.W.: A novel hybrid harmony search approach for the analysis of plane stress systems via total potential optimization. Appl. Sci. **10**(7), 2301 (2020)
139. Toklu, Y.C., Kayabekir, A.E., Bekdaş, G., Nigdeli, S.M., Yücel, M.: Analysis of plane-stress systems via total potential optimization method considering nonlinear behavior. J. Struct. Eng. **146**(11), 04020249 (2020)
140. Nigdeli, S.M., Bekdaş, G., Toklu, Y.C.: Total potential energy minimization using metaheuristic algorithms for spatial cable systems with increasing second order effects. In: 12th International Congress on Mechanics (HSTAM2019), pp. 22–25 (2019)
141. Toklu, Y.C., Bekdaş, G., Kayabekir, A.E., Nigdeli, S.M., Yücel, M.: Total potential optimization using metaheuristics: analysis of cantilever beam via plane-stress members. In: International Conference on Harmony Search Algorithm, pp. 127–138. Springer, Singapore (2020)
142. Toklu, Y.C., Bekdaş, G., Yücel, M., Nigdeli, S.M., Kayabekir, A.E., Kim, S., Geem, Z.W.: Total potential optimization using metaheuristic algorithms for solving nonlinear plane strain systems. Appl. Sci. **11**(7), 3220 (2021)
143. Toklu, Y.C., Bekdaş, G., Kayabekir, A.E., Nigdeli, S.M., Yücel, M.: Total potential optimization using hybrid metaheuristics: a tunnel problem solved via plane stress members. In: Advances in Structural Engineering—Optimization, pp. 221–236. Springer, Cham (2021)
144. Bekdaş, G., Kayabekir, A.E., Nigdeli, S.M., Toklu, Y.C.: Advanced energy-based analyses of trusses employing hybrid metaheuristics. Struct. Design Tall Spec. Build. **28**(9), e1609 (2019)

Bee Colony Optimization with Applications in Transportation Engineering

Dušan Teodorović, Miloš Nikolić, Milica Šelmić, and Ivana Jovanović

Abstract Metaheuristics turned out to be a dominant tool for solving difficult combinatorial optimization problems. Between them, a group of biologically motivated algorithms can be identified. The Bee Colony Optimization (BCO) method, which uses collective intelligence applied by honey bees, during the nectar gathering process, is one of them. The basic idea at the back of the BCO is to create a multi-agent system (a colony of artificial bees) competent to effectively discover solutions to difficult combinatorial optimization problems. The artificial bee colony performs to a certain degree alike, and in other parts differently, from bees in nature. The analysts specify the possible agents' behavior and simulate the relations between them (how artificial bees interact with each other). The model and related software imitate how artificial bees execute their actions. The BCO is a stochastic, random-search population-based method. The BCO uses likeness among how bees in nature search for nectar, and how optimization algorithms look for an optimum in combinatorial optimization problems under study. Two variations of the BCO algorithm can be differentiated easily: constructive BCO and the alternative based on the solutions improvement idea. This chapter offers a taxonomy and analysis of the research results accomplished using both variants of BCO to model transportation engineering processes. Thus far, BCO has effectively been used to various real-life traffic control and transportation planning problems: the vehicle routing problem, the ride-matching problem, the traffic sensor location problem on highways, inspection station location problem, the p-center problem, disruption mitigation on public transit, pre-timed control for isolated intersections, area-wide urban traffic control, etc. The main goal

D. Teodorović · M. Nikolić (✉) · M. Šelmić · I. Jovanović
University of Belgrade, Faculty of Transport and Traffic Engineering, Vojvode Stepe 305, 11000 Belgrade, Serbia
e-mail: m.nikolic@sf.bg.ac.rs
URL: http://www.sf.bg.ac.rs

M. Šelmić
e-mail: m.selmic@sf.bg.ac.rs

I. Jovanović
e-mail: ivana.jovanovic@sf.bg.ac.rs

A. Biswas et al. (eds.), *Advances in Swarm Intelligence*, Studies in Computational Intelligence 1054, https://doi.org/10.1007/978-3-031-09835-2_7

of this text is to notify colleagues with the key principles of Bee Colony Optimization, as well as to denote new possible BCO uses in traffic control and transportation planning.

Keywords Bee Colony Optimization · Traffic · Transportation

1 Introduction

We are witnessing climate change and emission problems. Also, an increasing number of cities in the world have a high population density. In the last few years, revolutionary technological developments in the transportation sector have also taken place. Electric vehicles and autonomous cars (self-driving cars, driverless cars) are now surrounding us. Intensive research is being conducted to develop alternative fuels. Autonomous vehicles can play a significant role in reducing the number of traffic accidents. Simultaneously, there is as well a belief that self-driving vehicles could decrease fuel consumption and the harmful consequences on the environment. A decrease in mean travel time, in addition to the decrease in areas' need for parking spaces, additionally signifies possible benefits that autonomous vehicles could bring. Autonomous vehicles have already been used in taxi services, dial-a-ride systems, and distribution systems. Smart cities, which involve total digital networking between things, have become a very important subject of research over the last decade. New transportation concepts are also emerging. The most important concept is definitely Transportation as a Service (TaaS). As an alternative to owning a car, people will buy trips and miles. TaaS vehicles could be available 24 h a day and can save a significant amount of money and time to their users. Because of TaaS, car ownership rates will finally start to fail.

Many current and future transportation engineering problems fit within the group of combinatorial optimization problems. These problems are characterized by a large number of feasible solutions (feasible solution space), as well as a large number of various constraints. Because of that, the exact algorithms, such as mixed-integer programming and dynamic programming, are not suitable approaches (primarily because the CPU time could be extremely large). This reason caused in recent decade's development of many applications based on diverse metaheuristic algorithms. A special group between them is a category of biologically motivated algorithms. The Bee Colony Optimization (BCO) algorithm is a biologically inspired metaheuristic considered in this paper. It goes to a cluster of swarm intelligence algorithms, inspired by the way how bees collect nectar in nature. Artificial bees look for the optimal solution within the solution space. During that process, they exchange some information and decide which area of the solution space they will continue to investigate.

This Chapter gives an analysis of the results attained by the BCO to model complex transportation engineering problems. Thus far, BCO has been fruitfully applied to a range of real-life traffic control and transportation planning problems: the vehicle routing problem, the ride-matching problem, the traffic sensor location problem on highways, inspection station location problem, the p-center problem, disruption mitigation in public transit, pre-timed control for isolated intersections, area-wide urban traffic control, etc.

The Chapter is arranged as follows. After the introductory section, we describe the BCO algorithm. The third part of this Chapter systematizes relevant literature - papers, dissertations and books in which this method is used to solve transportation engineering problems. Conclusions are given in the final section.

2 Bee Colony Optimization

BCO is motivated by the way how bees collect nectar in nature. In nature, some bee scouts usually explore the region. Finishing the search, scout bees go back to the hive. They report to their hive-mates about the sites, magnitude, and worth of obtainable food sources in the regions they have explored. If they find nectar, scout bees dance in the so-called "dance floor area" of the hive. In this way, they tried promoting food sites and engaging the other hive-mates to follow their lead. The information related to the food amount is given using a ritual called a "waggle danse". If a bee decides to depart the hive and gather food, it will move behind one of the scout bees to the earlier located area of flowers. After coming to the discovered area, the foraging bee grabs a load of nectar and flies back to the hive, relinquishing the nectar to food storage. These situations are then probable for a foraging bee: (1) it can discard the food site and become an uncommitted follower; (2) it can go on with the foraging behavior at the detected nectar source, with no promoting it to the bees; (3) it can try drafting free hive-mates with the dance ritual previous to the going back to the food site. The bee chooses one of these options. Several bees may be trying to recruit other bees at the dance floor area at the same time. "The recruitment among bees is always a function of the quality of the food source" [2]. This process goes over and over again, as the bees from a hive accumulate nectar and discover novel regions with prospective food supplies.

Inspired by the described process from nature Lučić and Teodorović [18–21] proposed the first version of the Bee Colony Optimization algorithm. The proposed algorithm authors called the Bee System. The algorithm has evolved. In the paper of Teodorović and Dell'Orco [43] it was called by the name it still has—Bee Colony Optimization.

There are two variants of the BCO algorithm: (1) constructive version (BCOc) and (2) version based on solution improvement (BCOi). BCOc was the first variant of this algorithm where at the start of every iteration the empty solution is allocated to each bee. The bees build up their solutions through iterations. A newer improvement variant of the algorithm, BCOi, was suggested more than 10 years ago by Davidović et al. [3]. The major innovation behind BCOi is that an entire solution should be allocated to each bee at the beginning of the search process, i.e. at the beginning of the first iteration. Bees later modify solutions during iterations to improve them. It has been shown through empirical analysis that the improvement variant has better performances, and nowadays it is more popular than constructive. Both BCO variants are described below in more detail.

2.1 Constructive Version of the Algorithm (BCOc)

BCOc was the original, first developed version of the BCO algorithm, proposed by Lučić and Teodorović [18–20]. The BCOc algorithm starts from an empty solution. Construction of a solution is carried out in forward passes in each iteration. In the backward pass, every bee chooses whether it will stay devoted to its partial solution or not. Once every bee creates a complete solution to a problem being studied, the iteration is completed. All the solutions are then compared, and the best is set as a global best solution. Then, all solutions are erased and another iteration starts.

The pseudo-code of the BCOc algorithm can be given in the following way:

```
1.  do
2.      Set empty solution to each bee
3.      for i = 1 to NP
4.          Forward pass
5.          Backward pass
6.      next
7.  while (stopping criteria is not satisfied)
```

where *NP* is number of forward/backward passes during one iteration.

In the forward pass, every bee expands previously created a partial solution. The choice of how bee will expand created partial solution is performed in a random manner. The *while* loop (steps from 1 to 7), is repeated during each iteration. The most frequently used stopping criteria are the number of iterations and the CPU time.

The main steps of the forward pass are:

```
for i = 1 to NC
    for b = 1 to B
        Evaluate possible construction steps for every bee b.
        Choose one construction step using random generator.
    next
next
```

where *NC* is number of construction steps used within one forward/backward pass, and *B* is number of bees.

In the backward pass, bees evaluate their partial/complete solutions and decide whether to maintain their own investigation and turn out to be a recruiter or to abandon the discovered solution and become a follower of some other bee. Pseudo-code of the backward pass can be presented as follows [30]:

```
1.  Normalization of the objective function values of the
    generated partial solutions.
2.  Loyalty decision for each bee.
3.  Recruitment (assigning every uncommitted bee to one of the
    loyal bees).
```

In the first step of the backward pass, the objective function value (that represents the quality of each bee's solution) should be normalized, i.e., scaled on the interval from 0 to 1. There are two possible ways of normalization depending on the character of the objective function (maximization or minimization).

In various transportation engineering problems, the objective function could represent total transportation costs, total travel time, average waiting time, the total delay, etc. All these objective functions should be minimized. Here, we compute the normalized values of the objective function in a next way [45]:

$$O_b = \frac{T_{\max} - T_b}{T_{\max} - T_{\min}} \tag{1}$$

where

B - is a number of bees in the colony,

T_b - is the objective function value of the solution generated by the bee b,

$T_{\max}$ - is the largest objective function value among all bees $T_{\max} = \max\limits_{b=1,\dots,B} T_b$,

$T_{\min}$ - is the smallest objective function value among all bees $T_{\min} = \min\limits_{b=1,\dots,B} T_b$.

There are lot of transportation engineering problems where objective function represents realized profit. In that case, objective function should be maximized, and the normalized values of the bees' solutions are calculated in the following way [45]:

$$O_b = \frac{T_b - T_{\min}}{T_{\max} - T_{\min}} \tag{2}$$

In the second step of the backward pass, we have to decide for each bee should it stay loyal to its solution or not. To make that decision, we calculate the probability that the bee will stay loyal to its solution. The probability that the bee b will continue to be loyal to its solution equals [45]:

$$p_b^{u+1} = e^{-\left(\frac{O_{\max} - O_b}{u}\right)} \tag{3}$$

where

$O_{\max}$ is maximal normalized value ($O_{\max} = \max\limits_{b=1,\dots,B} O_b$),

u is the number of current iteration.

Besides Eq. (3) researchers also frequently use other equations to represent this probability. One of the most commonly used is: $p_b = e^{-(O_{\max} - O_b)}$. The probability, calculated in this way, is independent on the number of iterations (u).

When we calculate the probability p_b^{u+1} (or p_b), we make a decision for the bee b will it stay loyal to its solution or not. In order to make such a decision, we generate a random number γ from the (0, 1) interval. The bee b will stay loyal to its solution if $\gamma \leq p_b^{u+1}$, otherwise it will not stay loyal. In this way, the loyal bees become the recruiters, while the bees that not stay loyal to their solutions become uncommitted bees.

In the third step of the backward pass, each uncommitted bee has to make a decision which one of the recruiter bees it will follow. In other words, each uncommitted bee will accept the solution of one of the recruiter bees.

The probability that uncommitted bee will follow the recruiter bee r is calculated in the following way [45]:

$$p_r = \frac{O_r}{\sum_{\forall k \in R} O_k}, \forall r \in R \tag{4}$$

where R is the set of the recruiter bees.

For each uncommitted bee, we generate a random number (from the (0, 1) interval), to decide which recruiter bee it will follow. Considering the probabilities p_r and the generated random numbers, we decide for each uncommitted bee which of the recruiter ones they will follow.

For example, let us denote by γ_1 the random number generated for the first uncommitted bee, and by p_1, p_2 and p_3 the probabilities that the first, second or third recruiter bee will be followed (in the case when we have three recruiter bees). The decision will be made in the following way: if $\gamma_1 \leq p_1$ the uncommitted bee will follow the first recruiter bee, if $p_1 \leq \gamma_1 \leq p_1 + p_2$ the uncommitted bee will follow the second recruiter bee, and if $p_1 + p_2 \leq \gamma_1 \leq p_1 + p_2 + p_3$ the uncommitted bee will follow the third recruiter bee. In the same manner, we make decisions for all uncommitted bees.

Illustration of forward and backward passes in BCOc algorithm is shown on Fig. 1.

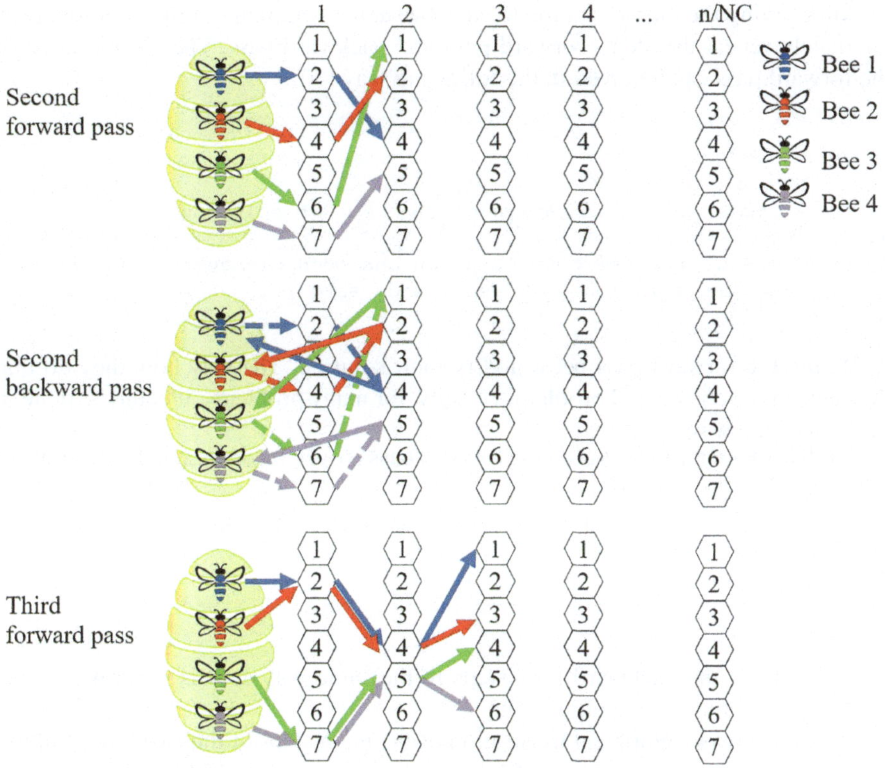

Fig. 1 Illustration of forward and backward passes in the BCOc algorithm

2.2 *Improvement Version of the Algorithm (BCOi)*

The pseudo-code of the BCOi algorithm can be described as follows [30]:

```
Generate initial solution.
do
  Set the solution to all bees.
  for i = 1 to NP
       Forward pass
       Backward pass
   next
while stopping criteria is not satisfied
```

In the first step of the algorithm, we have to determine the initial solution for the considered problem. That solution could be found in a random manner or by applying other heuristic algorithms. Usually, when the initial solution is better, a smaller number of iterations of the BCO algorithm is needed to achieve a satisfactory final result.

After setting the initial solution to each bee at the beginning of the iteration, bees alternately repeat the steps: forward pass and backward pass. The pseudo-code of the forward pass can be given in the following way [30]:

```
1.  for i = 1 to NC
2.     for b = 1 to B
3.        Make one modification of the bee's b solution.
4.     next
5.     Check if the new best solution has been discovered. If the
       new best solution has been found, save it.
6.  next
```

Within the forward pass bees modify their solution. The way how they do that depends on the considered problem. Usually, for many problems, it could be done in more than one way.

The backward pass starts after the forward pass is over. It consists of the subsequent major steps [30]:

```
1.  Normalization of the objective function values of the
    generated solutions.
2.  Loyalty decision for each bee.
3.  Recruitment (assigning every uncommitted bee to one of the
    loyal bees).
```

We perform the backward pass steps in the same way as it is described in the constructive variant of the algorithm.

Figure 2 shows a graphical illustration of the main steps of the BCOi algorithm.

3 Review of BCO Applications in Transportation Engineering

The BCO metaheuristic has been used, as a solution tool, when solving various engineering problems in the following transportation areas: urban and road (17 papers), logistics (8), air (4), railways (3), maritime and inland waterway transportation (3). In the text to follow, we analyze transportation engineering-related papers. For each paper, we emphasize the BCO variant and its specificity if it exists.

A famous problem in the field of logistics, the vehicle routing problem, was solved in the paper [21] by applying a basic variant of the algorithm called Bee system. The BCOi variant was used in the papers [27, 28], where the problem of vehicle routing with time windows was solved, and [29] in which the problem of determining new vehicle routes due to increased demand in nodes was considered. Marinelli et al. [23] considered the vehicle routing problem with simultaneous pickup/delivery. The authors proposed a two-stage approach. In the first stage, the authors applied Artificial Bee Colony, and in the second the BCOc metaheuristic.

Besides solving the vehicle routing problem, the BCO metaheuristic was applied in the following logistic problems.

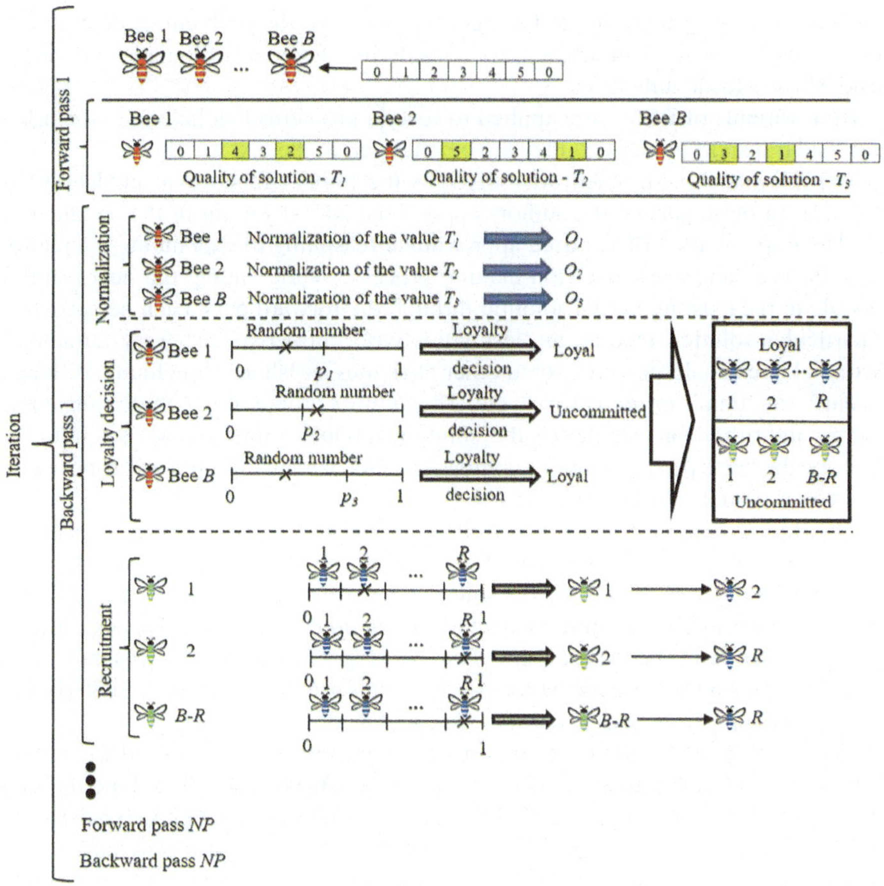

Fig. 2 A graphical illustration of the BCOi algorithm

Vukićević and Šelmić [53] discussed and solved the order picking routing problem in a warehouse with several blocks. To solve these tasks, the authors proposed an algorithm that combines Floyd's algorithm and the BCOc metaheuristic. The results were compared with results achieved using two standard heuristics (Largest Gap and S Shape strategies) that are used for solving the order picking routing problem.

Teodorović et al. [51] presented the model to discover the best locations of the p-hubs in the logistic network. Hubs operate as switching and transshipment points in transportation and communication networks and in logistic systems. Hub networks concentrate flows on the hub-to-hub links and benefit from economies of scale in interhub transportation. They used BCOi, tested it on various benchmark examples, and compared performances with results obtained by CPLEX.

Tadić et al. [42] used the BCOi algorithm to solve the problem of determining time access restrictions in urban areas considering the goals of logistics providers, residents and local authorities.

Both variants of BCO were applied to several problems that belong to the urban and road transport modes and they are emphasized in the text to follow.

The ride-matching problem was solved by the authors Teodorović and Dell'Orco [43, 44]. In these papers, the authors applied the BCOc variant of the metaheuristic. The implemented BCOc used approximate reasoning to explain the interaction between bees and bees' decision-making process. While adding the new solution module to the existing partial solution through the forward pass, each bee observes a particular solution module as 'less attractive', 'attractive', or 'very attractive'. Artificial bees could perceive some other attributes as 'short', 'medium' or 'long'; 'cheap', 'medium', or 'expensive'; etc. The authors proposed the approximate reasoning algorithm for calculating the solution module attractiveness. Intending to describe the bee's partial solutions comparison mechanism, the authors proposed the concept of partial solution badness.

The authors Nikolić et al. [36], as well as Nikolić and Teodorović [30], considered the problem of mitigation of traffic disturbances in public transport caused by bus shortages. In the first paper, [36], the authors try to find the best assignment of vacant buses on routes in a way to minimize total waiting time. In the second paper, [30], the authors minimize the negative effects of bus shortages by making new assignments of available buses and changing the topology of transit network routes. In both papers, the authors applied the BCOi variant of the metaheuristic.

The problem of locating sensors on transport networks was solved by authors Šelmić et al. [39] and Teodorović et al. [52] using BCOc, as well as Teodorović et al. [50] and Jovanović et al. [12, 13] by applying BCOi. In the first two papers, the authors compared BCOc with Genetic Algorithms. In the most recent papers, they compared the results obtained by BCOi metaheuristics with those achieved by the Simulated Annealing technique. BCOi demonstrated its competitiveness, especially on larger problems.

The problem of locating inspection facilities on transport networks was addressed by Šelmić et al. [40]. The authors applied the BCOc variant of metaheuristic. In this paper, for the first time, BCOc is combined with multi-criteria optimization, so the compromise programming in combination with metaheuristic was used.

During the past two decades, traffic authorities in many cities implemented various parking guidance and information systems. Teodorović and Šelmić [48] tried to get the answers to the following questions: (a) What should be the total number of parking guidance system signs in the traffic network? (b) Where the parking guidance system signs should be located? The model proposed is based on Fuzzy Sets Theory and BCOc. The model is supported by numerical examples.

Jovanović and Teodorović [9] considered the problem of urban traffic control at an isolated signalized intersection. The authors tried to minimize the average delay experienced by vehicles that arrive at the intersection. The authors extended their research in the papers [8, 10]. Jovanović et al. [8] applied BCOi to the problem of area-wide urban traffic control. Jovanović and Teodorović [10] considered the problem of optimal traffic control at a single intersection taking into account two objective functions: (1) average delay experienced by vehicle and (2) average delay experienced by all pedestrians that arrive at the intersection.

Dell'Orco et al. [5] used the BCO metaheuristic and a mesoscopic simulation when studying traffic assignment problem. The authors applied the constructive variant of the BCO algorithm. The authors tested the proposed approach on the Sioux-Falls network.

Nikolić and Teodorović [27, 28] studied the Transit Network Design problem. This problem aims to find a set of routes in the way to minimize the average total travel time of all users. In the first paper, [27], the total travel time consists of in-vehicle travel time and the penalties for the transfers, while in the second, [28], the total travel time also includes passengers' waiting time. The authors used the BCO algorithm to solve the considered problems. They have applied the improvement variant of the algorithm. The obtained results show that the developed approach is competitive with the other approaches in the literature.

Zhang et al. [54] applied the BCOi variant of metaheuristics to solve the transit network design problem. The authors solved the network design problem in order to find optimal bus stops and links that satisfy both passenger and transit operator. The passengers would appreciate having more transit lines, and transit operator, on contrary, would try to have fewer lines in order to reduce the cost. For a transit network design, all the city service zones should be covered with appropriate walking distances. BCOi algorithm is proposed to find the optimal transit sets, according to the origin-destination traffic demands.

Few authors used BCO algorithm to solve hard combinatorial optimization problems in the area of air transportation.

Marinelli et al. [24], Dell'Orco et al. [6], and Marinelli et al. [25] considered Flight-Gate Assignment Problem. In these papers, the authors solved the considered problem with the constructive variant of the BCO metaheuristic.

Marinelli et al. [24] defined two objective functions that should be minimized: (1) total walking distance and (2) the number of flights assigned to remote terminal gates (the remote terminal gates are not equipped with passenger bridges and because of that passengers should be transported to the terminal building by transfer buses). The proposed methodology, BCOc, was tested on the case study of Milano-Malpensa international airport.

Dell'Orco et al. [6] proposed the Fuzzy Bee Colony Optimization algorithm for the Flight-Gate Assignment Problem. The proposed algorithm is based on the BCOc algorithm and Fuzzy Inference System, where the artificial bees use fuzzy logic rules within the backward pass (to exchange information or make decisions). The proposed Fuzzy BCOc algorithm was tested on two real case studies from two Italian airports: (1) the Milan-Malpensa international airport, (2) the Turin international airport. The authors made a comparison with the results that were achieved by the Ant Colony Optimization metaheuristic. Obtained results show that the Fuzzy BCOc algorithm outperforms the ACO algorithm (on average, the Fuzzy BCOc has a better objective function value than the ACO algorithm for at least 4.8%).

Marinelli et al. [25] proposed a hybrid approach based on Biogeography-Based Optimization and the Bee Colony Optimization metaheuristic. This combined method was called Biogeography-Based Bee Colony Optimization (B-BCO). The Biogeography-Based Optimization algorithm was proposed by Simon [41]. It is an evolutionary algorithm, which includes two main operators: migration (immigration and emigration) and mutation. The proposed hybrid algorithm, B-BCOc, was applied to the case study of the Milano-Malpensa international airport. The obtained results were compared with the results got with the BCO metaheuristic. The authors concluded that B-BCOc outperformed the BCOc algorithm (the improvement of the objective function value is 9%).

Mijović et al. [26] developed a model based on the Particle Swarm Optimization and the BCO algorithm for tuning the fuzzy logic system. Within the presented model the BCOc algorithm is used for fine-tuning the membership functions. The authors used heterogeneous artificial bees (two types of bees are involved), which differ in the way how they modify solutions. The first type of bees makes small-scale changes, and the second type makes large-scale changes. The developed model was applied to the problem of airline market share on long-haul routes. The case study was developed for a set of routes originating at the London Airport system.

BCO algorithm also proved good performances when applied to various rail transportation problems.

The management of fees in railway transportation was considered by the authors Bugarinović et al. [1] using the BCOi variant of metaheuristic.

The problem of estimating electricity consumption was solved in the papers [32, 33] using the BCOi variant of metaheuristic. In these two papers, BCOi was used as a tool for fine-tuning the fuzzy membership functions of the variables that appeared in the paper. Also, the success of the BCOi technique was confirmed by comparing the achieved results with those obtained by applying the Simulated Annealing method.

Several authors have used BCO as a tool to solve problems that belong to waterway transport.

The problem of allocation of ship berths was considered by the authors Kovač [14], Kovač et al. [16], as well as Prencipe and Marinelli [37]. They all used the BCOc variant of the metaheuristic.

Kovač [14] and Kovač et al. [16] considered the Berth Allocation Problem with a discrete number of berths. Depending on the length, the ships can occupy one or more berths. The objective function, that should be minimized, represents the total penalty cost. The efficiency of the BCOc algorithm authors tested on the instances from the literature.

Prencipe and Marinelli [37] considered the dynamic Berth Allocation Problem with the discrete number of berths. The authors tested the proposed algorithm on real data instances from the Livorno port (in Italy). They compared the obtained results with the results obtained by CPLEX and Ant Colony Optimization metaheuristic.

The development of metaheuristics over the years has yielded seven doctoral dissertations dealing with problems in the field of traffic control and transportation planning using the Bee Colony Optimization metaheuristics.

In the first published dissertation, Lučić [22] applied the Bee System metaheuristic (the first name of the BCO metaheuristic) to solve the Traveling Salesman Problem and the Stochastic Vehicle Routing Problem. The dissertation of Šelmić [38] deals with the location theory problems in transportation (locating detectors on highways, locating inspection stations on the transport network, locating signs on the free parking spaces, as well as the general p-center location problem). Šelmić [38] approached the *p*-center problem with the constructive variant of the BCO algorithm. It performed well for small-scale problems but did not provide good results on relevant benchmark problems. Therefore, the variant of the BCO based on the improvement concept - BCOi was proposed, which proved itself, as a very good technique for solving the *p*-center problem. In his dissertation, Nikolić [31] considered the problems caused by disruptions in performing planned schedules in air transportation (disturbances within the gate assignment problem), the public transit (disturbances caused by shortages of buses in distribution systems), and distribution systems (the increased demand at the nodes makes distribution plan unfeasible, and a new one should be determined). For all considered problems the BCOi variant of metaheuristic was used. The problems of managing the fixed time control (selection of signal plans) at the isolated signalized intersection, as well as the area-wide urban traffic control, were solved in the dissertation Jovanović [7]. All problems are solved by BCOi. In the dissertation of Kovač [15], the problem of assigning berths to ships was solved, with the aim of minimizing total penalty costs. The problem was solved, among others, by BCOc and BCOi metaheuristics. The location of sensors on transport networks was solved by applying the BCOi variant of metaheuristic in the dissertation of Jovanović [11]. The solution to the problem was tested on real case study data on a section of the E-763 road in the Republic of Serbia. The BCOi metaheuristic has come up with good enough solutions for acceptable computer running times.

Teodorović and Šelmić [49] published a book "Computational Intelligence in Transportation." One part of the book is dedicated to swarm intelligence metaheuristics. One of the explained metaheuristics is the BCO.

In the book "Quantitative methods in transportation", Teodorović and Nikolić [45] presented widely used quantitative methods applied to the transportation engineering problems. One chapter is devoted to the metaheuristic algorithms. The authors explained Genetic Algorithm, Simulated Annealing, Tabu Search, Ant Colony Optimization and Bee Colony Optimization. The way how these techniques work is illustrated on the one simple traffic assignment problem.

Figure 3 represents a systematized presentation of analyzed papers by transport modes and by types of the BCO variant that is used as a tool.

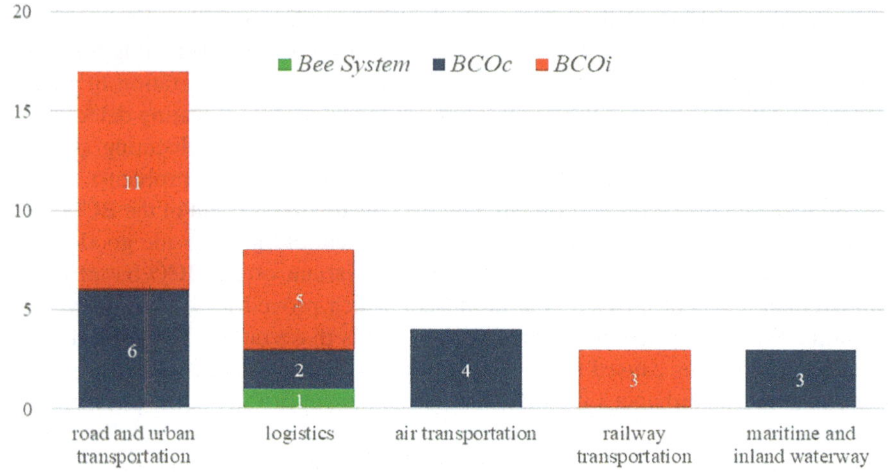

Fig. 3 The number of papers for different modes of transportation

4 Conclusions

Many current transportation engineering problems fit in the group of difficult combinatorial optimization problems. It has been demonstrated that metaheuristic algorithms are more suitable approaches for such problems than the exact approaches.

Electric vehicles and autonomous cars (self-driving cars, driverless cars) are by now on the roads and streets of several cities. This causes the need for intensive research in the field of area-wide traffic control of autonomous vehicles, dial-a-ride, and similar shared mobility systems, as well as distribution systems. There is a strong expectation in the scientific community that as an alternative to owning a car, people will buy trips and miles in the future. Transportation as a Service (TaaS) vehicles could be available 24 h a day, can save a considerable amount of money and time to their users, and decrease fuel consumption and the harmful effects on the environment.

Some of the mentioned, as well as many others, still unknown, future transportation engineering problems belong to the group of combinatorial optimization problems. The majority of them are characterized by large dimensions and the strong need to be solved very fast, or in real-time.

This Chapter presents the literature review of the applications of the Bee Colony Optimization metaheuristic on the transportation engineering problems whose combinatorial by their nature. Both BCO variants, BCOi and BCOc that are analyzed in this Chapter are tested on various numerical examples and are capable to produce high-quality solutions within small CPU times. In most of the analyzed papers, BCOi and BCOc performances are compared with other techniques from relevant literature and proved that they are very competitive.

The accomplished results show that the creation of new models founded on swarm intelligence principles could significantly contribute to the solution of complex transportation planning and traffic control engineering problems.

The Bee Colony Optimization metaheuristic originated in transportation engineering problems. Therefore, the authors aimed to systematize all problems in the field of transportation engineering that have so far been solved using the BCO metaheuristic, with the clearly emphasized type of used the BCO variant. The main contribution of this Chapter is to provide the readers interested in this particular area of research with the basics of the BCO metaheuristic, and possible good solutions that BCO can provide to their future research.

Acknowledgements This work was supported by the Serbian Ministry of Education, Science and Technological Development, through University of Belgrade, Faculty of Transport and Traffic Engineering.

References

1. Bugarinović, M., Davidović, T., Bošković, B.: Management of the access charges level for the use of railway infrastructure by bee colony optimization. In: Book of Abstracts 18th EWGT. p. 184 (2015)
2. Camazine, S., Sneyd, J.: A model of collective nectar source by honey bees: selforganization through simple rules. J. Theor. Biol. **149**, 547–571 (1991). https://doi.org/10.1016/S0022-5193(05)80098-0
3. Davidović, T., Ramljak, D., Šelmić, M., Teodorović, D.: Bee colony optimization for the p-center problem. Comput. Oper. Res. **38**(10), 1367–1376 (2011). https://doi.org/10.1016/j.cor.2010.12.002
4. Davidović, T., Teodorović, D., Šelmić, M.: Bee Colony Optimization Part I: The algorithm overview. Yugosl. J. Oper. Res. **25**(1), 33–56 (2015). https://doi.org/10.2298/YJOR131011017D
5. Dell'Orco, M., Marinelli, M., Ali Silgu, M.: Bee colony optimization for innovative travel time estimation, based on a mesoscopic traffic assignment model. Transp. Res. Part C Emerg. Technol. **66**, 48–60 (2016). https://doi.org/10.1016/j.trc.2015.10.001
6. Dell'Orco, M., Marinelli, M., Altieri, M.G.: Solving the gate assignment problem through the Fuzzy Bee Colony Optimization. Transp. Res. Part C Emerg. Technol. **80**, 424–438 (2017). https://doi.org/10.1016/j.trc.2017.03.019
7. Jovanović, A.: Choice of signal timing for traffic control by Bee Colony Optimization, doctoral dissertation, University of Belgrade, Faculty of Transport and Traffic Engineering, Belgrade (in Serbian) (2017)
8. Jovanović, A., Nikolić, M., Teodorović, D.: Area-wide urban traffic control: A Bee Colony Optimization approach. Transp. Res. Part C Emerg. Technol. **77**, 329–350 (2017). https://doi.org/10.1016/j.trc.2017.02.006
9. Jovanović, A., Teodorović, D.: Pre-timed control for an under-saturated and over-saturated isolated intersection: a Bee Colony Optimization approach. Transp. Plan. Technol. **40**(5), 556–576 (2017). https://doi.org/10.1080/03081060.2017.1314498
10. Jovanović, A., Teodorović, D.: Multi-objective optimization of a single intersection. Transp. Plan. Technol. **44**(2), 139–159 (2021). https://doi.org/10.1080/03081060.2020.1868083
11. Jovanović, I.: Sensors selection and deployment on transport networks using operations research methods, doctoral dissertation, University of Belgrade, Faculty of Transport and Traffic Engineering, Belgrade. (in Serbian) (2020)

12. Jovanović, I., Nikolić, M., Šelmić, M.: A comparative analysis of metaheuristic approaches for sensors deployment problem on transport networks. Int. J. Traffic Transp. Eng. (Online) **11**(2), 310–322 (2021). https://doi.org/10.7708/ijtte2021.11(2).10
13. Jovanović, I., Šelmić, M., Nikolić, M.: Metaheuristic approach to optimize placement of detectors in transport networks - Case study of Serbia. Can. J. Civ. Eng. **46**(3), 176–187 (2019). https://doi.org/10.1139/cjce-2018-0306
14. Kovač, N.: Bee colony optimization algorithm for the minimum cost berth allocation problem, In: XI Balkan Conference on Operational Research, BALCOR, pp. 245–254 (2013)
15. Kovač, N.: Metaheuristic approach for solving one class of optimization problems in transport. Doctoral dissertation, University of Belgrade, Faculty of Mathematics, Belgrade (in Serbian) (2018)
16. Kovač, N., Stanimirović, S., Davidović, T.: Metaheuristic approaches for the minimum cost hybrid berth allocation problem. In: Konstantopoulos C., Pantziou G. (eds.) Modeling, Computing and Data Handling Methodologies for Maritime Transportation, vol. 131, pp. 22–68. ISR Library (2018)
17. Leong, K.H., Abdul-Rahman, H., Wang, C., Onn, C.C., Loo, S.C.: Bee inspired novel optimization algorithm and mathematical model for effective and efficient route planning in railway system. PLoS ONE **11**(12), e0166064 (2016). https://doi.org/10.1371/journal.pone.0166064
18. Lučić, P., Teodorović, D.: Bee system: modeling combinatorial optimization transportation engineering problems by swarm intelligence. In: Preprints of the TRISTAN IV Triennial Symposium on Transportation Analysis. Sao Miguel, Azores Islands, Portugal, pp. 441–445 (2001)
19. Lučić, P., Teodorović, D.: Transportation modeling: an artificial life approach. In: Proceedings of the 14th IEEE International Conference on Tools with Artificial Intelligence, pp. 216–223. Washington, D.C. (2002). https://doi.org/10.1109/TAI.2002.1180807
20. Lučić, P., Teodorović, D.: Computing with bees: attacking complex transportation engineering problems. Int. J. Artif. Intell. Tools **12**, 375–394 (2003). https://doi.org/10.1142/S0218213003001289
21. Lučić, P., Teodorović, D.: Vehicle routing problem with uncertain demand at nodes: the bee system and fuzzy logic approach. In: Verdegay, J.L. (ed.) Fuzzy Sets in Optimization, pp. 67–82. Springer, Heidelberg (2003)
22. Lučić, P.: Modeling Transportation Problems Using Concepts of Swarm Intelligence and Soft Computing. doctoral dissertatioin, Virginia Polytechnic Institute and State University, Falls Church, Virginia, USA (2002)
23. Marinelli, M., Caggiani, L., Alnajajreh, A., Binetti, M.: A two-stage Metaheuristic approach for solving the Vehicle Routing Problem with Simultaneous Pickup/Delivery and Door-to-Door service. In: 6th International Conference on Models and Technologies for Intelligent Transportation Systems (MT-ITS), pp. 1-9 (2019). https://doi.org/10.1109/MTITS.2019.8883340
24. Marinelli, M., Dell' Orco, M., Sassanelli, D.: A metaheuristic approach to solve the flight gate assignment problem. Transp. Res. Procedia **5**, 211–220 (2015). https://doi.org/10.1016/j.trpro.2015.01.013
25. Marinelli, M., Palmisano, G., Dell'Orco, M., Ottomanelli. M.: Optimizing airport gate assignments through a hybrid metaheuristic, advanced concepts, methodologies and technologies for transportation and logistics. In: Book Series: Advances in Intelligence Systems and Computing,vol. 572, pp. 389–404. Springer International Publishing (2018). https://doi.org/10.1007/978-3-319-57105-8_19
26. Mijović, N., Kalić, M., Kuljanin, J.: Tuning the fuzzy logic system by two meta-heuristics: case study of airline market share on long-haul routes. Transp. Res. Procedia **52**, 453–460 (2021). https://doi.org/10.1016/j.trpro.2021.01.053
27. Nikolić, M., Teodorović, D.: Transit network design by Bee Colony Optimization. Expert Syst. Appl. **40**(15), 5945–5955 (2013). https://doi.org/10.1016/j.eswa.2013.05.002
28. Nikolić, M., Teodorović, D.: A simultaneous transit network design and frequency setting: computing with bees. Exp. Syst. Appl. **41**(16), 7200–7209 (2014). https://doi.org/10.1016/j.eswa.2014.05.034

29. Nikolić, M., Teodorović, D.: Vehicle rerouting in the case of unexpectedly high demand in distribution systems. Transp. Res. Part C Emerg. Technol. **55**, 535–545 (2015). https://doi.org/10.1016/j.trc.2015.03.002
30. Nikolić, M., Teodorović, D.: Mitigation of disruptions in public transit by Bee Colony Optimization. Transp. Plan. Technol. **42**(6), 573–586 (2019). https://doi.org/10.1080/03081060.2019.1622251
31. Nikolić, M.: Disruption management in transportation by the Bee Colony Optimization metaheuristic. doctoral dissertation. University of Belgrade, Faculty of Transport and Traffic Engineering, Belgrade (in Serbian) (2015)
32. Nikolić, M., Ćalić J., Šelmić, M., Macura, D.: Tuning fuzzy system for estimation of freight train energy consumption by the Bee Colony Optimization metaheuristic. In: Proceedings of conference SYM-OP-IS 2019, pp. 672–677. Kladovo, Serbia (2019)
33. Nikolić, M., Šelmić, M., Macura, D., Ćalić, J.: Bee Colony Optimization metaheuristic for fuzzy membership functions tuning. Exp. Syst. Appl. **158**, 113601 (2020). https://doi.org/10.1016/j.eswa.2020.113601
34. Nikolić, M., Teodorović, D., Šelmić, M.: Solving the vehicle routing problem with time windows by Bee Colony Optimization metaheuristic. In: Proceedings of the 1st Logistics International Conference LOGIC, pp. 44–48. Belgrade, Serbia (2013)
35. Nikolić, M., Teodorović, D., Šelmić, M.: Solving the vehicle routing problem with time windows by Bee Colony Optimization. In: Proceedings of conference SYM-OP-IS 2014, pp. 655–660. Divčibare, Serbia (2014)
36. Nikolić, M., Teodorović, D., Vukadinović, K.: Disruption management in public transit: the Bee Colony Optimization approach. Transp. Plan. Technol. **38**(2), 162–180 (2015). https://doi.org/10.1080/03081060.2014.997447
37. Prencipe, L.P., Marinelli, M.: A novel mathematical formulation for solving the dynamic and discrete berth allocation problem by using the Bee Colony Optimisation algorithm. Appl. Intell. **51**, 4127–4142 (2021). https://doi.org/10.1007/s10489-020-02062-y
38. Šelmić, M.: Locating facilities in the traffic networks using computational intelligence methods. doctoral dissertation. University of Belgrade, Faculty of Transport and Traffic Engineering, Belgrade (in Serbian) (2011)
39. Šelmić, M., Edara, P., Teodorović, D.: Bee colony optimization approach to optimize locations of traffic sensors on highways. Tehnika **6**, 9–15 (2008)
40. Šelmić, M., Teodorović, D., Vukadinović, K.: Locating inspection facilities in traffic networks: an artificial intelligence approach. Transp. Plan. Technol. **33**(6), 481–493 (2010). https://doi.org/10.1080/03081060.2010.505047
41. Simon, D.: Biogeography-based optimization. IEEE Trans. Evol. Comput. **12**, 702–713 (2008). https://doi.org/10.1109/TEVC.2008.919004
42. Tadić, S., Kovač, M., Krstić, M.: An approach for determining time access restriction measures in city logistics. In: Proceedings of Conference SYM-OP-IS 2020, pp. 195–200. Belgrade, Serbia (2020)
43. Teodorović, D., Dell'Orco, M.: Bee Colony Optimization—a cooperative learning approach to complex transportation problems. In: Proceedings of the 16th Mini-EURO Conference on Advanced OR and AI Methods in Transportation, pp. 51–60. Poznan, Poland (2005)
44. Teodorović, D., Dell'Orco, M.: Mitigating traffic congestion: solving the ride-matching problem by bee colony optimization. Transp. Plan. Technol. **31**(2), 135–152 (2008). https://doi.org/10.1080/03081060801948027
45. Teodorović, D., Nikolić, M.: Quantitative Methods in Transportation. CRC Press, Taylor and Francis Group (2021)9780367250539
46. Teodorović, D., Šelmić, M.: Bee colony optimization for the p median problem. In: Proceedings of Conference SYM-OP-IS 2007, pp. 417–420. Zlatibor, Serbia (in Serbian) (2007)
47. Teodorović, D., Šelmić, M.: Bee Colony Optimization for the p-center problem. In: Proceedings of Conference SYM-OP-IS 2008, pp. 591–594. Soko Banja, Serbia (2008)
48. Teodorović, D., Šelmić, M.: Locating parking guidance system signs in transportation networks, In: Proceedings of Conference SYM-OP-IS 2009, pp. 663–666. Ivanjica, Serbia (2009)

49. Teodorović, D., Šelmić, M.: Computational Intelligence in Transportation. University of Belgrade, Transport and Traffic Engineering Faculty, Belgrade, Serbia (in Serbian) (2012)978-86-7395-295-6
50. Teodorović D., Šelmić, M., Nikolić, M., Jovanović, I., Vidas M.: Metaheuristic approach for detector locations in transport networks, In: Proceedings of Conference SYM-OP-IS 2017, pp. 723–728. Zlatibor, Serbia (2017)
51. Teodorović, D., Šelmić, M., Vukićević, I.: Locating hubs in transport networks: an artificial intelligence approach. Int. J. Traffic Transp. Eng. (Online) **4**(3), 286–296 (2014). https://doi.org/10.7708/ijtte.2014.4(3).04
52. Teodorović, D., Šelmić, M., Praveen, E.: Bee Colony Optimization approach to optimize placement of traffic sensors on highways. In: Proceedings of the 13th International Conference on Traffic Science. Portorož, Slovenia (2010)
53. Vukićević, I., Šelmić, M.: Order picking routing using Bee colony optimization. In: Proceedings of Conference SYM-OP-IS 2011, pp. 376–379. Zlatibor, Serbia (2011)
54. Zhang, H., Zhao, P., Gao, J., Zhuge, C., Yao, X.: An effective intelligent method for optimal urban transit network design. J. Inf. Comput. Sci. **12**(6), 2177–2184 (2015). https://doi.org/10.12733/JICS20105667

Application of Swarm Based Approaches for Elastic Modulus Prediction of Recycled Aggregate Concrete

Harish Narayana and Prashanth Janardhan

Abstract In concrete, the elastic modulus plays a vital role as a design parameter. Most of the empirical formulas used to predict the elastic modulus are limited to natural aggregate concretes (NAC). The formulae developed for recycled aggregate concrete (RAC) are constrained by experimental conditions since there is a considerable variation in recycled aggregate properties from place to place. In the present study, swarm based soft computing technique is used to overcome this issue. Artificial Neural Network (ANN) is trained by both standalone Levenberg-Marquardt back-propagation (LM-BP) technique and a hybrid of LM-BP and elephant herding optimization (EHO). The developed model is trained by 400 datasets obtained from literature which include seven inputs (i.e., water to cement (w/c) ratio, replacement of NA by RA in volume (r), coarse aggregates to cement (CA/C) ratio, fine aggregate to total aggregate (FA/TA) ratio, saturated surface dry specific gravity (SGSSD) of mixed CA(NA + RA), water absorption (W_a) of mixed CA and cube compressive strength (F_c)) and one output as elastic modulus of RAC. The performances of the developed models are evaluated using standard statistical measures like MSE, r and MAE. The developed hybrid model yields 'r' of 0.9966 and 0.9935, MSE of 0.4765 and 0.4807, and MAE of 1.1080 and 1.1888 for training and testing respectively which outperformed standalone technique of LM-BP in every aspect for predicting the elastic modulus of RAC. This shows that the hybrid model can be used effectively to predict the elastic modulus of RAC.

Keywords Artificial neural network · Elastic modulus · Elephant herding optimization · Recycled aggregate concrete

The authors thank NIT Goa and NIT Silchar.

H. Narayana (✉)
National Institute of Technology Goa, Goa, India
e-mail: harishn@nitgoa.ac.in

P. Janardhan
National Institute of Technology Silchar, Assam, India
e-mail: prashanth@civil.nits.ac.in

A. Biswas et al. (eds.), *Advances in Swarm Intelligence*, Studies in Computational Intelligence 1054, https://doi.org/10.1007/978-3-031-09835-2_8

1 Introduction

The elastic modulus is one of the vital property of concrete since it shows ability of concrete to withstand deflection due to the applied stresses and also its stiffness. It is sensitive to mix proportions of concrete. A high or low modulus of elasticity requirement depends on whether deflection is to be avoided or to be allowed. Hence, it is vital for engineers to know the elastic modulus in order to understand the behavior of concrete.

In the past several decades, many experiments have been carried out to find the modulus of elasticity of different types of concrete [1–4]. However,the experimental procedure [5] to measure the elastic modulus of concrete is quite complicated and expensive, when compared to compressive strength tests. To overcome these difficulties, researchers have developed empirical formulae to estimate the elastic modulus of concrete [6–9] and these empirical formulas have been developed for natural aggregate concrete (NAC).

In recent years, Recycled Aggregates (RA) has been considered for producing concrete in order to conserve raw materials and facilitate utilization of demolition wastes. Whereas, RA have higher permeability and water absorption, which leads to lower performance compared to Natural Aggregates (NA) [4, 10]. With similar water to cement ratio, Recycled Aggregate Concrete (RAC) usually develops up to 45 percent lesser modulus of elasticity compared to NAC [11, 12]. RAC improved the properties of mixture used to build the base and sub-base [13]. Due to these complications, the empirical formulas developed for NAC are not entirely reliable for predicting the elastic modulus of RAC. Previously some researchers have proposed specific empirical formulas for predicting the elastic modulus of RAC [14–16]. However, the applicability of these relations has not been satisfactory since there is significant variation in the nature of recycled aggregates [17].

In the recent times, Soft computing is becoming more important in solving complex problems, including approximations, inaccuracies, uncertainties, and partial truths to solutions. Different soft computing techniques have been already used [18–22] since a large number of databases are accessible in the field of structural concrete. Golafshani and Behnood [23] have used Artificial Neural Network (ANN) for predicting the elastic modulus of RAC and further compared it to standalone techniques of Support Vector Regression (SVR), fuzzy logic and Radial Basis Function Neural Network (RBFNN). M5 model tree algorithm has been previously used for predicting the elastic modulus of RAC [24]. ANN trained by different types of Back-Propagation (BP) algorithms have been utilized to predict both compressive strength and elastic modulus of different types of concrete [25–27].

To reap the benefits of different techniques hybrid of Neuro-Fuzzy Inference System (ANFIS) has been utilized to predict the elastic modulus of normal and high strength concrete [28]. Pathak et al. [29] have also used ANFIS to predict the compressive strength of self-compacting concrete. Observing various drawbacks of standalone BP technique, [30] trained ANN using ant colony optimization (ACO) which is a global optimization technique. Further, they successfully hybridized ACO and BP for training ANN. Recently, [31] have hybridized ANN and Genetic Algorithm (GA) to predict and optimize the ultimate bond strength of reinforced concrete. Hence the hybrid techniques have proven to perform better than stand-alone techniques.

Wang et al. [32] being inspired by natural behaviour of elephant groups developed the Elephant Herding Optimization (EHO) methodology. Performance of the technique was benchmarked against Differential Evolution (DE), Bio-Geography Based Optimization (BBO), and GA. Tuba and Stanimirovic [33] employed EHO for Support Vector Machine (SVM) parameter tuning, and it proved to outperform GA concerning the accuracy of classification. EHO used as a training algorithm for ANN gave improved accuracy in classification when compared with crow search and LM-BP (Levenberg-Marquardt—Back Propagation) algorithm [34]. Recently, [35] suggested some modifications to existing EHO. These modifications have further enhanced the performance of EHO as a standalone technique and are yet to be tested for training ANN. Further, EHO has not been hybridized with other training algorithms in any field of engineering.

ANN trained by LM-BP has been previously used for prediction of elastic modulus and has proven to be a good prediction technique theoretically. Since elastic modulus is a very sensitive parameter in structural concrete, it would be hazardous to consider result obtained by ANN trained by LM-BP practically due to its tendency to get stuck in local optima from time to time. This drawback can be nullified by taking advantage of the complementary nature of soft computing techniques. From the past study, it is found that the EHO hybridized with LM-BP neural network in the prediction of the elastic modulus of RAC has not been investigated. In the present study, modified EHO is hybridized with LM-BP neural network to predict the elastic modulus of RAC. Performances of models are validated using statistical measures like Root Means Square (MSE), correlation factor (r) and Mean Absolute Error (MAE).

2 Data Collection

In this study, the ANN models have been developed using experimental data obtained from previously published literature [23, 24, 36]. Behnood et al. [24] have extracted 400 sets of data from 26 different kinds of literature, with the primary theme being RAC. They have excluded the effect of RCA as fine aggregate as well as supplementary cementitious material in mix proportion. Behnood et al. [24] observed that amount and type of the impurities in RAC parameters were not significant. In order to consider the effect of the mixture constituents, compressive strength at 28 days was considered as one of the input parameters by [24]. Further, to consider the effect of aggregate properties, saturated surface dry specific gravity (SGSSD) and water absorption of coarse aggregate were also considered as input parameters [24]. Based on the engineering judgement, they have chosen seven input variables as influencing parameters for the elastic modulus of RAC. These are water to cement (w/c) ratio, replacement of NA by RA in volume (r), coarse aggregates to cement (CA/C) ratio, fine aggregate to total aggregate (FA/TA) ratio, saturated surface dry specific gravity

Fig. 1 Variation of RAC data

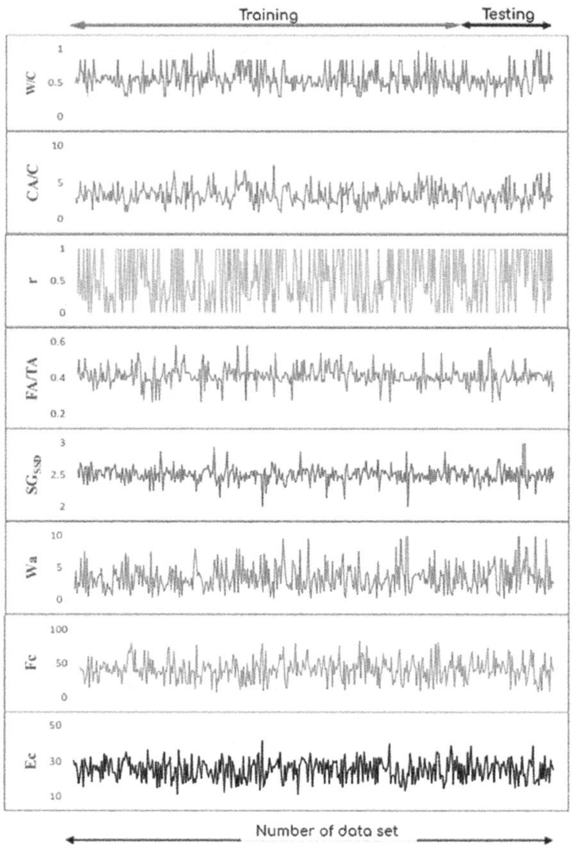

(SGSSD) of mixed CA(NA + RA), water absorption (W_a) of mixed CA and cube compressive strength (F_c) at 28th day. The underlying assumption by [24] is that these variables can reproduce the effects of different mix proportions and characteristics of the components used. Thus, a prediction model obtained using this dataset can be used for a wide range of RAC scenarios.

The resulting dataset was randomly divided into two groups, 80% and 20%, each used for training and testing the model. This way of grouping has been adopted based on the previous study conducted for the same data [23]. The variations in input and output data are shown in Fig. 1 for both training and testing datasets.

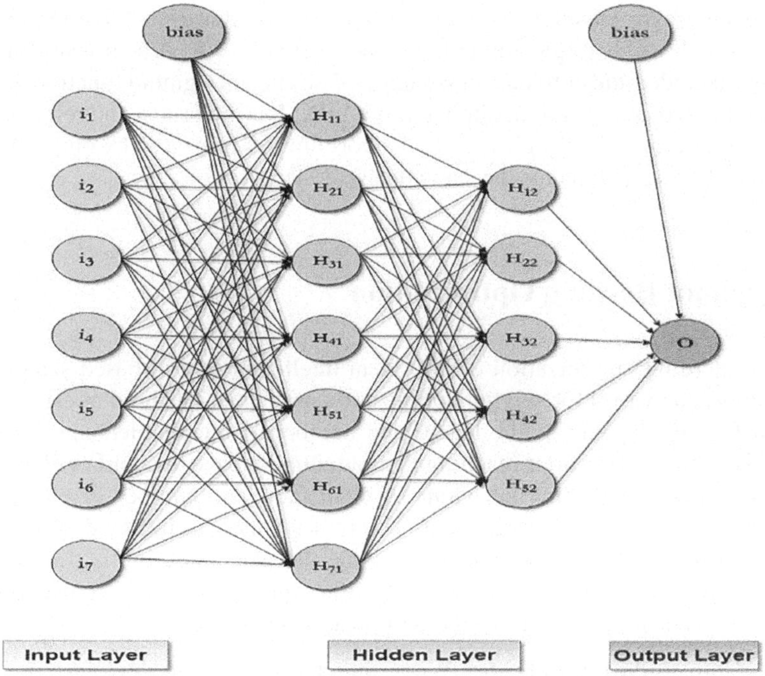

Fig. 2 Architecture of artificial neural network

3 Artificial Neural Network

Figure 2 shows the ANN flow diagram used in this study. Here, each of the nodes illustrates a neuron and together within each column represents a layer. Feedforward neural network (FFNN) is most commonly used, and it consists of 3 layers. The first layer is called the input layer, and the number of input nodes (I) depends upon the problem at hand. The second layer is called the hidden layer (H) just because it is a layer between the input and output layer and the number of hidden sub layers depend on the complexity of the problem at hand. The final layer being the output layer (O) typically represents end output. When an input is received in the input layer, that input is passed on to the next layer, via a connection. Thus, between the layers, each of the inputs is connected to every single node in the next layer. Each of these connections transfers the output from the previous node as input to the receiving node through a transfer function. Transfer functions are used to calculate the error when applying the network to new data, which can then be used to adjust weights accordingly so that it performs better on subsequent iterations of training according to iterative learning algorithms.

In the present study, the FFNN has seven input parameters (I=7), namely $i_1 =$ W/C, $i_2 =$ r, $i_3 =$ CA/C, $i_4 =$ FA/TA, $i_5 =$ SGSSD, $i_6 = W_a$ and $i_7 = f_c$. Three types of hidden layers are considered((10), (7 5) and (20 7)) and in each hidden layer H is a number of nodes. O is single output layer node, here $O_1 = E_c$. The transfer function used in both hidden and output layer is a hyperbolic tangent sigmoid function. LM-BP has been utilized as a default method to train ANN. It is the fastest and reliable ANN training algorithm [37–39]. Further, LM algorithm has proven to be outperforming eight other BP algorithms for the same dataset [23].

4 Elephant Herding Optimization

Elephant Herding Optimization (EHO) is an intelligent swarm-based search technique proposed by [32] for solving optimization problems. The herd or clan consists of adult female elephant as the leader and number of female elephants and their calves. The male calves after reaching their adulthood leave the herd or clan.

EHO makes use of this behaviour of elephants. The number of elephants is assumed to be the same in each clan. The best solution in the herd of elephants is assumed to be held by the matriarch, while the worst solution is decoded as a male elephant's position. Meena et al. [35] further improvised the [32] mathematical model used in this study. Steps of the optimization cycle are:

Step 1: Clan updating operator When the clan or herd of elephants p_k are on the move, the next position of each elephant in the clan p_k which depends on the matriarch is given by:

$$Z_{\text{new},pk,t} = Z_{pk,t} + \delta(Z_{\text{best},pk} - Z_{pk,t}) * r \ . \tag{1}$$

where $Z_{\text{new},pk,t}$ and $Z_{pk,t}$ are the new and old position of the elephant 't' in the p_k clan. δ is a scaling factor between 0 and 1 which determines how much old position of elephant $Z_{pk,t}$ is affected by matriarch of clan p_k. $Z_{\text{best},pk,t}$ is the best position in the p_k clan (matriarch), and r is uniform distribution which ranges between 0 and 1.

The best position of matriarch elephant is using Eq. 2.

$$Z_{\text{new},pk,t} = \gamma * Z_{\text{centre},pk} \ . \tag{2}$$

$$Z_{\text{centre},pk} = \sum_{t=1}^{pk} \frac{Z_{pk,t}}{n_z} \ . \tag{3}$$

where $Z_{\text{new},pk,t}$ is obtained from the information of all the elephants present in clan p_k and $Z_{\text{centre},pk}$ is the centre of clan p_k which can be obtained using Eq. 3. γ is a scale factor between 0 to 1 which determines how much $Z_{\text{new},pk,t}$ is affected by $Z_{\text{centre},pk}$ and n_z gives the total number of elephants in each clan.

Step 2: Clan separating operator In the clan, when the male elephant reaches puberty, it separates itself from the clan and lives alone. The worst fitness of the male elephant is used to improve the EHO model. The equation to find the worst fit male elephant is given by:

$$Z_{\text{worst},pk} = Z_{\min} + (Z_{\max} - Z_{\min} + 1) * \mu \ . \tag{4}$$

Where, $Z_{\text{worst},pk}$ is the worst or matured male elephant in the p_k clan. Z_max and Z_min is the upper and lower bound position of each elephant and μ is stochastic distribution factor between 0 to 1. In this step, the adult male or the worst performing elephant is removed and replaced by the younger calf.

Step 3: Steps 1 and 2 are repeated until the stopping condition is satisfied.

5 Hybrid of ANN and EHO

BP-algorithm is a gradient descent training technique in ANN, typically used to adjust connection weights. The gradient descent training technique performs trajectory searching, which is prone to get stuck in local optima easily [30]. Whereas, global optimization techniques such as EHO perform much longer jumps in the search environment. ANN trained by EHO has previously outperformed ANN trained by BP in classification problems [34] but are yet to be tested for prediction problems. Further in case of improved EHO, while its search is wide-ranging, it also makes sure to explore the neighborhood of the best solution by making herd elephants get closer to it every step. Each "elephant" is characterized by a "chromosome vector" which describes the state of all variables in the optimization problem (in our case, all the weights and biases of an ANN). Elephants are divided into "clans" of equal size, and each clan evolves independently. Here, each elephant's fitness is determined. In this case, fitness is a performance function of ANN with the weights and biases defined by the elephant's chromosome vector, calculated for all training datasets.

This study aims to employ EHO training to find the best initial weights which are then applied for LM-BP once the EHO is aborted. Gradient descent techniques like BP make small jumps in the search environment. As a result, it has an excellent ability to find the local optimum point nearby. Even though improved EHO makes sure to explore the neighborhood of the best answer by making herd elephants get closer to it every step, but during search it still makes large jumps due to which there may be some difficulty in locating local optima in the neighborhood. However, EHO can find more neighborhoods than BP technique.

In the present study, EHO is hybridized with LM-BP to train ANN as shown in Fig. 3. Initially, based on the defined number of populations, clans, and iterations, ANN is trained by EHO using Eqs. 1–4 appropriately. Once the EHO loop is completed, the obtained weights and biases are recorded and carried forward to LM-BP as default weights and biases. Thus, ANN is again trained using LM-BP algorithm

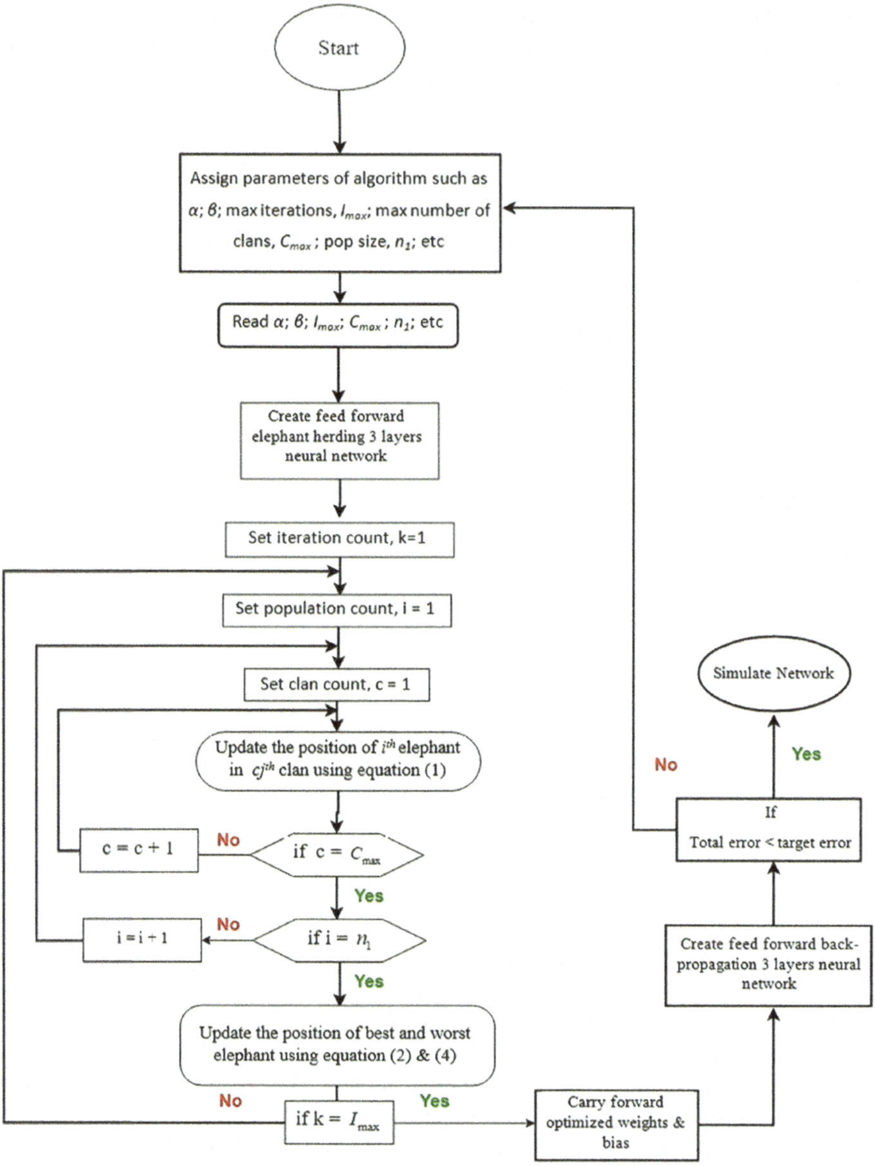

Fig. 3 Flowchart of ANN trained by a hybrid of EHO and LM-BP

with predefined weights and biases. The entire process is repeated if the results do not meet the stopping criteria by changing the number of clans or different population size or by altering the number of generations.

6 Results and Discussions

In the present study, the ANN model is trained by a hybrid of both LM-BP and EHO algorithm and compared with the ANN model trained by standalone LM-BP technique. The effectiveness of each approach is evaluated based on statistical parameters such as CC, MSE, and MAE, which are defined as:

$$CC = \frac{\sum_{i=1}^{n}(E_m - \overline{E_{mm}})(E_p - \overline{E_{pm}})}{\sqrt{\sum_{i=1}^{n}(E_m - \overline{E_{mm}})^2} * \sqrt{\sum_{i=1}^{n}(E_p - \overline{E_{pm}})^2}}. \tag{5}$$

$$MSE = \sqrt{\frac{1}{n}\sum_{i=1}^{n}(E_m - E_p)^2}. \tag{6}$$

$$MAE = \frac{\sum_{i=1}^{n}\left|\frac{E_m - E_p}{E_m}\right|}{n} * 100. \tag{7}$$

Here, E_m and E_p represents the measured and predicted modulus of elastic of concrete respectively, $\overline{E_{mm}}$ and $\overline{E_{pm}}$ are the mean value of measured and predicted observations, n is the number of observations.

The entire collected dataset is randomly divided into 80% for training and 20% for testing. All the developed models were run using HP Pavillion Intel® $Core^{TM}$ i5 CPU @ 2.30 GHz computer with 8GB RAM and MATlab® 2016b software.

6.1 *Performance of Artificial Neural Network*

Based on the study conducted by [23], ANN trained by LM-BP has proven to be a superior ANN training algorithm compared to 8 other different BP techniques. After a number of trials, they found an ANN with ten hidden nodes to be the best performing model for this dataset. In the present study, initially, the same LM-BP algorithm is utilized to train ANN with varying hidden layers to determine the performance of the model. Here each model is run for a minimum of 25 times since ANN assumes randomized weights on every occasion. The results obtained for each model regarding statistical measures are shown in Table 1.

From the obtained results, it is observed that the results from the previous study ([36]) and the present study are almost similar to the single hidden layer. The slight variation in training CC maybe due to different randomization of dataset compared to the previous study. The same training and testing data which is initially randomized for one hidden layer were used as base data for the rest of the ANN models.

As the number of hidden layers is increased, not much improvement in performance is observed. After running various trials, there was always mixed result for

Table 1 Statistical results of ANN trained by LM-BP

Network	Sample	CC	MSE	MAE
7-10-1 [24]	Train (80%)	0.9660	1.5156	–
	Test (20%)	0.9156	2.3463	–
7-10-1	Train (80%)	0.9414	1.9562	5.9921
	Test (20%)	0.9139	2.4903	7.2137
7-7-5-1	Train (80%)	0.9334	2.1217	6.5618
	Test (20%)	0.9116	2.3467	7.0960
7-20-7-1	Train (80%)	0.9794	1.1815	3.4690
	Test (20%)	0.8919	2.7560	8.4839

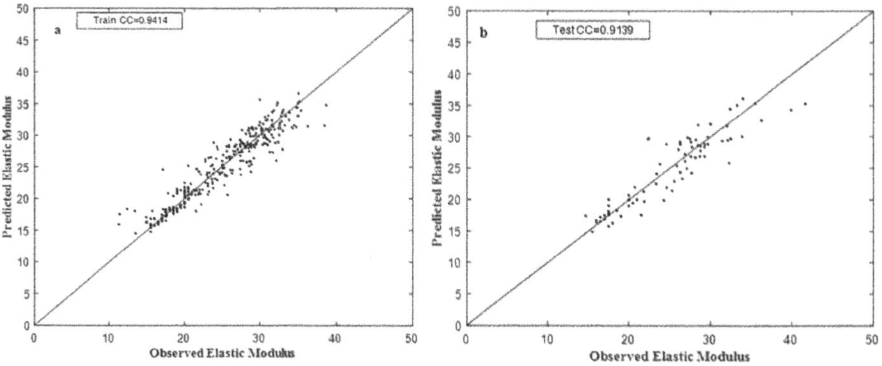

Fig. 4 Experimental and numerical elastic modulus for ANN (LM-BP) (7-10-1) **a** CCtrain and **b** CCtest

an increase in hidden layers, which is caused due to randomized weights used by ANN. Based on the tabulated result, there is a slight increase in training CC, but the testing CC has reduced further down. The decrease in CC may be due to overfitting which occurs as we make a neural network with higher hidden nodes. Figures 4 and 5 showed the CC between experimental and numerical elastic modulus for single and double hidden layers respectively.

Further, it was also observed that the performance of the model fluctuated over a wide range on each trial and also gave a poor performance at times, due to it getting stuck in local optima. This would be particularly worrisome and very risky to be adopted practically due to the sensitivity of elastic modulus in structural concrete. To bring more improvement to the results, EHO is introduced. It has already been proven to be better than BP and other algorithms to train ANN [34]. Further, it is observed that hybrid soft computing techniques have outperformed standalone models due to its complementary nature [30, 31, 40, 41]. In the present study, a LM-BP-EHO hybrid model is developed to overcome the drawbacks of each other and further improvise the results.

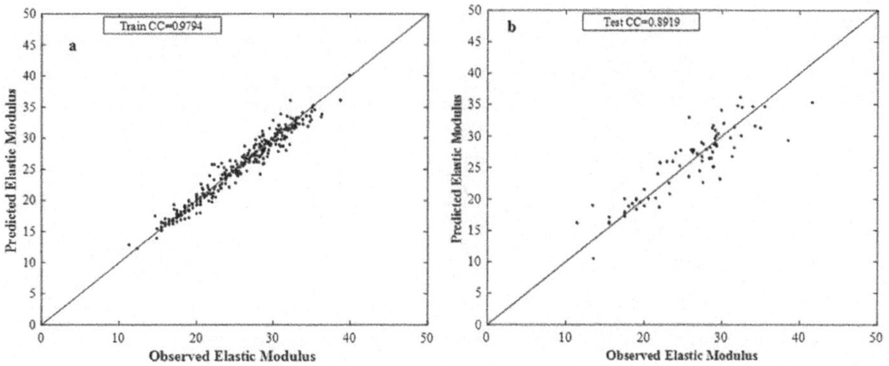

Fig. 5 Experimental and numerical elastic modulus for ANN (LM-BP) (7-20-7-1) **a** CCtrain and **b** CCtest

6.2 Performance of Artificial Neural Network-Elephant Herding Optimization

The weights are treated as state vectors and biases as a cost function when EHO trains ANN. EHO produces ANN weights and biases and adjusts them until there are no more generations with fixed values. As it runs out of generations, the obtained weights and biases are then passed on as default values for LM-BP to train ANN. In order to compare with the standalone LM-BP, the same number of hidden nodes has been used. Based on observations, three parameters namely the number of generations, total population, and numbers of clans play a vital role in increasing the accuracy of the results. Here generation is the number of steps required for optimization. It is observed that more the number of generations, longer the time required for optimization to complete. It is also observed that decreasing the number of generations lowered the performance and also there is not much improvement in results for generations over 30 for this particular dataset. Population size is the total number of elephants and changing this directly affects the solution time. A total of 100 populations yield the best result for the same dataset. Increasing the population further reduced performance. Also, the number of clans should always be the multiple of the population, an equal number of elephants is assumed in each clan, and the best result is obtained for ten clans for this particular dataset. Table 2 shows the values of CC, MSE, and MAE for ANN trained by a hybrid of EHO and LM-BP.

From Tables 1 and 2, it is observed that the performance of single hidden layer ANN trained by a hybrid of EHO and LM-BP has significantly improved with a test CC of 0.9652 compared to ANN trained by LM-BP alone, whose test CC was 0.9139. Further, as the hidden layers are increased there is a considerable increase in performance. Both hybrid models with a test CC of 0.9719 and 0.9975 have entirely outperformed the standalone model of BP whose CC are 0.9116 and 0.8919. Also based on running various trials for a hybrid model of EHO and LM-BP, it is observed

Table 2 Statistical results of ANN trained by a hybrid of EHO and LM-BP

Network	Sample	CC	MSE	MAE
7-10-1	Train (80%)	0.9692	1.4286	4.3858
	Test (20%)	0.9652	1.5900	4.5131
7-7-5-1	Train (80%)	0.9753	1.3038	4.0211
	Test (20%)	0.9719	1.3384	3.8966
7-20-7-1	Train (80%)	0.9966	0.4765	1.1080
	Test (20%)	0.9935	0.4807	1.1888

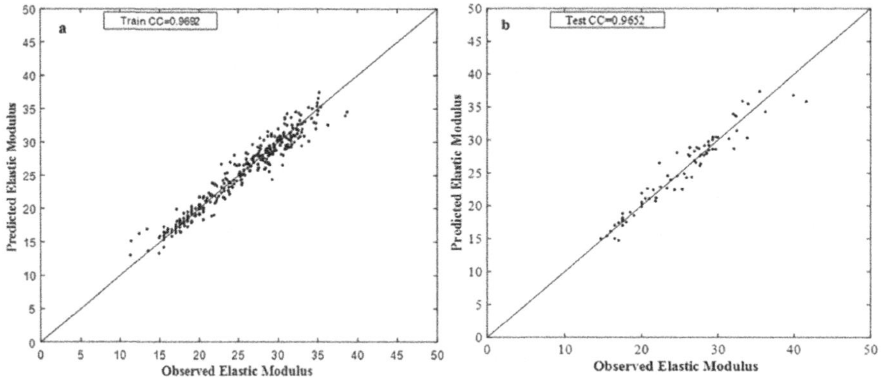

Fig. 6 Experimental and numerical elastic modulus for ANN (LM-BP) (7-20-7-1) **a** CCtrain and **b** CCtest

that the performance is consistent on each trial, unlike ANN trained by the standalone technique of LM-BP. This is because ANN utilizes already good sets of weights and biases obtained by EHO to further train by LM-BP. As a result, the hybrid technique makes use of EHO's strong global searching potential as well as LM-strong BP's local searching potential. Figure 6 and Fig. 7 show CCs between experimental and numerical elastic modulus for single and double hidden layer respectively.

7 Conclusions

Based on the results obtained for the present investigations and discussions, the following conclusions are drawn:

LM-BP initially trains ANN for predicting elastic modulus of RAC since it has been proved to be the best BP technique based on literature. The best results are obtained for one hidden layer with ten hidden nodes for the given dataset. Moving forward, as the hidden layers are increased to two, there are mixed results. Test CC reduces with increase in hidden nodes to (20 7), while the train CC has slightly

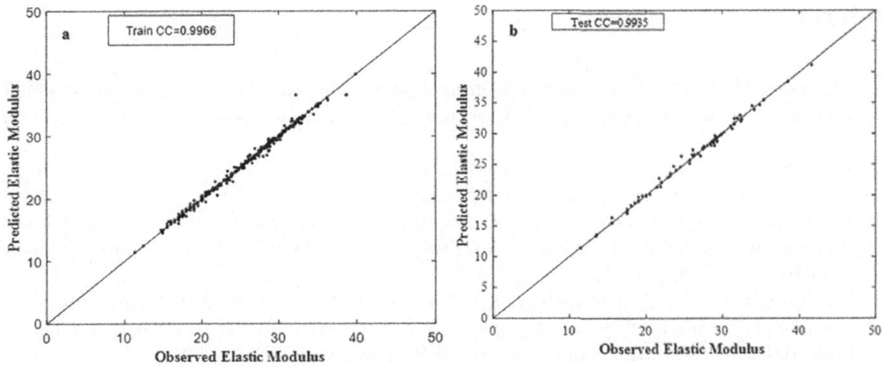

Fig. 7 Experimental and numerical elastic modulus for ANN (LM-BP) (7-20-7-1) **a** CCtrain and **b** CCtest

increased value. Thus, there are not many impactful changes with varying hidden layers for LM-BP technique. Further, performance fluctuates over a wide range across various trials and gives a poor performance at times owing to getting stuck in local minima. It would be very risky to adopt for practical purposes due to the sensitivity of elastic modulus in structural concrete.

Later, swarm inventive technique called EHO has been hybridized with LM-BP to train ANN. The aim here is to take advantage of complimentary features of either technique to train ANN and improvise the results. It is observed that the efficiency of the ANN model trained by the hybrid of EHO and LM-BP depends on population, number of clans and number of iterations. These values can keep varying depending on the dataset used. The obtained results for the hybrid technique completely outperformed the standalone BP technique. The tremendous local searching potential of BP and the strong global seeking potential of EHO appear to be substantially influencing this. Further, the performance is observed to be consistent over various trials. Thus, the hybrid model can be utilized as an alternative approach to obtain a reliable solution in predicting the elastic modulus of RAC.

Acknowledgements The authors are grateful to the Director, National Institute of Technology Goa, India; Director, National Institute of Technology Silchar, India for the support and encouragement provided by them and for permission to publish.

References

1. Sarıdemir, M.: Effect of silica fume and ground pumice on compressive strength and modulus of elasticity of high strength concrete. Constr. Build. Mater. **9**, 484–489 (2013). https://doi.org/10.1016/j.conbuildmat.2013.08.091
2. Mesbah, H.A., Lachemi, M., Aitkin, P.C.: Determination of elastic properties of high-performance concrete at early ages. ACI. Mater. J. **99**(1), 37–41 (2002)
3. Kocab, D., Barbara, K., Petr, M., Petr, Z., Monika, K.: Development of the elastic modulus of concrete under different curing conditions. Procedia Eng. **195**, 96–101 (2017). https://doi.org/10.1016/j.proeng.2017.04.529
4. Pedro, D., De Brito, J., Evangelista, L.: Mechanical characterization of high-performance concrete prepared with recycled aggregates and silica fume from precast industry. J. Cleaner. Prod. **164**, 939–949 (2017). https://doi.org/10.1016/j.jclepro.2017.06.249
5. ASTM C469-94: Test for Static Modulus of Elasticity and Poisson's Ratio of Concrete in Compression. ASTM, USA (2000)
6. IS 456: Indian standard plain and reinforced concrete - code of practice. Bureau of Indian Standards, New Delhi (2000)
7. ACI Committee 318: Building Code Requirements for Reinforced Concrete (ACI 318 M-95). American Concrete Institute (1995)
8. CSA Committee A23.3: Design of Concrete Structures: Structures (Design) - A National Standard of Canada. Canadian Standards Association, Rexdale, Canada (2014)
9. TS500: Betonarme Yapıların Tasarım ve Yapım Kuralları. Türk Standartları Enstitüsü, Ankara (2000)
10. McNeil, K., Kang, T.H.K.: Recycled concrete aggregates: a review. Int. J. Concr. Struct. Mater. **7**(1), 61–69 (2013)
11. Ajdukiewicz, A., Kliszczewicz, A.: Influence of recycled aggregates on mechanical properties of HS/HPC. Cem. Concr. Compos. **24**(2), 269–279 (2002). https://doi.org/10.1016/S0958-9465(01)00012-9
12. Rahal, K.: Mechanical properties of concrete with recycled coarse aggregate. Build. Envir. **42**(1), 407–415 (2007). https://doi.org/10.1016/j.buildenv.2005.07.033
13. Ebrahim Abu El-Maaty Behiry, A.: Utilization of cement treated recycled concrete aggregates as base or subbase layer in Egypt. Ain. Shams. Eng. J. **4**(4), 661–673 (2013)
14. Dhir, R., Limbachiya, M.C., Leelawat, T.: Suitability of recycled concrete aggregate for use in BS 5328 designated mixes. Proc. Inst. Civ. Eng.: Str. Build. **134**(3), 257–274 (1999)
15. Meinhold, U., Mellmann, G., Maultzsch, M.: Performance of high-grade concrete with full substitution of aggregates by recycled concrete. In: 3rd Canmet/ACI International Symposium: Sustainable Development of Cement and Concrete, p. 85 (2001)
16. Sri Ravindrarajah, R., Tam, C.T.: Properties of concrete made with crushed concrete as coarse aggregate. Mag. Concr. Res. **37**(130), 29–38 (1985)
17. Corinaldesi, V.: Mechanical and elastic behaviour of concretes made of recycled-concrete coarse aggregates. Constr. Build. Mater. **24**(9), 1616–1620 (2010)
18. Chithra, S., Kumar, S.S., Chinnaraju, K., Ashmita, F.A.: A comparative study on the compressive strength prediction models for high performance concrete containing nano silica and copper slag using regression analysis and artificial neural networks. Constr. Build. Mater. **114**, 528–535 (2016)
19. Naderpour, H., Mirrashid, M.: Application of soft computing to reinforced concrete beams strengthened with fibre reinforced polymers: a state-of-the-art review. Comput. Tech. Civ. Struct. Eng. **38**, 305–323 (2015)
20. Sadati, S., da Silva, L.E.B., Wunsch, I.I., Donald, C., Khayat, K.H.: Artificial intelligence to investigate modulus of elasticity of recycled aggregate concrete. ACI. Mater. J. **116**(1), 51–62 (2019)
21. Naderpour, H., Mirrashid, M.: Shear failure capacity prediction of concrete beam-column joints in terms of ANFIS and GMDH. Pract. Period. Struct. Des. Constr. **24**(2), 04019006 (2019)

22. Hemeida, A.M., Hassan, S.A., Mohamed, A.A.A., Alkhalaf, S., Mahmoud, M.M., Senjyu, T., El-Din, A.B., Alsayyari, A.: Nature-inspired algorithms for feed-forward neural network classifiers: a survey of one decade of research. Ain. Shams. Eng. J. **11**(3), 659–675 (2020)
23. Golafshani, E.M., Behnood, A.: Automatic regression methods for formulation of elastic modulus of recycled aggregate concrete. Appl. Soft. Comput. J. **64**, 377–400 (2018)
24. Behnood, A., Olek, J., Glinicki, M.A.: Predicting modulus elasticity of recycled aggregate concrete using M5‘ model tree algorithm. Constr. Build. Mater. **94**, 137–147 (2015)
25. Chopra, P., Sharma, R.K., Kumar, M.: Prediction of compressive strength of concrete using artificial neural network and genetic programming. Adv. Mater. Sci. Eng. 1–10 (2016)
26. Duan, Z.H., Kou, S.C., Poon, C.S.: Using artificial neural networks for predicting the elastic modulus of recycled aggregate concrete. Constr. Build. Mater. **44**, 524–532 (2013)
27. Moretti, J.F., Minussi, C.R., Akasaki, J.L., Fioriti, C.F., Pinheiro Melges, J.L., Mitsuuchi Tashima, M.: Prediction of modulus of elasticity and compressive strength of concrete specimens by means of artificial neural networks. Acta. Scientiarum. Tech. **38**(1), 65–70 (2016)
28. Ahmadi-Nedushan, B.: Prediction of elastic modulus of normal and high strength concrete using ANFIS and optimal nonlinear regression models. Constr. Build. Mater. **36**, 665–673 (2012)
29. Pathak, S.S., Sharma, S., Sood, H., Khitoliya, R.K.: Prediction of compressive strength of self-compacting concrete with flyash and rice husk ash using adaptive neuro-fuzzy inference system. Int. J. Adv. Comput. Sci. Appl. **3**(10), 119–122 (2012)
30. Mavrovouniotis, M., Yang, S.: Training neural networks with ant colony optimization algorithms for pattern classification. Soft. Comput. **19**(6), 1511–1522 (2015)
31. Rinchon, J.P.M., Concha, N.C., Calilung, M.G.V.: Reinforced concrete ultimate bond strength model using hybrid neural network-genetic algorithm. In: 9th International Conference on Humanoid, Nanotechnology, Information Technology, Communication and Control, Environment and Management, pp. 1–6. Manila (2017)
32. Wang, G.G., Deb, S., Coelho, L.D.S.: Elephant herding optimization. In: 3rd International Symposium on Computational and Business Intelligence, pp. 1–5. Bali (2015)
33. Tuba, E., Stanimirovic, Z.: Elephant herding optimization algorithm for support vector machine parameters tuning. In: 9th International Conference on Electronics, Computers and Artificial Intelligence, pp. 1–4. Romania (2017)
34. Sahlol, A.T., Ismail, F.H., Abdeldaim, A., Hassanien, A.E.: Elephant herd optimization with neural networks: a case study on acute lymphoblastic leukemia diagnosis. In: 12th International Conference on Computer Engineering and Systems, pp. 657–662. Cairo (2017)
35. Meena, N.K., Parashar, S., Swarnkar, A., Gupta, N., Niazi, K.R.: Improved elephant herding optimization for multiobjective DER accommodation in distribution systems. IEEE Trans. Ind. Inf. **14**(3), 1029–1039 (2018)
36. Golafshani, E.M., Behnood, A.: Application of soft computing methods for predicting the elastic modulus of recycled aggregate concrete. J. Cleaner. Prod. **176**, 1163–1176 (2018)
37. Levenberg, K.: A method for the solution of certain non-linear problems in least squares. Q. Appl. Math. **2**(2), 164–168 (1944)
38. Bal, L., Buyle-Bodin, F.: Artificial neural network for predicting drying shrinkage of concrete. Constr. Build. Mater. **38**, 248–254 (2013)
39. Mandal, S., Rao, S., Harish, N.: Lokesha,: Damage level prediction of non-reshaped berm breakwater using ANN, SVM, and ANFIS models. Int. J. Naval. Arch. Ocean. Eng. **4**(2), 112–122 (2012)
40. Zhang, J.R., Zhang, J., Lok, T.M., Lyu, M.R.: A hybrid particle swarm optimization-back-propagation algorithm for feedforward neural network training. Appl. Math. Comput. **185**(2), 1026–1037 (2007)
41. Ethaib, S., Omar, R., Mazlina, M.K.S., Radiah, A.B.D., Syafiie, S.: Development of a hybrid PSO-ANN model for estimating glucose and xylose yields for microwave-assisted pretreatment and the enzymatic hydrolysis of lignocellulosic biomass. Neural. Comput. Appl. **30**(4), 1111–1121 (2018)

Grey Wolf Optimizer, Whale Optimization Algorithm, and Moth Flame Optimization for Optimizing Photonics Crystals

Seyed Mohammad Mirjalili, Seyedeh Zahra Mirjalili, Nima Khodadadi, Vaclav Snasel, and Seyedali Mirjalili

Abstract In this chapter, three recent swarm intelligence algorithms are used to solve a challenging optimization problem in the field of photonics, including Grey Wolf Optimizer, Whale Optimization Algorithm, and Moth Flame Optimization Algorithm. The problem is to optimize the radii of several rods in a photonics crystal to minimize light wave loss when there is a bend corner. This problem is first presented and formulated in details. It is discussed that due to the use of complex simulations, analytics equations are ill-defined for this problem thereby justifying the use of black-box optimization algorithms. The above-mentioned algorithms are then employed to estimate the global optimal for this problem by finding the optimal values for its structural parameters. The results show that the GWO algorithm provides the best results. The chapter also considers a convergence analysis of all algorithms that led to interesting insights about the process of solution improved during the course of optimization. It is observed that GWO shows constant improvement while others tend to show steady and slow improvement.

Keywords Optimization · Grey Wolf optimizer · Whale optimization algorithm · Moth flame optimization algorithm · Photonic crystal

---

S. M. Mirjalili
Department of Engineering Physics, Polytechnique Montréal, Montreal, QC H3C 3A7, Canada

S. Z. Mirjalili · S. Mirjalili (✉)
Torrens University Australia, Brisbane, QLD, Australia
e-mail: ali.mirjalili@gmail.com

N. Khodadadi
Department of Civil and Environmental Engineering, Florida International University, Miami, USA

V. Snasel
VSB Technical University of Ostrava, 17. listopadu 2172/15, Ostrava-Poruba 708 00, Czech Republic

S. Mirjalili
Yonsei Frontier Lab, Yonsei University, Seoul, South Korea

University Research and Innovation Center, Obuda University, Budapest 1034, Hungary

A. Biswas et al. (eds.), *Advances in Swarm Intelligence*, Studies in Computational Intelligence 1054, https://doi.org/10.1007/978-3-031-09835-2_9

1 Introduction

Engineering optimization, which is often called design optimization, is the process of optimization in engineering design to achieve a goal. Engineers usually create engineering systems that operate in optimal matter to ensure reaching highest potential and applicability. For instance, they design such systems for maximizing efficiency, maximizing performance, minimization cost, or minimizing waste. Optimization problems can be found everywhere. Despite the diverse range of such problems, at the foundational level, they are quite similar key components, including decision variables, objectives functions, and constraints.

Decision variables are the inputs to the objective functions and constraints. They are independent elements of an optimization problems that impact the value that an objective function or a constraint return during the optimization process. Changing the decision variables within the allowed domain results in generating different solutions to an optimization problem. The set of all possible, unique permutations of the values for the decision variable is called solution set or search space in an optimization problem. It should be noted that in engineering problems, a system typically operates in a real environment. Some examples are temperature, humidity, acidity, etc. Such inputs are not considered as decision variables and can be considered as secondary inputs to the system. Engineers need to use them during the calculation of the objective values and constraints if it is considered necessary depending on the nature of the optimization problem.

An objective function is the second key element of an optimization problem. It takes inputs as independent decision variables and produce dependent, objective values. Since it is a function, it produces values for every input and many-to-one mapping is allowed (but not one-to-many). The set of all objective values creates the objective space in an optimization problems. The goal in an optimization problem is to find the best solution in the search space that provides the best value (minimum or maximum depending on the problem) in the objective space.

The last key component of an optimization problem is a constraint. It serves as a function too to map the decision variables to values that can be later used to decide whether a solution is feasible or not. Therefore, constraints are not directly minimized or maximized as opposed to objective functions. Instead, we use them to filter the solutions in the search space that are not desirable for any reason defined based on the nature of the optimization problem. For example, a constraint may limit feasible solution to those with equal values for two given decision variables. For a solution to be considered optimally good and feasible, it needs to provide the best objective value and does not violate any of the constraints.

The first step in solving any optimization problem is to identify the three components of decision variables, objectives, and constraints. The next step is formulate problem. Without the loss of generality, an optimization problem can be formulated as a minimization problem as follows:

$$\text{Minimize:} f(x) \tag{1}$$

$$\text{Subject to:} g(x) \geq 0 \tag{2}$$

where x is the decision variable, f shows the objective function, g is a constraint.

It should be noted that in this formulate we have one decision variable, one objective function, and one constraints. In real-world problems, we typically have more than one of each. Constraints are also divided into two classes of equality and inequality. To have a unified problem formulation, those constraint can be expressed as inequality. For instance, $g(x) = 0$ can be written as $g(x) \geq 0 \wedge g(x) \leq 0$.

The identification of variables, constraint, and objectives as well as problem formulated define a search space and a search landscape. The search space is a set that includes all the possible solutions for a given optimization problems. In case of using discrete values, the size of this set is finite, while continuous variables lead to an infinite size. This set is often called solution, decision variable set, parameter set. Another set includes all the corresponding objective values for the solutions in the solutions set. The ultimate goal in a minimization problem is the find the solution in the solution set that gives the least value in the objective values set. A combination of these two sets will create what is called search landscape.

Defining an optimization problem allows us to find an optimization algorithm to solve it. One classification of optimization algorithms is based on the type of optimization problems that they can solve. Figure 1 shows a classification.

Figure 1 shows that in terms of variables, optimization algorithms can be divided into three sub-categories of continuous, discrete, and mixed-integer. The nature of

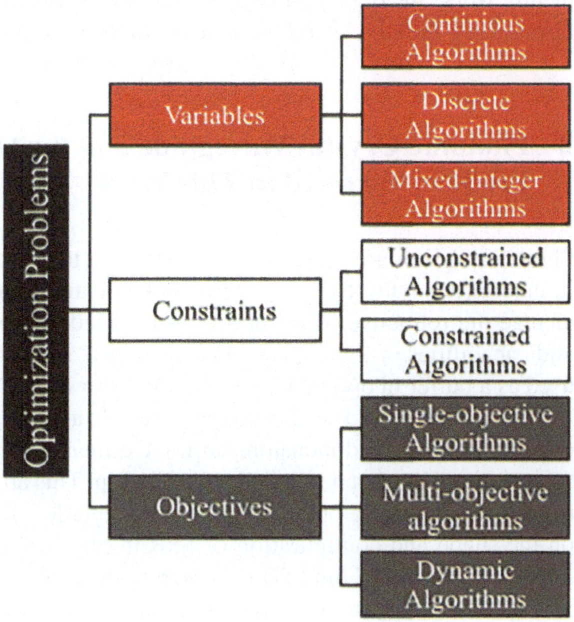

Fig. 1 Classification of optimization algorithms based on the three components of optimization problems

variables will dictate the shape and size of the search space, so different algorithms are required. In regards to the constraints, a problem without a constraint is considered unconstrained. If case of even a single constraint, there should be special mechanisms to handle them (e.g. penalty functions).

Finally, an optimization algorithm may have single or multiple-objectives. In case of a single-objective problem, there is one global optimum that represents the best solution for the problem. In multi-objective problems, however, we deal with several, often conflicting objectives to address. Due to the nature of such problems, an optimization algorithm typically finds a set of solutions that represents the best trade-offs between the objectives.

Solving challenging optimization problem requires addressing several challenges in each of the above elements. The case study chosen in this work is to optimize the shape of a bend photonic crystal waveguide. The properties of this optimization problem are as follows:

- Variables: ten continuous
- Objectives: single, computationally expensive (each simulation takes nearly one min), and calculating the gradient is impossible due to the use of simulator
- Constraint: none

The above characteristics make this problem impossible to solve using conventional, gradient-based algorithms. Therefore, we have chosen a set of recent swarm intelligence algorithms to solve it. The rest of chapter is organized as follows:

Section 2 provides the details of the optimization problem solved in this chapter. The three swarm intelligence algorithms (GWO, WOA, and MFO) algorithms used in this work are briefly presented in Sect. 3. Section 4 provides the results and discussions. Finally, Sect. 5 concludes the work and suggests future works.

2 Photonic Crystal Waveguide and the Optimization Problem Targeted in This Work

Photonic crystals are popular nanostructure in the field of photonics. They are found in nature (e.g. butterfly wigs or peacock's feature) and laboratories. This allows to change the refractive index profile of the medium to achieve different behaviours and capabilities of light waves. For insurance, a photonic crystal waveguide can be used as a buffer in optical CPUs. Figure 2 shows a 3D model of a photonic crystal:

As can be seen in Fig. 2, Photonic crystal can be a periodic dielectric structure in one dimension, two dimensions, or three dimensions. In all these cases, light waves "get trapped" in what it is called photonic gap. This allows us to guide waves through the lattice and create photonic crystal waveguides. During this process, depending on the shape and configuration of rods in photonic crystal waveguide (PCW), we lose some portion of light. This is exactly the problem targeted in this work: to find and optimal shape to minimize light wave loss. The PCW used in this work has a

Fig. 2 Different type of photonic crystal structures

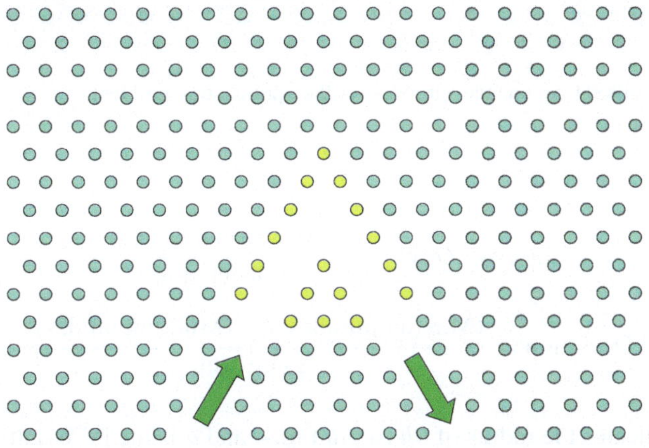

Fig. 3 A 2D PCW with angle bend used in this work

sharp bend as can be seen in Fig. 3 [1, 2]. The sharp bend is evident that leads to light loss and will be minimized in this work.

In Fig. 3, the rods are made of Si which shows a refractive index of 3.6. The radius of all rods are 114 nm, and the lattice constant is 614 nm. To minimize the wave loss, the radius of rods can be changed. We can formulate an optimization problem with all the radii of rods as the parameters. However, this work considers the radius of 17 rods but due to the symmetrical nature of PCW, seven pair of them will have the same radius. Therefore, this problem has 10 variables shown in Fig. 4. It can be seen that the rods form two triangular shapes and with changing their radii, the output of PCW will be different.

With identifying the variables and objectives, this optimization problem can be formulated as follows:

$$\text{Parameters:}\vec{x} = [R_1, R_2, R_3, R_4, R_5, R_6, R_7, R_8, R_9, R_{10}]$$

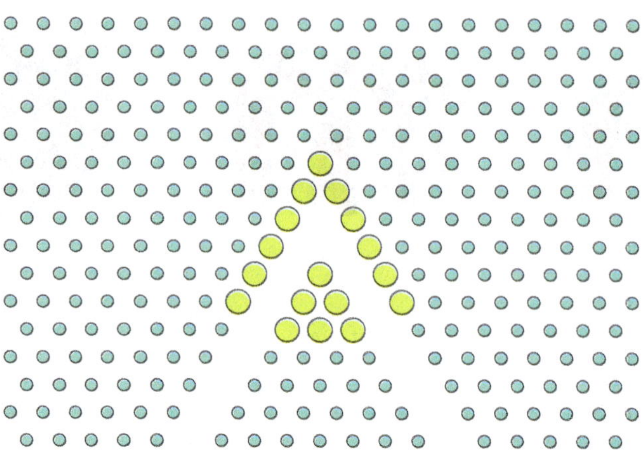

Fig. 4 The rods used to define the variables for the optimization problem

where: $0 \leq R_x \leq 0.5a$,

$$\text{Objective:} O(\vec{x}) = -\frac{\text{Average of amplitudes}}{\text{Deviation}}$$

$$\text{Deviation} = \frac{(\text{Max(amplitudes)}) - (\text{Min(amplitudes)})}{\text{Bandwidth}}$$

where R_i indicates the radius of *ith* rod in Fig. 4 and a is the PCW lattice constant.

The problem formulation shows that the parameters are 10 and stored in the vector x. This vector is then an input to the objective function, which is calculated by running a simulator. Each simulation will take around one minute to complete. In the next sections, this problem will be optimized using several swarm intelligence algorithms.

3 Swarm Intelligence Algorithm

Swarm Intelligence methods mimic collective intelligence of creatures in nature [1]. The evolution has led to amazing collective behaviour for survival of species that are not dependent on individual for reasons such as being smarter, faster, or better able to camouflage. One of the well-regarded swarm intelligence methods is Particle Swarm Optimization (PSO) [2]. This algorithm mimic the foraging behaviour of birds in a flock or fish in a school. To perform the optimization process (also known as search), each particle is updated considering its current speed, the distance to the best solution obtained so far personally, and the best solution obtained so far globally in the swarm. This allows search the most promising regions of the search space.

Grey Wolf Optimizer (GWO) [3] is another recent but very popular swarm intelligence method (see Fig. 5). This algorithm mimic two key behaviours in grey wolves:

Fig. 5 Main mechanisms of search implemented in (left) GWO (middle) WOA (right) MFO

the hierarchy level of dominance in a pack (alpha, beta, delta, and omega wolves) and the hunting mechanism of a pack (chasing, harassing, attacking, and killing a prey). An in-depth review of this algorithm can be found in [4] for interested readers.

Whale Optimization Algorithm (WOA) was proposed in 2016 [5] (see Fig. 5). This algorithm mainly simulates the interesting bubble-net feeding mechanism of humpback whales in oceans. The spiral position updating of a group of humpback whales is mathematically models to estimate the global optimum for a given optimization algorithm. For more details on this algorithm, interested readers are referred to this great survey paper [6].

Moths are known as being gravitated towards source of lights (see Fig. 5). They maintain a certain angle with a source of light to migrate between from one region to another. When the light source is very far, it leads to flyting on a straight line. It is like walking while keeping the moon on your right shoulder. When the source of light is close, this mechanism list to spiral movement and convergence towards it. This is the main mechanism that the MFO algorithm [7] is equipped with to solve optimization problems. For more details on this algorithm, interested readers are referred to this great survey paper [8].

In the next section, these three algorithms are employed to optimize the shape of bend photonic crystal waveguides.

4 Results

This section employs GWO, WOA, and MFO to find an optimal solution for the bend PCW descripted above. For verification of results, these algorithms are compared with Sine Cosine Algorithm (SCA) [9]. For all algorithms, the number of search agents and maximum iterations are set to 50 and 200 respectively. A sensitivity analysis to these parameters were conducted. It was observed that no significant improvement is achieved when increasing these numbers over the two figures on this problem. The best results obtained by all algorithms are presented in Table 1.

This table shows that optimal values obtained for R1 to R10 and the objective value. The first row of this table shows a typical bend PCW with default radii for

Table 1 Obtained optimal bend PCW structures

Label	R1	R2	R3	R4	R5	R6	R7	R8	R9	R10	Average of amplitude
Typical bend PCW	114	114	114	114	114	114	114	114	114	114	−0.62
Optimized bend PCW using SCA	0	0	0	0	0	118	0	45	32	0	−1
Optimized bend PCW using GWO	113	71	75	153	100	111	23	48	144	2	−1.68
Optimized bend PCW using WOA	205	112	149	135	124	154	170	151	113	216	−1.31
Optimized bend PCW using MFO	54	119	76	124	110	107	122	113	127	126	−1.64

Unit of R_i is in nanometre

the rods. The average of amplitude (objective) is −0.62. The first observation by looking at the results of all algorithms is that the average of amplitude has been improved. The benchmark, SCA, improves it nearly 61%. The solution obtained by this algorithm shows several 0 s for the radii of rods. This means that those rods should be completely removed.

The solution obtained by GWO, WOA, and MFO, however, show no removal of any of the rods. The radii of rods are quite diverse and different from the typical bend PCW. Looking at the last column of Table 1, it can be seen that GWO, WOA, and MFO found solutions that provide average amplitude of −1.68, −1.31, and −1.64. In comparison with the typical bend PCW, these solutions provide nearly 170, 111, and 164% improvements. These results are also substantially better (over 60%) better than those obtained by the SCA algorithm. This shows how swarm intelligence algorithms better suits this problem.

The convergence curves of the algorithms are provided in Fig. 6. This figure shows that the slowest convergence rate is given by SCA. This algorithm provides a very steady convergence and not much significant improvement during the optimization problems. This shows that SCA got trapped in a locally optimal solution. WOA shows better improved compared to SCA. Around the 50th iteration, there is a substantial drop in the WOA's convergence curve, which might be an indication of finding a promising "valley" towards the global optimum. However, this algorithm did not manage to substantially improve the accuracy since then.

Figure 6 shows that both MFO and GWO provide the fastest convergence. One interesting differences between these curves is that the GWO delivers constants improvements up until the last iteration. However, the MFO does the behaviour until nearly 100th iteration followed by very slow but steady decline. This shows

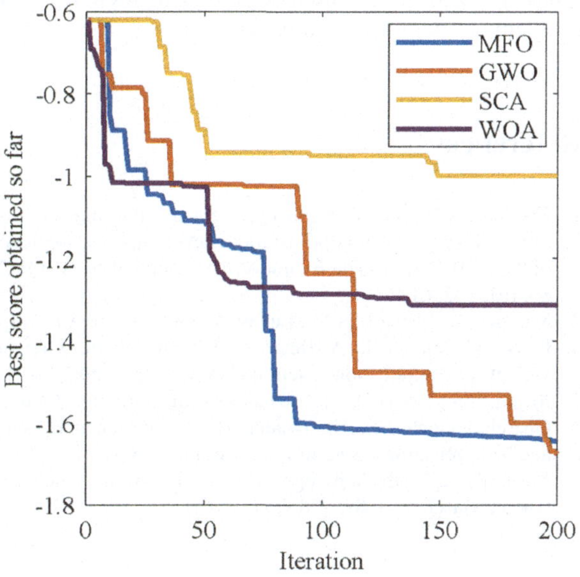

Fig. 6 Convergence curve of the optimization algorithms for the bend PCW designing problem

the superiority of GWO in constantly finding promising regions of the search space by keeping the exploration high. This is probably due to the parameter C in this algorithm that always provides random values and promotes exploratory behaviour regardless of optimization progression.

5 Conclusion

This chapter employed three recent swarm intelligence algorithms to solve a challenging optimization problem in the field of photonics. A problem on the design of a PCW was first formulated by identifying the variables and the objective. It was discussed that the optimization problem is to find optimal values for the radii of several rods to minimize the loss of light in a PCW with angle corners. After the problem formulation, GWO, WOA, and MFO were employed to solve this problem. The resulted were verified in a comparative study with the SCA algorithm.

The results showed that all the swarm intelligence algorithms utilized in this work provide better results than a typical PCW and the one optimized by SCA. The best results were given by the GWO algorithm followed by MFO and WOA. The chapter also considered an convergence analysis of all algorithms that led to interesting insights about the process of solution improved during the course of optimization. It was observed that GWO shows constant improvement while others tend to show steady and slow improvement.

The problem formulated with single objective, but there are other objectives too. For future works, it is recommended to consider and simultaneously optimize all the objectives of this problem. Also applying swarm intelligence algorithms to other types of PCW is recommended due to their high efficiency in such challenging problems as demonstrated in this work.

References

1. Eberhart, R.C., Shi, Y., Kennedy, J.: Swarm Intelligence. Elsevier (2001)
2. Shi, Y.: Particle swarm optimization: developments, applications and resources. In: Proceedings of the 2001 Congress on Evolutionary Computation (IEEE Cat. No. 01TH8546), vol. 1, pp. 81–86. IEEE (2001)
3. Mirjalili, S., Mirjalili, S.M., Lewis, A.: Grey wolf optimizer. Adv. Eng. Softw. **69**, 46–61 (2014)
4. Faris, H., Aljarah, I., Al-Betar, M.A., Mirjalili, S.: Grey wolf optimizer: a review of recent variants and applications. Neural Comput. Appl. **30**(2), 413–435 (2018)
5. Mirjalili, S., Lewis, A.: The whale optimization algorithm. Adv. Eng. Softw. **95**, 51–67 (2016)
6. Gharehchopogh, F.S., Gholizadeh, H.: A comprehensive survey: Whale optimization algorithm and its applications. Swarm Evol. Comput. **48**, 1–24 (2019)
7. Mirjalili, S.: Moth-flame optimization algorithm: A novel nature-inspired heuristic paradigm. Knowl. Based Syst. **89**, 228–249 (2015)

8. Shehab, M., Abualigah, L., Al Hamad, H., Alabool, H., Alshinwan, M., Khasawneh, A.M.: Moth–flame optimization algorithm: variants and applications. Neural Comput. Appl. **32**(14), 9859–9884 (2020)
9. Mirjalili, S.: SCA: a sine cosine algorithm for solving optimization problems. Knowl. Based Syst. **96**, 120–133 (2016)

Intelligent and Reliable Cognitive 5G Networks Using Whale Optimization Techniques

Jasiya Bashir, Javaid Ahmad Sheikh, and Zahid A. Bhat

Abstract In forthcoming networks for high definition radio large bandwidth, low latency and several emergence applications like e-health, Industrial IOT, smart transportation etc. will be conquered by 5G networks. Therefore, more capacity and consequently efficient spectrum sensing will be an awful prerequisite for huge demand for certain applications. In this field of research, attempts have been made to develop new techniques to improve the reliability and channel capacity in 5G Networks using WOA, LDPC and Cognitive Concepts. The results have been presented in the form of various plots and graphs.

Keywords Cognitive Radio (CR) · Fifth Generation (5G) · Low-Density Parity Check (LDPC) · Whale Optimization Algorithm (WOA)

1 Introduction

Communication systems are of utmost importance in a world we are living today and wireless communication is the fastest growing technology in the field of communication. Even with the cellular network's evolution and growth, dearth of spectrum and the scarcity of intelligent and independent capabilities still continue to be a reason for concern [1]. These complications have ensued in capacity thirsty networks, high overhead signaling, data forwarding inefficiency, and low scalability [2]. To overawed these issues there is an awful need of additional capacity and efficient spectrum sensing which will be in ore demand [3].

The striking operational features of fifth-generation (5G) cellular networking can meet such curbs to gratify upcoming situations. The characterization of highly assorted, ultra-dense, and highly mutable environments will shape up the above-mentioned networks. Although 5G give the impression of offering virtually every service, the prominence of cognitive resource management and LDPC are projected to support those extraordinary requirements and influence the arrival of caching,

J. Bashir · J. A. Sheikh (✉) · Z. A. Bhat
P.G. Department of Electronics and I.T, University of Kashmir, Hazratbal, Srinagar, Kashmir, India
e-mail: sheikhjavaid@uok.edu.in

A. Biswas et al. (eds.), *Advances in Swarm Intelligence*, Studies in Computational Intelligence 1054, https://doi.org/10.1007/978-3-031-09835-2_10

mobile edge computing, smart cities and context-aware networking [4]. In order to enable potential wireless applications, ratings of technological challenges must be focused on. Such problems cover all aspects relevant to device design. Multiple modes of operation must be included in these small devices in order to allow various applications and media, provided that wireless terminals provide additional features. Computers allow text, images, voice and data processing in the form of video. It will take advances in the circuit layout for an affordable and lightweight portable gadget that offers the same functionality. This allows the smallest amount of power to be used by these systems, as consumers do not welcome the use of large batteries that need to be recharged very frequently. For that purpose, signal processing for supporting networking and multimedia applications may be power-intensive. For various applications due to their versatility and vigorousness, ad hoc wireless networks lacking infrastructure are extremely desirable. In such networks, all transmission and control must be achieved with the help of nodes in a disseminated fashion, making it impossible to achieve energy efficiency. For sensing applications, the nodes in the network will not recharge their batteries, thereby making energy an essential resource in such networks. Robust applications that gracefully degrade are involved in finite bandwidth and arbitrary wireless channel deviations as network efficiency becomes corrupt. The essence of the wireless channel separates wireless network design from wired network design. Wireless channels are an unpredictable and dynamic means of interacting. The radio spectrum is a meager resource to be allotted to several different systems and applications and is therefore regulated by regulatory bodies regionally and internationally. For a prearranged frequency band, the operating system (regional or global) must work in compliance with the restrictions set by the relevant regulatory body for that band. Spectral licenses are auctioned to the maximum bidder in different nations, which makes bandwidth expensive. The spectrum obtained by means of these auctions can be used to achieve a logical return on their venture in an extremely efficient way. They must be constantly reused in the same geographical area, requiring the cellular system designs to have greater capacity and high-quality performance. Wireless radio modules of acceptable scale, power expenditure and costs are provided at frequencies approximately close to several GHz. Yet the bandwidth in this frequency range is awfully jam-packed. The issue of spectrum scarcity will be minimized by technical developments that are compatible with the same investment and efficiency for higher frequency systems. The path loss is therefore greater at these higher frequencies, restricting the range if directional antennas are not in place. When a signal is transmitted via a wireless channel, as soon as the sender, receiver or contiguous objects move about, haphazard variations are encountered due to altered reflections and attenuation. Consequently, it becomes challenging to build steadfast systems with assured efficiency as the characteristics of the channel appear to change arbitrarily over time. Security implementation in wireless systems is complex because airwaves are at risk of being pried on by anyone with an RF antenna. Protection is virtually negligible in analog cellular systems. By scanning the analog cellular frequency band; one can easily drop the conversations. In comparison, the degree of encryption is related to all digital cellular networks. Nevertheless, most of these encryption methods can be broken with sufficient evidence, time and determination, and many of them have been compromised in practice. Such networks must be secured against such audiences in order to retain multiple applications (electronic commerce and credit card transactions). The major challenge in these scenarios is indeed Wireless networking. A specific user must be intelligent to discover the network between billions of mobile terminals that are distributed globally. At speeds of up to 100 km/h, it must then call the recipient. In relation to changing locations and user demands, the scarce resources of the network must be disseminated equally and effectively. There is also a large infrastructure for physically guided networks: the POTS, the Internet and the optic fiber cables, which can be exploited to link wireless systems to a global network. data rates and reliability of guided systems can always be more as compared to wireless networks for mobile users. The interface between wired networks and wireless and with enormously different presentation competences is a difficult issue. Reviewing the design process itself is probably the most important technological challenge in the design of wireless networks. Wired networks are typically constructed as per layered method, whereby protocols related with dissimilar device operation layers are separated between layers using simple interface mechanisms. Since the link or physical layer layers in wireless systems handles bit transmissions through the medium of communication, the access layer that switches mutual access to the medium of communication, the network and the transport layer that routes data crossways the network and safeguards comprehensive data delivery and connectivity the application layer which dictates the data rate from end to end. Although a layering approach decreases complexity and promotes standardization and modularity because of the lack of global design optimization, it also contributes to performance loss and inefficiency. For many wired network applications. The inherent miracles such as large capacity and good reliability of wired networks make these inefficiencies relatively benevolent, while the good presentation of delay-restricted applications such as video and voice is prevented. Due to the nature of radio broadcasting and transmission in fact, the very idea of a wireless link is somewhat unclear. The complex flora and deprived performance of the fundamental wireless communication channel means that it is important to optimize high-performance networks for this channel and to be stable and adaptable to their variations and network dynamics. Unified and adapt protocols on entire layers are therefore needed in these networks both from the connection layer and the application layer. This cross-layer protocol enterprise includes interdisciplinary communication skills, network theory and design signal processing.

2 Challenges in 5G

1. **Device capacity and data rate**: Mobile networks are projected to accommodate mobile traffic several times over current levels and to have an improved data rate of 10–100 times even in conditions of high mobility. The data rate of 5G is estimated to be about 10 Gbps. With rising demand for high speeds, a network

congestion issue is anticipated. In order to accommodate multiple devices on a cell site, system capability should be increased.

2. **E2E latency**: E2E latencies would be essential to help real-time applications. Remote controlled robots, for example, need a fast feedback service. Enhanced reality and applications for virtual reality require immediate feedback and reaction time to make sense in real time. Applications such as cars and vehicles to communicating networks often demand rapid answers to ensure safety. The latencies in 5G should be around 1 ms in order to accommodate the previously cited applications, while the E2E latency in the existing LTE-A systems is about 10 ms.
3. **Massive number of connected devices**: The invention of IoT has provided many wearable and intelligent home devices with connectivity. As a consequence, there will be 10–100 times more interconnected devices after 2020. Some of these linked devices (for example, sensors) need intermittent connectivity to send data to devices which are constantly connected (e.g., cameras). The challenge is to connect thousands of different devices. The solutions must also be scalable and functional
4. **Cost**: Mobile telephones and their connectivity are an important part of human lives. The cost of infrastructure and their deployment must therefore be reduced. Infrastructure implementation, repair and operating costs must be reduced in order to make connectivity a universal and affordable service. To satisfy the 5G needs tremendous changes that increase the cost of the services provided by 5G. However, consumers would not pay for the resulting rise in operating costs. 5G network architecture should also be built so that it offers consumers sustainable services.

3 Cognitive Radio

Due to a growing number of mobile devices, the rise in traffic has multifold in the recent past. Effective usage of bandwidth is a key to fulfilling users' expectations for high data rates. The spectrum, however, is inefficiently used due to static spectrum allocations and fixed regulations. CR is recommended as a new approach for optimizing the use of spectrum. CR is a technology that allows consumers to use the spectrum effectively by tracking the spectrum on an ongoing basis. As established by the Federal Commission for Communication (FCC), "Cognitive radio: a radio or device in which electromagnetic after sensing the operating environment and has the capability to change its radio operating parameters enthusiastically and unconventionally to alter interference, promote interoperability, and access secondary markets.

The major applications of CR are: dynamic spectrum access and coexistence of heterogeneous networks (interference management).

Dynamic spectrum access: In CR words, the spectrum user is called main user (PU) and the unlicensed user is called secondary user in the absence of a PU (SU). When the spectrum is vacant, an SU may use the PU spectrum. Whenever the PU needs an SU, using the spectrum of a PU, must release the spectrum. One of the difficulties in CR is to reduce the intervention in order not to deteriorate the efficiency of the PU. The SUs use the diversity of various Pus opportunistically to meet the demands on bandwidth. The 'dynamic Spectrum Access (DSA)' is considered an opportunistic use of spectrum. To solve the problem of spectrum scarcity, DSA is very important. Three important components of DSA are: Spectrum awareness, spectrum access and cognitive processing.

- *Spectrum awareness*: Spectrum sensing is responsible for selecting frequency bands which are complex in time, frequency and space on an opportunistic basis. Spectrum consciousness is primarily concerned with finding whether or not a certain frequency band is open. If it is occupied, the transmitter parameters such as position, power etc. are important to know. Spectrum sensitivity can be achieved using geolocation databases or through local spectrum sensing on CR.
- *Spectrum access*: The SUs use the current spectrum in the Spectrum access based on the conducted spectrum knowledge. Since many CR users attempt to access the un extended spectrum, spectrum management strategies are essential to limit interference between CR users in the sense of spectrum access. There are four critical steps in the management of spectrum to effectively minimize interference: the sensing of spectrum, spectrum judgment, spectrum share and spectrum mobility.
- *Cognitive processing*: Cognitive processing involves taking decisions by carrying out such tasks intelligently such as radio environment knowledge, effective sensing of the spectrum and interference with the co-existence of SU and PU.

Interference management: Many heterogeneous networks coexist, can contribute to interference in 5G surroundings [5]. Network levels with various cell sizes, transmission power levels, back-to-house connections are projected to be 5G. In addition to these enhanced technologies, large MIMO technology will be adopted in 5G networks to improve cell ability to concurrently support numerous users [6, 7]. Nevertheless, such MIMO-based networks appear to compete with other groups and interfere interference, which ultimately reduces the efficiency of intended users. In the event of a heterogeneous and multi-level 5G network interference cannot be mitigated by the existing state of the art cancelation strategies such as channel distribution, power management and charges. Effective interference management strategies for 5G networks must also be established, considering that the specifications, waveforms and transmission power used for various networks are different [8, 9]. CR is a new approach to handle interference in 5G networks. Cognitive skills allow the use of a spectrum sensing to detect the presence or lack of other users, to minimize the interference to the other existing users. The CR recommends efficient distribution of resources and management strategies. There is also an attempt to minimize interference by allowing the dynamic frequency selection and optimum power allocation with CR in heterogeneous networks. CR is thus regarded as an efficient way to handle interference with 5G.

4 Low-Density Parity-Check (LDPC) in 5G Channel

the Low-density parity-check (LDPC) first proposed by Gallage [10] in the initial 1960s and revived by MacKay and Neal [28] in 1996. These codes provide near to Shannon error have excellent error improvement capabilities. Further, Very Large-Scale Integration (VLSI) advancements LDPC have gained extensive attention. Amongst all forward error correction (FEC), LDPC codes are amid the most widely used types codes in numerous communications standards like wireless local area network (WLAN, IEEE 802.11n), IEEE 802.22, Advanced Television System Committee (ATSC), digital video broadcast (DVB) and the recently, the fifth generation (5G) communication has been a talk of research and development. More specifically, enhanced Mobile Broad (eMBB) for the 5G data channel band LDPC have been selected as the coding scheme. Such codes have huge impact on role in 5G communication. Moreover, to support scalable data transmission and compatible rate, 3rd Generation Partnership Project (3GPP) has settled to consider two BG1 and BG2 the rate-compatible base graphs for the channel coding [11, 12]. It is estimated that greater parallelism can be more effectively achieved in hardware after LDPC codes is decomposed into a larger number of smaller independent atomic units. compared to turbo code decoders.

As 5G is well-thought-out to include the key characteristics of approximately 10 Gbps data transmission rate, delay less than 1 ms, higher energy and spectral efficiency and higher as compared to 4G [10]. The use of LDPC as the 5G channel coding will provide the following advantages [5].

1. **Parallel decoders and shorter code-word lengths: Design perspective**: Since with the addition of extra bits and need for advanced data rate expectedly up to 10 Gbps [8]. Additionally, in 5G channel coding may decrease transmission rate requirement. To overcome this encounter, M. Khan et al. showed the usage of LDPC codes which ultimately solves issues of both code-word length and throughput. Since data is passed after reception through the channel-decoder therefore for convergence N = 10 iterations are needed. This is true whenever processing units in the channel decoder are operating at frequency f = 400 MHz. For attaining 20 Gbps maximum data rate (N*throughput/f) i.e. 500 parallel processing units are needed [13, 14]. Further, the inherent parallelism of LDPC decoder as described by F. Kahn et al. can assist in intensifying the transmission rate to approximate level. It is also possible that joint detector can be made if detection and demodulation is combined to form a Joint detector and demodulator which can also prove beneficial for shorter-code words.
2. **Channel decoding delay**: It is known that round-trip delay of 15 ms is integral for 4G technologies. It is estimated that in future endeavors for 5G such as Google glass, could have latency of about 1 ms. Moreover, physical layer delay of approximatively 50 microseconds in case of 5G the is decreased to around. Eventually, this value of delay is pooled amid all the mechanisms prevailing in physical layer with few gears like synchronization block having higher delay requirements. So lower channel delay must possess the delay in the decoder by

operating their processing units in co-exist and this an inherent feature of LDPC decoders.

3. **Lower decoding error rate**: The m-MIMO system is considered for utilization in 5G which is basically use of huge number of antennas and vast base-stations. Therefore, in this context is that the cost and data rate should not work complementary to each other. Moreover, the cost and power/bit should reduce by 100 times as the data rate need to be enhanced by 100 times. Moreover, the channel coding adopted in 5G must ensure good energy efficiency and less cost [13]. For non-coherent UWB communication systems Liang et al. exploited low-density parity-check codes such that energy efficiency is improved. This is a decoding process restrained with the number of translated bits per nano-Joule of the energy wasted in translating.

   In the proposed field of research, LDPC codes has been exploited for achieving efficient channel coding in 5G since the channel interpreter must have the lower delay by operating their processing units in parallelism. Further, the inherent parallelism of LDPC decoder as described by F. Kahn et al. can assist in intensifying the transmission rate to the partiality [14]. Also, the demodulation and detection can be combined to form a Joint detector and demodulator and therefore shorter-code words can also work at.

5 Proposed Approach: Integration of CR, LDPC in 5G with Whale Optimization Algorithm

5G, and Cognitive Radio (CR) are the evolving technologies to meet the challenges of heavy mobile data traffic of upcoming wireless networks. The use of more spectrums is essential for more capacity resulting in the integration of Cognitive Radio in 5G networks and a reliable channel modeling technique which forces the use of LDPC. Hence in our field of research we have combined the advantages of 5G networks, Cognitive Radio (CR) technology and LDPC (Low-density parity-check) where 5G provide high data rate and good quality of service. Cognitive radio provides efficient spectrum sharing, flexibility and adaptability to 5G while LDPC provides reliable and efficient channel modeling.

5.1 Whale Optimization Technique

The main challenge faced is the design of wireless networks because the environmental conditions are highly dynamic, due to which parameter optimization is becoming even more difficult and complex. The requirements for wireless networking are often highly dependent on machine learning techniques and artificial intelligence (AI) algorithms since the operational conditions are unknown and the environmental conditions frequently change.

In this segment the proposed method and its inspiration is discussed.

Regarding humpback whales, the most fascinating thing is their form of special hunting. Bubble-net is the name of this foraging behavior. It is a method for feeding. Humpback whales tend to go hunting at school, near to the surface of krill or small fish. It is also noted that this scavenging is achieved by generating characteristic bubbles along a circular path or '9'-shape. This behavior before 2011was only tested on the basis of surface observation. However, this activity was studied using tags by Goldbogen et al. With 300 bubble-net feeding sensors they captured events derived from tags of 9 individual whales of the humpback. It is indeed a unique behavior which can only be experiential in humpback whales.

5.1.1 WOA Algorithm's Fundamentals

In this section, the WOA algorithm's fundamentals are outlined including bubble-net feeding method, encircling prey and a prey search.

- **Encircling Prey**

  The location of prey can be recognized by humpback whales (e.g., krill) and can be enclosed completely. In the WOA, it is expected that the target prey represents the present finest search mediator. Humpback whales are upgraded to the finest search over the iterations of the agent. The following equations need to be formulated using behavior mathematically.

$$\vec{d} = \left|\vec{c} \cdot \vec{x}^*(t) - \vec{x}(t)\right| \tag{1}$$

$$\vec{x}(t+1) = \left|\vec{x}^*(t) - \vec{a} \cdot \vec{d}\right| \tag{2}$$

where $\vec{a}$ and $\vec{c}$ are vector coefficients, latest iteration is given by t, x ∗ (t)2 represents the best search agent position, | · | signifies the absolute value, and · denotes the element-wise multiplication.

The $\vec{a}$ and $\vec{c}$ coefficient vectors are determined as follows:

$$\vec{a} = 2\vec{l} \cdot \vec{r} - \vec{a} \tag{3}$$

$$\vec{c} = 2 \cdot \vec{r} \tag{4}$$

From above we can decrease l from 2 to 0 linearly over the progression of iterations in both exploitation and exploration phases, and r is a random vector over [0, 1].

The control parameter _a can be updated as _a = 2(1 − t/Imax) after we denote by t and Imax the iteration index and maximum number of iterations, respectively,

The control parameter $\overrightarrow{a}$ can be updated as:

$$\vec{a} = 2(1 - t / i_{\max})$$

where t indicates the index iteration and Imax denotes number of iterations maximally.

The main aim of (3) and (4) is to balance exploration and exploitation. The both of the above equations and perimeter r is random. These equations also provide a stochastic behavior for location updating of the population. The selection of random numbers decreases from 2 to 0 in Eq. 3.

For a ≥ 1, exploration is achieved, and for a < 1 the WOA performs exploitation. When the WOA performs exploitation, we need to minimize the likelihood of permanently being stuck in local solutions, the parameter c is basically a random number in [0, 2]. These mains to improving discovery of exploitation at any phases of optimization.

- **Bubble-Net Attacking Method**
  The simultaneous process of the humpback whales shrinking encircling and spiral updating location mechanisms is used for modelling of the bubble-net attacking method. Moreover, by making the coefficient vector $\overrightarrow{a}$ in [−1, 1] the shrinking encircling process is accomplished though linearly dropping the value of a over the classification of iterations. Current location of the agent and the position of the best search agent and the new position will be located in this fashion. Between the position of the prey and the whale and the spiral equation can be followed to imitate the movement of humpback which is helix-shaped whales as follows:

$$\vec{d}' = \left|\vec{x}^*(t) - \vec{x}(t)\right| \tag{5}$$

$$\vec{x}(t+1) = \vec{d}' \cdot e^{bs} \cdot \cos(2\pi s) + \vec{x}^*(t) \tag{6}$$

where the constant is b and that is used to define the logarithmic spiral shape, and s is a random number in [−1, 1].

The spiral approach and shrinking encircling method are simultaneously used at the same time. This is due to the fact that humpback whales swim within a dwindling circle around the prey. For modeling purpose, it is expected that each process is accomplished with 50% probability as:

$$\begin{cases} \vec{x}(t+1) = \left|\vec{x}^*(t) - \vec{x}(t)\right| & \text{if } p < 0.5 \\ \vec{d}' \cdot e^{bs} \cdot \cos(2\pi s) + \vec{x}^*(t) & \text{if } p \geq 0.5 \end{cases} \tag{7}$$

- **Search for Prey**
  Since same shrinking encircling mechanism can be utilized for the prey search. Though, the coefficient vector a with |a| > 1 is utilized and xrand of a whale nominated randomly from the existing population the position is used in place of

$\overrightarrow{x}^*$(t) of the suited search. Alternatively, the humpback whales are being enforced to travel gone from a random whale, this way WOA algorithm has ability to lengthen the search space and accomplish the global search. For the prey search the mathematical model is given as:

$$\vec{d} = \left|\vec{c} \cdot \overrightarrow{\text{xrand}} - \vec{x}(t)\right| \tag{8}$$

$$\vec{x}(t+1) = \left|\overrightarrow{\text{xrand}} - \vec{a} \cdot \vec{d}\right| \tag{9}$$

It should be noted that in any meta-heuristic algorithm, the bubble-net assault process and search for prey are precisely two stages: exploration and exploitation respectively. The Bubble-net Attack approach focuses on local searches by using the best solution available, while the search for prey is designed to maximize the variety of the explanations to realize a global solution. With the increasing number of iterations, it is more desired to use exploration while experimentation at very early iterations is preferred. Adopting this algorithm in our field of research optimizes channel encoding in 5G networks in an enticing potential direction and with the strong balance between exploitation and discovery, the WOA algorithm can be observed as a successful global optimizer.

5.1.2 Algorithm Outline

Flowchart of the algorithm WOA is shown in Fig. 1.

6 Results and Discussion

This chapter emphasis is on the spectrum sharing techniques in cognitive radio network where the unoccupied spectrum holes is shared by various number of secondary users. Moreover this chapter is focussed on two cooperative cognitive radio networks wherein the secondary users collaborate with the primary user to provide the primary's data. Further in our field of research, we have used LDPC Codes to escalate spectrum efficiency. This paper provides a framework for increasing the spectral efficiency, network coverage and of the next generation systems while diminishing end to end delay. Multiple antennas are used at transmitter and receiver so that the data traffic at base stations is trimmed down to facilitate users to communicate directly. This has been accomplished by using the LDPC codes and bio-inspired Whale Optimization algorithm in Matlab software and the corresponding simulation results are discussed.

In our proposed work, we have demonstrated different scenarios wherein the number of Secondary Users (SUs) in attendance vary due to absenteeism of primary user (PU)resulting in the occupancy of different slots in spectrum as per

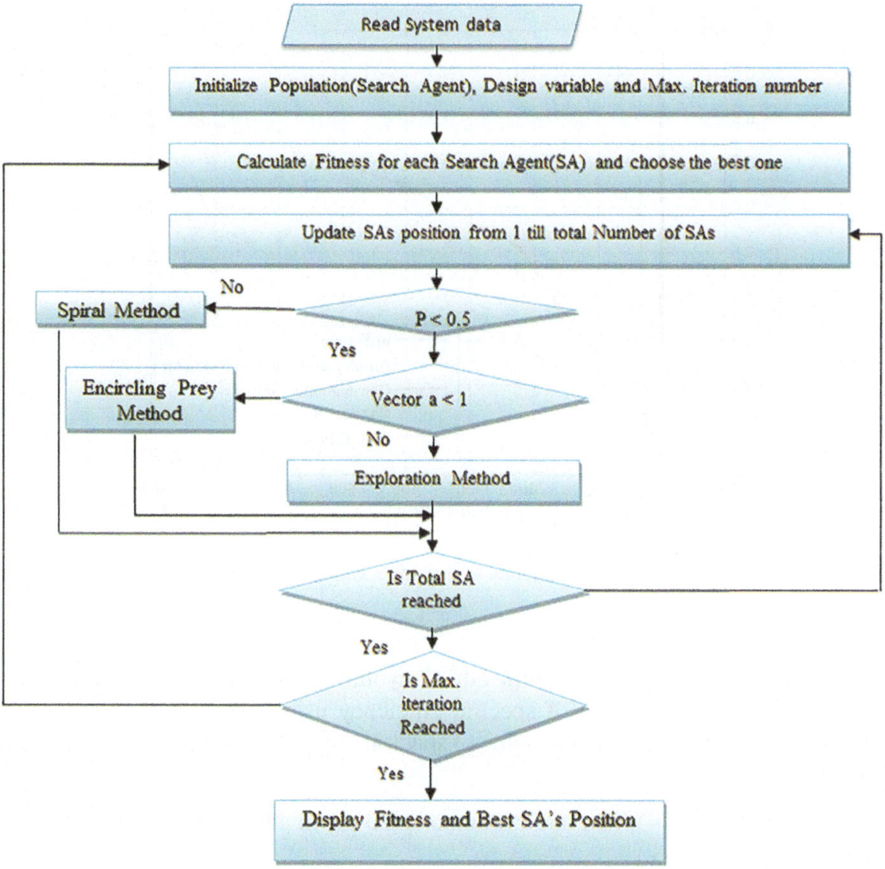

Fig. 1 Flowchart for WOA

their availability. To understand the effect of whale optimization on spectral efficiency we analyzed the eNB (enhanced nodeB), amplify and forward relaying (AF relaying), Nakagami Fading channel and Convolution coding and calculated numerous percentiles of cdf.

The different scenarios are given below:

Scenario1: PAAAA: –When only one PU is in attendance and all other slots are vacant.

In this scenario, we determined our analysis by taking only one PU in attendance while keeping all other slots vacant. The results obtained, Fig. 2, demonstrates that the eNB (enhanced nodeB) provides the least value of spectral efficiency. When amplify and forward relaying (AF relaying) is applied, the spectral efficiency still remains at lower levels. Convolution coding shows slight increase in spectral efficiency. A higher value for spectral efficiency can be obtained by using Nakagami. The Whale algorithm provides the superior value at numerous percentiles of cdf for

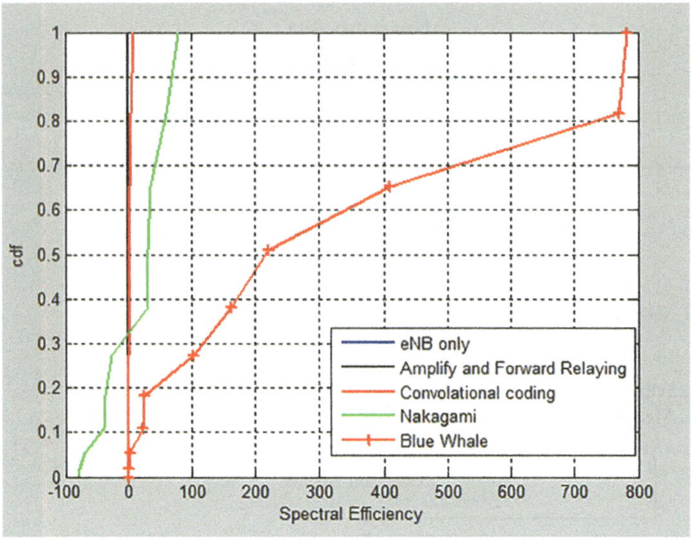

Fig. 2 Plot of cdf versus spectral efficiency when only a single primary user is present and rest of slots are vacant

spectral efficiency. At 50th percentile cdf, Convolution provides a spectral efficiency 30 bits/s/Hz while the value of spectral efficiency using whale Algorithm is 220 bits/s/Hz. The net increase in the value of spectral efficiency at 50th percentile of cdf is 633%.

Scenario 2: PPAAA: –When one PU and one SU are in attendance and remaining 3 slots are vacant.

This scenario encompasses only one PU with all other slots as vacant, the only SU in attendance senses the empty slot and occupies it. As shown in Fig. 3, the eNB (enhanced nodeB), and Nakagami amplify and forward relaying all provides the least value of spectral efficiency. However, the convolution shows slight increase in spectral efficiency. The superior efficiency is shown by whale optimization algorithm at numerous percentiles of cdf for spectral efficiency. At 50th percentile of cdf, convolution coding provides a spectral efficiency 40bits/s/Hz while the value of spectral efficiency using WOA is 1381 bits/s/Hz. The net increase in the value of spectral efficiency at 50th percentile of cdf is 3352%.

Scenario 3: PPPAA: –When one PU and two secondary users (SU) are in attendance and remaining 2 slots are vacant.

Here, the two SU in attendance senses the empty slot and occupies it. As shown in Fig. 4, the eNB (enhanced nodeB), Nakagami and amplify and forward relaying all provides the least value of spectral efficiency. However, the convolution shows slight increase in spectral efficiency. The superior efficiency is shown by whale optimization algorithm at numerous percentiles of cdf for spectral efficiency. At 50th percentile

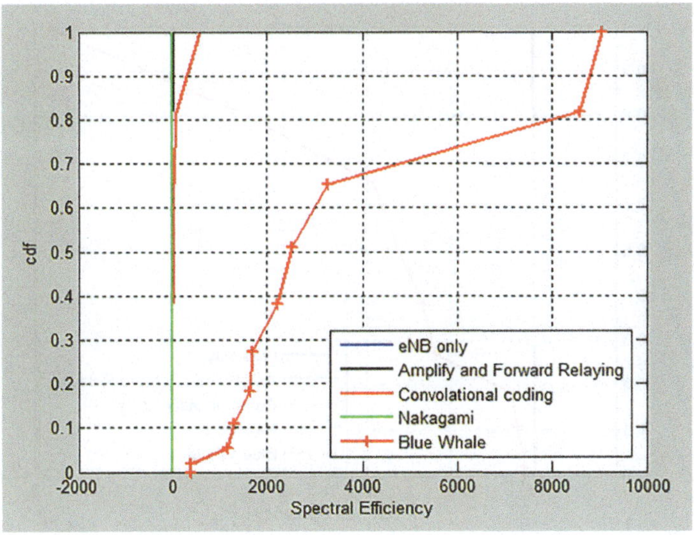

Fig. 3 Plot of cdf versus spectral efficiency when one PU and one SU are present and rest of slots are vacant

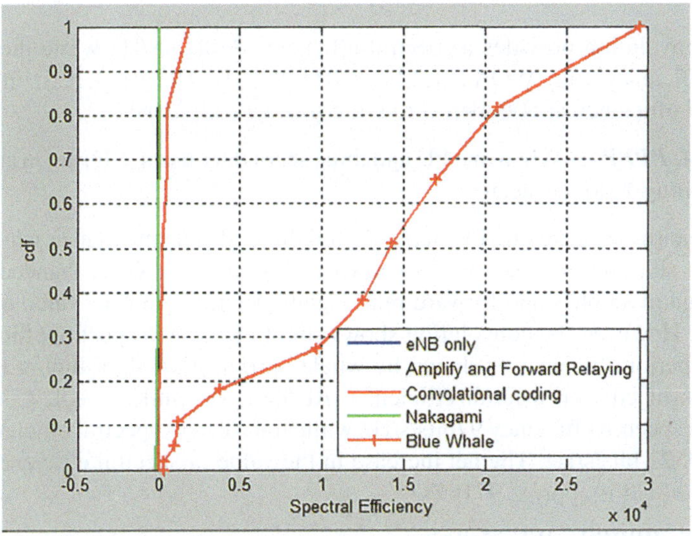

Fig. 4 Plot of cdf versus spectral efficiency when one PU and two SU are present and rest of slots are vacant

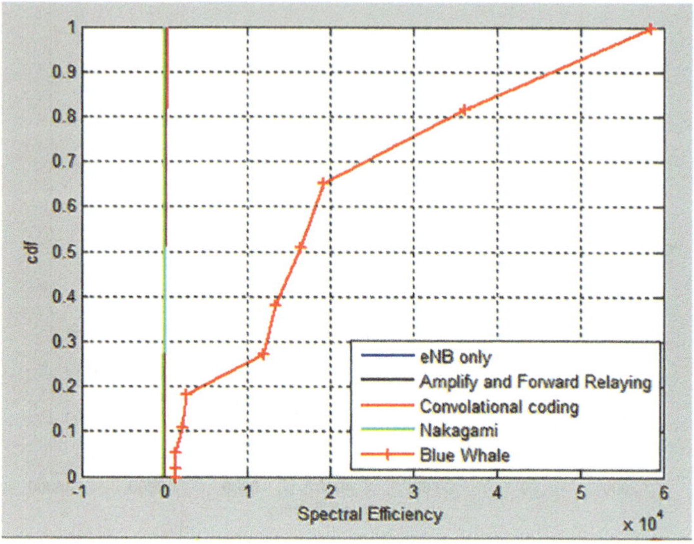

Fig. 5 Plot of cdf versus spectral efficiency when one PU and three SU are present and one slots is vacant

of cdf, Convolution provides a spectral efficiency 230bits/s/Hz while the value of spectral efficiency using WOA is 1.428e + 04 bits/s/Hz. The net increase in the value of spectral efficiency at 50th percentile of cdf is approx. 6108%.

Scenario 4: PPPPA: –When one PU and three secondary users (SU) are in attendance and remaining 1 slot is vacant.

In this scenario we have one PU in place, and three SUs in attendance which senses the empty slot and occupies it. As shown in Fig. 5, the eNB (enhanced nodeB), and Nakagami amplify and forward relaying all provides the least value of spectral efficiency. However, the convolution shows slight increase in spectral efficiency.

The superior efficiency is shown by whale optimization algorithm at numerous percentiles of cdf for spectral efficiency. At 50th percentile of cdf, Convolution provides a spectral efficiency 97bits/s/Hz while the value of spectral efficiency using WOA is 9522 bits/s/Hz. The net increase in the value of spectral efficiency at 50th percentile of cdf is approx. 9716%.

Scenario 5: PPPPP: –When all slots are occupied.

In this scenario we have one PU in place and four SUs in attendance which after sensing the empty slots in spectrum occupy them. Hence all the slots are occupied. As shown in Fig. 6, the eNB (enhanced nodeB) provides the least value of spectral efficiency. When amplify and forward relaying (AF relaying) is applied, the spectral efficiency still remains low. A higher value for spectral efficiency can be obtained by using convolution coding The WOA provides the superior value at numerous percentiles of cdf for spectral efficiency. At 50th percentile of cdf, convolution coding

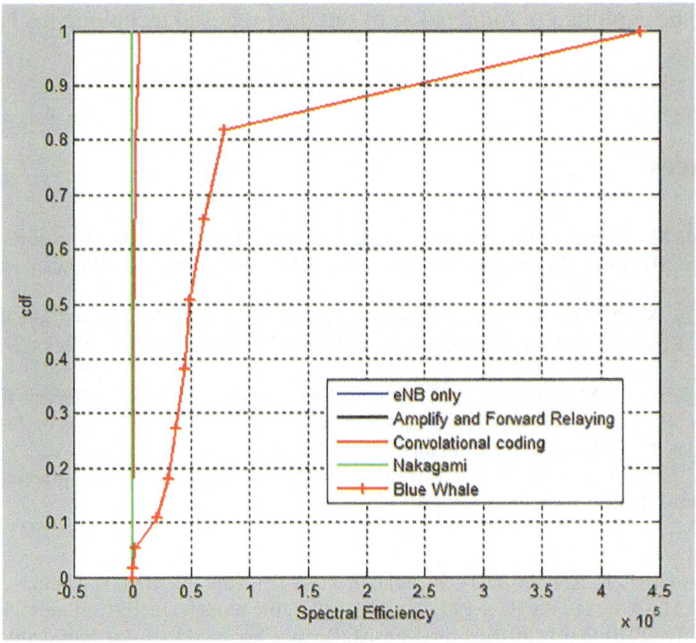

Fig. 6 Plot of cdf versus spectral efficiency when all the slots are occupied

provides a spectral efficiency 1072bits/s/Hz while the value of spectral efficiency using WOA is $4.97e^{+04}$ bits/s/Hz. The net increase in the value of spectral efficiency at 50th percentile of cdf is approx.4536%.

7 Conclusion

In order to overcome the demand and challenges of future 5G networks, an indulgent and amalgamation of Cognitive Radio with LDPC in the 5G the futures wireless networks is proposed in this chapter. The proposed technique has the grave ability to utilize the network resources in an effective way while diminishing end to end delay and can be proved to be intelligent and reliable. Additionally, WAO has been used to distribute resources and to address the issue of optimization to a greater extent. Since the work demonstrated in this chapter integrates the cognitive radio and LDPC with the 5G networks using Whale Optimization Techniques, consequently this will prove to be the ideal candidate for deploying the 5th generation wireless networks with IOT platform. Besides, the projected method will also act as benchmark for further research to extend the work for Internet of Medical Things (IOMT) for better and efficient transmission of medical over 5G and IOT networks. A quantum concept could also be exploited in future so as to suit the proposed technique for beyond 5G networks.

References

1. Haddad, M., Hayar, A.M.: Mobile Communications Group, Institute Eurecom, Merqueen Debbah ,SUPELEC,Plateau de Moulon, 3 rue Joliot-Curie Spectral efficiency of Cognitive Radio systems. IEEE (2007)
2. Zhang, S., Xu, X., Wu, Y., Lu, L.: Huawei Technologies, Co. Ltd., Shanghai, Chin 5G: Towards Energy-Efficient, Low-Latency and High-Reliable Communications Networks. ©2014 IEEE (2014)
3. Cayamcela, M.E.M., Lim, W.: Artificial Intelligence in 5G Technology: A Survey. IEEE (2018)
4. Pirinen, P.: Centre for wireless communications. In: A Brief Overview of 5G Research Activities. University of Oulu, Finland. ICST (2014)
5. Niu, K., Chen, K., Lin, J., Zhang, Q.T.: Polar codes: primary concepts and practical decoding algorithms. IEEE Commun. Mag. (2014)
6. Yao, M., Sohul, M., Marojevic, V., Reed, J.H.: Artificial intelligence-defined 5G radio access networks. IEEE Commun. Mag. (2019)
7. Navarro-Ortiz, J., Romero-Diaz, P., Sendra, S., Ameigeiras, P., Ramos-Munoz, J.J., Lopez-Soler, J.M.: A survey on 5G usage scenarios and traffic models. IEEE Commun. Surv. Tutor.
8. Strutz, T.: 2010–2014, 2016. Low-Density Parity-Check codes—An introduction, TStrutz Tutorial on LDPC codes (2016)
9. Pham, Q.-V., Mirjalili, S., Kumar, N., Alazab, M., Hwang, W.-J.: Whale Optimization Algorithm With Applications to Resource Allocation in Wireless Networks. IEEE (2020)
10. Arora, K., Singh, J., Randhawa, Y.S.: A Survey on Channel Coding Techniques for 5G Wireless Networks. Springer Science Business Media, LLC, Part of Springer Nature (2019)
11. Bae, J.H., Abotabl, A., Lin, H.-P., Song, K.-B., Lee, J.: An overview of channel coding for 5G NR cellular communications. SIP 8(e17), 1 of 14 © The Authors (2019). https://creativecommons.org/licenses/by/4.0/
12. Richardson, T., Kudekar, S.: Design of low-density, parity check codes for 5G new radio. IEEE Commun. Mag. (2018)
13. Andrews, J.G., et al.: What will 5G be? IEEE J. Sel. Areas Commun. **32**(6), 1065–1082 (2014)
14. Shukurillaevich, U.B., Sattorivich, R.O., Amrillojonovich, R.U.: 5G Technology Evolution. IEEE (2019)
15. Mirjalili, S.: A School of Information and Communication Technology. Griffith University, Nathan Campus, Brisbane, QLD 4111, Australia, Andrew Lewis Griffith College, Mt Gravatt, Brisbane, QLD 4122, Australia The Whale Optimization Algorithm 2016 Elsevier (2016)

Machine Learning

Automatic Data Clustering Using Farmland Fertility Metaheuristic Algorithm

Farhad Soleimanian Gharehchopogh and Human Shayanfar

Abstract Data clustering is a data mining task, and it means finding clusters from among data whose labels are not predetermined. It is a popular analytics tool for statistical data in various domains. The k-means algorithm is a basic algorithm for data clustering, which has initial problems such as dependence on the cluster centers' initial value, sensitivity to outliers, and non-guaranteed optimal solutions to unbalanced cluster formation. This book chapter uses Farmland Fertility Algorithm (FFA) for the data clustering algorithm. Ten standard data sets used to evaluate the effectiveness of FFA are compared Harmony Search (HS), Monarch Butterfly Optimization (MBO), Artificial Bee Colony (ABC), Symbiotic Organism Search (SOS), Differential Evolution (DE), and Crow Search Algorithm (CSA) in terms of statistical criteria such as analysis of variance (ANOVA) and the convergence rate. Experimental results demonstrate that FFA has better performance than other optimization algorithms and is more stable than these algorithms.

Keywords Farmland fertility algorithm · Clustering analysis · Metaheuristic algorithm · Optimization · Data clustering

1 Introduction

One of the important actions in data mining is the data clustering of a data set. This technique explores the structures that led to grouping samples available in a data set. So, similar samples are placed in the most similar categories, while they have an acceptable difference from other groups. The data clustering aims to display large data sets in the form of a smaller number of primary samples or primary clusters, which simplifies modeling data. Therefore, data clustering is an attractive and essential task in data mining used in many applications. It refers to groups with a collective score similar to each other based on some criteria. A metaheuristic

---

F. S. Gharehchopogh (✉) · H. Shayanfar
Department of Computer Engineering, Urmia Branch, Islamic Azad University, Urmia, Iran
e-mail: bonab.farhad@gmail.com

A. Biswas et al. (eds.), *Advances in Swarm Intelligence*, Studies in Computational Intelligence 1054, https://doi.org/10.1007/978-3-031-09835-2_11

algorithm is used in data clustering problems. The traditional and basic algorithm for clustering is the popular K-means algorithm, which this algorithm has been widely used in several applications such as data analysis to logistical applications. However, the K-means algorithm can be easily influenced by factors such as the initial choice of geometric centers, and it can be easily trapped in a local optimum. In summary, data clustering is a prominent data mining approach for clustering a collection of objects into multiple clusters. Entities in separate clusters are unique, whereas entities inside a cluster are very similar. In unsupervised learning, data clustering is perhaps the most critical problem. When it comes to data clustering, we divide the data into clusters or groups with the maximum similarity between each cluster's data and the minimum similarity between different clusters' data. The clustering process divides into supervised learning and unsupervised learning parts. From the beginning, categories are distinct in supervised learning, and each of the educational data is attributed to specific data. Indeed, during the learning process, a supervisor provides information (such as the number of clusters, etc.) to the learner in addition to the educational data. However, there is information on educational data available only to the learner in unsupervised learning, and this learner should look for a particular structure in the data. Clustering methods determine similar data and label them implicitly. Among the desirable characteristics considered for a clustering algorithm, we can mention the following [1].

Scalability: the high volume of data is usually the default available in data mining techniques. In massive data, using fewer samples will result in directional results. For this reason, we need high scalability in clustering algorithms. *Ability to deal with data types*: Clustering algorithms and working with numeric data should encounter other data types such as nominal data. *Extraction of clusters to the desired shape*: Some criteria to measure the distance, such as the Euclidean distance, find clusters with a particular shape (spherical). However, the clustering algorithm is expected to distinguish groups with unequal sizes and densities. *The ability to confront noisy, incorrect, and incomplete data*: Although the data preparation and pre-processing stage, detecting this type of data is essential not to enter the next steps. However, this work is not done with 100% certainty, and one of its disadvantages is the excessive sensitivity of the clustering algorithm to this kind of data. *Insensitivity to data entry sequence*: A single algorithm with different lines of identical data produces very different results, which is not acceptable. After executing and obtaining clusters, an algorithm that fails to increase the new data is not good. *Lack of need for input parameters*: Many clustering algorithms require parameters to start their work. The results obtained from these algorithms can depend on these parameters, and this subject is one of their disadvantages. *Acceptance of high-dimensional data*: Finding a cluster in high-dimensional data is a challenge in clustering algorithms. *Comprehensibility of the algorithm results*: Interpreting the output of clustering algorithms for the user with applications' help.

A new metaheuristic algorithm called the FFA was presented in 2018 [3]. The simulation results and various experiments of this algorithm show that this algorithm has a higher performance in solving optimization problems than other metaheuristic algorithms [3]. We use FFA to solve the data clustering problem. It was compared with optimization algorithms Including DE [4], HS [5], ABC [21], and MBO [2] in terms of ANOVA [6] and statistical criteria, and these algorithms will be briefly explained in the next section.

In [3], the authors compared this algorithm with different benchmark functions on several mathematical algorithms. The result shows that the FFA algorithm is due to internal memory (local solutions) and external (global solution) having better results than another algorithm. This algorithm in various problems such as traveling salesman problems [7], Photovoltaic model parameters [8], constrained engineering problems [9], Optimal design of interconnected multimachine power system [10], feature selection in analysis of the text psychology [11], Improved Intrusion Detection Systems [12] has a significant advantage over other basic algorithms.

In the rest of this book chapter, Sect. 2 investigates the types of metaheuristic algorithms in solving the data clustering problem. The theory and formulation of FFA are introduced in Sect. 3. Section 4 presents the proposed FFA-based approach to the data clustering problem. The proposed approach's efficiency and performance are tested in Sect. 5, and the results obtained in the figures and tables are displayed. In the final section, the conclusion and future works are presented.

2 Related Works

The k-means algorithm most commonly used method for data clustering that which the data are divided into k separate clusters. In the first phase, the center point of the cluster is obtained, and in the next phase, each point of the dataset is attributed to a cluster closer to its center. Wang et al. proposed a new hybrid method based on a flower pollination algorithm (FPA) and pollen operator for solving data clustering problems [13]. This paper applies pollen and crossover operators to increase population diversity. Also, the ability to search locally using the elite-based mutation operator is increased. The experimental results show that the proposed method has higher accuracy, higher stability level, and faster convergence speed than another algorithm. In another study, Zhou et al. proposed a social spider optimization algorithm based on a simple method called SMSSO for solving data clustering problems [14]. In this paper, to overcome the problem of optimal local trapping and poor convergence rates, the simplex method is a random variable strategy that increases population variability while increasing the algorithm's local search capability. The results of various experiments show that the proposed algorithm has performed better than the k-means algorithm and other comparison algorithms in terms of convergence speed and stability. Boushaki et al. proposed a quantum chaotic cuckoo search algorithm to solve the clustering problem [15]. Turbulence maps and boundary management strategies are used to improve the cuckoo search algorithm, and heterogeneous quantum updating also enhances the global search capability of the cuckoo search algorithm. The results of various experiments were performed on six data sets. The results of multiple experiments show that the quantum turbulent cuckoo search algorithm performs better in all data sets in terms of internal and external clustering quality. Elaziz et al. proposed a hybridization of atom search optimization and the cosine sine algorithm(SCA) called ASOSCA to solve the clustering problem [16]. ASOSCA uses SCA as a local search operator to improve the quality of ASO algorithm solutions. The primary purpose of the proposed it is to automatically find the optimal number of centers and their positions to minimize the internal distance. The results of various experiments show that the proposed ASOSCA method leads to a high advantage compared to other types of hybrid meta-innovation in terms of clustering criteria. Han et al. proposed a modified algorithm based on the gradient search algorithm(GSA) and Bird Flock called BFGSA to solve the clustering problem [17]. A new mechanism for increasing population diversity in the GSA algorithm is performed through three main steps: initialization, identifying nearest neighbors, and changing direction. The simulation results show that BFGSA can be used effectively for data clustering. In another study, to use the strengths of the nature of the two artificial bee colony (ABC) algorithms and the K-means algorithm, Kumar et al. proposed a hybrid algorithm for these two algorithms called the MABCKM algorithm [18]. In this algorithm, the solutions generated by ABC are modified and considered as initial solutions for the K-means algorithm. Zhou et al. [19] presented an improved version using the symbiotic organism search (SOS) algorithm to solve clustering problems. The SOS algorithm simulates the coexistence strategies of living organisms for survival and reproduction in an ecosystem. In another study, Mageshkumar et al. proposed a hybrid method based on the ant lion algorithm and the ant colony algorithm called ACO-ALO to solve the clustering [20]. The results of various experiments show that the proposed ACO-ALO hybrid method has achieved better results than the other data clustering algorithms. Rahnema and Gharehchopogh have proposed an improved artificial bee colony (ABC) optimization algorithm based on the whale optimization algorithm for data clustering [21]. In this paper, for the problem of late detection and convergence, two memories called random memory, and elite memory in the ABC is used for solving data clustering. Ahmadi et al. [22] have presented an improved version of the Grey Wolf Optimization (GWO) algorithm for solving the data clustering problem. This improvement includes balancing the exploration and exploitation of the GWO and a local search for the best solution found. The results show that the intra-cluster distance of the proposed algorithm is less than other algorithms and has an average error of about 11%, which is the lowest among all comparative algorithms. A modified Sine Cosine Algorithm (SCA) called AMSCAC is proposed for data clustering [23]. Both local and global search methods have increased the relatively slow SCA convergence rate. In another study, a new meta-heuristic algorithm based on water wave optimization (WWO) for data clustering was proposed by Kaur et al. [24]. In recent years, various metaheuristic algorithms have been proposed for data clustering problems, such as Multi-verse Optimizer [25], krill herd [26], Chaotic cuckoo Search [15], a hybrid of GWO and WOA [27], memetic DE [28], Variance-based DE with an optional rossover [29], Critical Distance Methodology [30], etc. And in this book chapter, we used an FFA algorithm to solve a data clustering problem, which results are satisfactory.

3 Farmland Fertility Algorithm

FFA is inspired by farmland fertility in nature and tries to optimize each sector's solutions based on the division of farmland and the desired productivity of both external and internal memory [3]. The FFA algorithm has a better convergence and exploration approach due to internal memory (local solutions) and external memory (global solutions) in optimization problems such as data clustering. In addition, the FFA algorithm uses global solutions that can ensure the convergence of the algorithm. On the other hand, local solutions prevent trapping in the local trap in this algorithm. Therefore, unlike the methods reviewed in the background, this algorithm does not need to be fundamentally changed. It is enough to use it in the problem of data clustering by setting the correct parameter. The results also prove that the solutions of this algorithm to the problem of data clustering are of better quality than other methods. Of course, the positive point is that as the number of iterations of the FFA algorithm increases, the variety and quality of solutions of the FFA algorithm increases. This shows that this algorithm maintains the balance between exploration and efficiency in the problem of data clustering and can be effectively used to solve data clustering and other machine learning problems. But the disadvantage of the FFA algorithm in the data clustering problem is that, like all meta-heuristic algorithms, with increasing the dimensions of the problem, the algorithm may lose its efficiency to some extent, in which case the parameters of the FFA algorithm must be selected more carefully. We introduced the formulation and step by step of FFA algorithm. Generating an initial population is the first step of each metaheuristic algorithm. In this algorithm, Eq. (1) determines the number of the initial population:

$$N = k * n \tag{1}$$

In Eq. (1), k determines the number of sections of land or space, and n specifies the number of solutions available in each area of farmland. A separate section is considered to determine k's value that. The optimal value is specified as k between 2 and 8 [3]. Each area of farmland's quality is obtained based on the average of the solutions available in each area of farmland. Initially, Eq. (2) assigns solutions to every area.

$$\text{Section}_s = x(aj), a = n * (s - 1) : n * s \qquad s = 1, 2, \ldots k, \qquad j = 1, 2, \ldots D \tag{2}$$

In Eq. (2), s represents the number of sections, and j = 1, 2, …, D indicates the dimension of the variable x(aj) that all solutions in the search space. This equation separates each segment's present solutions in a simple way to calculate the average of each section separately. After segmentation, Eq. (3) determines the quality of each section.

$$\text{Fit\_Section}_s = \text{Mean}(\text{all Fit}(x_{ji}) \text{ in Section}_s) \quad s = 1.2\ldots k, \quad i = 1.2\ldots n \tag{3}$$

In Eq. (3), $Fit\_Section_s$ determines the value that determines the quality of the outreach section of farmland's solutions. The number of best memory determines the local and global memory of each section according to Eq. (4) and the number of best local memory to Eq. (5).

$$M\_\text{Global} = \text{round}(t * N) \qquad 0.1 < t < 1 \tag{4}$$

$$M\_\text{Local} = \text{round}(t * n) \qquad 0.1 < t < 1 \tag{5}$$

In Eqs. (4) and (5), $M_$Global and $M_$Local represent the number of solutions in global memory and local memory, respectively. Moreover, at this stage, both local and international memories are updated. All solutions available in the worst section of farmland are hybridized with global memory solutions according to Eqs. (6) and (7).

$$h = \alpha * rand(-1, 1) \tag{6}$$

$$X_{\text{new}} = h * (X_{ij} - X_{M\_\text{Global}}) + X_{ij} \tag{7}$$

where α is a random number between zero and one. X_{M_Global} is a random solution among the solutions available in the global memory, X_{ij} is a solution selected in the worst section of the farmland to must make changes, and h is also a decimal number. The solutions available in the other sections change based on Eqs. (8) and (9).

$$h = \beta * rand(-1, 1) \tag{8}$$

$$X_{\text{new}} = h * (X_{ij} - X_{uj}) + X_{ij} \tag{9}$$

where, X_{uj} is a random path selected between all the solutions in the sections, and β is a number between zero and one. X_{ij} is a solution from the sections selected to apply the changes, and h is also a decimal number calculated by Eq. (9). For updating solutions based on local and global memory, The combination of the considered solution with $Best_{\text{Local}}$ or $\text{Best}_{\text{Global}}$ is determined by Eq. (10).

$$H = \begin{cases} X_{\text{new}} = X_{ij} + \omega_1 * (X_{ij} - \text{Best}_{\text{Local}}(b)) & : \ Q > rand(0, 1) \\ X_{\text{new}} = X_{ij} + rand(0, 1) * (X_{ij} - \text{Best}_{\text{Local}}(b)) & : \ \text{else} \end{cases} \tag{10}$$

In Eq. (10), Q is between zero and one that determines the amount of the combination of the solutions with $\text{Best}_{\text{Global}}$. The ω_1 is parameters related to FFA, which is an integer. At first, the algorithm should be specified, and its amount decreases based on the iteration of the algorithm during the time (Eq. 11).

$$\omega_1 = \omega_1 * R_v \qquad 0 < R_v < 1 \tag{11}$$

lb_1, ub_1	lb_2, ub_2	lb_3, ub_3	lb_4, ub_4	lb_1, ub_1	lb_2, ub_2	lb_3, ub_3	lb_4, ub_4
A_1	A_2	A_3	A_4	A_5	A_6	A_7	A_8
Cluster 1				Cluster 2			

Fig. 1 FFA solution for data clustering

The conditions of termination are investigated at the final stage of the FFA. If the last situation is present, the algorithm ends. Otherwise, the algorithm continues its task until the termination condition is completed.

4 Proposed Method

The FFA is implemented to solve the data clustering problem in this section. This solution is considered for an optimization problem as follows: $A = A_1, A_2 \ldots A_n$ and if we want to use the solutions of this algorithm for a dataset with Dn×m and with k cluster, it is better to produce the k × m solution. Since the matrix's production and operation are complicated in the FFA, we first considered each solution of FFA for more convenience and speed (Fig. 1). The number of clusters (k) = 4, and the number of features(m) = 2.

According to Fig. 1, we do not need to produce and perform operations on the (K*M Matrix), Therefore, instead of a 2 × 4 matrix, we used an array of 1 × 8, and we considered its dimensions for the desired cluster, respectively. Also, we need to determine the minimum and maximum limit for each dimension of the solution so that we can obtain lower bounds (lb) and upper bounds (ub) based on Eq. (12) in the dataset D_{n*m} , and apply to any solution. Thus, in FFA, a series of operations are initially performed according to Fig. 1 and Eq. 12.

$$D_{n*m} = \text{load}(\text{DataSet}), ub = \max(D_1, D_2, \ldots, D_m), lb = \min(D_1, D_2, \ldots, D_m) \tag{12}$$

Therefore, we accomplished a series of initial operations to produce suitable solutions for solving the FFA's clustering problem; moreover, we tried not to change the structure of this algorithm's solutions. At the initial stage of the FFA, each key for the dataset D_{n*m} is generated as Eq. (13):

$$X_{ij} = rand(1, k * n) * (ub - lb) + ub \tag{13}$$

In Eq. (13), a solution size equal to $k * m$ is randomly generated between each data set's minimum and maximum limit. Also, for each metaheuristic algorithm, we need a proper objective function for clustering, which defines the objective function according to the clustering problem as Eq. (14):

$$f(D, C) = \sum_{i=0}^{N} \min ||d_i - C_z||^2, z = 1, 2, \ldots, k \tag{14}$$

In Eq. (14), N is the number of objects, and d_i represents the distance of data to the centers of each cluster. The goal of clustering is to minimize the distance between the subset of data $d_i(i = 1 \ldots n)$ and the centers of cluster$C_z(z = 1, 2, \ldots, k)$ that belongs to them. We replaced this equation more precisely in FFA, and its pseudo-code is shown in Algorithm (1) to understand better.

Algorithm (1): Pseudo-code of FFA Clustering Algorithm

```
Initialize parameters: K, beta, alpha, Q, N
Determining the number of sections of farmland according to Equation(1)
Load DataSet and Determining ub and lb according to Equation(12)
Generate initial population Xi (i=1... N) according to Equation(13)
Calculate Objective function f(Xi) according to Equation(14)
While (t < maximum number of iterations)
  Determining quality according to equations (2) and (3)
  Update Global Memory and Local Memory according to equations (4) and (5)
  Worst Sections: Changes according to equations (6) and (7)
  Other sections: Changes according to equations (8) and (9)
  Evaluation of all New Solutions according to Equation (14)
   For all solutions
       If (Q>rand)
          Changes of Solution according to Equation (10).{part: Q>rand }
        else
        Changes of Solution according to Equation (10).{part: else}
      End if
    End for
    Evaluation of all New Solutions
End while
Print Best Solution
```

5 Result and Discussion

All algorithms implemented in this section are simulated in a 5 core system, 8 GB of memory, and Windows 10 operating system and are implemented in MATLAB 2021. We use ten standard clustering data sets (Table 1) to evaluate the proposed approach. The details of implementation, simulation, and implementation of FFA for data clustering will be tested and reviewed. FFA's initial parameters and other comparative metaheuristic algorithms are determined here so that all algorithms are implemented and then compared in similar conditions. The proposed algorithm is compared with another robust metaheuristic algorithm, Including DE, HS, ABC, SOS, MBO, and CSA, in terms of the ANOVA test and statistical criteria, which initializes the FFA parameters and other algorithms that have been shown in Table 2. After determining the simulation environment and the FFA's primary parameters and other comparative metaheuristic algorithms and according to the parameters specified in Table 2, clustering will be implemented on the dataset. We used ten

Table 1 Dataset for clustering

Dataset	Number of Samples	Number of attribution	Number of class
Iris	150	4	3
Wine	178	13	3
CMC	1473	9	3
Glass	214	10	6
Vowel	871	3	6
Wisconsin Breast Canver	683	9	2
Ionosphere	351	31	2
Spect	267	22	2
Blood	748	4	2
Zoo	101	17	7

Table 2 Initialization of the FFA parameters and other algorithms

Algorithm	The value of parameters
ABC	limit = 5D, Population size is 50, Nonlooker = 20, and MaxIt = 50
SOS	Population size is 50, and MaxIt = 50
DE	CR = 0.9 , F = 0.5, Population size is 50, and MaxIt = 50
HS	HMCR = 0.5, PAR = 0.3, bw = 0.2, population size is 50, and MaxIt = 50
MBO	Population size is 50, and MaxIt = 50
CSA	AP = 0.1, population size is 50, and MaxIt = 50
FFA	k = 2; n = 25, NPop = k*n, alpha = 0.6, betha = 0.4, W = 1, Q = 0.5 , and MaxIt = 50

standard clustering datasets with the same population and iteration to compare DE, HS, ABC, SOS, MBO, and CSA algorithms.

In this part of the book chapter, our goal is to find the optimal parameters for the FFA algorithm. However, the results prove that the effect of the parameters on the final output of the algorithm was minimal. However, in this experiment, we aim to investigate the effective parameters of the FFA algorithm, including alpha, beta, and Q, which are quantified in Table 3 as different ranges, and the results are shown in different criteria. Of course, other values were also tested, of which seven examples of optimal parameters were named FFA1 to FFA7.

The results of determining the optimal parameter for the FFA algorithm with different ranges for the alpha, beta, and Q parameters in Table 3 show that the optimal values of these parameters, including 0.6, 0.4, and 0.5, respectively, have achieved much more promising results than the other parameters. The results show that the Q values are between 0.1 and 0.9 due to the imbalance between the discovery and the efficiency of the worst solutions. The results of many parameters of the FFA algorithm

Table 3 Determine optimal parameter values for FFA

Data Set	Criterion	FFA1	FFA2	FFA3	FFA4	FFA5	FFA6	FFA7
–	alpha	0.6	0.5	0.5	0.8	0.6	0.3	0.7
	beta	0.4	0.5	0.6	0.2	0.4	0.7	0.3
	Q	0.5	0.8	0.4	0.3	0.2	0.1	0.9
Iris	Best	96.66	96.67	97.54	96.66	97.1	96.66	97.24
	Worst	97.5	97.5	98.39	97.52	97.95	97.52	98.09
	Mean	97.06	97.06	97.88	97.11	97.53	97.11	97.69
	Std	0.22	0.25	0.24	0.24	0.26	0.31	0.24
Wine	Best	16300.82	16300.75	16296.32	16296.68	16300.72	16301.46	16299.39
	Worst	16446.3	16441.17	16441.73	16438.63	16446.71	16445.65	16445.41
	Mean	16372.05	16366.3	16374.2	16362.65	16393.78	16368.8	16362.57
	Std	48.44	43.35	44.15	44.18	41.21	43.35	40.17
CMC	Best	5534.82	5545.96	5554.52	5547.43	5550.68	5543.06	5546.54
	Worst	5579.2	5595.01	5603.41	5596.84	5597.44	5592.74	5595.95
	Mean	5558.13	5569.51	5579.63	5573.83	5573.55	5566.01	5572.26
	Std	13.85	14.92	15.98	16.33	13.92	13.29	13.47
Glass	Best	2964.57	2968.83	2964.84	2966.77	2965.01	2965.47	2965.81
	Worst	2990.41	2995.43	2991.18	2992.78	2991.56	2991.01	2992.43
	Mean	2975.10	2983.56	2975.16	2979.51	2978.84	2976.34	2978.6
	Std	7.21	7.84	7.29	8.63	7.84	7.61	8.04
Vowel	Best	251.2	260.03	239.65	257.67	277.8	266.92	276.06
	Worst	253.38	262.37	241.72	259.97	280.27	269.31	278.52
	Mean	252.35	261.24	240.66	258.87	279.07	268.22	277.38
	Std	0.64	0.73	0.52	0.71	0.85	0.7	0.82
cancer	Best	156116.6	155832.3	161891.6	153058.6	156513.2	161775.1	155989.4
	Worst	157498	157225.7	163329.3	154429.5	157913.1	163226	157360.8
	Mean	156894.3	156499.4	162627.7	153683.2	157226	162514.5	156748
	Std	399.79	459.14	461.74	385.51	452.27	446.62	420.72
ionosphere	Best	181.74	181.76	181.74	182.18	181.66	181.67	182.48
	Worst	183.35	183.35	183.36	183.81	183.23	183.27	184.1
	Mean	182.46	182.6	182.50	183.06	182.48	182.48	183.11
	Std	0.55	0.49	0.48	0.45	0.48	0.49	0.41
spect	Best	498.1	506.6	501.04	504.46	501.41	502.88	506.11
	Worst	502.57	510.95	505.47	508.98	505.92	507.36	510.61
	Mean	500.39	508.61	502.88	506.61	503.67	505.09	508.52
	Std	1.35	1.42	1.26	1.29	1.4	1.32	1.4
Blood	Best	934.89	919.95	919.31	920.21	942.25	949.83	962.9
	Worst	943.18	928.17	927.57	928.48	950.65	958.35	971.52
	Mean	939.02	924.24	923.46	924.41	946.58	953.42	966.75
	Std	2.47	2.52	2.43	2.51	2.53	2.61	2.33
Zoo	Best	407714.2	407714.2	407714.2	407714.2	407714.2	407714.2	407714.2
	Worst	411228.6	411348.3	411381.9	411348.5	411378.6	411362.7	411338.9
	Mean	409330	409410.7	409350.4	409590.8	409831.5	409382.8	409544
	Std	1049.61	1025.12	1092.35	1087.02	1055.35	1077.68	1111.01

have been satisfactory. We showed the results in graphical charts. We considered a separate section for each clustering data set and analyzed FFA and other comparative algorithms to illustrate the results better. Finally, we compared the FFA with different algorithms for a more accurate assessment (Fig. 2).

Figure 3 shows that the FFA algorithm exhibits the best performance. These experiments have proven that the FFA algorithm's superiority is not random. It has performed better than other algorithms in most implementations and has shown an excellent convergence in all performances. The FFA algorithm has consistently demonstrated a good convergence in the final iterations due to better local and global memory. As a result, the comparison of FFA with DE, HS, ABC, SOS, MBO, and CSA algorithms on ten standard clustering data sets has proven this algorithm's superiority in terms of the rate of convergence. To further investigate the FFA, ANOVA was displayed to measure the difference between clusters in Fig. 4.

Figure 4 shows that the FFA algorithm with other algorithms has been implemented on ten different data sets in ANOVA. It is shown in the boxplots, in which the height of each boxplot confirms the algorithm's stability. As a result, whatever the measurement of boxplots is more significant than others, its corresponding algorithm has higher noise, which indicates that the algorithm has not worked well. Therefore, the FFA algorithm's implementation on the ten datasets in Fig. 4 shows that the FFA algorithm in all datasets has been displayed a lower height and better results than other algorithms. The experiments also prove that the FFA algorithm positively impacts the entire population due to local and global memory and can create the convergence of the whole population towards the goal. Finally, the FFA algorithm in ANOVA has proven its superiority over the algorithms on ten clustering standard data sets. This algorithm has the minimum possible limit for ANOVA, even in most data sets. The FFA algorithm's clustering algorithm on the ten standard clustering data sets is in Fig. 4.

The FFA algorithm's clustering algorithm has been shown on the ten standard clustering data sets in Fig. 4. To illustrate the clustering process, we considered each dataset with several iterations 10, 20, and we showed selectively 2 and 3 features in two-dimensional and three-dimensional formats. This experiment shows that the FFA algorithm has performed the data clustering process well and placed it in a suitable cluster. The FFA algorithm has presented better clustering with an increasing number of iterations. Therefore, for further comparison, in Table 3, the FFA algorithm has been compared with other algorithms with different statistical criteria (Table 4).

Table 3 shows that FFA has been compared with other optimization algorithms with statistical criteria(best, worst, mean, and standard deviation) in 100 iterations. This experiment shows that the FFA algorithm performs better than DE, HS, ABC, SOS, MBO, and CSA algorithms. This experiment's remarkable point is that the FFA algorithm has performed well in the best criterion. On the other hand, the lower standard deviation indicates that this algorithm positively impacts the entire population using global and local memory operators and displays a good convergence.

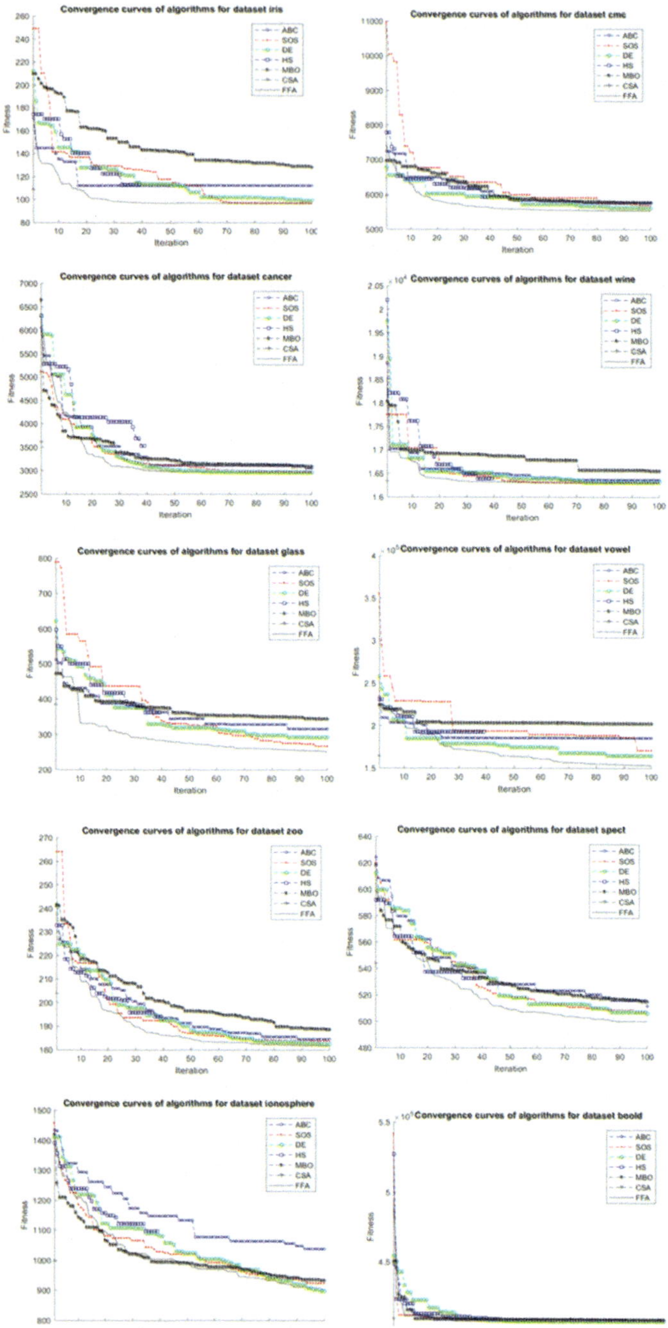

Fig. 2 Convergence curves of the Proposed method and other algorithms

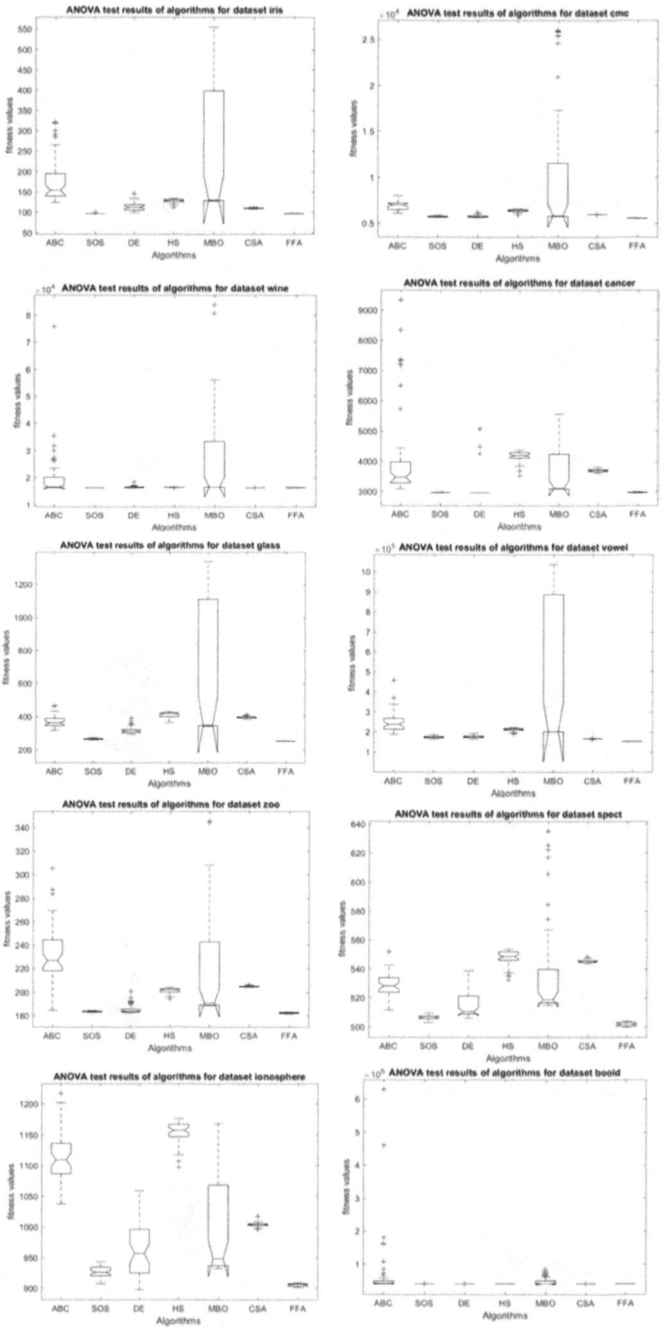

Fig. 3 ANOVA test results of Proposed method and other algorithms

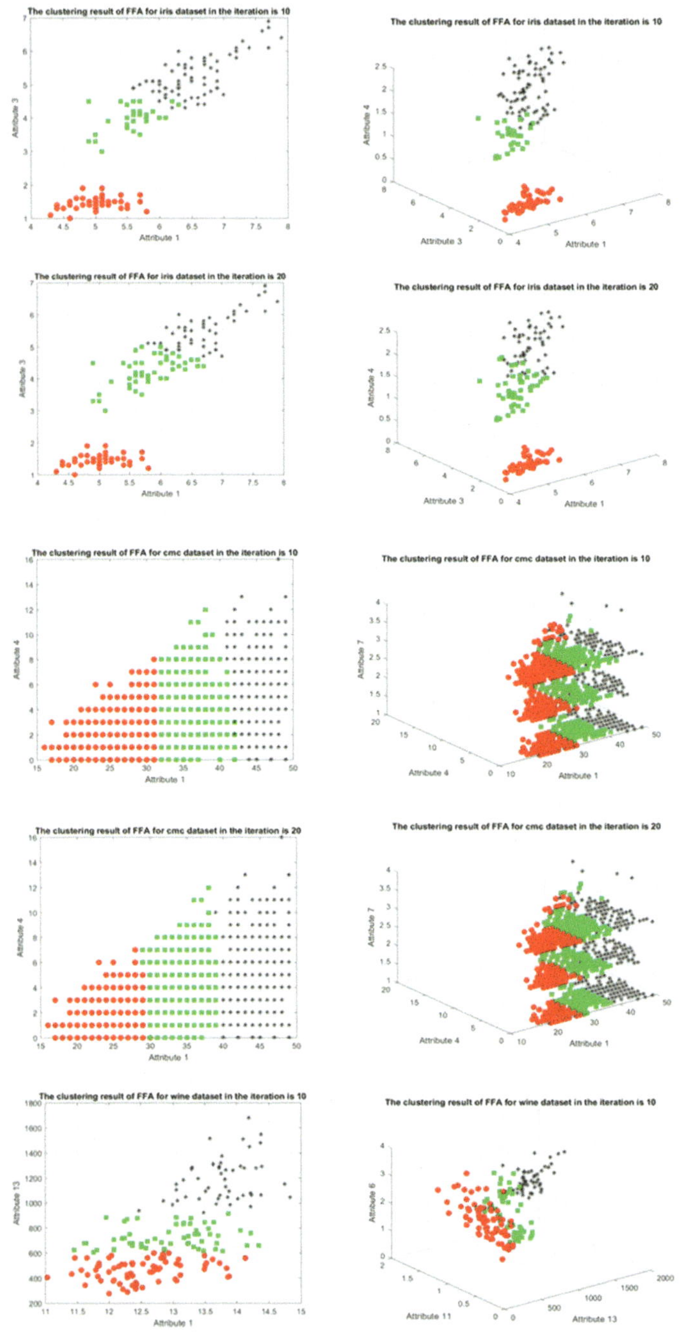

Fig. 4 Clustering process of FFA

Table 4 FFA algorithm with other algorithms

Data set	Criterion	MBO	ABC	SOS	DE	HS	CSA	FFA
Iris	Best	128.04	112.21	96.87	96.87	113.1	108.82	96.66
	Worst	554.75	321.07	99.39	145.78	134.38	112.06	97.52
	Mean	259.1	177.89	96.85	113.15	128.22	110.15	97.07
	Std	172.14	58.67	0.37	9.5	4.71	0.72	0.24
Wine	Best	16549.03	16362.81	16298.65	16315.85	16390.17	16336.47	16296.46
	Worst	83800.82	75751.08	16301.18	18435.63	16561.39	16369.74	16442.32
	Mean	30073.12	20014.58	16299.59	16541.28	16495.91	16343.08	16388.76
	Std	22253.97	9279.02	0.93	341.34	38.55	7.13	41.06
CMC	Best	5764.14	5767.54	5673.54	5615.99	5930.26	5952.65	5534.82
	Worst	26000.1	7987.49	5899.65	6212.56	6580.45	6012.53	5579.2
	Mean	10446.75	6884.03	5733.71	5772.28	6395.07	5987.36	5558.13
	Std	7038.26	458.22	60.14	121.87	140.02	9.34	13.85
Glass	Best	3091.63	3061.66	2971.4	2964.98	3529.57	3611.66	2964.76
	Worst	5554.28	9339.74	2982.46	5067.45	4364.93	3806.48	2991
	Mean	3672.04	4198.69	2974.01	3064.71	4167.7	3701.65	2977.05
	Std	824.12	1590.27	3.37	400.89	175.35	51.55	7.34
Vowel	Best	340.22	316.03	267.61	292.74	362.38	384.07	251.2
	Worst	1336.26	465.17	270.82	388.92	428.84	411	253.38
	Mean	690.9	367.96	265.49	316.56	409.64	395.42	252.35
	Std	395.4	34.33	2.55	20.75	16.45	5.1	0.64
Cancer	Best	201945.78	185428.07	171050.6	164829.9	192702.5	164363.3	153290.8
	Worst	1036524	457456.32	185785.7	192302.6	221241.2	169540.4	154660.1
	Mean	483680.5	247565.26	174714.4	177506.8	213025.7	166986.1	153990.8
	Std	344629.6	49715.4	5162.84	6437.18	7200.97	801.23	410.54
Ionosphere	Best	188.69	184.56	183.97	182.49	194.18	203.98	181.74
	Worst	345.32	305.21	184.37	200.71	203.85	206.08	183.35
	Mean	218.65	230.81	183.62	185.66	201.14	204.86	182.49
	Std	42.58	24.4	0.34	4.08	2.33	0.38	0.52
Spect	Best	515.03	511.77	507.34	506.17	532.79	543.59	499.95
	Worst	635.11	551.75	509.64	538.69	553.59	548.57	504.27
	Mean	502.09	545.43	547.96	515.19	506.71	529.19	536.41
	Std	33.23	7.49	1.48	8.51	5.05	1.08	1.39
Blood	Best	932.28	1037.77	924.29	897.98	1096.97	996.42	901.66
	Worst	1168.18	1216.94	943.49	1059.02	1176.1	1017.23	909.65
	Mean	990.7	1115.77	927.21	960.44	1154.2	1004.02	906.15
	Std	74.06	42.77	8.62	41.62	17.68	3.28	2.39
Zoo	Best	408416.71	408312.6	407714.38	407714.28	409330.32	409651.4	407714
	Worst	846895.1	6288648	407714.3	407714.8	411124.2	414418.2	411300.6
	Mean	474224.9	762296.03	407714.2	407714.3	410527.8	410156.2	409636.5
	Std	113302.1	1041264.3	0.03	0.1	442.96	694.71	1060.02

6 Conclusion and Future Works

This book chapter proposes FFA based on clustering to solve the data clustering problem. This algorithm has strong operators to optimize optimal issues. We first described the FFA based on clustering and showed how this algorithm presents a clustering solution. After giving the proposed method, we told ten valid clustering datasets for evaluating the proposed method. Then, we compared FFA with optimization algorithms Including DE, HS, ABC, SOS, MBO, and CSA in terms of statistical criteria such as ANOVA and the rate of convergence. Initially, FFA with other algorithms was implemented on ten different databases in terms of convergence curves; this test proved that the FFA algorithm's superiority was recent. This algorithm worked better than other algorithms in m of the implementation and showed an excellent convergence in all the performances. The ANOVA test results showed that the FFA algorithm in all data sets presents a lower height of boxplots (boxplots with less size) and better results. Finally, the statistical criteria test shows that the FFA algorithm has an excellent performance in different criteria in the ten standard clustering data sets compared to the above-mentioned comparative algorithms. Experimental results show that the FFA algorithm has correctly performed the data clustering process and put the data in a suitable cluster.

References

1. Rahnema, N., Gharehchopogh, F.S.: An improved artificial bee colony algorithm based on whale optimization algorithm for data clustering. Multimed. Tools Appl. **79**(43), 32169–32194 (2020)
2. Wang, G.-G., et al.: A new monarch butterfly optimization with an improved crossover operator. Oper. Res. **18**(3), 731–755 (2018)
3. Shayanfar, H., Gharehchopogh, F.S.: Farmland fertility: a new metaheuristic algorithm for solving continuous optimization problems. Appl. Soft Comput. **71**, 728–746 (2018)
4. Price, K.V.: Differential evolution: a fast and simple numerical optimizer. In: Proceedings of North American Fuzzy Information Processing. IEEE (1996)
5. Geem, Z.W., Kim, J.H., Loganathan, G.V.: A new heuristic optimization algorithm: harmony search. Simulation **76**(2), 60–68 (2001)
6. Shapiro, S.S., Wilk, M.B.: An analysis of variance test for normality (complete samples). Biometrika **52**(3/4), 591–611 (1965)
7. Benyamin, A., Farhad, S.G., Saeid, B.: Discrete farmland fertility optimization algorithm with metropolis acceptance criterion for traveling salesman problems. Int. J. Intell. Syst. **36**(3), 1270–1303 (2021)
8. Agwa, A.M., El-Fergany, A.A., Maksoud, H.A.: Electrical characterization of photovoltaic modules using farmland fertility optimizer. Energy Convers. Manag. **217**, 112990 (2020)
9. Gharehchopogh, F.S., Farnad, B., Alizadeh, A.: A modified farmland fertility algorithm for solving constrained engineering problems. Concurr. Comput.: Pract. Exp. **33**(17), e6310 (2021)
10. Sabo, A., et al.: Optimal design of power system stabilizer for multimachine power system using farmland fertility algorithm. Int. Trans. Electr. Energy Syst. **30**(12), e12657 (2020)
11. Hosseinalipour, A., et al.: A novel binary farmland fertility algorithm for feature selection in analysis of the text psychology. Appl. Intell. **51**(7), 4824–4859 (2021)

12. Naseri, T.S., Gharehchopogh, F.S.: A feature selection based on the farmland fertility algorithm for improved intrusion detection systems. J. Netw. Syst. Manag. **30**(3), 1–27 (2022)
13. Wang, R., et al.: Flower pollination algorithm with bee pollinator for cluster analysis. Inf. Process. Lett. **116**(1), 1–14 (2016)
14. Zhou, Y., et al.: A simplex method-based social spider optimization algorithm for clustering analysis. Eng. Appl. Artif. Intell. **64**, 67–82 (2017)
15. Boushaki, S.I., Kamel, N., Bendjeghaba, O.: A new quantum chaotic cuckoo search algorithm for data clustering. Expert. Syst. Appl. **96**, 358–372 (2018)
16. Abd Elaziz, M., et al.: Automatic data clustering based on hybrid atom search optimization and sine-cosine algorithm. In: 2019 IEEE Congress on Evolutionary Computation (CEC). IEEE (2019)
17. Han, X., et al.: A novel data clustering algorithm based on modified gravitational search algorithm. Eng. Appl. Artif. Intell. **61**, 1–7 (2017)
18. Kumar, A., Kumar, D., Jarial, S.: A novel hybrid K-means and artificial bee colony algorithm approach for data clustering. Decis. Sci. Lett. **7**(1), 65–76 (2018)
19. Zhou, Y., et al.: Automatic data clustering using nature-inspired symbiotic organism search algorithm. Knowl.-Based Syst. **163**, 546–557 (2019)
20. Mageshkumar, C., Karthik, S., Arunachalam, V.: Hybrid metaheuristic algorithm for improving the efficiency of data clustering. Clust. Comput. **22**(1), 435–442 (2019)
21. Karaboga, D., Ozturk, C.: A novel clustering approach: Artificial Bee Colony (ABC) algorithm. Appl. Soft Comput. **11**(1), 652–657 (2011)
22. Ahmadi, R., Ekbatanifard, G., Bayat, P.: A modified grey wolf optimizer based data clustering algorithm. Appl. Artif. Intell. **35**(1), 63–79 (2021)
23. Boushaki, S.I., Bendjeghaba, O., Brakta, N.: Accelerated modified sine cosine algorithm for data clustering. In: 2021 IEEE 11th Annual Computing and Communication Workshop and Conference (CCWC). IEEE (2021)
24. Kaur, A., Kumar, Y.: A new metaheuristic algorithm based on water wave optimization for data clustering. Evol. Intell. **15**(1), 759–783 (2022)
25. Aljarah, I., et al.: Multi-verse optimizer: theory, literature review, and application in data clustering. In: Nature-Inspired Optimizers, pp. 123–141. Springer, Berlin (2020)
26. Abualigah, L.M., et al.: A novel hybridization strategy for krill herd algorithm applied to clustering techniques. Appl. Soft Comput. **60**, 423–435 (2017)
27. Jadhav, A.N., Gomathi, N.: WGC: Hybridization of exponential grey wolf optimizer with whale optimization for data clustering. Alex. Eng. J. (2017)
28. Mustafa, H.M., et al.: An improved adaptive memetic differential evolution optimization algorithms for data clustering problems. PloS one **14**(5), e0216906 (2019)
29. Alswaitti, M., Albughdadi, M., Isa, N.A.M.: Variance-based differential evolution algorithm with an optional crossover for data clustering. Appl. Soft Comput. (2019)
30. Kuwil, F.H., et al.: A new data clustering based on critical distance methodology. Exp. Syst. Appl. (2019)

A Comprehensive Review of the Firefly Algorithms for Data Clustering

MKA Ariyaratne and TGI Fernando

Abstract Separating a given data set into groups (clusters) based on their natural similar characteristics is one of the main concerns in data clustering. A cluster can be defined as a collection of objects which are "homogeneous" between them and are "heterogeneous" to the objects belonging to other clusters. Many areas encounter clustering based applications including the fields of medical, image processing, engineering, economics, social sciences, biology, machine learning and data mining. Clustering goes under unsupervised learning where no labels are given to the learning algorithm, leaving it on its own to find structure in its input. Even though many classical clustering algorithms can be found, most of such suffer from severe drawbacks such as sensitivity over initial cluster centroids and hence can be easily trapped in local optimum solutions. The other main problem with the data clustering algorithms is that it cannot be standardized. On the other hand, clustering can be considered under optimizations which goes to the category of NP hard optimization making more difficult in solving. Addressing such NP hard problems, meta-heuristics play a remarkable role in optimization. Since its appearance from more than a decade ago, Firefly Algorithm (FA), a stochastic meta- heuristic in nature inspired algorithms has shown significant performance in giving solutions to many optimization problems. Hence FA has been used in research addressing the problem of clustering optimization. This chapter forestalls the ability of firefly algorithm in solving data clustering problem. It presents an introduction to clustering and the performance of FA, briefly reviews and summarizes some of the recent firefly-based algorithms used for data clustering with the emphasis on how FA has been combined/ hybridized with other methods to contribute to the problem of data clustering. Further it discusses on different representations, initializations, and the used cluster validation criteria in FA based clustering methods. The chapter also discusses why FA is to be more useful

M. Ariyaratne (✉) · T. Fernando
Department of Computer Science, Faculty of Applied Sciences University of Sri Jayewardenepura, Nugegoda, Sri Lanka
e-mail: mkanuradha@sjp.ac.lk

T. Fernando
e-mail: tgi@sjp.ac.lk
URL: https://www.sjp.ac.lk

A. Biswas et al. (eds.), *Advances in Swarm Intelligence*, Studies in Computational Intelligence 1054, https://doi.org/10.1007/978-3-031-09835-2_12

for clustering over other methods and what features made it more suitable for handling the clustering problem compared with other meta-heuristics. Finally, it focuses on the limitations that have been found in the literature on clustering grounded on FA-based applications and discusses possible avenues in future.

Keywords Firefly algorithm · Clustering · Meta-heuristics · Unsupervised

1 Introduction

The necessity of optimization is an often transpiring situation in many fields including science, engineering, economics, computing, biology etc. Most of the time these optimizations need to be handled with several constraints and in large scales. When it comes to real world problem scenarios, the traditional optimization techniques such as gradient based methods frequently find it difficult to tackle the problems due to the inherited drawbacks in many of such methods. Some of these drawbacks such as dependability on the initial guesses, need of continuous differentiable problem representations have made their usability on real world problems somewhat unrealistic [1].

The current state of the art solution for this can be stated as the use of meta-heuristics; basically, the nature inspired algorithms [2]. The robust, flexible nature of these algorithms along with their effectiveness have made them popular using for practical optimization problems. These nature inspired algorithms mainly are of two types: evolutionary computing algorithms such as genetic algorithms (GA) [3] and differential evolution (DE) [4], and swarm intelligence algorithms such as particle swarm optimization (PSO) [5], ant colony optimization (ACO) [6], firefly algorithm (FA) [7], cuckoo search (CS) [8] and many other. For the last five decades, the emergence of the swarm intelligence algorithms can be stated as rapid, although many of such algorithms have to be revisited in order to see the power and the dangers of the metaphor, how good in dealing benchmarks and the scalability and usability of the algorithms [9]. Therefore, selecting suitable algorithms for the problem has become another crucial task when using swarm intelligence algorithms.

Clustering can be considered as a process of dividing a given data set into homogeneous sets based on the given characteristics where similar data are kept in a group and of the dissimilar data placed in different groups. It is the most important unsupervised learning problem. It deals with finding structure in a collection of unlabeled data. Cluster analysis has become an emerging research problem in data mining due to its various applications. Various fields in science need clustering in various situations. With the advent of many data clustering algorithms in recent years and their extensive use in wide variety of applications such as computational biology, mobile communication, medicine and economics and many other, these algorithms became popular. Main problem with the data clustering algorithms is that it cannot be standardized. An algorithm may give best result with one type of data set but may fail or give poor result with data set of other types. Although many attempts were taken in standardizing the algorithms which can perform well in all case of scenarios, to date no major accomplishment has been achieved. Many clustering algorithms have been proposed so far. However, each algorithm has its own merits and inferiorities and cannot work for all real situations [10].

There are only a few clustering problems which are polynomial bounded. Such algorithms have the property that there exists a method where the number of computational steps are bounded by a polynomial in the length of the input of the problem [11]. On the other hand, many clustering problems that encounter in real life are NP hard in nature [12], where the computational time grow exponentially with the size of the input.

Clustering is usually done when there is no information about the membership of the data items for class pre-definition. Because of that, clustering is traditionally regarded as part of non-supervised/unsupervised learning. Conventional clustering algorithms such as K-means, fuzzy c-means, K-medoids, x-means and Nelder Mead simplex search have several drawbacks. Dependency over the initial parameter values, inability in escaping from local optimum points and relatively low scalability are some of them. Clustering big data sets is beyond the ability of these conventional methods. Therefore, the need of searching for efficient and accurate big data clustering algorithms has come to the stage. Because of the complexity of the problem, it has been observed that there is a tendency in using meta-heuristics such as nature inspired algorithms in recent research to solve the problem. Tabu search, simulated annealing, genetic algorithms, differential evolution, firefly algorithm, gray-wolf swarm optimization (GWO), particle swarm optimization are some of the meta-heuristics that have been tried on clustering [13–19].

The firefly algorithm, which was introduced to the world of meta-heuristics in early 2008 can be considered as one of the powerful swarm intelligence algorithms. It has proven capabilities in solving multi-model optimization problems [7]. Many studies have demonstrated the different abilities of the FA including software testing [20], modeling to generate design alternatives [21] and scheduling problems [22]. In many other studies, FA has been effectively modified to solve a variety of problems in optimization, enhancing the performance as well [23–25].

In a handful of latest research, FA has been effectively adapted to solve the problems in clustering. In some of the research FA was used to modify the existing standard clustering algorithms for instance like K-means [17] whereas in others, new fitness measures and search mechanisms have introduced [26]. However, there is a great need of taking all these attempts into one place and discuss the potential of firefly algorithm in solving the data clustering problem. Hence The aim of this paper is twofold: to present and discuss the research where FA has been successfully applied to solve the problem of data clustering, and thus to broaden the range of its potential users. This paper is a preliminary attempt to review the capability of FA in addressing the data clustering problem. For this, 23 research papers were examined and a comprehensive review is presented. We believe that we included almost all articles regarding data clustering solved using FA, to date.

This paper is organized as follows. Section 2 briefly reviews and summarizes some of the recent firefly based algorithms used for clustering data with the emphasis on

Table 1 Summary of the clustering problems solved with FA based algorithms. Background colors represent different combinations of algorithms with FA: FA only, FA + K means, FA + GA, FA + DE, FA + PSO, FA + Fuzzy C Means, FA + density peaks clustering, FA + LEACH, FA + Markov clustering algorithms

Clustering problem	Literature
Bench mark clustering problems (Standard and synthetic data sets)	[26–35]
Clustering Nodes in Wireless sensor networks	[36–40]
Image segmentation	[41–44]
Forecasting Applications	[45]
Information Retrieval	[46]
Clustering solutions to ensure privacy in social networks	[47]
Clustering dynamic protein-protein interaction networks	[48]

how FA has been combined/hybridized with different other methods to contribute to the problem of data clustering. Section 3 focuses on representations, fitness evaluations, search mechanisms and the cluster evaluation criteria used in FA based clustering methods. Section 3.2 strives to highlight some of the main issues in the existing FA based clustering algorithms and suggest some improvements that are possible.

2 Contribution of the FA for Data Clustering

Use of Nature inspired algorithms for data clustering is not a new topic in the optimization literature and has been accomplished a decade or two ago. Several variants of firefly algorithms used to solve the clustering problem exist in the literature. This section will basically review the existing research work carried out based on the firefly algorithm in solving different clustering problems. Although different algorithms can be described in different ways, if the description of the algorithm can be divided into categories, according to the way the FA has been combined / hybridized with other methods in order to solve various clustering problems, then it will be easier to discuss. The summary of the review in Sect. 2.1 is presented in Table 1.

2.1 Use of Different Variants of FA Without Hybridization

First, we will be focusing on studies which haven't done any combination or hybridization, but modifications to the FA itself to solve different clustering problems.

The earliest of such work was carried out by J. Senthilnath and the fellows [26]. The authors used the firefly algorithm to solve benchmark clustering problems and the performance were compared with several other nature inspired and other clustering algorithms. A given dataset was divided as training (75%) and testing (25%) datasets. The classification error percentage which is defined as the ratio of number of misclassified samples in the test data set and total number of samples in the test data set and was used to measure the performance of the algorithm. The clustering objective function used is the sum of error squared. Thirteen benchmark problems were solved to obtain the results. The parameter values used for the algorithm are encouraging and the average error percentages of FA over the other two nature inspired algorithms are low. These findings have helped to establish firefly algorithm's high capability in successfully identifying the best cluster centroids.

Further in [37], FA is used to implement an energy efficient clustering in wireless sensor networks. They have defined a new cost / fitness function to minimize the intra-cluster distance to optimize the energy consumption of the network. In a wireless sensor network, there are partially distributed small sensor nodes to communicate among themselves. There is a need to cluster these sensor nodes into small groups which are the clusters addressed in the problem. The cluster centroid is responsible in monitoring the others in the same cluster. Because of this, the energy consumption and number of messages transmitted to the base station will be reduced and number of active nodes in communication is also reduced. That increases the lifetime of network. Among the three basic methods that are available for the network clustering: centralized, distributed and hybrid, a centralized, energy concerned protocol to extend the lifetime of the sensor network based on FA has been developed. The performance of the FA has been compared with a previous work done using PSO and LEACH (Low Energy Adaptive Clustering Hierarchy [49]); a well-known distributed clustering technique. The main concern was the lifetime of network and the results indicate that the clustering done based on FA has prolonged the life time of network than the other two.

In [36], the authors presented FA to select cluster heads in a Wireless Sensor Network (WSN). They formed a fitness function based on the values of residual energy and the distance between the nodes. This fitness function with the canonical firefly algorithm gave the solution to the problem of cluster head selection. Comparisons are made with the Energy-Aware Clustering algorithm [50] where the cluster heads were selected based on the energy and determined after every transaction cycle. According to the results obtained, the lifetime of the network and the packet delivery ratio of the proposed method is higher than that of the energy aware method. In [40], the same problem of Wireless Sensor Network (WSN) is addressed in a different way where a data aggregation solution to the WSN is provided using the FA. The cluster heads were selected using FA and the LEACH model and distribution of sensor nodes among clusters is done by the FA. The main purpose of this research is to reduce the duplicate data transmitted by the sensor nodes. The FA mainly used here as an aggregator for removing duplicate data and converting them into unified data. The efficiency of the proposed model was compared with that of LEACH [49] and SFA [51] models and evaluated as efficient than those.

2.2 FA Combined with K-Means Algorithm

In [27], the authors offered a hybrid clustering algorithm which in cooperates the firefly algorithm and the K-means clustering algorithm. The anytime behavior of the FA which helps the meta-heuristic algorithms to stand along over the provided initial guesses has been used to overcome the initial sensibility problem of the K-means algorithm. They used the FA to find the initial optimal cluster centroid and then used the K-means algorithm with the optimal centroid to further refine it and improve cluster accuracy. Appropriate modifications were provided in chaining the classical FA such that in-cooperating global optimum value of each iteration to its movement function. Instead of moving fireflies towards their neighbors, like in the PSO, this proposed modification used the global best firefly in each iteration to influence others and affect in their movement. Therefore, the movement function in the standard FA is slightly modified. Experiments were performed on five data sets from UC Irvine Machine Learning Repository [52], including Iris, WDBC, Sonar, Glass and Wine. Several performance measures including intra-cluste distance and clustering error were obtained. According to the results obtained, the discussed method is able to reduce the intra-cluster distance of each cluster in all cases and helped to do appropriate initialization of the K-means algorithm and reduces the clustering error.

The authors in [41], implemented a firefly variant with the K-means algorithm for the purpose of image segmentation. In the problem of image segmentation, similar pixels are grouped together into clusters. The proposed KFA used the exploration property of the FA and calculated the centroids according to the defined formula and the fitness of a centroid. When the FA achieved the best centroids, the K-means algorithm can run smoothly without trapped in local optima taking them as the initial centroids for the K-means. The authors applied the KFA on three types of gray scale images cameraman, coins, and toyobjects. Correlation coefficient is used as a valid image segmentation measure where higher values of it indicate the goodness of the segmentation. They showed how the proposed KFA obtained higher correlation coefficients during different number of clusters.

An interesting forecasting application is presented in [45] using FA, K-means clustering, and Support Vector Regression (SVR) with wavelet transform. A prediction system for Taiwanese export trade values is developed with the help of FA, K-means, and SVM. The presented model consists of three stages. Initially, Wavelet transform is used to reduce the noise in data preprocessing. Then the FA based K-means clustering is applied for the cluster analysis and finally the FA based SVM is used to develop a forecasting model for each of the clusters. Standard firefly algorithm with Euclidean distance as the distance measure to calculate distance between two centroids is taken. The K-means algorithm is in cooperated with FA by evaluating the fitness function to update the cluster centroids. The experimentations conducted indicated positive feedback on performance of the new forecasting algorithm over the existing ones. To avoid the famous issues in K-means on initialization and getting locally optimum solutions, [28] proposed a hybrid solution for K-means with FA. Implementing FA to find a suitable optimum solution to be an initial seed of the K-means is the core of the study. For the evaluations, Iris, Glass, Cancer-Int and Haberman data sets from UCI Machine Learning repository [52] were used. The conclusions supported the idea that the proposed method is successful in finding more suitable clusters than K-means alone.

Addressing the drawbacks in both [27, 28], two variants of hybrid FA-K-means algorithms were proposed in [29] to improve the limitations of the previously proposed hybrid versions. The developed probabilistic firefly k-means (PFK) algorithm and the greedy probabilistic firefly k-means (GPFK) algorithm improved the FK algorithm [28] in terms of the percentage of correct answer (PCA). The PFK algorithm calculated the probability that each data point belonging to a cluster and use the highest probability to select the final cluster. To avoid the computational complexity in the newly introduced movement step of proposed PFK, GPFK is introduced. In GPFK, fireflies move only towards the best firefly in the population similar to the classical greedy search. The results indicated that GPFK worked best for the testing data sets compared to the PCA.

Also, in [30], two variants of the firefly algorithm are implemented to reduce the existing drawbacks of the K-means clustering algorithm. The solution addressed the inherited problems in K-means such as dependability over the initial guess and the tendency in falling on local optimum solutions. The firefly algorithm is also modified to enhance its capabilities in exploration, exploitation, and efficiency. In the proposed variants, two novel strategies are formulated to increase search diversification and the efficiency of FA. In the first variant, inward intensified exploration FA (IIEFA), a randomized control matrix is replaced the attractiveness coefficient in the FA to increase the exploitation diversity. The second variant, compound intensified exploration FA (CIEFA), also increased the diversity of the exploration by scattering fireflies with high similarities to different directions. To combine these two algorithms with K-means, initially a solution (centroids to the predefined number of clusters) is generated using K-means algorithm. This solution replaced the first firefly in the initial population of either IIEFA or CIEFA. Then these two variants produce the quality cluster centroids. The results are compared with several meta-heuristics including some firefly variants as well. Examples from IDB2 database [53] and the UCI machine learning repository [54] are tested and compared. The results and the comparisons have proved statistically significant superiority of the two proposed variants over the other methods.

2.3 FA Combined with Evolutionary Computing Algorithms

Kaushik and the team presented a hybrid data clustering method using a firefly algorithm based improved genetic algorithm [31]. In this study firefly algorithm is used to boost the initialization process. Therefore, a good population of centroids selected by the FA are feed to the GA for further processing. The performance of the hybrid is tested with four benchmark datasets: Iris, Glass, Brest cancer and Wine taken from UCI repository [52]. Inner cluster distances and the intra cluster distances are measured and the goodness of the results are visible in the new algorithm compared with the FA and GA along. However, the method adapted questions the initialization advises that are given by the anytime behavior of meta-heuristics like GA or FA. However, in this study the initialization approch can be admired as it succeeded.

Further, cluster head selection of a wireless sensor network is done in [38], using a modified firefly algorithm combined with the concepts in GA, named as synchronous firefly algorithm. WSNs are typically organized in a hierarchically and for the purpose of energy saving some nodes are selected to be cluster heads based on a fitness function. They proposed a fitness function based on energy, end to end delay, and packet loss rate in the nodes in the network. The fireflies are ranked based on the fitness and selected for mating based on tournament selection. Binary representation is used to represent the nodes in each solution. Selected fireflies are taken to crossover and then to mutate. Comparisons are made with LEACH, EEHC and the classical firefly algorithm. They summarized the results with the statements that the proposed synchronous firefly algorithm minimizes the packet loss and the end-to-end delay and increases the number of clusters hence reducing the energy consumption significantly.

In addition, [32] proposed an improved firefly algorithm for data clustering. The FA is coupled with differential evolution in order to build the proposed hybrid. Initially the population is sorted according to the fitness and divided in to two parts as the best and worst parts. These two sub parts then are manipulated separately by the two algorithms: the best part with FA and the worst part with the DE. Then best values are selected from each sub-population and among all of those findings, the best value is obtained as the representations for centers. Three fitness measures were used to develop the fitness function: distance within the cluster, average distance between the cluster centers and the third one as a ratio of the previous two. The data sets used included Glass, Wine, Breast cancer, Iris which were extracted from the popular database UCI [52]. The results are compared with K-means, FA and DE. The results showed that the proposed algorithm has average performance when compared with DE but can be considered as better than K-means and FA.

2.4 FA Combined with the Particle Swarm Optimization Algorithm

Authors in [46] implemented FA to solve the clustering problem of information retrieval from the web. Extraction of information specially from the web, highly depends on the efficiency of the methods that use. The authors recast the work of [55] which has been done for continuous constrained optimization tasks to make it capable in clustering of data. Instead of moving a particular firefly towards a brighter firefly, which is the idea in the original concept of FA, they identify the most attractive firefly for a particular firefly. Then the centroid of that firefly is updated using the centroids of the most attractive firefly. As in the PSO, if the updated one is better / fitter than the existing one, the centroids are then replaced for that firefly. Else the replacement of random centroids of the firefly will be done with uniform random centroids. To evaluate the performance of the proposed methods, mean best fitness value of TRW [16], VRC [16], percentage of number of successive runs and run length distribution (RLD) [56] were measured. They used two simulated datasets generated from multivariate normal distributions and four benchmark datasets: Fisher's iris [57], Wine [58], Thyroid [59] and Wisconsin breast cancer [60] for the testing. The results obtained point out that the FClust has higher possibility to achieve the optimal results as compared to PSO algorithm and DE algorithm. The results of the RLD plots and the run length distributions further revealed the fast convergence and the stability of the FClust compared to PSO and DE.

As one of the latest works, the authors in [33], used the firefly algorithm hybridized with the particle swarm optimization for the use of automatic data clustering. By hybridizing FA with PSO, the authors tried to reduce few drawbacks resides in the canonical firefly algorithm such as need of proper parameter tuning and the reduction of speed and convergence when the dimension of the problem grows.

One advantage of the proposed hybrids (FAPSO) method is the ability to determine number of cluster for any given set of data based on common cluster validity indices such as the compact-separated (CS) validity index [61] and Davies-Bouldin (DB) validity index [62]. Euclidean distance between two solutions connected with the above two indices of validity that computed the fitness of a cluster. The proposed algorithm validated with 12 benchmark datasets from UCI repository [52] and the two moons datasets [63]. They compared the new FAPSO with six meta-heuristic algorithms; automatic clustering DE (ACDE) [64], genetic clustering with an unknown number of clusters K (GCUK) [65], dynamic clustering particle swarm optimization (DCPSO) [66], classical differential evolution (DE) [4], canonical FA [7] and canonical PSO [5]. The users combined the exploitation property of the FA and the exploration property of the PSO in order to gain the effectiveness of the hybrid. The indices (CS, DB) were also used to decide the proper number of clusters and for them, to observe the best possible partitioning.

The authors in [39] proposed a firefly algorithm coupled with particle swarm optimization (HFAPSO) to modify the LEACH-C algorithm [67] for effective cluster head selection in WSNs. In selecting the optimal cluster header, the authors considered the remaining strength and distance between the cluster header and cluster members in the WSN.

It is essential to select an efficient and optimal cluster head for coordinating the data transmission activities of the cluster nodes for a specific cluster and gathering information to the base station of the WSN.FA and PSO were combined to improve the global search effort, when considering/finding the best firefly in the population. The movement function of the FA is modified accordingly by in cooperating P_{best} and g_{best} values to it. Authors proposed a new fitness function mainly based on the energy levels and the distance between the cluster heads. The comparisons encouraged the idea that the proposed HFAPSO outperformed LEACH-C and the FA, depending on the lifetime of the network, residual power, number of active nodes, and throughput of the network.

2.5 FA Combined with Fuzzy C-Means Algorithm

One of the earliest works, addresses the limitations in the Fuzzy C-Means algorithm (FCM), carried out for the segmentation of brain tumors using a hybrid firefly algorithm with FCM known as FAFCM [42]. The FA determines the better cluster centers in the initial phase and they were used for the initialization of the Fuzzy C-Mean algorithm which is used to determine the final optimum cluster centers. It is implemented to be a dynamic clustering method to segment the tumors. The FA initially started with predefined number of clusters and then based on the pixel colors the algorithm will increase the number of clusters where the FCM will then use the best centroids for further improvements. To evaluate the implemented algorithm, real 3 dimensional3-dimensional brain images and simulated brain images were used. The simulated MRI images of the brain were obtained from the brain-web simulated brain database (SBD) repository and the actual data from the Morphometric Analysis Center at the Massachusetts General Hospital Storage. As a result, FAFCM was able to determine the types of the brain image, tumor clusters in the abnormal brain image and the number of tumor pixels in the abnormal image without prior information, while avoiding the hereditary deficiencies of FCM.

Addressing the same limitations in FCM, Janmenjoy Nayak and others implemented a hybrid with the use of firefly algorithm with a similar name FAFCM [34] as in the previous study had [42]. This study has a similar approach of hybridizing FA with FCM but when calculating the intensity or the fitness of the fireflies they have used the objective function of the fuzzy c-means algorithm which is computed by using membership value and Euclidian distance. Glass, Iris, Lung cancer and Single outlier datasets were used to test the algorithm [54]. Comparisons were done with limited studies; however, the results indicate better performance in the proposed method providing faster convergence and optimum objective function value.

Suggesting an improvement to all fuzzy clustering related applications and mainly focusing the MRI brain image segmentation, Alomoush and his team introduced a FA based fuzzy clustering algorithm [43]. The algorithm is two phased; the first phase works with the firefly algorithm obtaining the near optimum cluster centers based on the clustering validity index which is an enhanced version of the conventional FCM proposed in [68]. Then as in the [34], those cluster centers are used as initial data set for the fuzzy C means algorithm. The main difference with [34] is the way that FA evaluate the cluster centers, simply the fitness function. Brain MRI data sets from'BrainWeb:' simulated brain database and Morphometric Analysis and IBSR Center for Morphometric analysis, Massachusetts General Hospital data storage were used, and the results are compared with the standard FCM with random initialization technique. The results showed a significance improvement over the conventional FCM with regards to the fitness value/efficiency. A chaos theory based firefly algorithm combined with fuzzy C means algorithm is used in a paper to segment brain Magnetic Resonance (MR) images (C-FAFCM) [44]. A chaotic map is used for the initialization purpose of the fireflies and to tune the absorption coefficient in FA in order to increase the global search property. The proposed algorithm is two phase; as FA has been used in the previously discussed studies [34, 43], here the initial better cluster centers are selected employing the FA. FCM objective function is used to evaluate the goodness of the obtained cluster centers from FA. The same approach of evaluating cluster centers determined by FA in [34] is used here. These outcomes are taken to set the initial values of the Fuzzy C-Means algorithm as seen in most similar studies. By doing so, reducing the chances of trapping in local optimum solutions is achieved. The concept of chaos is used to overcome the limitations in the standard FA, where the study moved steps ahead from the previously discussed same approaches based on FA and fuzzy C means. Using the Chebyshev Chaotic map [69], the parameters of the firefly algorithm is optimized with a view of improving the global search mechanism. The parameters/coefficients to adjust attractiveness and the light absorption in FA are tuned using the chaotic map. The results of the study are compared with stochastic gradient descent based fuzzy clustering (SGFC) method [70]. Synthetically generated data sets with two different classes were used. Clustering accuracy and the total execution time were measured to compare the performances. Satisfactorily better results are obtained from the proposed algorithm for both accuracy and the time.

In a latest study combining fuzzy clustering and firefly algorithm, a solution is proposed to ensure the privacy in the social networks [47]. Contrast to all the studies that has been discussed based on fuzzy clustering, firefly algorithms and their combinations, the two algorithms are used here in a different order.

A K-member version of fuzzy c-means algorithm is modified to generate balanced clusters. The method ensures including t least K members in each cluster. FA is then used to perform further optimization of the primary clusters and anonymizing the network graph and data. The algorithm is named as K-member Fuzzy Clustering Firefly Algorithm (KFCFA). Combining K member clustering with Fuzzy C means, the algorithm is capable of generating clusters that are balanced, with at the minimum of K members within every cluster. These clusters are then used by the FA to start with a set of better solutions. The proposed FA's multi-objective fitness function is used to reduce the ratio of inter-cluster inter-cluster distances, to reduce the average distortion rate in the graph and the matrix, and to minimize unbalanced clusters.

Comparisons were made with existing methods [71–74] over sub databases of popular social net- work sets of data: Facebook, Google+, Twitter and YouTube. Simulation results illustrate the effectiveness of the KFCFA algorithm against existing anonymity technologies.

2.6 FA Combined with Density Peaks Clustering Algorithms

Modifications to the density peaks clustering (DPC) algorithm [75] using the firefly algorithm is suggested in [35]. Some drawbacks of DPC such as

- manual selection of cut off distance
- lower cluster effect of the algorithm, the greater the variability of the data
- need of different local density definition methods for different data sets

are addressed. The proposed FA-based DCP algorithm merges the cut-off kernel and the Gaussian kernel defined by the DPC algorithm and the effects of the two kernels are balanced by the weight factor. Synthetic and real data sets are tested with the modified algorithm. Test results shows the superiority of the cluster performance of IDPC-FA algorithm than the DPC algorithm and its variants.

2.7 FA Combined with Markov Clustering Algorithms

To improve the clustering performance in Markov clustering (MCL), FA is used in a recent research specifically to optimize the parameters of MCL [48]. The specialty is that, here FA does not directly contribute to the process of optimizing the clusters but by optimizing the inflation parameter (rp) of MCL, it improved the ability of the MCL in better clustering. Therefore, the fireflies represented the values for the 'rp' and the brightness/fitness is determined by the clustering performance at each 'rp' value. The obtained results are compared with seven different methods where some of them used fixed parameter values and some of the others used fixed parameter values except the values for 'rp'. For the 'rp' value, they all have used a kind of an automatic tuning method. When compared with the other swarm intelligence methods used to tune 'rp', FA has shown better performances.

3 Representations, Initialization of Fireflies and Cluster Validation Measures Used by the FA Based Methods for Data Clustering

Meta-heuristic algorithms such as nature inspired algorithms are famous for nonlinear optimization due to their robustness in solving wide variety of problems. When

Fig. 1 A solution containing K centroids for a 2D data set

C_{11}	C_{12}	C_{21}	•	•	•	C_{K1}	C_{K2}

using a meta-heuristic algorithm to solve an optimization problem, several considerations are there for the success of the algorithm in solving the problem. Representation of a solution in the problem, objective function used to evaluate the fitness and the parameter adjustment techniques used are some of them. This section will present a detailed description on the solution representations, initialization procedures, performance measures and the cluster selection methods used in the discussed literature (Sects. 2.1–2.7) using FA to solve different clustering problems.

The clustering problem addressed in the studies here can be defined as follows.

The given dataset $X = \{x_1, x_2, \ldots, x_n\}$ needs to be grouped into non overlapping set of classes $P = \{p_1, p_2, \ldots, p_K\}$ where the dimension of x_i, $\{i = 1, 2, \ldots, n\}$ is D. For each cluster/class, there exist a centroid c_j, $j = \{1, 2, \ldots K\}$ which represents the centers of P.

Further, the following conditions should hold.

$$p_i \cap p_j = \emptyset \quad i, j = \{1, 2, \ldots, K\} \quad and \quad i \neq j$$
$$p_1 \cup p_2 \cup \cdots \cup p_K = X$$
$$p_i \subseteq X \quad p_i \neq 0, \quad i = 1, 2, \ldots, K$$

3.1 *Definition of Individuals (Representations)*

Representations or definition of individuals is the first step of most of all nature inspired algorithms. It can be considered as the way of linking the real world to the world of meta-heuristics. This process can be viewed as simple, but expertise knowledge is needed when simplifying or abstracting a real-world complex problem. In evolutionary computing this is known as the phenotype - genotype mapping where phenotype is the original representation and the genotype is the encoded representation. In most of nature inspired algorithms, an individual represents a possible solution to the given problem. Therefore, the representation should be better enough to represent all possible feasible solutions for the desired optimization problem. It should also not/minimize represent infeasible or wrong solutions out of the solution space.

For the clustering problem, the most common way of representing a solution is to arrange cluster centroids to the positions of each of the individual solution. That means each solution can be considered as an array containing $K \times D$ elements where the number of clusters is represented by K and D is the number of features (dimension) in the set of data. Figure 1, a feasible solution with K centroids for a 2D data set is demonstrated. In this figure C_{ij} is the j_{th} dimension of the i_{th} centroid.

Fig. 2 Representation of Fireflies dividing data points to 5 clusters, number of data points in the data set is N

Data ID	1	2	3	•	•	N-1	N
firefly 01	3	5	2	1	5	3	4
firefly 02	2	5	3	2	1	4	3

Fig. 3 Representation of Fireflies dividing data points be cluster heads or not, number of data points in the data set is n

Data	D1	D2	D3	.	.	.	Dn
firefly 1	1	0	1	1	0	1	0
firefly 2	1	1	0	0	0	1	1

Table 2 Summary of the different firefly representations used in clustering problems solved with FA based algorithms. Background colors represent different combinations of algorithms with FA: FA only, FA + K means, FA + GA, FA + DE, FA + PSO, FA + Fuzzy C Means, FA + density peaks clustering, FA + LEACH, FA + Markov clustering algorithms

Representation	Literature
A firefly represents set of cluster centroids	[26, 27, 30–34, 36, 37, 39–47]
A firefly represents set of cluster numbers assigned to each data point	[28, 29]
A firefly is a binary string showing whether each data point is a cluster head or not	[38]
A firefly represent parameter values of another clustering algorithm	[35, 48]

This representation is easy for the meta-heuristics to show a solution in a clustering problem and hence the following studies that discussed here [26, 27, 30–34, 36, 37, 39–47] used this method of solution representation.

Another type of representation assigned a cluster number for each data item and it is represented by a firefly. Hence, the length of a vector representing a firefly is equal to the data points to be clustered (see Fig. 2). This kind of representation is adopted in [28, 29].

In some of the studies, FA is not used as the primary algorithm to find the perfect clusters, instead it is used as a parameter tuning method of another clustering mechanism. There a firefly is used to represent the parameter values of the second algorithm, which has done the clustering job. References [35, 48] have used FA in that way.

Binary representation is also used in FA based clustering algorithms where each node's (data point) suitability of being a cluster head is denoted by 1 and not by 0. In [38], they have approached in that way since FA is combined with GA and hence use of crossover and mutation operators are simple with that way (Fig. 3).

A summary on the representations of fireflies used in different studies is presented in Table 2.

3.2 *Initializations*

Initialization of population/solutions is the first step of any nature inspired algorithm. Initialization is kept simple in most applications that deal with meta-heuristics. That is, the first population is seeded by randomly generated individuals. However, there is no rule of thumb and hence problem-specific heuristics can be used in this step, to create an initial population with higher fitness. However, the anytime behavior of these meta-heuristics [76] questions the need of such heuristics but some studies used them in order to accelerate the convergence of the algorithms.

The studies that we have selected for the review here, have used random initialization process on the firefly algorithm. However, in many of these studies that combines FA with another clustering mechanism, firefly algorithm itself worked as a heuristic initializer which generate better initial population that is later fed to the next algorithm to carry out the clustering task [27–29, 31, 34, 41–44].

3.3 *Performance Measures*

The algorithms implemented based on firefly algorithm to solve data clustering problems have used different types of performance measures to evaluate the goodness of the algorithms. Evaluating the goodness of a clustering algorithm is essential and the validation indices used should be carefully selected. An extensive review on evaluations and the measures of fitness for evolutionary data clustering can be found in [77]. There, they have divided the measures mainly as internal, external and fitness functions. Basically, cluster validation indices are numerical measures that are applied to judge various aspects of cluster validity. They are classified as external index which is used to measure the extent to which cluster labels match externally supplied class labels, internal index which is used to measure the goodness of a clustering structure without respect to external information and the relative index that compares two different clusterings or clusters.

This section summarizes different clustering validation indices or performance measuring techniques used with FA based clustering.

3.3.1 Accuracy

Measuring the accuracy of the algorithms, several studies have evaluated percentage of correct answers (PCA) in order to measure the performances. [26, 28–30, 35, 42, 44, 47, 48] .

3.3.2 Determining the 'Correct' Number of Clusters

When evaluating benchmark and synthetic datasets, some studies have used number of clusters as a performance measure to compare the FA based algorithm with other clustering algorithms [38, 47] .

3.3.3 External Measures

External cluster validation compares cluster results with externally known results, such as externally provided class labels. It measures the extent to which cluster labels match externally supplied class labels. Because we know the "real" cluster number in advance, this approach is primarily used to select the correct cluster algorithm for a specific set of data. The following external measures were used in the selected studies in measuring performances.

External measure	Literature
Rand Index (RI) [78]	[42]
Adjusted Rand Index (ARI) [79]	[35]
Adjusted Mutual Information (AMI) [80]	[35]
Fowlkes-Mallows index (FMI) [81]	[35]
Jaccard Index [82]	[42]
Davies-Bouldin Index (DB) [62]	[33]
F-measure [83]	[30, 42, 48]
Rogers-Tanimoto index [77]	[44]
Czekanowski-Dice Index [77]	[44]

3.3.4 Internal Measures

In contrast to the external validation measures, that use external information which are not present in the data, internal validation measures depend on information in the data. Practically, external information like class labels is often not available in many practical situations. And so, in a situation where there is no external information present, internal validation measures are the only option for cluster validation. In the discussed FA based algorithms, only two internal indices were used; CS-index and the Average Between Group Sum of Squares (ABGSS).

Internal measure	Literature
CS- index [84]	[33]
Average Between Group Sum of Squares [77]	[43]

3.3.5 Measuring Cluster Validity via Correlation

This approach computes the correlation between the proximity matrix and the incidence matrix. High correlation indicates the points that belong to the same cluster are close to each other. To measure the quality of the image segmentation results, in [41], the authors have used correlation coefficient as the measuring criteria. A higher value of correlation coefficient signifies better segmentation.

3.3.6 Fitness Functions

There can be found many studies which has been used to evaluate the cluster quality as the fitness measure of the algorithm. Since FA based algorithms are optimization algorithms, the clustering problems are converted to optimization problems and the fitness functions evaluate the goodness of the clustering via different problem related/not measures. The following different fitness measures were used in the selected studies in measuring the clustering performances.

Fitness measure	Literature
Intra cluster distance	[27, 30–32]
Inter cluster distance	[27, 31, 32]
The objective function of the fuzzy c-means algorithm (Base on membership value and Euclidian distance)	[34]
Network Lifetime (in WSN)	[36–40]
Packet Delivery Ratio of the Network (in WSN)	[36, 38, 40]
Mean Square Error	[45]
Trace Within Criteria (TRW)	[46]
Variance Ratio Criteria (VRC)	[46]

Apart from these, few studies recorded running time of the algorithm as a performance measure [29, 47].

Figure 4 summarizes the found results based on clustering validation measures used in the studies explained here.

4 Possible Further Enhancements to FA Based Clustering Algorithms

This section addresses the nature of the firefly algorithm and why it has been used for the clustering problem. As discussed in [85] about the features of FA, it has been recognized as a combination of several other meta-heuristics such as PSO and HS and hence it can perform better than such algorithms solving many problems. It is remarkable here the powerfulness of short distance attractions of FA over the long-distance attraction, which made it more capable of subdividing to swarms. This multi-swarm feature enables enormous avenues to the FA to be an effective optimizer

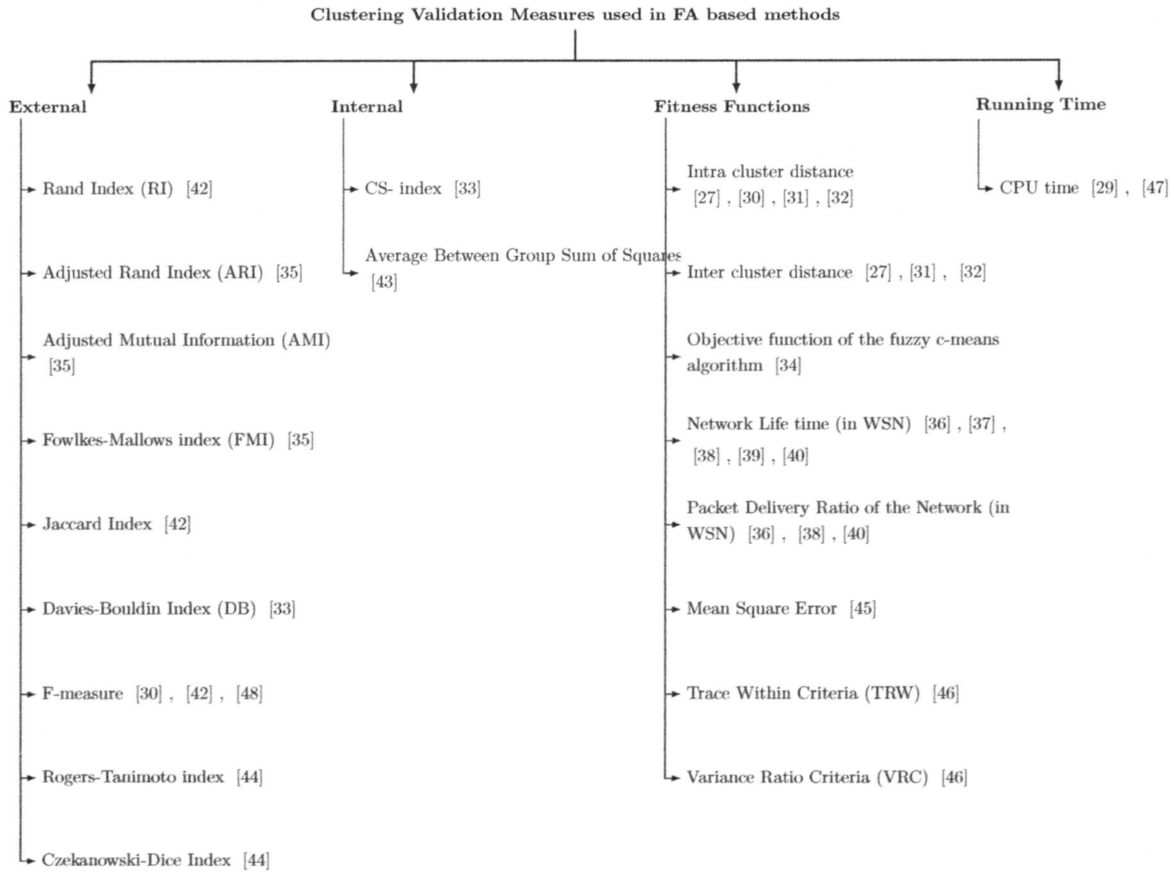

Fig. 4 A figure illustrating internal and external cluster validation measures along with the fitness functions used in FA-based clustering techniques

for multi-model problems where data clustering can also be considered to be one of such. The swarm is sub dividing into sub swarms where each sub swarm is a cluster and each cluster is having a best centroid which is the global optimum of the sub swarm. This capability is used in the studies [26, 36, 37] where FA has been used as the primary method to solve the data clustering problem. In most of the FA based studies, FA has been used as the primary algorithm to find the better cluster centroids because of this subdivision capability and then another algorithm is hybridized to the further optimization of cluster centroids.

However, by reviewing the FA based methods on data clustering, following possible enhancements to the algorithms are identified.

- The performance of the most meta-heuristics including FA, highly depend on the selected parameter values and hence a proper parameter tuning should be used in order to have the full potential of the algorithm. Yet none of these studies have used a parameter tuning mechanism to select suitable parameters for the algorithm. It is worth pointing out here as a further enhancement that can be carried out on the algorithms.
- Another important aspect regarding data clustering is the capability of algorithms in automatically detecting the $'K'$ or the effective value for the number of groups in a given dataset. Among the reviewed studies based of FA, only one method approached this automatic clustering [33]. By using the subdivision capability of the FA, this can be tried and can be touched on as an enhancement to be incorporated in the FA based studies.

5 Conclusion

This review summarizes different FA based algorithms used for data clustering. First we briefly introduced the clustering problem and the firefly algorithm. Next, we analyzed and discussed studies applying FA based algorithms to solve data clustering tasks. Different methods are discussed in Sect. 2.1 through Sect. 2.7 and later classified them in different perspectives such as representations, generation of initial solutions and cluster validation criteria. The studies considered in this paper demonstrate the powerfulness of FA in solving the data clustering problem.

Through this comprehensive study on FA based data clustering methods, following conclusions are possible.

- FA can be effectively used to solve data clustering problems.
- FA can be effectively hybridized with other meta-heuristics as well as existing clustering algorithms in order to improve the performance of such algorithms.
- FA based algorithms have been used to solve different types of clustering problems in different domains.

The aim of this review was twofold. First it summarized the research where FA has been successfully used to solve data clustering problems. The review manifested the practical capability of the FA solving data clustering problem. Secondly the review has shown the different domains in clustering where FA based methods have been applied and different representation methods, initialization methods and cluster evaluation criteria used in the studies, allowing the users to broaden their research ideas. Further, the paper provides some limitations found and possible enhancements to improve the limitations by opening more research avenues in the area of data clustering.

References

1. Boyd, S., Boyd, S.P., Vandenberghe, L.: Convex Optimization. Cambridge university press, Cambridge (2004)
2. Yang, X.S.: Nature-inspired optimization algorithms: challenges and open problems. J. Comput. Sci. **46**, 101104 (2020)
3. Holland, J.H., et al.: Adaptation in Natural and Artificial Systems: An Introductory Analysis with Applications to Biology, Control, and Artificial Intelligence. MIT press (1992)
4. Storn, R., Price, K.: Differential evolution-a simple and efficient heuristic for global optimization over continuous spaces. J. Glob. Optim. **11**(4), 341–359 (1997)
5. Kennedy, J., Eberhart, R.: Particle swarm optimization. In: Proceedings of ICNN'95-International Conference on Neural Networks, vol. 4, pp. 1942–1948. IEEE (1995)
6. Dorigo, M., Di Caro, G.: Ant colony optimization: a new meta-heuristic. In: Proceedings of the 1999 Congress on Evolutionary Computation-CEC99 (Cat. No. 99TH8406), vol. 2, pp. 1470–1477. IEEE (1999)
7. Yang, X.S.: Firefly algorithms for multimodal optimization. In: International Symposium on Stochastic Algorithms, pp. 169–178. Springer, Berlin (2009)
8. Yang, X.S., Deb, S.: Cuckoo search via lévy flights. In: World Congress on Nature & Biologically Inspired Computing (NaBIC), pp. 210–214. IEEE (2009)
9. Brabazon, A., O'Neill, M., McGarraghy, S.: Natural Computing Algorithms, vol. 554. Springer, Berlin (2015)
10. Jain, A.K., Murty, M.N., Flynn, P.J.: Data clustering: a review. ACM Comput. Surv. (CSUR) **31**(3), 264–323 (1999)
11. Brucker, P.: On the complexity of clustering problems. In: Henn, R., Korte, B., Oettli, W. (eds.) Optimization and Operations Research, pp. 45–54. Springer, Berlin (1978)
12. Welch, W.J.: Algorithmic complexity: three np-hard problems in computational statistics. J. Stat. Comput. Simul. **15**(1), 17–25 (1982)
13. Al-Sultan, K.S.: A tabu search approach to the clustering problem. Pattern Recognit. **28**(9), 1443–1451 (1995)
14. Selim, S.Z., Alsultan, K.: A simulated annealing algorithm for the clustering problem. Pattern Recognit. **24**(10), 1003–1008 (1991)
15. Akay, Ö, Tekeli, E., Yüksel, G.: Genetic algorithm with new fitness function for clustering. Iran. J. Sci. Technol., Trans. A: Sci. **44**, 865–874 (2020)
16. Paterlini, S., Krink, T.: Differential evolution and particle swarm optimisation in partitional clustering. Comput. Stat. Data Anal. **50**(5), 1220–1247 (2006)
17. Al Radhwani, A.M.N., Algamal, Z.Y.: Improving k-means clustering based on firefly algorithm. J. Phys.: Conf. Ser. **1897**, 012004 (2021). IOP Publishing
18. Jadhav, A.N., Gomathi, N.: Hybridization of exponential grey wolf optimizer with whale optimization for data clustering. Alex. Eng. J. **57**(3), 1569–1584 (2018)
19. Behravan, I., Zahiri, S.H., Razavi, S.M., Trasarti, R.: Finding roles of players in football using automatic particle swarm optimization-clustering algorithm. Big Data **7**(1), 35–56 (2019)

20. Srivatsava, P.R., Mallikarjun, B., Yang, X.S.: Optimal test sequence generation using firefly algorithm. Swarm Evol. Comput. **8**, 44–53 (2013)
21. Imanirad, R., Yang, X.S., Yeomans, J.S.: Modelling-to-generate-alternatives via the firefly algorithm. J. Appl. Oper. Res. **5**(1), 14–21 (2013)
22. Udaiyakumar, K., Chandrasekaran, M.: Application of firefly algorithm in job shop scheduling problem for minimization of makespan. Procedia Eng. **97**, 1798–1807 (2014)
23. Kavousi-Fard, A., Samet, H., Marzbani, F.: A new hybrid modified firefly algorithm and support vector regression model for accurate short term load forecasting. Exp. Syst. Appl. **41**(13), 6047–6056 (2014)
24. Ariyaratne, M., Fernando, T., Weerakoon, S.: Solving systems of nonlinear equations using a modified firefly algorithm (modfa). Swarm Evol. Comput. **48**, 72–92 (2019)
25. Yelghi, A., Köse, C.: A modified firefly algorithm for global minimum optimization. Appl. Soft Comput. **62**, 29–44 (2018)
26. Senthilnath, J., Omkar, S.N., Mani, V.: Clustering using firefly algorithm: performance study. Swarm Evol. Comput. **1**(3), 164–171 (2011)
27. Hassanzadeh, T., Meybodi, M.R.: A new hybrid approach for data clustering using firefly algorithm and k-means. In: The 16th CSI International Symposium on Artificial Intelligence and Signal Processing (AISP 2012), pp. 007–011. IEEE (2012)
28. Mizuno, K., Takamatsu, S., Shimoyama, T., Nishihara, S.: Fireflies can find groups for data clustering. In: 2016 IEEE International Conference on Industrial Technology (ICIT), pp. 746–751. IEEE (2016)
29. Zhou, L., Li, L.: Improvement of the firefly-based k-means clustering algorithm. In: Proceedings of the 2018 International Conference on Data Science, pp. 157–162 (2018)
30. Xie, H., Zhang, L., Lim, C.P., Yu, Y., Liu, C., Liu, H., et al.: Improving k-means clustering with enhanced firefly algorithms. Appl. Soft Comput. **84**, 105763 (2019)
31. Kaushik, K., Arora, V., et al.: A hybrid data clustering using firefly algorithm based improved genetic algorithm. Procedia Comput. Sci. **58**, 249–256 (2015)
32. Sadeghzadeh M.: Data clustering using improved fire fly algorithm. In: Information Technology: New Generations, pp. 801–809. Springer, Berlin (2016)
33. Agbaje, M.B., Ezugwu, A.E., Els, R.: Automatic data clustering using hybrid firefly particle swarm optimization algorithm. IEEE Access **7**, 184963–184984 (2019)
34. Nayak, J., Nanda, M., Nayak, K., Naik, B., Behera, H.S.: An improved firefly fuzzy c-means (fafcm) algorithm for clustering real world data sets. In: Advanced Computing, Networking and Informatics, Vol. 1, pp. 339–348. Springer, Berlin (2014)
35. Zhao, J., Tang, J., Shi, A., Fan, T., Xu, L.: Improved density peaks clustering based on firefly algorithm. Int. J. Bio-Inspired Comput. **15**(1), 24–42 (2020)
36. Manshahia, M.S., Dave, M., Singh, S.: Firefly algorithm based clustering technique for wireless sensor networks. In: 2016 International Conference on Wireless Communications, Signal Processing and Networking (WiSPNET), pp. 1273–1276. IEEE (2016)
37. Sarma, N., Gopi, M.: Implementation of energy efficient clustering using firefly algorithm in wireless sensor networks. Int. Proc. Comput. Sci. Inf. Technol. **59**, 1 (2014)
38. Baskaran, M., Sadagopan, C.: Synchronous firefly algorithm for cluster head selection in WSN. Sci. World J. (2015)
39. Pitchaimanickam, B., Murugaboopathi, G.: A hybrid firefly algorithm with particle swarm optimization for energy efficient optimal cluster head selection in wireless sensor networks. Neural Comput. Appl. **32**(12), 7709–7723 (2020)
40. Mosavvar, I., Ghaffari, A.: Data aggregation in wireless sensor networks using firefly algorithm. Wirel. Pers. Commun. **104**(1), 307–324 (2019)
41. Sharma, A., Sehgal, S.: Image segmentation using firefly algorithm. In: 2016 International Conference on Information Technology (InCITe)-The Next Generation IT Summit on the Theme-Internet of Things: Connect your Worlds, pp. 99–102. IEEE (2016)
42. Alsmadi, M.K.: A hybrid firefly algorithm with fuzzy-c mean algorithm for mri brain segmentation. Am. J. Appl. Sci. **11**(9), 1676–1691 (2014)

43. Alomoush, W., Sheikh Abdullah, S., Sahran, S., Hussain, R.: Segmentation of mri brain images using fcm improved by firefly algorithms. J. Appl. Sci. **14**, 66–71 (2014)
44. Ghosh, P., Mali, K., Das, S.K.: Chaotic firefly algorithm-based fuzzy c-means algorithm for segmentation of brain tissues in magnetic resonance images. J. Vis. Commun. Image Represent. **54**, 63–79 (2018)
45. Kuo, R., Li, P.: Taiwanese export trade forecasting using firefly algorithm based k-means algorithm and svr with wavelet transform. Comput. Indus. Eng. **99**, 153–161 (2016)
46. Banati, H., Bajaj, M.: Performance analysis of firefly algorithm for data clustering. Int. J. Swarm Intell. **1**(1), 19–35 (2013)
47. Langari, R.K., Sardar, S., Mousavi, S.A.A., Radfar, R.: Combined fuzzy clustering and firefly algorithm for privacy preserving in social networks. Exp. Syst. Appl. **141**, 112968 (2020)
48. Lei, X., Wang, F., Wu, F.X., Zhang, A., Pedrycz, W.: Protein complex identification through markov clustering with firefly algorithm on dynamic protein-protein interaction networks. Inf. Sci. **329**, 303–316 (2016)
49. Heinzelman, W.R., Chandrakasan, A., Balakrishnan, H.: Energy-efficient communication protocol for wireless microsensor networks. In: Proceedings of the 33rd Annual Hawaii International Conference on System Sciences, p. 10. IEEE (2000)
50. Lou, C., Gao, X., Wu, F., Chen, G.: Energy-aware clustering and routing scheme in wireless sensor network. In: International Conference on Wireless Algorithms, Systems, and Applications, pp. 386–395. Springer, Berlin (2015)
51. Abirami, T., Anandamurugan, S.: Data aggregation in wireless sensor network using shuffled frog algorithm. Wirel. Pers. Commun. **90**(2), 537–549 (2016)
52. Dua D., Graff C.: UCI Machine Learning Repository (2017)
53. Labati, R.D., Piuri, V., Scotti, F.: All-idb: The acute lymphoblastic leukemia image database for image processing. In: 2011 18th IEEE International Conference on Image Processing, pp. 2045–2048. IEEE (2011)
54. Blake C.: Uci Repository of Machine Learning Databases (1998). https://www.icsuciedu/mlearn/MLRepository.html
55. Łukasik, S., Zak, S.: Firefly algorithm for continuous constrained optimization tasks. In: International Conference on Computational Collective Intelligence, pp. 97–106. Springer, Berlin (2009)
56. Hoos, H.H., Stützle T.: Stochastic Local Search: Foundations and Applications. Elsevier (2004)
57. Fisher, R.A.: The use of multiple measurements in taxonomic problems. Ann. Eugen. **7**(2), 179–188 (1936)
58. Aeberhard, S., Coomans, D., De Vel, O.: Comparative analysis of statistical pattern recognition methods in high dimensional settings. Pattern Recognit. **27**(8), 1065–1077 (1994)
59. Coomans, D., Jonckheer, M., Massart, D.L., Broeckaert, I., Blockx, P.: The application of linear discriminant analysis in the diagnosis of thyroid diseases. Anal. Chim. Acta **103**(4), 409–415 (1978)
60. Street, W.N., Wolberg, W.H., Mangasarian, O.L.: Nuclear feature extraction for breast tumor diagnosis. In: International Society for Optics and Photonics Biomedical Image Processing and Biomedical Visualization, vol. 1905, pp. 861–870 (1993)
61. Chou, C.H., Su, M.C., Lai, E.: A new cluster validity measure and its application to image compression. Pattern Anal. Appl. **7**(2), 205–220 (2004)
62. Davies, D.L., Bouldin, D.W.: A cluster separation measure. IEEE Trans. Pattern Anal. Mach. Intell. **2**, 224–227 (1979)
63. Zhang, X., Li, J., Yu, H.: Local density adaptive similarity measurement for spectral clustering. Pattern Recognit. Lett. **32**(2), 352–358 (2011)
64. Das, S., Abraham, A., Konar, A.: Automatic clustering using an improved differential evolution algorithm. IEEE Trans. Syst. Man Cybern.-Part A: Syst. Hum **38**(1), 218–237 (2007)
65. Bandyopadhyay, S., Maulik, U.: Genetic clustering for automatic evolution of clusters and application to image classification. Pattern Recognit. **35**(6), 1197–1208 (2002)
66. Omran, M., Salman, A., Engelbrecht, A.: Dynamic clustering using particle swarm optimization with application in unsupervised image classification. In: Fifth World Enformatika Conference (ICCI 2005), pp. 199–204. Prague, Czech Republic (2005)

67. Heinzelman, W.B., Chandrakasan, A.P., Balakrishnan, H.: An application-specific protocol architecture for wireless microsensor networks. IEEE Trans. Wirel. Commun. **1**(4), 660–670 (2002)
68. Hathaway, R.J., Bezdek, J.C.: Optimization of clustering criteria by reformulation. IEEE Trans. Fuzzy Syst. **3**(2), 241–245 (1995)
69. Gandomi, A.H., Yang, X.S., Talatahari, S., Alavi, A.H.: Firefly algorithm with chaos. Commun. Nonlinear Sci. Numer. Simul. **18**(1), 89–98 (2013)
70. Wang, Y., Chen, L., Mei, J.P.: Stochastic gradient descent based fuzzy clustering for large data. In: 2014 IEEE International Conference on Fuzzy Systems (FUZZ-IEEE), pp. 2511–2518. IEEE (2014)
71. Sweeney, L.: k-anonymity: a model for protecting privacy. Int. J. Uncertain. Fuzziness Knowl. Based Syst. **10**(05), 557–570 (2002)
72. Sun, X., Sun, L., Wang, H.: Extended k-anonymity models against sensitive attribute disclosure. Comput. Commun. **34**(4), 526–535 (2011)
73. Honda, K., Kawano, A., Notsu, A., Ichihashi, H.: A fuzzy variant of k-member clustering for collaborative filtering with data anonymization. In: 2012 IEEE International Conference on Fuzzy Systems, pp. 1–6. IEEE (2012)
74. Rahimi, M., Bateni, M., Mohammadinejad, H.: Extended k-anonymity model for privacy preserving on micro data. Int. J. Comput. Netw. Inf. Secur. **7**(12), 42–51 (2015)
75. Rodriguez, A., Laio, A.: Clustering by fast search and find of density peaks. Science **344**(6191), 1492–1496 (2014)
76. Zilberstein, S.: Using anytime algorithms in intelligent systems. AI Mag. **17**(3), 73–73 (1996)
77. Aljarah, I., Habib, M., Nujoom, R., Faris, H., Mirjalili, S.: A comprehensive review of evaluation and fitness measures for evolutionary data clustering. Algorithms Appl. Evol. Data Clust. **23** (2021)
78. Rand, W.M.: Objective criteria for the evaluation of clustering methods. J. Am. Stat. Assoc. **66**(336), 846–850 (1971)
79. Hubert, L., Arabie, P.: Comparing partitions. J. Classification **2**(1), 193–218 (1985)
80. Vinh, N.X., Epps, J., Bailey, J.: Information theoretic measures for clusterings comparison: variants, properties, normalization and correction for chance. J. Mach. Learn. Res. **11**, 2837–2854 (2010)
81. Fowlkes, E.B., Mallows, C.L.: A method for comparing two hierarchical clusterings. J. Am. Stat. Assoc. **78**(383), 553–569 (1983)
82. Jaccard, P.: Distribution comparée de la flore alpine dans quelques régions des alpes occidentales et orientales. Bulletin de la Murithienne **31**, 81–92 (1902)
83. Chinchor, N., Sundheim, B.M.: Muc-5 evaluation metrics. In: Fifth Message Understanding Conference (MUC-5): Proceedings of a Conference Held in Baltimore. Maryland (1993)
84. Chou, C.H., Su, M.C., Lai, E.: A new cluster validity measure for clusters with different densities. In: IASTED International Conference on Intelligent Systems and Control, pp. 276–281 (2003)
85. Yang, X.S., He, X.S.: Why the firefly algorithm works? Stud. Comput. Intell. 245–259 (2017)

A Hybrid African Vulture Optimization Algorithm and Harmony Search: Algorithm and Application in Clustering

Farhad Soleimanian Gharehchopogh, Benyamin Abdollahzadeh, Nima Khodadadi, and Seyedali Mirjalili

Abstract Data clustering is one of the necessary research fields in data analysis. Clustering is an unsupervised classification method for assigning data objects to separate groups, which are called clusters. So that the similarity of the data within each cluster and the difference between the cluster data is high, a variety of meta-heuristic algorithms can be used to solve this problem. In this paper, a new algorithm created using a combination of African Vulture Optimization Algorithm (AVOA) and Harmony Search (HA) is used. The proposed algorithm is implemented on the clustering dataset of the UCI machine learning repository. Furthermore, the results obtained from the proposed algorithm are compared with other meta-heuristic algorithms. The experiments show that the proposed method has good and better performance than other optimization algorithms.

Keywords Data clustering · African vulture optimization algorithm · Harmony search algorithm · Hybridization · Optimization

1 Introduction

Data clustering is one of the most important research fields in data analysis. Data clustering can find helpful information and has been widely used in market research, data analysis, pattern recognition, image processing and web document classification [1]. Data clustering is an unsupervised classification method for assigning data

F. S. Gharehchopogh · B. Abdollahzadeh
Department of Computer Engineering, Urmia Branch, Islamic Azad University, Urmia, Iran

N. Khodadadi
Department of Civil and Environmental Engineering, Florida International University, Florida, USA

S. Mirjalili (✉)
Centre for Artificial Intelligence Research and Optimisation, Torrens University Australia, Fortitude Valley, Brisbane, QLD 4006, Australia
e-mail: ali.mirjalili@gmail.com

Yonsei Frontier Lab, Yonsei University, Seoul, South Korea

A. Biswas et al. (eds.), *Advances in Swarm Intelligence*, Studies in Computational Intelligence 1054, https://doi.org/10.1007/978-3-031-09835-2_13

objects to separator groups called clusters so that the points with many similarities fall in the same cluster. Data clustering is used to classify data and is obtained to identify distinctive features and diagnose abnormalities [2]. It can be said that data clustering is one of the most challenging tasks, especially in pattern recognition, image analysis and other complex applications. K-means algorithm is one of the most well-known data clustering algorithms, a fast, simple and center-based algorithm. The primary task of the K-means algorithm is to find these clusters so that the square error between the points in the cluster and the overall clustering average is reduced [3].

Most clustering algorithms face challenges such as the problem of cluster center determination, low clustering accuracy, inadequate clustering efficiency of different data sets, and the dependence of tangible parameters. Of course, the problem of data clustering can be considered an optimization problem. Moreover, meta-heuristic methods can be used to solve it. In other words, meta-heuristic algorithms can obtain optimal or near-optimal solutions in a reasonable time for the data clustering problem. The critical issue in these algorithms is defining the objective function and producing the appropriate solution to the problem. In the case of data clustering, these algorithms can generate solutions according to the data set and minimize or maximize the objective function of clustering. So far, many meta-heuristic algorithms have been proposed by researchers. This paper presents a new hybrid algorithm using AVOA and HS algorithms to solve the clustering problem.

The structure of this paper is as follows: in the next section to the research done by meta-heuristic algorithms that have been done in the field of data clustering and Sect. 3, to the basic concepts and in Sect. 4, the proposed method is explained in detail. Furthermore, in Sect. 5, the proposed method is reviewed and evaluated, and finally, in the last section, conclusions and future works are discussed.

2 Related Work

In this section, the work done in the field of clustering with meta-heuristic algorithms is discussed. Of course, here we will look at works that are relevant to the last few years. Ahmadi et al. [4] have provided an improved version of the Grey Wolf Optimization (GWO) algorithm for the clustering problem. The improved GWO performance is compared with seven other clustering methods in nine datasets of the UCI Machine Learning Lab. The results of various experiments show that the proposed algorithm performs better in solving data clustering problems. The results show that the intra-cluster distance of the proposed algorithm is less than other algorithms.

Ashish et al. [5] have provided a fast and efficient parallel bat algorithm for data clustering using mapping reduction architecture. Also, Shufen et al. [6] have proposed a space-based quantum step clustering algorithm using quantum step for clustering analysis. In the space-based quantum step clustering algorithm, data points are considered as participants in the march. Furthermore, similar data points are clustered according to a particular rule using the step function in the pay-off matrix. Simulation experiments have tested the Quantum step clustering algorithm. Furthermore, the results show that the quantum step clustering algorithm has a superior performance to the existing clustering algorithms, namely K-means, PCA + K-means and LDA-Km.

Saida et al. [7] introduced the Quantum *chaotic* Cuckoo Search Algorithm in 2017. The results of various experiments were performed on six data sets. The results of various experiments show that the quantum *chaotic* cuckoo search algorithm performs better in all data sets in terms of internal and external clustering quality than the comparative algorithm.

Geng et al. [8], In 2018, to improve the K-means algorithm's accuracy and stability and determine the most appropriate number of K clusters and the best initial data, they have presented an improved K-means algorithm based on density canopy. Ahmad Tajudeen et al. [9] investigated the application of multidimensional optimization algorithms in data clustering. This paper discusses the theoretical foundations, operations, and main strengths of the multidimensional optimization algorithm. Finally, the multidimensional optimization algorithm was evaluated qualitatively and quantitatively by several datasets. The results show that the clustering algorithm based on multidimensional optimizer performs better than several meta-heuristic algorithms in terms of intra-cluster distance, cluster homogeneity and clustering completeness.

Qaddoura et al. [10] provide an improved version of the evolutionary behavior algorithm of the genetic algorithm and an advanced version of the nearest neighbor search technique based on allocation and selection mechanisms for the data clustering problem. The proposed algorithm aims to improve the quality of clustering results by finding a solution that maximizes the separation between different clusters and maximizes coherence between data points in an identical cluster. The results of various experiments show that the proposed algorithm works well with the objective function of the Silhouette coefficient. Furthermore, the proposed performance of the data set works better than other algorithms. Zhou and et al. [11] provide an improved version of the coexistence organism search algorithm to solve clustering problems. In this paper, the proposed method is implemented on ten standard datasets of the UCI machine learning repository and compared with other optimization algorithms. The results of various experiments show that the coexistence organism search algorithm has performed better in terms of accuracy and precision criteria and has better performance than other algorithms.

Rahnema and Gharehchopogh present an improved artificial bee algorithm based on the whale optimization algorithm for data clustering in 2020 [12]. In this paper, two memories called random memory and elite memory in artificial bee algorithm have been used to solve the problem of late detection and convergence. Finally, the proposed method was implemented on ten standard datasets of the UCI machine learning repository for evaluation. It has also been compared with other meta-heuristic algorithms in terms of statistical criteria and ANOVA tests. The results of the simulation of the proposed algorithm show that the proposed algorithm performed better than other meta-heuristic algorithms.

3 Background Knowledge

This section provides the essential knowledge required for creating the proposed metaheuristic algorithm. For this purpose, the following items will be discussed in this section: Data clustering problem, AVOA, and HS.

3.1 Data Clustering Problem

Data clustering can be expressed by segmenting or segmenting a data set (with the number of N samples and D feature) to the number of K in clusters. This segmentation is done in a dimensional D space based on similarity criteria such as Euclidean distance, cosine, etc. So that data patterns belonging to a cluster should have more coherency. However, there must be more differences or distances between different data clusters. So, according to the above definitions if $S = (x_1, x_2, x_3, \ldots, x_n)$ be considered as a data set with N samples, then it can be said that each instance in this data set has a D dimension which can be expressed as $(x_{i1}, x_{i2}, x_{i3}, \ldots, x_{id})$. Finally, if we want to divide this data set into K sections or clusters, it can be expressed as Eqs. (1) and (2).

$$C = (C_1, C_2, C_3, \ldots, C_k) \tag{1}$$

$$\begin{aligned} & C_j \neq \emptyset, \forall j\{1, 2, 3, \ldots, k\} \\ & \cup_{j=1}^{k} C_j = S \\ & C_j \cap C_i = \emptyset, \forall j \neq i \text{ and } j, i \in \{1, 2, 3, \ldots, k\} \end{aligned} \tag{2}$$

According to Eqs. (1) and (2), the data set is divided into K clusters. Furthermore, none of the clusters are empty. Also, the sum of all clusters is equal to the total data set S, that there is no commonality between the samples in each cluster. This clustering problem can be defined as an optimization problem. The most commonly used objective function can be expressed using Eq. (3):

$$f(S, C) = \sum_{j=1}^{n} \min\left\{ \left\| x_j - C_j \right\|^2 \right\} \quad i = \{1, 2, 3, \ldots, k\} \tag{3}$$

According to Eq. (3), each data sample is assigned to a header with the least distance. This distance is based on various criteria, which is the most critical and standard Euclidean method. Furthermore, it is shown mathematically in Eq. (4).

$$d(x_i, x_j) = \sqrt{\sum_{m=1}^{d} \left| x_{im} - x_{jm} \right|^2} \tag{4}$$

In Eq. (4), the distance of each property from the header and each property from the data is obtained. In the proposed algorithm, the goal is to select the heads of clusters that minimize this distance.

3.2 African Vulture Optimization Algorithm

The AVOA proposed by Abdollahzadeh et al. [13] is inspired and formulated based on how African vultures eat and live together. The AVOA algorithm uses two different processes in the exploration phase shown in Eqs. (5) and (6).

$$P(i+1) = R(i) - D(i) \times F \tag{5}$$

$$D(i) = |X \times R(i) - P(i)| \tag{6}$$

In Eq. (6), $R(i)$ is one of the two best vultures which is selected using the roulette cycle. $D(i)$ is also calculated using Eq. (7). F is also calculated using Eq. (8). $P(i)$ is also the position of the current vulture vector. X is a random number between 0 and 2.

$$t = h \times \left(\sin^{w}\left(\frac{\pi}{2} \times \frac{\text{iteration}_i}{\text{max iterations}}\right) + \cos\left(\frac{\pi}{2} \times \frac{\text{iteration}_i}{\text{max iterations}}\right)\right) \tag{7}$$

$$F = (2 \times rand_1 + 1) \times z \times \left(1 - \frac{\text{iteration}_i}{\text{max iterations}}\right) + t \tag{8}$$

In Eq. (7), h is a random number between -2 and 2. sin and cos represent the sine and cosine functions. $iteration_i$ is the current iteration value. $maxiterations$ is the total amount of iterations to perform the optimization operation. w is set to a constant value and $rand_1$ is a random number between 0 and 1.

$$P(i+1) = R(i) - F + rand_2 \times ((ub - lb) \times rand_3 + lb) \tag{9}$$

In Eq. (9), ub and lb are the upper and lower bounds of the problem, respectively. $rand_2$ and $rand_3$ are random numbers between 0 and 1.

Exploitation: The AVOA uses four different processes for optimization operations in the exploitation phase, which are: Competition for Food, Rotating flight of Vultures, The accumulation of several types of vultures over the food source, aggressive Competition for Food.

The Competition for Food mechanism is used to simulate vultures competing for food. In Eq. (10), $D(i)$ is calculated using Eq. (7) and F also using Eq. (8). $rand_4$ is a random number between 0 and 1. Finally, $d(t)$ is calculated using Eq. (11).

$$P(i+1) = D(i) \times (F + rand_4) - d(t) \tag{10}$$

$$d(t) = R(i) - P(i) \tag{11}$$

The Rotating flight of Vultures mechanism is used to simulate the rotating flight of vultures in nature.

In Eq. (12), $rand_5$ and $rand_6$ are random numbers between 0 and 1. cos and sin represent the functions of sine and cosine, respectively. Finally, the new position of the vultures is calculated using Eq. (13).

$$\begin{aligned} S_1 &= R(i) \times \left(\frac{rand_5 \times P(i)}{2\pi}\right) \times \cos(P(i)) \\ S_2 &= R(i) \times \left(\frac{rand_6 \times P(i)}{2\pi}\right) \times \sin(P(i)) \end{aligned} \tag{12}$$

$$P(i+1) = R(i) - (S_1 + S_2) \tag{13}$$

In the wild, large numbers of vultures congregate on food sources to obtain food and compete. Equations (14) and (15) have been used to simulate this behavior.

$$\begin{aligned} A_1 &= \text{Best Vulture}_1(i) - \frac{\text{Best Vulture}_1(i) \times P(i)}{\text{Best Vulture}_1(i) \times P(i)^2} \times F \\ A_2 &= \text{Best Vulture}_2(i) - \frac{\text{Best Vulture}_2(i) \times P(i)}{\text{Best Vulture}_2(i) \times P(i)^2} \times F \end{aligned} \tag{14}$$

$$P(i+1) = \frac{A_1 + A_2}{2} \tag{15}$$

In Eq. (14), BestVulture_1 and BestVulture_2 are the first and second vultures of the total population, respectively. F is also calculated using Eq. (8). Finally, the position of the new vulture vector is calculated using Eq. (15). Sometimes vultures compete to obtain food, and this competition can lead to much violence.

$$P(i+1) = R(i) - |d(t)| \times F \times Levy(d) \tag{16}$$

Equation (16) is used to simulate this behavior of vultures. In Eq. 16, d(t) t is calculated using Eq. (11). F is also calculated using Eq. (8). $Levy(d)$ represents a random number in the Levy distribution [14], which is calculated using Eq. (17).

$$LF(x) = 0.01 \times \frac{u \times \sigma}{|v|^{\frac{1}{\beta}}}, \sigma = \left(\frac{\Gamma(1+\beta) \times \sin\left(\frac{\pi\beta}{2}\right)}{\Gamma(1+\beta 2) \times \beta \times 2\left(\frac{\beta-1}{2}\right)}\right)^{\frac{1}{\beta}} \tag{17}$$

In Eq. (17), d indicates the dimensions of the problem. u and v are random numbers in the range 0 and 1. β is also a fixed number with a value of 1.5.

3.3 Harmony Search Algorithm

HS algorithm with the purpose of harmony, to achieve the best answer, is inspired by music. Trying to find this harmony in music is like finding the optimal conditions in the optimization process. The HS algorithm is the best strategy for transforming qualitatively studied processes into quantitative and tangible optimization processes. A process with some ideal rules will turn beautiful pieces of music into a suitable solution for solving various optimization problems. The optimization steps in the harmonic algorithm are described below:

Initialization of Harmony Memory (HM): HM is created according to the solution matrix in the dimensions of the problem in HMS. All the elements in the harmonic memory matrix represent a solution. Solutions are created randomly and are sorted according to the defined cost function.

Improvise New Harmony: The new harmonic vector improves according to the HS process.

$$a_i = a_1, a_2, a_3, \ldots, a_N \tag{18}$$

HMCR, $PAR_{\max}$ and $PAR_{\min}$ are being processed, and new values are being updated.

$$a_i \in \left\{a_i \in \left\{a_i^1, a_i^2, a_i^3, \ldots, a_i^{HMS}\right\} w \cdot pHMCRa_i \in Ai\right\} \tag{19}$$

Due to the adjustment of all decision variables, a search process is added to the new harmonic vector. The PAR operator is taken from HM.

$$a_i \leftarrow \{\text{Pitch adjusted } w \cdot p\,PAR \text{ No change } w \cdot p(1 - PAR)\} \tag{20}$$

Each number produced $rd\ \varepsilon[0, 1]$ is in the probability range PAR. And a new a_i evaluation variable is computed in Eq. (21):

$$a_i = (a_i + rd()) \times bw \tag{21}$$

In Eq. 21, bw is a random bandwidth of distance. Updating the harmony memory: After generating the new position vector by HM, the cost of the new harmonic vector is calculated according to the cost function f(a). If the value of the objective function of the new vector is better than the current harmonic vector, then it is stored in HM; otherwise, the new vector is ignored.

4 AVOAHS Algorithm

To hybridize, the two AVOA and the HS algorithm, the main framework of the HS algorithm, have been used. In the way that the initial solutions are created, and then the harmonic memory is also created. However, in the production of new harmonies, the whole population is divided into two parts. The first half of the population is updated by the AVOA algorithm. Moreover, the second half of the search engine population is updated using the mechanisms in HS. Then the cost of the new solutions produced is then calculated using the objective function. If the cost of new solutions is better than the cost of existing solutions, it is stored in HM. Otherwise, new solutions will be ignored. The above process is shown in Fig. 1.

The main reason for combining the two AVOA and HS algorithms is to balance the resonance and diversity components. The AVOA algorithm focuses more on the resonant component during the optimization operation, which can lead to a drop in local optimality. Because there is no guarantee of being close to the global optimum in the early stages of optimization, however, if combined with the HS algorithm in all optimization stages, the diversity component can be increased. Because HS has good capabilities in the diversity component and early convergence and escapes from the optimal local trap can help.

On the other hand, the temporal complexity of the algorithm does not increase. Because for each search factor, only one update operation is performed with one of the AVOA and HS algorithms. The first step in the AVOAHS algorithm to solve the clustering problem is related to the structure of each solution. The structure of any solution must be adjusted so that it can solve a clustering problem.

According to Eqs. (1)–(4), if the data set contains N samples with D dimensions, the AVOAHS solution must be a one-dimensional representation of size K* to divide that K cluster. Each solution in the AVOAHS algorithm population represents a clustering solution, calculated by the objective function using Eq. (3).

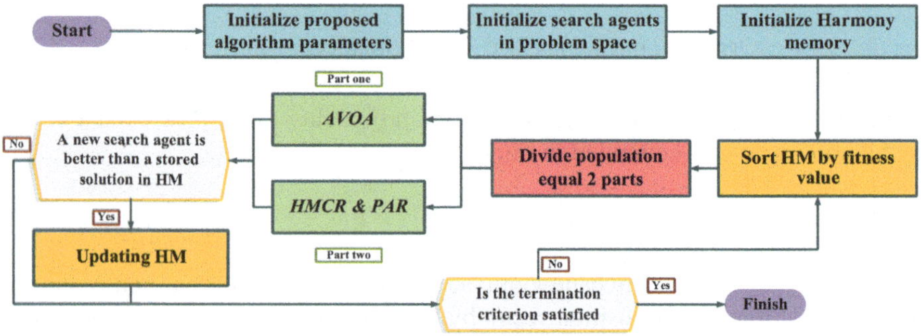

Fig. 1 Flowchart of AVOAHS

5 Result and Discussion

In this section, the proposed algorithm and other optimization algorithms, AVOA[13], CSA [15], MBO [16], HS [17], and ABC [18] on ten standard datasets taken from the UCI repository, are being performed. The descriptions of the datasets used are shown in Table 1. Moreover, its results are shown graphically in Table 2.

In order to compare and evaluate the population number of all algorithms, 30 and in a maximum of 100 repetitions in 30 independent performances have been performed. Also, the proposed algorithm and other comparative algorithms have been used for statistical comparison using the ANOVA test. On the other hand, the parameter settings of all the comparable algorithms are by using the settings used in the main works AVOA [15], CSA [17], MBO [18], HS [19], ABC [20]. The results of the experiments are shown in Table 2 and Figs. 1, 2 and 3.

According to the results shown in Table 2, it is clear that the proposed AVOAHS algorithm has a better and significant performance in all data sets compared to other compared optimization algorithms, and it has been able to achieve good results with quality. On the other hand, AVOA and HS algorithms have shown good performance after the proposed model. According to the results obtained from the tested algorithms, it is simply evident that performance is significantly increased when combining AVOA and HS algorithms. In the following, Fig. 1 shows the convergence diagrams obtained from the optimized algorithms tested.

Figure 2, which shows the convergence diagrams obtained from the experiments of the proposed algorithm and other optimization algorithms, shows that the AVOAHS algorithm has a rapidly decreasing pattern and compared to other optimization algorithms, in all stages of optimization, it has tried to find better solutions and has performed better.

Figure 1 shows the Plat box samples related to the ANOVA statistical test. To perform this experiment, we used each Pilot box separately based on the population of each algorithm to show the effect of the operators of each algorithm on the total

Table 1 Clustering dataset

Data set	Instance number	Number of classes	Number of features
Iris	150	3	4
Wine	178	3	13
CMC	1473	3	9
Glass	214	6	9
Vowel	871	6	3
Cancer	683	2	9
Zoo	101	2	16
Spect	267	2	22
Ionosphere	351	2	34
Blood	748	2	4

Table 2 Comparison of the AVOAHS with another optimization algorithm

AVOAHS	AVOA	CSA	MBO	HS	ABC	Metric	Dataset
96.89	99.16	113.05	98.47	97.38	114.73	Best	Iris
96.92	99.16	117.37	556.7	99.56	267.78	Worst	
96.9	99.16	115.52	246.78	98.85	179.86	Mean	
0.01	0	0.94	191.65	0.69	37.04	STD	
16,300.79	16,389.91	16,622.06	16,838.48	16,300.58	16,490.8	Best	Wine
16,301.06	16,389.93	16,700.86	83,800.87	16,310.16	18,714.79	Worst	
16,300.97	16,389.91	16,655.29	33,257.94	16,307.65	17,153.03	Mean	
0.07	0	19.27	24,385.37	2.06	611.16	STD	
5576.43	5715.63	6057.6	6168.17	5647.56	5797.84	Best	CMC
5577.68	5716.19	6161.42	25,810.73	5712.11	9802.07	Worst	
5576.95	5715.83	6126.03	10,366.29	5690.41	7812.89	Mean	
0.37	0.19	22.8	6216.96	20.71	941.4	STD	
2987.24	2992.58	4087.99	3369.17	3018.05	3298.68	Best	Cancer
2989.07	2992.62	4260.14	4818.06	3112.89	9104.87	Worst	
2988.3	2992.59	4188.14	3664.76	3085.85	5359	Mean	
0.77	0.02	42.96	441.6	22.2	1706.3	STD	
271.49	278.63	371.63	371.47	291.96	311.75	Best	Glass
271.83	278.73	401.52	1330.38	316.96	811.82	Worst	
271.7	278.65	386.91	680.38	311.2	544.87	Mean	
0.09	0.04	5.89	363.7	6.45	94.95	STD	

(continued)

Table 2 (continued)

AVOAHS	AVOA	CSA	MBO	HS	ABC	Metric	Dataset
153,945.25	165,157.23	174,967.43	222,940.36	160,885.97	190,568.92	Best	Vowel
154,106.09	165,166.66	182,674.49	1,036,524.94	173,665.14	407,745.63	Worst	
154,024.66	165,157.97	179,942.24	455,771.77	169,874.97	251,353.97	Mean	
50.21	2.24	2211.77	312,769.94	3090.51	43,406.54	STD	
185.15	193.98	206.32	193.29	186.44	191.03	Best	Zoo
186.52	194.03	208.54	404.39	189.64	265.25	Worst	
185.68	193.99	207.33	227.96	188.94	207.06	Mean	
0.45	0.01	0.6	64.91	0.74	16.96	STD	
504.3	506.33	557.8	516.77	514.34	515.16	Best	Spect
504.52	506.33	563.52	596.06	529.3	605.76	Worst	
504.45	506.33	561.52	529.5	525.57	564.23	Mean	
0.05	0	1.29	20.73	3.92	16.67	STD	
910.33	912.06	994.12	933.79	1011.86	1033.76	Best	Ionosphere
911.15	912.06	1023.12	1197.13	1073.38	1274.09	Worst	
910.8	912.06	1005.12	993.15	1056.93	1103.56	Mean	
0.24	0	5.05	82.13	15.33	77.22	STD	
407,718.92	408,433.78	415,172.41	408,815.76	407,941.38	409,881.03	Best	Boold
407,719.72	408,436.08	415,718.98	810,579.69	407,951.59	3,928,094.48	Worst	
407,719.27	408,434.02	415,548.1	471,076.45	407,947.11	732,412.75	Mean	
0.29	0.5	137.6	112,445.32	2.94	787,222.38	STD	

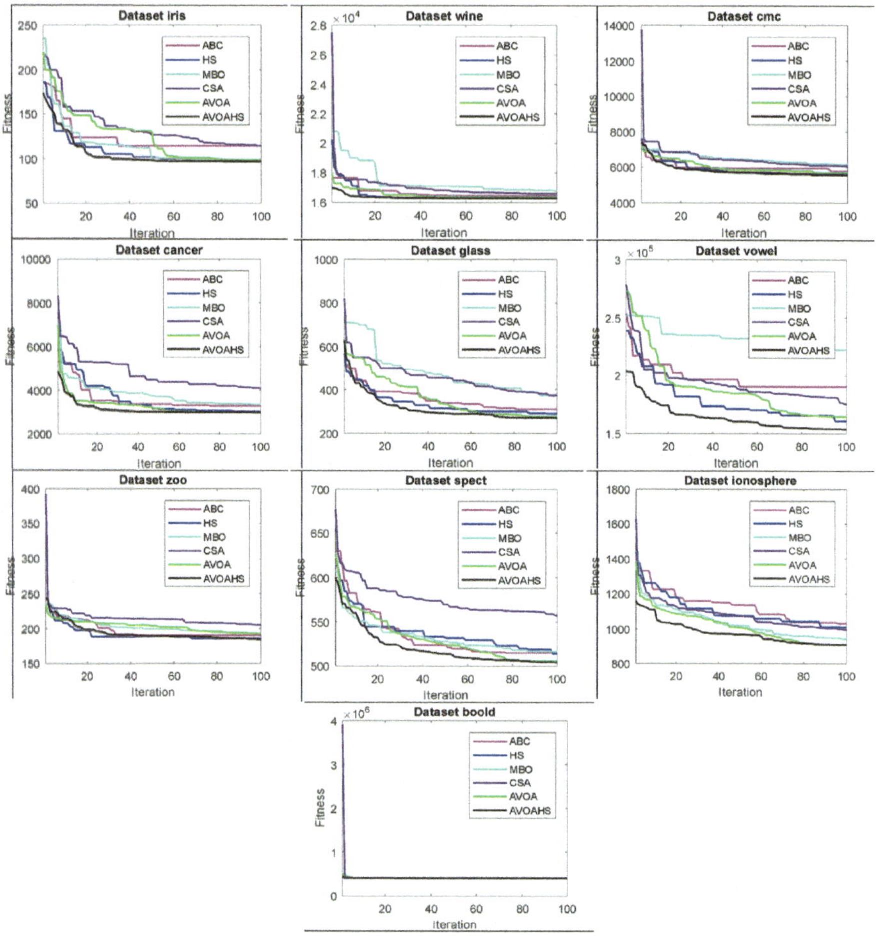

Fig. 2 Convergence curves of each optimization algorithm

population of each algorithm. Therefore, if the height of each boxplot is high, it indicates more noise and shows that operators have not been able to improve the entire population. On the other hand, the lower the height, the less noise and the better the performance. According to the description given, the results of this experiment show that AVOAHS has had a positive effect on the whole population, and it has been able to show good performance in all datasets and boxes of the plot with low height and with the same pattern.

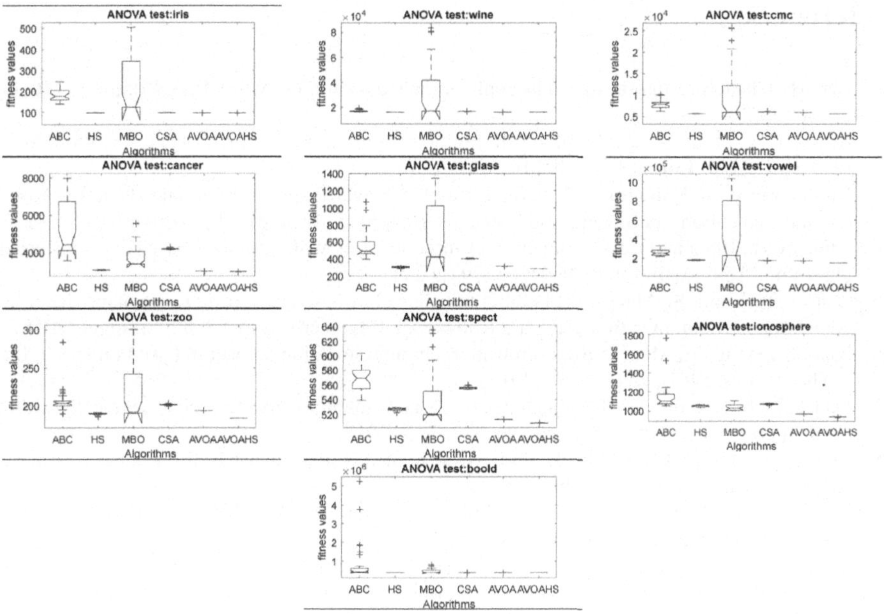

Fig. 3 ANOVA test result of each optimization algorithm

6 Conclusion and Future Work

Data clustering is one of the essential machine learning methods to recognize similar data in different groups. On the other hand, clustering is usually used to detect unlabeled data and analyzing this data is also very challenging for humans. In this paper, we use an algorithm based on a combination of two AVOA and HS algorithms, which we named AVOAHS, to solve the clustering problem. The purpose of combining these two algorithms was to use the useful advantages of both algorithms simultaneously and to cover the weaknesses of each algorithm without increasing the computational complexity. The results also indicate the positive impact of this type of approach. In addition to evaluating and testing optimization algorithms, ten standard datasets that have been taken from the UCI machine learning repository are being used. Also, to evaluate the results obtained from AVOAHS, a comparison has been made with AVOA, CSA, MBO, HS, ABC algorithms. ANOVA statistical test was used to evaluate the results statistically. In all evaluations and tests performed from the proposed model, it has been determined that it has performed very well.

Furthermore, in terms of components of diversity and exploration, creating a balance in these two components has a very high and good ability. Due to the excellent performance of AVOAHS and its unique capabilities and abilities, in the future, it can be used and evaluated to solve a wide range of different problems in different dimensions. On the other hand, a multi-objective version of AVOAHS could be introduced in the future to solve multi-objective problems.

References

1. Zou, H.: Clustering algorithm and its application in data mining. Wirel. Pers. Commun. **110**(1), 21–30 (2020)
2. Kuwil, F.H., et al.: A novel data clustering algorithm based on gravity center methodology. Exp. Syst. Appl. **156**, 113435 (2020)
3. Krishnasamy, G., Kulkarni, A.J., Paramesran, R.: A hybrid approach for data clustering based on modified cohort intelligence and K-means. Exp. Syst. Appl. **41**(13), 6009–6016 (2014)
4. Ahmadi, R., Ekbatanifard, G., Bayat, P.: A modified grey wolf optimizer based data clustering algorithm. Appl. Artif. Intell. **35**(1), 63–79 (2021)
5. Ashish, T., Kapil, S., Manju, B.: Parallel bat algorithm-based clustering using mapreduce. In: Networking Communication and Data Knowledge Engineering, pp. 73–82. Springer (2018)
6. Xiao, S., Dong, Y., Ma, H.: Random walk quantum clustering algorithm based on space. Int. J. Theor. Phys. **57**(5), 1344–1355 (2018)
7. Boushaki, S.I., Kamel, N., Bendjeghaba, O.: A new quantum chaotic cuckoo search algorithm for data clustering. Exp. Syst. Appl. **96**, 358–372 (2018)
8. Zhang, G., Zhang, C., Zhang, H.: Improved K-means algorithm based on density Canopy. Knowl. Based Syst. **145**, 289–297 (2018)
9. Abualigah, L.M., et al.: A novel hybridization strategy for krill herd algorithm applied to clustering techniques. Appl. Soft Comput. **60**, 423–435 (2017)
10. Qaddoura, R., Faris, H., Aljarah, I.: An efficient evolutionary algorithm with a nearest neighbor search technique for clustering analysis. J. Ambient. Intell. Humaniz. Comput. **12**(8), 8387–8412 (2021)
11. Zhou, Y., et al.: Automatic data clustering using nature-inspired symbiotic organism search algorithm. Knowl. Based Syst. **163**, 546–557 (2019)
12. Rahnema, N., Gharehchopogh, F.S.: An improved artificial bee colony algorithm based on whale optimization algorithm for data clustering. Multimed. Tools Appl. **79**(43), 32169–32194 (2020)
13. Abdollahzadeh, B., Gharehchopogh, F.S., Mirjalili, S.: African vultures optimization algorithm: a new nature-inspired metaheuristic algorithm for global optimization problems. Comput. Ind. Eng. **158**, 107408 (2021)
14. Yang, X.-S.: Nature-Inspired Metaheuristic Algorithms. Luniver Press (2010)
15. Askarzadeh, A.: A novel metaheuristic method for solving constrained engineering optimization problems: crow search algorithm. Comput. Struct. **169**, 1–12 (2016)
16. Wang, G.-G., Deb, S., Cui, Z.: Monarch butterfly optimization. Neural Comput. Appl. **31**(7), 1995–2014 (2019)
17. Geem, Z.W., Kim, J.H., Loganathan, G.V.: A new heuristic optimization algorithm: harmony search. Simulation **76**(2), 60–68 (2001)
18. Karaboga, D., Basturk, B.: A powerful and efficient algorithm for numerical function optimization: artificial bee colony (ABC) algorithm. J. Global Optim. **39**(3), 459–471 (2007)

Estimation Models for Optimum Design of Structural Engineering Problems via Swarm-Intelligence Based Algorithms and Artificial Neural Networks

Melda Yücel, Sinan Melih Nigdeli, and Gebrail Bekdaş

Abstract In structural engineering, iterative optimization processes may take a long time and several cases of the same problem need the process of the same optimization run. For that reason, machine learning can be used to generate artificial intelligence estimation models. This process includes the steps of generating data including optimum results of several cases, machine learning via using the generated data, and providing an estimation model via artificial intelligence methods. Artificial Neural Networks (ANNs) have been successfully used in developing estimation models for solving and optimization of structural engineering problems. In optimization, swarm intelligence-based algorithms have been generally used to provide machine learning data. In this chapter, a review of ANNs applications in civil engineering is presented. Then, examples of structural optimization that propose an estimation model are presented. The problems include benchmark structural optimization problems, tuned mass dampers, and reinforced concrete retaining walls.

Keywords Structural systems · Optimization · Swarm-intelligence algorithms · Machine learning · Artificial Neural Networks (ANNs)

1 Introduction

The concept of machine learning mainly means that software besides computers, calculators, or various technical devices have an idea about the events by gaining experience and taking lessons and can give reaction intended for this case. The mentioned machine learning is accepted as a sub-discipline of the technological area, which is called artificial intelligence, and made progress gradually from the 1950s, makes possible by such these types of devices to perform some activities that can be realized by only a human under normal circumstances. If it is a must to give

M. Yücel · S. M. Nigdeli · G. Bekdaş (✉)
Department of Civil Engineering, Istanbul University-Cerrahpaşa, 34320 Avcılar, Istanbul, Turkey
e-mail: bekdas@iuc.edu.tr

S. M. Nigdeli
e-mail: melihnig@iuc.edu.tr

A. Biswas et al. (eds.), *Advances in Swarm Intelligence*, Studies in Computational Intelligence 1054, https://doi.org/10.1007/978-3-031-09835-2_14

examples to these activities, for instance, the cases such as matching of the voice and face belonging different people, completing of the missing parts in a puzzle or picture piece, playing games such as chess, chequer, etc. and even winning, recognition the tone of voice and speech accent possessed to a person can be also performed by these systems learning by mimicking the human intelligence.

In this respect, within the various and many fields of sciences and natural life, it has been benefited from the learning systems developed with the usage of performing such these activities or similar together with different ones. Moreover, these tools, which were generated through the utilizing of different machine learning applications such as pattern recognition, classification of the labels, numerical prediction, etc. can be created with various algorithms. However, it can be said and recognized that especially artificial neural networks (ANNs) and several kinds of them have been more preferred than other methodologies due to that ANNs can carry out all of the mentioned activities successfully and effectively within a short time and using less effort. For this reason, in Table 1 [1–26], summary literature for many applications of several fields is represented that they were conducted in the way of benefiting from ANNs and some types of it.

On the other hand, in most of the disciplines of engineering, which is one of the major science fields, it can be seen that different machine learning applications were carried out with various purposes by benefiting from ANNs. Also, for the structural engineering branch, which is a field of civil engineering covering the main context of this chapter, the mentioned applications turned to more frequently preferred and continued to utilize. In this scope, for example, the rapid calculation tools providing the design parameters for members of the structural systems were generated [27–32], the creation of the optimum design for structures (such as truss or frame models, retaining walls, etc.) were performed [33–38], early processes were made possible for forecasting the material ingredients/properties such as concrete strength, required amount cement mortar, etc. [39–42], the effective performance was ensured for seismically designed devices like base isolation systems, tuned mass dampers (TMDs), vibrators by observing dynamic properties [43–46], etc.

In this chapter, different machine learning applications carried out with the usage of ANNs concerning the structural engineering area were presented. Here, all of the applications are based on the designing of the best structural model namely optimum design, and the prediction process for the desired parameters or outcomes. In this scope, different structural models were handled like several steel benchmark systems, retaining wall, and tuned mass damper (TMD), and estimation models were provided in the way of finding these parameters via ANNs. For this respect, after an introduction to machine learning, artificial neural networks (ANNs), and a literature summary for various applications are given in Sect. 1, general representation for ANNs was ensured in Sect. 2. As to Sect. 3, the estimation applications concerning the structural engineering problems, which were conducted with ANNs, were reflected, and then the numerical investigations in terms of numerical prediction processes for all of the designs were presented in Sect. 4, besides that Sect. 5 was concluded with general comments and results.

Table 1 The applications realized via different types of ANNs within several fields

The field	Application	Performed by	Year	Cite
Medicine	Detection of cancer or tumor cells	A. Sawant et al.	2018	[1]
		Y. Fujisawa et al.	2019	[2]
	Foresee of the disease occurrence possibility	T. Ayer et al.	2010	[3]
		A. K. Sahoo et al.	2020	[4]
	Determination of the best medicine dose for a disease treatment	E. Grossi et al.	2014	[5]
Social media applications	Detection of the cyberbullying cases	J. Hani et al.	2019	[6]
	Revealing the fake news, contents, or e-mails	A. K. Sharma et al.	2014	[7]
		D. V. B. de Oliveira and U. P. Albuquerque	2021	[8]
	Detection of the possible frauds and suspicious operations	E. M. Carneiro et al.	2015	[9]
		R. Mouawi et al.	2018	[10]
Transportation services	Prediction of the traffic flow	E. Doğan	2020	[11]
		C. Chen et al.	2020	[12]
	Estimation of the vehicular traffic accident possibility	S. Sikka	2014	[13]
		L. Wenqi et al.	2017	[14]
	Prediction of the pedestrian number within the current traffic	J. Song et al.	2020	[15]
Finance and marketing activities	Detection of the financial risk by early warning	Z. Chen et al.	2018	[16]
	Pre-determination whether give the credit to the customer by a company	G. Teles et al.	2020	[17]
	Estimation of buying behaviors of consumers	L. M. Badea	2014	[18]
		Y. Xu et al.	2019	[19]
Energy, agriculture, and production industry	Developing the energy conservation models for the constructions	F. Ascione et al.	2017	[20]
		M. Zekić-Sušac et al.	2021	[21]
	Foresee the soil properties and humid ingredients	S. Prakash and S. S. Sahu	2020	[22]
	Prediction of the crop yield and management of the crop quality	T. Islam et al.	2018	[23]
		M. Alagurajan and C. Vijayakumaran	2020	[24]
	Determination of the parameters affecting the production efficiency	B. Najafi et al.	2018	[25]
		S. Francik et al.	2020	[26]

2 Artificial Neural Networks (ANNs)

Artificial neural networks (ANNs), which are one of the machine learning algorithms that provide the direct prediction of numerical parameters or classification of categorical labels, have a structure similar to the central neural system together with cells of the human brain. In this regard, ANNs can carry out some activities like a human such as learning of data samples, extraction of the features of parameters, producing suitable responses to the taken information, etc. If the working principle and structural mechanism of ANNs must be explained, they have different nodes simulating the neural cells and they are generally parted into three different classes like input, hidden and output nodes. Also, these can be named as layers, which contain parameter points and their numerical or alpha-numerical/label values. Furthermore, the structure of ANNs is determined according to whether a hidden layer exists or not. In this meaning, in the case of existing one or more hidden layers and nodes, the kind of ANNs turn into multilayer perceptrons (MLPs), and the exact opposite situation also causes single layer ANNs. The basic structure of ANNs is formed via constituents of the central neural system of humans. As an example, neural cells in this biologic structure, and axons that provide the transferring of information between cells within a neural system, represent the connection nodes of input, hidden, and output and node connections in the training algorithm.

Here, neural nodes take the information from input layers and transmit the processed data towards the output layer by operating with hidden nodes. While these processes are carried out, different parameters come into prominence. For example, a parameter called connection weight is utilized for transferring data coming from the input layer to the hidden layer, and it is also handled between hidden and output layers. Due to benefiting from the weight connection, it is provided that the creature of ANNs is trained and the relationships among input and output data are learned by this formation. Moreover, different activation functions can be utilized for processing the ultimate total value given to the sequent layers from the previous one. In this respect, training of ANNs is provided in the way of realizing the learning and then directly determining the outcomes.

3 Applications via ANNs

In this section, machine learning applications based on the prediction of the optimum design properties for several structural systems were presented. For this respect, optimization and machine learning processes in terms of benchmark structures, which contain tubular column, I-beam, 3-bar, and 10-bar trusses together with a simply supported reinforced concrete (RC) beam, tuned mass damper (TMD), and RC cantilever retaining wall were applied. Here, initially, the required design properties were provided by realizing the optimization applications via metaheuristic algorithms. Then, to directly predict the optimum design parameters and also the ultimate objectives with respect to the mentioned structures, machine learning applications were carried out with the usage of ANNs.

The first optimization problem is related to a benchmark structure that is a tubular column. To minimize the total cost for column structure was targeted by optimizing the design parameters as central diameter (d) and the thickness of the tubular column (Fig. 1). Also, there are two design limitations related to the capacity of column compression and buckling force.

Another benchmark problem is related to the I-beam structure (Fig. 2). The optimization objective is to minimize the vertical deflection of the beam under the design constraints based on cross-section area and allowable moment stress. On the other hand, there are four different design variables to optimize that are beam height (h) and beam width (b) together with the thickness of flange (t_f) and beam web (t_w).

Fig. 1 Details of structural model and cross-section of tubular column [28]

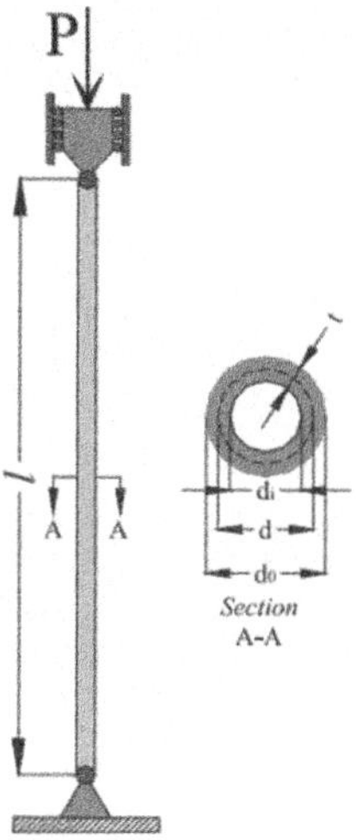

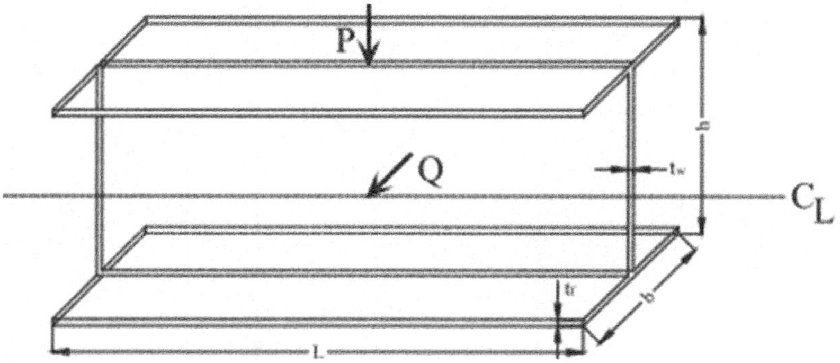

Fig. 2 Structural model of I-beam [29]

While the previous benchmark structures were designed as optimal models by utilizing of flower pollination algorithm (FPA), the last models, which include 3-bar and 10-bar truss structures by finding optimum bar sections (A_s) with RC beam by ensuring the optimum section sizes together with reinforcement bar area, were analyzed with harmony search (HS) for optimization processes. Here, the main optimization aims are the minimization of structural volume, weight, and total cost for these three structures, respectively (Figs. 3, 4 and 5).

One of the other structural problems where the designs were tried to determine with prediction applications, is a vibration control device as TMD. For this structural problem, the acceleration transfer function (TF) in the frequency domain was considered to minimize. In this process, an SDOF structure was utilized for generating optimum TMD designs by adjusting the mechanical parameters as period (T_d)

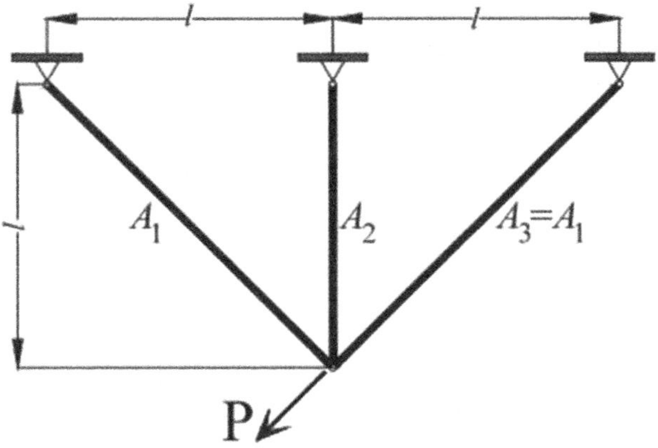

Fig. 3 3-bar truss model [36]

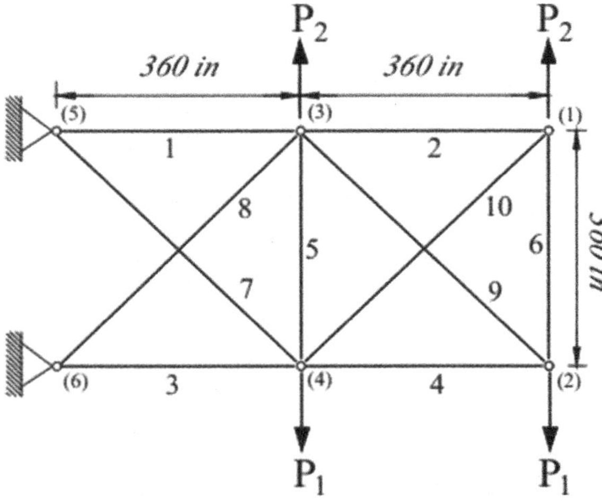

Fig. 4 10-bar truss model [36]

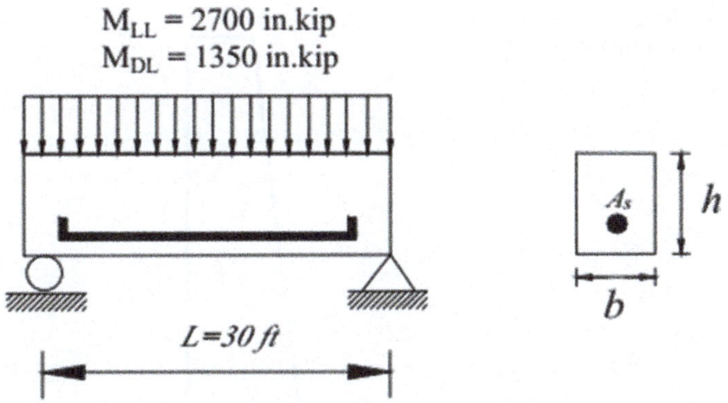

Fig. 5 The structure of simply supported RC beam [36]

and damping ratio (ξ_d). The structural model of the main structure with TMD can be shown in Fig. 6.

Finally, an RC cantilever retaining wall model has been handled with the aim of minimization the total cost by determining the optimized section sizes ($X_{1\text{-}5}$). This structure was optimized via FPA and the wall model is subjected to some design constraints as geotechnical and structural limit states (Fig. 7).

Here, the second process of the applications is a prediction of some properties as mentioned previously. In this respect, for all of the structural designs presented in this chapter, the basic structural variables such as optimum section sizes, mechanical parameters, and also the main optimization objectives were directly determined through the generated prediction tools, which were ensured with the usage of ANNs. Also, different and multiple test designs were investigated to approve the performance and efficiency of these tools.

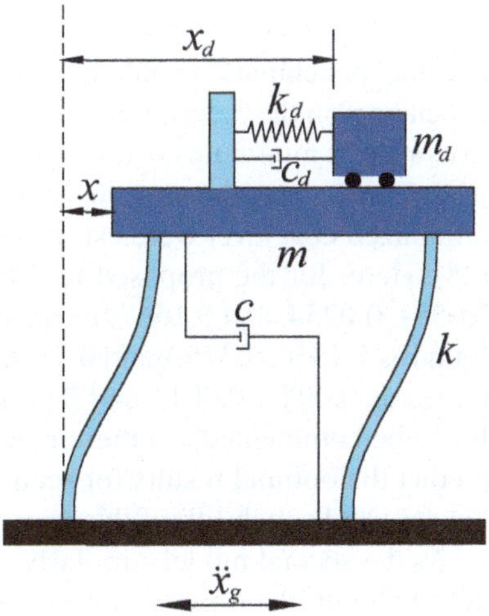

Fig. 6 SDOF structure with a TMD [45]

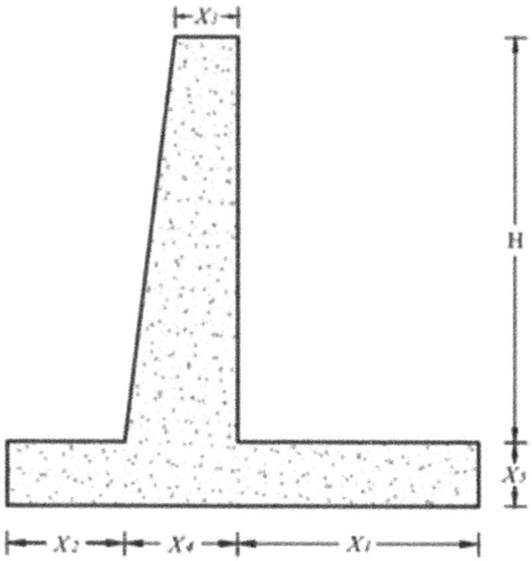

Fig. 7 Structure of RC cantilever retaining wall [38]

Besides, if it must be pointed out, the optimized structures can be designed in an extremely short time, consuming less effort and time thanks to the prediction models. The reason for this is based on the realization of direct prediction processes via created ANNs tool without extra optimization analyses. On the other side, before the construction of generating the large-scale or multi number structural models, these tools provide to make a decision or form an opinion whether they have the sufficient/required strength, safety, cost-efficiency, etc.

4 Numerical Results

For the benchmark problems, the realized prediction processes via ANNs are depending on the determination of the optimum section properties with main objectives such as minimum cost, weight, volume, etc. Thus, in terms of the tubular column model, both optimum section sizes as central diameter (d) with the thickness (t) and minimized cost level were estimated by benefiting from an ANNs prediction model [28]. Here, for the proposed test designs, mean absolute error (MAE) metrics are 0.0554, 0.0254, and 0.1667; mean absolute percentage (%) error (MAPE) values are found as 1.2356, 8.7760, and 0.9378; and mean squared error (MSE) values are determined as 0.0049, 0.0011 and 0.0502 for d, t and minimum cost level, respectively. It can be commented via these results that the ANNs model is extremely proper to predict the optimal results for section sizes together with minimum cost in terms of lowness and suitability of the error values.

As the second model, similarly, optimized section properties of the I-beam were detected with the created ANNs structure. Here, as a difference compared with the tubular column, the minimum vertical deflection was calculated via the estimated optimum properties, and design constraints were also detected with these results to control whether limitation values are exceeded or not [29]. While the performance of the estimation model was investigated, it can be recognized that the correlation of the main ANNs model is a high level for the fitness of actual and predicted outputs (correlation coefficient as R = 99%). Thus, a test model could be generated to examine the prediction model. Here, all of the error values are quite low (smaller than 0.08) for h and b, these values are a bit greater for t_w and t_f (not exceeding 2.35). Also, for the minimum vertical deflection, calculated errors are found at a very low rate, besides the design constraints are exceeded for only three test designs decimally.

In the last benchmark structures as 3-bar and 10-bar truss structures with RC beam model, optimum section areas of bars with minimum structural volume and weight together minimum cost were tried to determine via ANNs tool, respectively [36]. Here, in terms of 3-bar truss, for the training model of the main dataset, the ensured error rates were found maximum values as 3.82e−04, 3.9182, and 0.3072 for a specific parameter of each structural model (in terms of MSE). On the other side, while these successes were evaluated for the test designs, all of the values of error metrics as MAE, MSE, and root mean squared error (RMSE) are observed between 4.95e−05 and 0.0911; 7.01e−05 and 1.8410; 0.0186 and 0.6036 rates as minimum and maximum level respect to optimum design parameters and minimum objective function values for each truss and RC beam.

Concerning the other structural problem, which is based on the generation optimum design model of TMDs, the optimum values of mechanical parameters and transfer function belonging to the presented vehicle control device were provided by applying the prediction process [45]. For this reason, different combinations for the structural parameters were handled and combined, then the main dataset was created for the training of ANNs. In the second phase, by proposing some test designs, the performance of the prediction tool can be seen and validated thanks to the ANNs performance. Accordingly, both mechanic parameters (T_d and ξ_d) and minimum transfer functions are predicted via ANNs, and error metrics are calculated in comparison to optimum data. Whole MAPE values are determined at quite low levels (0.02–6.70 for T_d; 0.17–5.15 for ξ_d; 0.02–4.21 for minimum transfer function). Thanks to the good performance of the main prediction model, an extra-large test dataset was also generated to propose some generalized formulations for determining the values of the parameters directly.

For the last structure, optimized design properties as section sizes together with the minimum total cost of the RC retaining wall model were ensured by operating ANNs. In this respect, some parameters of the wall and earth soil were handled to create a dataset in terms of the formation of the ANNs estimation model [38]. According to this, the general performance of the model was found at the level of 99% in terms of correlation between actual and estimated values of outputs. On the other side, for the test designs, the predicted values can be determined in a pretty close way to actual optimum results by deviating with extremely small error rates in terms of $X_{1\text{-}5}$ together minimum cost. As to MAE, MSE, and RMSE, the values of metrics are observed in the level of 0.5432 as the maximum rate. In this meaning, section sizes and also cost values can be detected via ANNs with very effective and reliable results. Also, the design constraints can be provided with the predicted parameters in the desired and required limitations.

5 Results and Conclusion

In this chapter, machine learning technology, and several applications together with developments realized within the structural engineering area by benefiting from the advantages of this area were mentioned. In this respect, the main aim was determined as the creation of the best design models for various structural systems through realizing both optimization process and numerical prediction investigations via artificial neural networks (ANNs). In summary, for all of the design structures, the most suitable and valid models were tried to find directly in the way of the usage of a quick, time-saving, and effective prediction tool.

According to the presented numerical results and investigations, it can be noticed that all of the values belonging to error measurements and also correlation levels are extremely suitable, and reliable in terms of the prediction performance and efficiency. The cause of this is that the generated ANNs models have deviated from the actual output results with really small rates generally. Moreover, it can be said that ANNs models are quite suitable to design the structural models in a short time by not using so much effort and thus cost. Additionally, due to the success of these prediction models, most of the design parameters can be determined directly and rapidly for the mentioned structural designs.

In conclusion, it can be understood and mentioned that ANNs algorithm is very effective, quick, suitable, and sensitive to determine the optimum parameter values for any design parameters presented in this chapter. Also, these models provide the saving of time besides the effort of the designers. In this meaning, ANNs prediction models can decide for the detection of the real/actual optimum parameter values concerning the given models as truss systems, I-beam, tubular column, tuned mass damper, and also RC retaining wall without performing extra optimization analyses.

References

1. Sawant, A., Bhandari, M., Yadav, R., Yele, R., Bendale, M.S.: Brain cancer detection from mri: a machine learning approach (tensorflow). Brain **5**(04), 2089–2094 (2018)
2. Fujisawa, Y., Inoue, S., Nakamura, Y.: The possibility of deep learning-based, computer-aided skin tumor classifiers. Front. Med. **6**, 191 (2019)
3. Ayer, T., Alagoz, O., Chhatwal, J., Shavlik, J.W., Kahn, C.E., Jr., Burnside, E.S.: Breast cancer risk estimation with artificial neural networks revisited: discrimination and calibration. Cancer **116**(14), 3310–3321 (2010)

4. Sahoo, A.K., Pradhan, C., Das, H.: Performance evaluation of different machine learning methods and deep-learning based convolutional neural network for health decision making. In: Nature Inspired Computing for Data Science, pp. 201–212. Springer, Cham (2020)
5. Grossi, E., Podda, G.M., Pugliano, M., Gabba, S., Verri, A., Carpani, G., Buscema, M., Casazza, G., Cattaneo, M.: Prediction of optimal warfarin maintenance dose using advanced artificial neural networks. Pharmacogenomics **15**(1), 29–37 (2014)
6. Hani, J., Nashaat, M., Ahmed, M., Emad, Z., Amer, E., Mohammed, A.: Social media cyberbullying detection using machine learning. Int. J. Adv. Comput. Sci. Appl. **10**(5), 703–707 (2019)
7. Sharma, A.K., Prajapat, S.K., Aslam, M.: A comparative study between naïve Bayes and neural network (MLP) classifier for spam email detection. Int. J. Comput. Appl. 12–16 (2014)
8. de Oliveira, D.V.B., Albuquerque, U.P.: Cultural evolution and digital media: diffusion of fake news about COVID-19 on Twitter. SN Comput. Sci. **2**(6), 1–12 (2021)
9. Carneiro, E.M., Dias, L.A.V., Da Cunha, A.M., Mialaret, L.F.S.: Cluster analysis and artificial neural networks: a case study in credit card fraud detection. In: 2015 12th International Conference on Information Technology-New Generations, pp. 122–126. IEEE (2015)
10. Mouawi, R., Awad, M., Chehab, A., El Hajj, I.H., Kayssi, A.: Towards a machine learning approach for detecting click fraud in mobile advertizing. In: 2018 International Conference on Innovations in Information Technology (IIT), pp. 88–92. IEEE (2018)
11. Doğan, E.: Short-term traffic flow prediction using artificial intelligence with periodic clustering and elected set. Promet-Traffic Transp. **32**(1), 65–78 (2020)
12. Chen, C., Li, K., Teo, S.G., Zou, X., Li, K., Zeng, Z.: Citywide traffic flow prediction based on multiple gated spatio-temporal convolutional neural networks. ACM Trans. Knowl. Discov. Data (TKDD) **14**(4), 1–23 (2020)
13. Sikka, S.: Prediction of road accidents in Delhi using back propagation neural network model. Int. J. Comput. Sci. Eng. Technol. (IJCSET) **5**(08), 798–804 (2014)
14. Wenqi, L., Dongyu, L., Menghua, Y.: A model of traffic accident prediction based on convolutional neural network. In: 2017 2nd IEEE International Conference on Intelligent Transportation Engineering (ICITE), pp. 198–202. IEEE, Singapore (2017)
15. Song, J., Qiu, Z., Ren, G., Li, X.: Prediction of pedestrian exposure to traffic particulate matters (PMs) at urban signalized intersection. Sustain. Cities Soc. **60**, 102153 (2020)
16. Chen, Z., Van Khoa, L.D., Teoh, E.N., Nazir, A., Karuppiah, E.K., Lam, K.S.: Machine learning techniques for anti-money laundering (AML) solutions in suspicious transaction detection: a review. Knowl. Inf. Syst. **57**(2), 245–285 (2018)
17. Teles, G., Rodrigues, J.J.P.C., Rabê, R.A., Kozlov, S.A.: Artificial neural network and Bayesian network models for credit risk prediction. J. Artif. Intell. Syst. **2**(1), 118–132 (2020)
18. Badea, L.M.: Predicting consumer behavior with artificial neural networks. Procedia Econ. Financ. **15**, 238–246 (2014)
19. Xu, Y., Zhang, W., Bao, H., Zhang, S., Xiang, Y.: A SEM–neural network approach to predict customers' intention to purchase battery electric vehicles in China's Zhejiang province. Sustainability **11**(11), 3164 (2019)
20. Ascione, F., Bianco, N., De Stasio, C., Mauro, G.M., Vanoli, G.P.: Artificial neural networks to predict energy performance and retrofit scenarios for any member of a building category: a novel approach. Energy **118**, 999–1017 (2017)
21. Zekić-Sušac, M., Mitrović, S., Has, A.: Machine learning based system for managing energy efficiency of public sector as an approach towards smart cities. Int. J. Inf. Manag. **58**, 102074 (2021)
22. Prakash, S., Sahu, S.S.: Soil moisture prediction using shallow neural network. Int. J. Adv. Res. Eng. Technol. **11**(6), 426–435 (2020)
23. Islam, T., Chisty, T.A., Chakrabarty, A.: A deep neural network approach for crop selection and yield prediction in Bangladesh. In: 2018 IEEE Region 10 Humanitarian Technology Conference (R10-HTC), pp. 1–6. IEEE (2018)
24. Alagurajan, M., Vijayakumaran, C.: ML methods for crop yield prediction and estimation: an exploration. Int. J. Eng. Adv. Technol. **9**(3), 3506–3508 (2020)

25. Najafi, B., Faizollahzadeh Ardabili, S., Shamshirband, S., Chau, K.W., Rabczuk, T.: Application of ANNs, ANFIS and RSM to estimating and optimizing the parameters that affect the yield and cost of biodiesel production. Eng. Appl. Comput. Fluid Mech. **12**(1), 611–624 (2018)
26. Francik, S., Knapczyk, A., Knapczyk, A., Francik, R.: Decision support system for the production of Miscanthus and willow briquettes. Energies **13**(6), 1364 (2020)
27. Du, Y., Chen, Z., Zhang, C., Cao, X.: Research on axial bearing capacity of rectangular concrete-filled steel tubular columns based on artificial neural networks. Front. Comput. Sci. **11**(5), 863–873 (2017)
28. Yucel, M., Bekdas, G., Nigdeli, S.M., Sevgen, S.: Artificial neural network model for optimum design of tubular columns. Int. J. Theor. Appl. Mech. **3**, 82–86 (2018)
29. Yucel, M., Nigdeli, S.M., Bekdas, G.: Estimation model for generation optimization of design variables for I-beam vertical deflection minimization. In: IV. Eurasian Conference on Civil and Environmental Engineering (ECOCEE), pp. 17–18 (2019)
30. Zhang, G., Ali, Z.H., Aldlemy, M.S., Mussa, M.H., Salih, S.Q., Hameed, M.M., Al-Khafaji, Z.S., Yaseen, Z.M.: Reinforced concrete deep beam shear strength capacity modelling using an integrative bio-inspired algorithm with an artificial intelligence model. Eng. Comput. 1–14 (2020)
31. Yucel, M., Kayabekir, A.E., Nigdeli, S.M., Bekdaş, G: Optimum design of carbon fiber-reinforced polymer (CFRP) beams for shear capacity via machine learning methods: optimum prediction methods on advance ensemble algorithms–bagging combinations. In: Artificial Intelligence and Machine Learning Applications in Civil, Mechanical, and Industrial Engineering, pp. 85–103. IGI Global (2020)
32. Yücel, M., Nigdeli, S.M., Kayabekir, A.E., Bekdaş, G.: Optimization and artificial neural network models for reinforced concrete members. In: Nature-Inspired Metaheuristic Algorithms for Engineering Optimization Applications, pp. 181–199. Springer, Singapore (2021)
33. Gholizadeh, S.: Performance-based optimum seismic design of steel structures by a modified firefly algorithm and a new neural network. Adv. Eng. Softw. **81**, 50–65 (2015)
34. Asteris, P.G., Nikoo, M.: Artificial bee colony-based neural network for the prediction of the fundamental period of infilled frame structures. Neural Comput. Appl. **31**(9), 4837–4847 (2019)
35. Yücel, M., Bekdaş, G., Nigdeli, S.M.: Prediction of optimum 3-bar truss model parameters with an ANN model. In: International Conference on Harmony Search Algorithm, pp. 317–324. Springer, Singapore (2020)
36. Bekdaş, G., Yücel, M., Nigdeli, S.M.: Estimation of optimum design of structural systems via machine learning. Front. Struct. Civ. Eng. 1–12 (2021)
37. Momeni, E., Yarivand, A., Dowlatshahi, M.B., Armaghani, D.J.: An efficient optimal neural network based on gravitational search algorithm in predicting the deformation of geogrid-reinforced soil structures. Transp. Geotech. **26**, 100446 (2021)
38. Yücel, M., Bekdaş, G., Nigdeli, S.M., Kayabekir, A.E.: An artificial intelligence-based prediction model for optimum design variables of reinforced concrete retaining walls. Int. J. Geomech. **21**(12), 04021244 (2021)
39. Behnood, A., Golafshani, E.M.: Predicting the compressive strength of silica fume concrete using hybrid artificial neural network with multi-objective grey wolves. J. Clean. Prod. **202**, 54–64 (2018)
40. Yucel, M., Namlı, E.: High performance concrete (HPC) compressive strength prediction with advanced machine learning methods: combinations of machine learning algorithms with bagging, rotation forest, and additive regression. In: Artificial Intelligence and Machine Learning Applications in Civil, Mechanical, and Industrial Engineering, pp. 118–140. IGI Global (2020)
41. Abellán García, J., Fernández Gómez, J., Torres Castellanos, N.: Properties prediction of environmentally friendly ultra-high-performance concrete using artificial neural networks. Eur. J. Environ. Civ. Eng. 1–25 (2020)
42. Armaghani, D.J., Asteris, P.G.: A comparative study of ANN and ANFIS models for the prediction of cement-based mortar materials compressive strength. Neural Comput. Appl. **33**(9), 4501–4532 (2021)

43. Yucel, M., Öncü-Davas, S., Nigdeli, S.M., Bekdas, G., Sevgen, S.: Estimating of analysis results for structures with linear base isolation systems using artificial neural network model. Int. J. Control Syst. Robot. **3**, 50–56 (2018)
44. Ramezani, M., Bathaei, A., Ghorbani-Tanha, A.K.: Application of artificial neural networks in optimal tuning of tuned mass dampers implemented in high-rise buildings subjected to wind load. Earthq. Eng. Eng. Vib. **17**(4), 903–915 (2018)
45. Yucel, M., Bekdaş, G., Nigdeli, S.M., Sevgen, S.: Estimation of optimum tuned mass damper parameters via machine learning. J. Build. Eng. **26**, 100847 (2019)
46. Farrokhi, F., Rahimi, S.: Supervised probabilistic failure prediction of tuned mass damper-equipped high steel frames using machine learning methods. Studia Geotechnica et Mechanica **42**(3), 179–190 (2020)

A Novel Codebook Generation by Lévy Flight Based Firefly Algorithm

Ilker Kilic

Abstract Nature inspired metaheuristic algorithms are become more powerful and useful in image processing algorithms especially by the developments in microprocessor technology. In the last several decades the Linde-Buzo-Gray algorithm is a powerful technique for local optimum codebook generation in image compression. Fuzzy C-Means and C-Means are the alternative ones for the same process. On the other hand nature-inspired metaheuristic algorithms have also become other alternate technics for designing the optimum codebook. In this paper the Firefly technique is enhanced by the Lévy flight function to achieve the global optimum codebook. The Firefly technique contains two sub search mechanisms to reach the global minimum solution. The first one is the attraction of any firefly by a brighter one. This process strongly guides to the firefly on the global minimum way especially if it is attracted by the brightest one. The second one is the random search of a firefly in a circle with a radius of α. On the other hand if α is determined so big, the firefly may lose its way and come to a location that is much far away from the firefly group and possibly there is no brighter firefly to be followed. On the contrary, if α is determined so small, this time the fireflies fall into a local minimum and can not escape. Therefore we need to have such an α that in most of the iterations its value changes in a small random value interval, but in rare iterations its value must be relatively big in order to escape local minimums. Therefore if a firefly is captured by a local minimum point by accidentally, Lévy Flight step provides an opportunity to escape from it easily. Numerical results suggest that the new introduced Lévy Flight based Firefly Algorithm is better than the classical techniques and provides the global optimum codebook for image compression.

Keywords Fire Fly algorithm · Metaheuristic algorithm · Levy flight · Image compression · Vector quantization

---

I. Kilic (✉)
Pamukkale University, Electical and Electronics Department, Kinikli Campus, 20160, Denizli, Turkey
e-mail: ilkerk@pau.edu.tr
URL: https://www.pau.edu.tr/eem

A. Biswas et al. (eds.), *Advances in Swarm Intelligence*, Studies in Computational Intelligence 1054, https://doi.org/10.1007/978-3-031-09835-2_15

1 Introduction

Vector quantization (VQ) [1] is a lossy image compression algorithm combined by a codebook generation technique named LBG [2]. A codebook is a combination of vectors that represents the image best. The LBG technique is the well known codebook producing technique which uses local search procedures and satisfies a local minimum solution. A faster and robust LBG technique is improved by Lin [3]. In the new algorithm by Patane [4], the codewords are designed by an error minimization strategy such that badly positioned codewords are shifted in order to escape local minimum points. The VQ technique is also improved by the help of reinforced learning strategy, competitive learning and codevector structure [5]. An improved and faster version of VQ, which is a pattern reduction technique for codebook generation process to reduce the computation time is presented [6]. Fuzzy vector quantization is the similar version of the fuzzy c-means (FCM) technique [7]. The FCM locates each vector to a cluster with a new membership degrees. So, all codewords can change itself and become more dynamic. Therefore, the FCM reduces the dependence of VQ initial start [8]. A fuzzy based VQ clustering VQ is produced to obtain the benefits of fuzzy decision process. Therefore the computational performance of crisp decision is maintained [9]. This is achieved by improving an effective way for the transition between fuzzy and crisp decisions. Therefore this strategy reduces the random initialization effect on the codebook. The first form of k-means algorithm is improved by Ant Colony Optimization (ACO) method [10] by clustering the vectors by the help of pheromone variable probability. The total within the cluster variance is the criteria for updating the ant pheromone. The ACO based codebook generation algorithm is speeded up by Preaco algorithm [11]. Genetic algorithm (GA) [12] is a branch of evolutionary algorithms that simulates the natural evolution process in order to overcome the optimization problems. GA is used for codebook generation for the first time [13]. Then, GA is combined with Simulated Annealing in order to both generate better codebooks and decrease the run time [14]. A Codebook design using GA is proposed for speaker identification [15]. In another study, the researchers introduced a constrained storage based genetic codebook for image compression [16]. Evolution based tabu search approach [17] and fuzzy particle swarm optimization [18] are also used for codebook design. A codebook based image compression is proposed by honey bee optimization [19], firefly algorithm [20], bat algorithm [21]. Fruit Fly Algorithm (FFA) is an algorithm that imitates the foraging behaviors of drosophila fruit flies [22]. The FFA is better than the other nature inspired algorithms, easy to understand and implement. Hence, FFA technique has been implemented in different kinds of optimization problems. Some of them can be listed as financial distress models [22], the hybrid annual power load forecasting [23], a combination of common regression neural network and fruit fly algorithm [24], annual electric load forecasting [25], PID controller tuning [26], fractional PID controller [27], artificial neural network [28], knapsack problem [29], parameter definition of synchronous generator [30], joint search strategies [31], FFA based fuzzy-pid controller for electronic throttle [32], solving the steelmaking casting problem [33]. Recently, FFA has been specifically implemented in image watermarking using a deep learning method through wavelet transform [34]. An optimum medical image compression technique based on vq is improved by firefly algorithm [35]. In this paper Lévy Flight technique is combined by Firefly Algorithm(FRA) for codebook improving process named as Lévy Flight based Firefly algorithm (LFRA). In PSO, each particle determines its route by using only its and swarm's best positions. This restricts the search area, increases the probabilty of captured by a local minimum point and causes a delay in reaching the global minimum point. On the contrary, Lévy Flight is a kind of power law distribution that in most of the scale it generates small random numbers whereas rarely produces big values. Therefore if a firefly is captured by a local minimum point, Lévy Flight distribution provides an opportunity to escape from it easily. The paper is composed of six sections. Following the introduction the popular clustering algorithms are explained briefly in Sect. 2. The proposed LFA for codebook generation is introduced by third section. The parameters of the algorithms are explained in Sect. 4. Application results of the proposed new technique declared in Sect. 5. In Sect. 6 a breaf conclusion is written.

2 Codebook Generation Algorithms

In this section some fundamental features of clustering methods in the literature are explained.

2.1 *Vector Quantization and LBG Algorithm*

Vector quantization (VQ) is a well known image processing algorithm. The most important part of VQ is the codebook generation. Lets determine the size of the original image as $Y = Xij$. N $\times$ N pixels of the image divided into mxm pixel sized sub blocks. Therefore the whole image is defined by N_b = (N/m $\times$ N/m) number of sub blocks that represents the collection of original image blocks $X = Xi$. Here i = 1,2,..., N_b. The L is the dimension of the blocks which is assumed as mxm pixel size. The original sub blocks xi is defined as L dimensional Euclidean space. The number of Nc codeword grup collection is called as a codebook. $C = C_1, C_2, \ldots, C_j$, j = 1,2,...,Nc. The original image vectors are defined as row vectors. $Xi = (Xi_1, Xi_2, Xi_3, \ldots, Xi_L)$ The ith codeword is defined as $Ci = (Ci_1, Ci_2, Ci_3, \ldots, Ci_L)$. In VQ, the most suitable codeword from the codebook which has the minimum error, is assigned to each image block. Therefore instead of using original image block the number of the codeword from the codebook is used for the lossy compression. Determination of the C is achieved by minimizing the following mean square error (MSE) criteria.

$$MSE(C) = \left(\frac{1}{N_b}\right)\sum_{j=1}^{N_c}\sum_{i=1}^{N_b}\mu_{ij} \parallel (x_i - c_j) \parallel^2 \quad (1)$$

$$\sum_{j=1}^{N_c}\mu_{ij} = 1, \qquad i \in 1, 2, ..., N_b \quad (2)$$

$$\mu_{ij} = 1 \quad if \quad x_i \in j\text{th cluster}, \quad \text{otherwise} \quad \mu_{ij} = 0 \quad (3)$$

The Euclidean distance between the codeword c and image vector xi is calculated in Eq. (1). Following two rules exist for an local optimum codebook.

The group of vector R_j, j = 1,2,3,…,Nc must obey the rule given below.

$$R_j \supset \{x \in X : d(x, c_j) < d(x, c_k), \quad \forall k \neq j\} \quad (4)$$

The center of the R_j called c_j is calculated as

$$c_j = \left(\frac{1}{N_j}\right)\sum_{j=1}^{N_j} x_i, \quad x_i \in R_j \quad (5)$$

Here, N_j identifies the number of elements in R_j

Image vectors are determined as $x_i =, i = 1, 2, \ldots,$ Nb. The distance between the vectors is called d. The initial codewords are determined as $c_j(0), j = 1, 2, \ldots,$ Nc. The LBG algorithm executes the following three steps in order to obtain the local optimal codebook:

(a) Partition the original image blocks into several clusters using Euclidean distance. The result of clustering is saved in a Nb × Nc matrix U. The elements of U determined as

$$\mu_{ij} = 1 \quad if \quad d(x_i, c_j(k)) = \min \quad d(x_i, c_j(k)), \quad \text{otherwise} \quad \mu_{ij} = 0 \quad (6)$$

(b) Define the center of each cluster. Replace the old centers by the following new ones.

$$c_j(k+1) = \frac{\sum_{i=1}^{N_b}\mu_{ij}x_i}{\sum_{i=1}^{N_b}\mu_{ij}}, \quad j = 1, 2, \ldots, Nc \quad (7)$$

(c) Repeat the steps a and b until there is no change of centroid c_j values.

2.2 PSO Algorithm

Particle swarm optimization (PSO) is a metaheuristic technique originally proposed by Karayiannis [7]. In PSO, a particle determines a solution of the problem in the search area. Each particle has a velocity, a position and a fitness function value. In every iteration a particle records it's and the swarm's best solutions named "pbest" and "gbest" respectively and tries to reach the global optimum solution by using them. In image processing, each particle denotes a possible codebook C, whose error is determined by Eq. (1). Therefore PSO algorithm minimizes the codebook error to achieve the best representation. The fireflies in the population change their velocity and position by using Eqs. (8) and (9).

$$v_{ik}^{n+1} = k_w v_{ik}^n + k_p \text{random}_1 (pbest_{ik}^n - x_{ik}^n) + k_g \text{random}_2 (gbest_{ik}^n - x_{ik}^n) \quad (8)$$

$$x_{ik}^{n+1} = x_{ik}^n + v_{ik}^{n+1} \quad (9)$$

where i and k represent the number of particles and number of dimensions (k = 1, 2, 3, …, L) respectively. X and V are also named as the particle's position and velocity. The constants of k_w, k_p and k_g are the user defined parameters generally change in [0, 2]. The random1 and random2 are the uniform randoms change in [0, 1].

2.3 Fruit Fly Algorithm

The Fruit Fly Algorithm (FFA) is an optimization technique that imitates the Drosophila [25–33]. A fruit fly follows two steps for food; firstly, smells the food location, and flies towards that source. Then the drosophila uses its sensitive vision feature in order to determine if there is a better food source by looking around to the other fruit flies. Then all drosophilas fly toward the best food location using their vision feature by leaving their food storage. Fruit fly foraging procedure can be grouped into three parts;

(i) initial population generation, parameter setting,
(ii) search phase of osphresis,
(iii) search phase of vision.

The steps of FFA are determined as the following,
Step 1. Determine the iteration number, population size of the swarm and the search diameter value.
Step 2. Randomly define the locations of the fruit flies.
Step 3. Osphresis search phase; randomly search in the search space.
Step 4. Determine the smell intensity values of fruit flies.
Step 5. Phase of vision search: Define the best smell concentration location. Then,

all drosophilas flies to the best smell location. The next iteration starts from this location.
Step 6. Repeat the steps 3 to 6 sequentally until the maximum iteration number is reached.

2.4 Firefly Algorithm

The Firefly Algorithm (FRA) is a technique that inspired by nature and that imitates the flashing mechanism of the firefy. The two basic purposes of such flashing are attracting the prey and protective warning mechanism. The rate and pattern of flashing attract both sexes and brings them together. Females replies to the male's special pattern of flashing and come closer to their partners in the same species. The brightness of the flashing firefly can be determined by the objective function.

The FRA assumes three rules: (1) Each firefly is unisex, therefore any firefly may be attracted by other firefly without looking at their sex. (2) Attractiveness depends on the brightness of the fireflies. Therefore more bright one attracts the less bright one. If the brightness of each firefly are equal then search for another one. If there is not any brighter firefly than fireflies fly randomly. (3) The fitness function values are determined by the brightness values of the fireflies.

Two features of the FRA are important. First one is brightness and the second one is the attractiveness. The brightness value I is defined as the attractiveness of a firefly. Consequently the brightness of a firefly is proportional to the objective function as I(x)$\approx$ Objective Function(x). On the other hand the attractiveness β depends on the Euclidean distance r_{ij} between two fireflies numbered i and j. Since the light is absorbed by the air, then its intensity decreases when the interval from the source increases. Therefore the attractiveness should be changed according to the margin from source and absorption. The light intensity changes by an exponential function.

$$I = I_o \exp(-\gamma.r) \tag{10}$$

Here, I_0 is the source light intensity, where the γ is the light absorption parameter that is related to the attractiveness, determines the speed of the convergence and generally defined as $\gamma \in [0.01,10]$. The interval between two fireflies is represented by the value r. The attractiveness of a firefly β is defined by the Eq. (11) that is given below.

$$\beta = \beta_o \exp(-\gamma.r^2) \tag{11}$$

Here, β_o is the attractiveness at r=0. This situation may occur if the two fireflies are at the same position. The parameter β_o is generally defined as $\beta_o \in [0,1]$. The brightest firefly attracts the others at a maximum rate when β_o=1. On the contrast, if β_o=0 there is no attraction and cooperative study between the brightest one and the others. Therefore firefly flies randomly. The Euclidean distance between ith and jth fireflies i and j at the coordinate of x_i and x_j can be expressed as

$$r_{ij} = \| x_i - x_j \| = \sqrt{\sum_{k=1}^{d} (x_{ik} - x_{jk})^2} \tag{12}$$

where d is the total element number of firefly coordinate, x_{ik} is the kth element of the x_i. The ith firefly is attracted by more brighter jth one and flies towards it by

$$x_i = x_i + \beta_o(x_j - x_i)exp(-\gamma.r^2) + \alpha(\text{RandomNumber} - \frac{1}{2}) \tag{13}$$

where random number is distributed uniformly in [0, 1] and $\alpha \in [0, 1]$ is user defined parameter. Here, in Eq. (13), the second term represents the attraction by the brighter firefly when the third term corresponds to a random fly in a circle with radius α in order to find a possible brighter partner.

3 Lévy Flight Based Firefly Algorithm for Codebook Generation

The FRA contains two search mechanisms to reach the global minimum location. First one is the attraction of any firefly by a brighter one. This process strongly guides to the firefly on the global minimum way especially if it is attracted by the brightest one. The second one is the random search of a firefly in a circle with a radius of α. On the other hand if α is determined so big, the firefly may lose its way and come to a location that is much far away from the firefly group and possibly there is no brighter firefly to be followed. On the contrary, if α is determined so small, this time the firefly falls into a local minimum and can not escape. Therefore we need to have such an α that in most of the iterations its value changes in a small random value interval, but in rare iterations its value must be relatively big in order to escape local minimums. The Lévy distribution satisfies this need. Let f(x) be a Normal distribution,

$$f(x) = \frac{1}{\sigma\sqrt{2\pi}} \exp\left(-\frac{(x-\mu)^2}{2\sigma^2}\right) \tag{14}$$

Let μ=0 and standard deviation σ be a function of x,

$$\sigma = \sqrt{\frac{x^3}{c}} \tag{15}$$

Then the normal distribution is converted to the Lévy distribution,

$$f_L(x) = \sqrt{\frac{c}{2\pi}}\left(\frac{\exp(-\frac{c}{2x})}{x^{3/2}}\right) \tag{16}$$

By increasing x, exponential term decreases to 1, and the Lévy distribution becomes like a power law function with long tail,

$$f_L(x) \approx k.x^{-\lambda} \tag{17}$$

which has an infinite mean and variance. The approximated version of Lévy distribution of Eq. (17) is plotted in Fig. 1. It is seen that the Lévy function has higher values at the small values of x, on the contrary it has too much lower values of Lévy function at the big values of x. The approximated Lévy distributions for the values of $\lambda_1 = -1.0$, $\lambda_2 = -1.25$, $\lambda_3 = -1.5$, $\lambda_4 = -2.0$ are plotted in Fig. 1. The maximum value of $f_L(x)$ is constructed by k = 2.0 value. Lévy Flight based Firefly Algorithm (LFRA) can be obtained by using Lévy distribution function $f_L(x)$ in the firefly movement equation. Therefore, if Eq. (17) is embedded into Eq. (13) the firefly movement is determined as Eq. (18) where the Rand1 is uniformly distributed in [0, 1], k and λ are the user defined coefficients and Rand2 is a uniform distributed value in [a, b] interval. The a and b are determined from the horizontal axes. Since the third term of Eq. (18) is a power law function, in most of the iterations its value is small whereas the value is rarely relatively big. Consequently, the third term of Eq. (18) is called Lévy Flight step guides to the firefly in order to escape from local minimums.

$$x_i = x_i + \beta_o(x_j - x_i)\exp(-\gamma.r^2) + k.\text{sign}\left(Rand1 - \frac{1}{2}\right)(Rand2^{-\lambda}) \tag{18}$$

The pseudo code of the proposed LFRA is given below;

```
Program Pseudo Code (Output)
   Generate initial population of fireflies
   Objective function MSE(C) by Eq(1)
   Define the light absorption coefficient gamma
   Calculate the light intensities of each firefly  by 1/MSE(C)
   Define iteration number n
   while (x <Max Generation)
     while (n<Max Iteration Number)
       i = Uniform random number
       j = Uniform random number
        if (Ij > Ii)
         LÃ©vy Flight step is applied to the firefly i towards j
          end if
             Attractiveness changes exponentially
             Calculate locations and light intensity
```

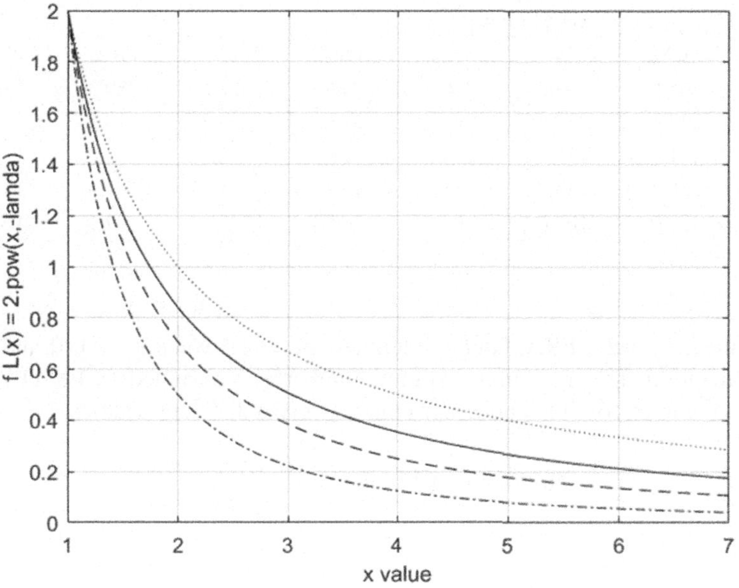

Fig. 1 Approximated Lévy distributions for different λ values

```
    end while
    Order the fireflies according to MSE(C) and define the best
    Move the best  firefly by L\'{e}vy Flight
  end while
  Further process results and visualization
  end
```

4 Parameters

In this paper, since the LBG algorithm is adapted to the FA then each firefly is represented by a codebook defined as $C = C_1, C_2, \ldots, C_{N_c} j = 1, 2, \ldots, N_c$. The number of codewords are selected as 8, 16, 32, 64 and 128 for the 256 gray level and 256 × 256 pixel sized standard images. In FCM, fuzzifier m is selected as 1.4. In both K-Means and VQ ϵ is determined as 0.001. The MSE performance of K-Means is calculated as the best of 10 K-Means runs. In PSO, user defined parameters kw, kp and kg are defined as 0.9, 0.9 and 0.9 respectively when the random1 and random2 are the uniform randoms change in [0, 1]. In Firefly algorithm, the α is determined as 0.5 when the γ is 0.01. I_o is determined as $MSE^{-1}(C)$. On the other

Table 1 MSE performances of Lena image

Cb Size	FCM	CM	V.Q.	PSO	FFA	FRA	LFRA
8	305.8	303.8	296.3	293.6	322.6	289.3	286.3
16	245.8	232.6	221.8	219.8	268.8	216.5	214.4
32	191.7	175.1	173.2	173.0	202.0	172.0	169.7
64	150.3	138.8	135.5	135.2	171.9	134.9	133.5
128	121.0	109.6	108.2	108.0	133.9	107.7	104.2

hand in the proposed LFRA, lamda is defined as $\lambda = 1.5$ when $\gamma = 0.01$ and k = 2.0 for optimization. The maximum iteration number is determined by Eq. (19) which indicates the total possible encounter number of two different fireflies.

$$\text{Max Iteration Number} = \frac{N_c(N_c - 1)}{2} \tag{19}$$

5 Simulations and Results

The codebook generation process is performed on 256 gray levels and 256 $\times$ 256 pixel size of standard images. First of all, standard images are dived into 4 $\times$ 4 pixel sized of 4096 subvectors. The standard VQ, PSO, FFA, FRA and the proposed LFRA techniques are used for codebook generation. In the proposed LFRA, the values of k and α parameters belong to the approximated Lévy Flight function $f_L(x)$ are important for determining the convergence speed. The k determines the maximum Lévy Flight step value whereas the λ defines the percentile of the long Lévy Flight step rate. High k and λ values cause a delay in the convergence whereas the small values may cause to be caught by a local minimum. Therefore k and λ are defined as 2.0 and 1.5 respectively for the optimization. Since $f_L(x)$ approaches too high values when x is nearly zero, the interval of a random variable is determined as x$\in$ [1, 7]. Consequently, Lévy Flight values change from $f_L(1)$ = 2.0 to $f_L(7)$ = 0.108, generally smaller than 0.5 but rarely bigger than 0.5 whereas in the classical FRA, the random step of α is changed uniformly in [0, 2.0]. The mean square error MSE(C) performance values for the codebooks of different and proposed techniques are given in Tables 1 and 2. The number of 8, 16, 32, 64 and 128 codewords are used in the simulation. In addition to these results the convergence curves of the algorithms according to MSE versus iteration number are presented in Figs. 2 and 3 for the high contrast 256 $\times$ 256 pixels size Lena and low contrast 256 $\times$ 256 pixel size Airplane standard images.

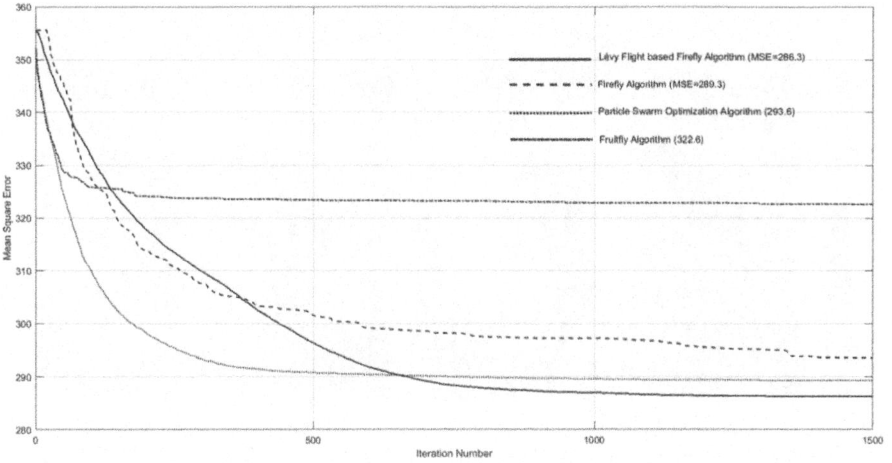

Fig. 2 Convergence results of LFRA, FRA, FFA and PSO algorithms for LENA image and 8 codewords of codebook

Table 2 MSE performances of airplane image

Cb Size	FCM	CM	VQ	PSO	FFA	FRA	LFRA
8	157.1	123.4	122.8	122.5	167.4	120.0	118.5
16	105.5	86.3	85.7	83.3	121.0	81.1	79.4
32	99.5	61.5	60.8	57.8	109.9	52.0	49.5
64	54.2	44.8	40.9	38.5	65.5	34.3	32.1
128	40.1	33.3	31.8	24.0	53.3	21.1	19.8

Table 3 MSE performances of different techniques

0.122bpp	FCM	CM	VQ	PSO	FFA	FRA	LFRA
Peppers	365.7	333.9	332.4	330.1	335.8	325.0	320.2
Clock	357.4	311.0	309.9	304.4	306.3	300.3	295.1
Airplane	157.1	123.4	122.8	122.5	130.2	120.0	118.5
Boat	400.4	379.0	376.7	372.8	373.3	369.1	363.2
ChemicalP.	402.3	371.5	369.9	362.2	364.0	359.8	355.9
Cameraman	444.5	417.6	416.6	410.0	411.3	408.7	405.4
Barbara	382.2	351.1	349.7	345.5	348.7	343.5	338.4
Einstein	270.5	245.7	242.1	238.8	240.1	235.6	229.2
Couple	340.0	315.9	312.8	308.1	309.7	307.2	303.0
Lena	305.8	303.8	296.3	293.6	322.6	289.3	286.3
Moon	188.4	152.1	147.7	142.2	145.6	140.0	137.4
Baboon	465.0	439.0	433.3	427.7	432.3	424.4	418.7
Aerial	562.7	538.2	532.2	526.6	528.5	524.3	518.4

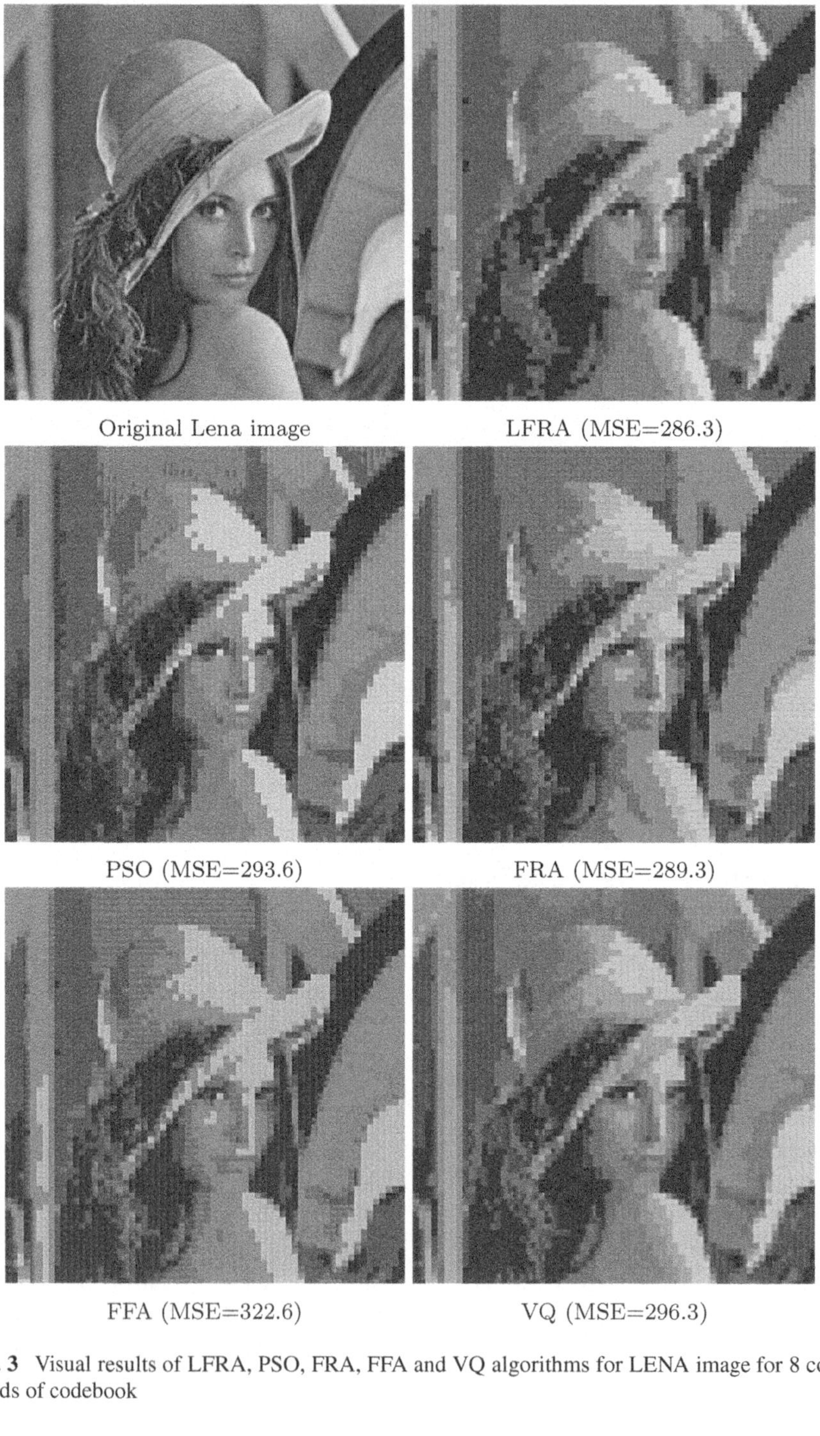

Fig. 3 Visual results of LFRA, PSO, FRA, FFA and VQ algorithms for LENA image for 8 codewords of codebook

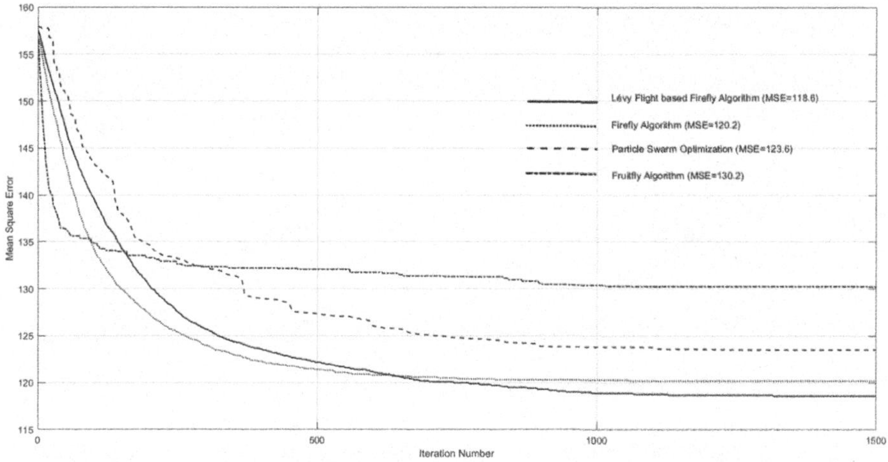

Fig. 4 Convergence results of LFRA, FRA, FFA and PSO algorithms for AIRPLANE image and 8 codewords of codebook

The results of other standard images of 256 × 256 pixel size are given in Table 3. It is seen from Tables 1, 2 and 3 and Figs. 3, 4, 5 and 6 that the proposed LFRA codebook MSE performances are better than both classical techniques and metaheuristic algorithms. On the other hand FFA, FRA and PSO classical metaheuristic algorithms are superior to the VQ, FCM, C-Means techniques. The proposed LFRA algorithm performs better and reaches to the global optimum codebook whereas VQ, CM, FCM and FFA techniques are seem to be captured local minimums. Consequently in the proposed LFRA, a firefly moves to the more brighter one and meanwhile searches around by Lévy Flight, which is a small random step in most of the iterations but rarely a big random step. Therefore this mechanism protects a firefly to be captured by a possible local minimum and also leads to it on the way of global minimum solution.

6 Conclusion

In this paper, the FRA metaheuristic algorithm is improved by using Lévy Flight distribution for codebook generation in image processing. In the proposed LFRA, besides the movement to the brighter one, a special random Lévy Flight step opportunity is given to a firefly. Lévy Flight satisfies to a firefly an opportunity of special random movement, which is generally small random steps in most of the iterations but rarely a big one. Therefore this mechanism protects a firefly to be captured by a possible local minimum and also leads to it on the way of global minimum solution. The achieved solutions suggest that the new introduced LFRA is better than both the classical techniques and the metaheuristic algorithms and provides the global optimum codebook.

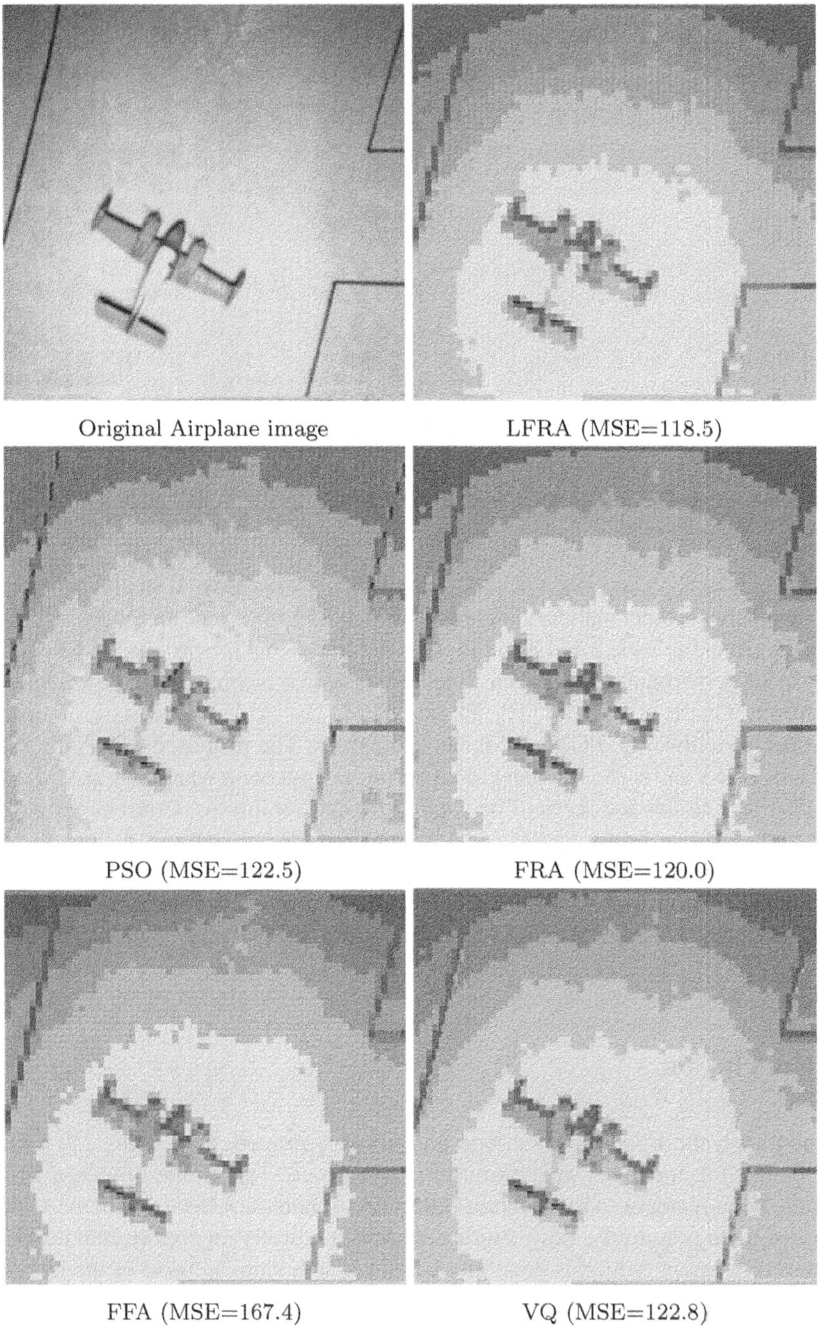

Fig. 5 Visual results of LFRA, PSO, FRA, FFA and VQ algorithms for AIRPLANE image for 8 codewords of codebook

References

1. Gray, T., R.M.: Vector quantization. IEEE ASSP Mag. **1**, 4–29 (1984)
2. Linde, Y., Buzo, A., Gray, R.M.: An algorithm for vector quantizer design. IEEE Trans. Commun. **1**, 84–95 (1980)
3. Lin, Y.C., Tai, S.C.: A fast Linde-Buzo-Gray algorithmin image vector quantization. IEEE Trans. Circuits Syst.-II Analog. Digit. Signal Processing. **45**(3), 432–435 (1998)
4. Patane, G., Russo, M.: The anhanced LBG algorithm. Neural Netw. **14**, 1219–1237 (2001)
5. Xu, W., Nandi, A.K., et al.: Novel vector quantiser design using reinforced learning. Signal Process. **85**, 1315–1333 (2005)
6. Tsai, C.W., Lee, C.Y., et al.: A fast VQ codebook generation algorithm via pattern reduction. Pattern Recognit. Lett. **30**, 653–660 (2009)
7. Karayiannis, N.B., Pai, P.I.: Fuzzy vector quantization algorithms and their application in image compression. IEEE Trans. Image Process. **4**(9), 1193–1201 (1995)
8. Karayiannis, N.B., Bezdek, J.C.: An integrated approach to fuzzy learning vector quantization and fuzzy c-means clustering. IEEE Trans. Fuzzy Syst. **5**(4), 622–628 (1997)
9. Tsekouras, G.E.: A fuzzy vector quantization approach to image compression. Appl. Math. Comput. **167**, 539–560 (2005)
10. Kuo, R.C., Wang, H.S., et al.: Application of ant K-Means on clustering analysis. Comput. Math. Appl. **50**, 1709–1724 (2005)
11. Tsai, C.W., Tseng, S.P., et al.: PREACO: a fast ant colony optimization for codebook generation. Appl. Soft Comput. **13**, 3008–3020 (2013)
12. Goldberg D.E.: Genetic Algorithms in Search Optimization and Machine Learning. Addison Wesley (1989)
13. Pan, J.S., McInnes, F.R., et al.: VQ codebook design using genetic algorithms. Electron. Lett. **31**(17), 1418–1419 (1995)
14. Huang, H.C., Pan, J.S., et al.: Vector quantization based on genetic simulated annealing. Signal Process. **81**, 1513–1523 (2001)
15. Zhang, L., Zheng, B., Yang, Z.: Codebook design using genetic algorithm and its application to speaker identification. Electron. Lett. **41**(10), 619–620 (2005)
16. Yang, S.B.: Constrained-storage multistage vector quantization based on genetic algorithms. Pattern Recognit. **41**, 689–700 (2008)
17. Pana, S.M., Chenga, K.S.: An evolution-based tabu search approach to codebook design. Pattern Recognit. **40**, 476–491 (2007)
18. Feng, H.M., Chen, C.Y., Ye, F.: Evolutionary fuzzy particle swarm optimization vector quantization learning scheme in image compression. Exp. Syst. Appl. **32**, 213–222 (2007)
19. Horng, M.H., Jiang, T.W.: Image vector quantization algorithm via honey bee mating optimization. Exp. Syst. Appl. **38**, 1382–1392 (2011)
20. Rani, M.L.P., Rao, G.S., Rao, B.P.: An efficient codebook generation using firefly algorithm for optimum medical image compression. J. Ambient. Intell. Hum. Comput. 1–13 (2020)
21. Kumar, S.N., Fred, A.L., Kumar, H.A., et.al.: Bat optimization-based vector quantization algorithm for compression of CT medical images. In: ICTMI, pp. 53–64. Springer, Singapore (2017)
22. Dai, H., Zhao, G., Lu, J., Dai, S.: Comment and improvement on A new fruit fly optimization algorithm: taking the financial distress model as an example. Knowl. Based Syst. **59**, 159–160 (2014)
23. Li, H., Guo, S., Li, C., Sun, J.: A hybrid annual power load forecasting model based on generalized regression neural network with fruit fly optimization algorithm. Knowl. Based Syst. **37**, 378–387 (2013)
24. Lin, S.M.: Analysis of service satisfaction in web auction logistics service using a combination of fruit fly optimization algorithm and general regression neural network. Neural Comput. Appl. **7**, 459–465 (2013)
25. Jiang, W., Wu, X., Gong, Y., et al.: Holt-Winters smoothing enhanced by fruit fly optimization algorithm to forecast monthly electricity consumption. Energy **193**, 116779 (2020)

26. Li, C., Xu, S., Li, W., L. Hu, L.: A novel modified fruit fly optimization algorithm for designing the self-tuning proportional integral derivative controller. J. Converg. Inf. Technol. **7**, 69–77 (2012)
27. Sheng, W., Bao, Y.: Fruit fly optimization algorithm based fractional order fuzzy-PID controller for electronic throttle. Nonlinear Dyn. **73**, 611–619 (2013)
28. Chen, P.W., Lin, W.Y., Huang, T.H., et al.: Using fruit fly optimization algorithm optimized grey model neural network to perform satisfaction analysis for e-business service. Appl. Math. Inf. Sci. **7**(21), 459–465 (2013)
29. Meng, T., Pan, Q.K.: An improved fruit fly optimization algorithm for solving the multidimensional knapsack problem. Appl. Soft Comput. **50**, 79–93 (2017)
30. Yuan, X., Dai, X., Zhao, J., et al.: On a novel multi-swarm fruit fly optimization algorithm and its application. Appl. Math. Comput. **233**, 260–271 (2014)
31. Wang, L., Xiong, Y., Li, S., Zeng, Y.R.: New fruit fly optimization algorithm with joint search strategies for function optimization problems. Knowl. Based Syst. **176**, 77–96 (2019)
32. Sheng, W., Bao, Y.: Fruit fly optimization algorithm based fractional order fuzzy—pid controller for electronic throttle. Nonlinear Dyn. **73**(1), 611–619 (2013)
33. Li, J.Q., Pan, Q.K., Mao, K.: A hybrid fruit fly optimization algorithm for the realistic hybrid flowshop rescheduling problem in steelmaking systems. IEEE Trans. Autom. Sci. Eng. **13**(2), 932–949 (2015)
34. Ingaleshwar, S., Dharwadkar, N.V., Jayadevappa, D.: Water chaotic fruit fly optimization-based deep convolutional neural network for image watermarking using wavelet transform. Multimed. Tools Appl. 1–25 (2021)
35. Rani, M.L.P., Rao, G.S., Rao, G.S.: An efficient codebook generation using firefly algorithm for optimum medical image compression. J. Ambient. Intell. Hum. Comput. **12**(1), 1–13 (2021)

Novel Chaotic Best Firefly Algorithm: COVID-19 Fake News Detection Application

Miodrag Zivkovic, Aleksandar Petrovic, K. Venkatachalam, Ivana Strumberger, Hothefa Shaker Jassim, and Nebojsa Bacanin

Abstract The COVID-19 pandemic period is approaching the two-year mark of its lifetime. During the pandemic, the people experienced various restrictions and different problems were raised unlike before. While the people stayed in their homes, electronic device usage had a decisive spike. This increased time spent on the internet and the amount of misinformation followed the same trend. The problem with the information on the COVID-19 is that there are too many different sources presenting different data. While some are just trying to help and do not have the latest or the most accurate information, some sources are malicious. Either way, everything but the latest and the most accurate data needs to be filtered when regarding a serious matter as such. The research proposed in this manuscript is aimed to identify COVID-19 fake news by performing wrapper-based feature by using an improved version of the well-known firefly algorithm. Practical simulations were done against a well-known dataset used in the domain of this problem, Koirala. The proposed method managed to achieve high accuracy of classification by using a smaller number of features in comparison with the state-of-the-art methods tested in the same experimental environment.

M. Zivkovic · A. Petrovic · I. Strumberger · N. Bacanin (✉)
Singidunum University, Danijelova 32, 11000, Belgrade, Serbia
e-mail: nbacanin@singidunum.ac.rs

M. Zivkovic
e-mail: mzivkovic@singidunum.ac.rs

A. Petrovic
e-mail: aleksandar.petrovic@singidunum.ac.rs

I. Strumberger
e-mail: istrumberger@singidunum.ac.rs

K. Venkatachalam
Faculty of Science, University of Hradec Králové, Hradec Kralove, Czech Republic
e-mail: venkatachalam.k@ieee.org

H. S. Jassim
Modern College of Business and Science, AL-Khuwair 133, Muscat, Oman
e-mail: hothefa.shaker@mcbs.edu.om

A. Biswas et al. (eds.), *Advances in Swarm Intelligence*, Studies in Computational Intelligence 1054, https://doi.org/10.1007/978-3-031-09835-2_16

Keywords COVID-19 · Fake news · Swarm intelligence · Firefly algorithm · Optimization · Metaheuristics

1 Introduction

The world was presented with novel problems as the global pandemic started gaining momentum while the effects of some of the known problems became amplified. This was the case with the problem of fake news. The characteristics of the internet platform make it susceptible to inaccuracies of any information that it holds. In consequence, the individuals that are to gain something from spreading erroneous information are capable to do so. These individuals can have different motives that range from personal amusement to elaborate frauds. The methods for solving this problem are present in the field of artificial intelligence but require further advancements since the problem is still relevant. This work presents improvements with the algorithm used as the foundation for such artificially intelligent solutions. The effects of misleading information such as fake news are more often than not harmless easy to recognize. On the other hand, when health is in question the consequences of such actions can lead even to death. The threat is created not only by malicious individuals but also by those who are incompetent to discuss such serious topics. It is paramount that the information easily available to large masses is of the highest accuracy and up-to-date. The trouble in the beginning of the COVID-19 pandemic was that the professionals did not have enough time to research the threat and immediately provide an unambiguous opinion for the measures of prevention, treatment, and even symptoms that still are in high variety. This lead to inconsistencies even with the trusted mediums and later on gave space to those forwarding their agendas regarding the treatment and vaccination. All of the above brings to the conclusion that is preeminent that all of the information regarding unpredictable health topics goes through a filter for harmful and obsolete information. To battle the mentioned issues, the authors propose an improved version of a population-based metaheuristic that is utilized as the wrapper method for the process of feature selection. The algorithm that was experimented on is the Firefly Algorithm with the goal of classification improvement. The proposed algorithm enhances the basic FA implementation by utilizing the chaotic-based approach.

The contributions of this manuscript are two-fold:

- New improved FA method has been devised by focusing on the known shortcomings with the FA version;
- Developed algorithm was used for the task of detecting COVID-19 fake-news and compared to other cutting-edge approaches;

This manuscript's organization is described in the following paragraph. The 2 Section explains the basis of technologies that the research relies on, explaining in detail the feature selection and the metaheuristic process. Following is the Sect. 3 that is used to explain the FA for the reasons of better understanding of the modifications that follow in the same section. Section 4 presents the experimental setup consisting of the environment in which the tests were performed, the experimental dataset, and the model. The research's results are displayed in Sect. 5 that additionally presents the discussion and comparative analysis with similar efforts. Final observations are given in Sect. 6 that finalizes this research.

2 Background

This section first gives a literature survey on the feature selection task. After that, a brief survey on swarm intelligence metaheuristics is provided.

2.1 Feature Selection

The classification accuracy is the main target of the Feature Selection (FS) method which leads to the general improvement of performance. The goal of the procedure is to reduce the number of features by recognizing those of the least quality which is determined by the amount of information they carry and their uniqueness. Similar features do not contribute greatly to the quality of the batch hence they should be removed. Optimal subset selection of features (solutions) is categorized as an NP-hard problem. FS process consists of three methods.

Firstly, the wrapper method is utilized that assigns scores for the features in the given subset. The score is affected by the number of mistakes that happened on the hold-out set that test every fresh subset with the role of model training. Note that this method yields the most computational costs to execute, but it is also the best-performing one. Following is the filter method with different attributes and performance. FS attempts to diminish the execution complexity while retaining performance quality. To achieve simplicity over the previous method the fine-tuning of features is not possible with the use of the filter method. Consequentially, it is more general than the wrapper method. Lastly, the third method is the embedded method that exploits the previous processes for the reason of constructing models. Therefore, the embedded method is the most general one out of the three. The embedded methods are in between the other two methods regarding the computational costs.

The problem of the COVID-19 fake news detection has been addressed recently by several models [28, 45, 52]. The performances of the models depend heavily of the appropriate feature selection process. Considering the previously stated, it is clear that the FS relies on search and evaluation. For the search process and traversing of the space of solutions, the algorithms are utilized with that goal. The result should be the best possible feature subset [3]. Suitable algorithms for this process proved to be those of a metaheuristic nature. Such algorithms generate random populations and iterate until they obtain the near-optimal sub-set of features [21, 44]. The subject of this works optimization, the Firefly Algorithm has also proven to yield high-quality results with the improvements of initialization strategy [15] as well as the Grey Wolf Optimizer (GWO) [4], variations of the Particle Swarm Optimization (PSO) [51], the Salp Swarm Algorithm [3], etc.

2.2 *Metaheuristic Algorithms*

The nature-inspired population-based algorithms are considered metaheuristic. This sort of algorithm iterates the same steps for searching but with a random population that is generated at the beginning of each iteration. As mentioned, the focus of this work is on nature-inspired metaheuristics and such solutions are nature inspired. Swarm Intelligence (SI) algorithms also belong to this group of algorithms since they take inspiration from the swarm-like behavior exhibited in the animal species that live in large groups. These algorithms found various purposes in the optimization problems a have successfully optimized the solving of NP-hard problems. Following this ability, the FA performs exceptionally well with the same sort of problems. It is becoming a standard to further modified these metaheuristics in the search of the method that provides the best possible solutions. The common practice in these modifications became the process of hybridization in which the advantages of two different SI solutions are combined. The unmodified natural metaheuristics favor either the process of exploration or exploitation. Hence, the process is with the goal of optimizing these two phases to both perform well. The process of SI is stochastic in nature, because of the random population initialized each time, and iteration-based. The object is the global optimum. The main obstruction in the process are these solutions occurrences providing results that are not the global, rater a local optimum. The search si done during phase of exploration, while the exploitation focuses on accuracy improvement.

The SI solutions became among the most popular subgroup of metaheuristic family finding the use for them in the solving of NP-hard problems across a wide variety of usages. Notable algorithms that have given best results include the Ant Colony Optimization (ACO) [25], artificial bee colony (ABC) [33], PSO [34], as well as the FA [57]. Some of the later solutions that have distinguished themselves among the other similar solutions are the GWO [43], moth search (MS) [53], monarch butterfly algorithm (MBA) [54], whale optimization algorithm (WOA) [42], and the harris hawks optimization (HHO) [29]. Some of the most recent metaheuristics algorithms include sine cosine algorithm (SCA) [40], arithmetic optimization algorithm (AOA) [2], jellyfish optimization algorithm [23] and salp swarm algorithm (SSA) [41]. In addition, the following algorithms exhibited leading performance in this category: the Differential Evolution Algorithm [32] and the Covariance Matrix Adaptation [30].

The application of the metaheuristics discussed take on a wide spectrum of different problems with NP-hardness in the information technologies field. Some of the such applications are with the problem of global numerical optimization [17], scheduling of tasks in the cloud reliant systems [7, 20], the problems of wireless sensors networks such as localization of nodes and the network lifetime prolonging [11, 60, 62], artificial neural networks optimization [6, 8–10, 13, 18, 24, 27, 31, 39, 48, 50], histological slides or MRI classifier optimization [14, 16, 19, 38, 46], and last but not least the COVID-19 case prediction [61, 63].

3 Proposed Method

This section first describes the basic implementation of the FA metaheuristics, followed by the known and observed drawbacks of the algorithm. At the end, a modified version of FA is proposed, that specifically targets the deficiencies of the basic FA.

3.1 The Firefly Algorithm

The basis of the algorithm created by Yang [58], that is FA, relies on the behavior of the firefly insect that the species use to communicate. To better understand this process a brief analysis of the species is given. The bioluminescence, the natural production of light in living organisms, is the main characteristic of fireflies and as with every evolutionary aspect it was not developed for no reason. The fireflies use this rare ability of nature to communicate. The way in which this communication is achieved is based on the intensity of an individual's light. The fireflies will always move towards the stronger light-emitting units. The brightness is directly proportional to the strength of attractiveness. In the case of equal brightness, the group will move randomly. It is important to emphasize that the firefly kind does not recognize sex, and hence this is not an additional requirement that would further complicate and possibly negatively influence the metaheuristic process. The amount of light that a single firefly emits is influenced by the objective function's setting which is the item for optimization.

Taking previously stated into account, three rules can be derived and they are described in the following text.

The first rule focuses on the modeling of the fitness function and it takes advantage of the brightness as well as attractiveness for this process. The objective function provides the value of brightness that typically affects the brightness in most of the different developed variations of the FA. The minimization problems require the following implementation of such formula:

$$I(x) = \begin{cases} \frac{1}{f(x)} & \text{, if } f(x) > 0 \\ 1 + \mid f(x) \mid & \text{, otherwise} \end{cases} \tag{1}$$

and the attractiveness is displayed with $I(x)$, while the objective function's value in x location is shown as $f(x)$.

The intensity of the emitted light falls off with distance and this directly influences the attractiveness decreasing its value of it as well. This behavior has been mathematically translated into the following equation:

$$I(r) = \frac{I_0}{1 + \gamma r^2} \tag{2}$$

and the intensity of light is represented by $I(r)$ and with the r distance, and its intensity at the source is I_0. Moreover, γ parameter is used for the modeling of the coefficient of absorption when the surroundings absorb the portion of the light. The combined impact of the γ coefficient and the inverse square for distance is approximately determined by the Gaussian form that is described below:

$$I(r) = I_0 \cdot e^{-\gamma r^2} \tag{3}$$

As discussed earlier, every unit in the swarm uses the attractiveness β that varies in intensity-based value of the subject's light intensity. The effects that are the consequences of distance in this scenario are described in the next equation:

$$\beta(r) = \beta_0 \cdot e^{-\gamma r^2} \tag{4}$$

that exploits the β_0 denomination as the attractiveness of the distance $r = 0$. The emphasis is brought to the replacement of Eq. (4) with the Eq. (5).

$$\beta(r) = \frac{\beta_0}{1 + \gamma r^2} \tag{5}$$

Considering the previous equations the calculation that performs the search method of a random unit i, that also changes locations in the new iteration $t + 1$ with the goal of approaching the stronger light emitting individual j is given:

$$x_i^{t+1} = x_i^t + \beta_0 \cdot e^{-\gamma r_{i,j}^2}(x_j^t - x_i^t) + \alpha^t(\kappa - 0.5) \tag{6}$$

and the parameter of randomization is denoted by α, uniform distribution or the Gaussian provided random number is κ. The distance from firefly i to firefly j is denoted as $r_{i,j}$. The values that have been previously employed for substantial results with the majority of problems for α and β_0 are [0, 1] and 1, respectively.

The Cartesian distance $r_{i,j}$ is measured with the return of the next equation:

$$r_{i,j} = ||x_i - x_j|| = \sqrt{\sum_{k=1}^{D}(x_{i,k} - x_{j,k})^2} \tag{7}$$

while the D denotes specific problem parameters number.

3.2 Drawbacks of the FA

With acknowledgment of the FA's excellent performances that were proven on many benchmarks [59] and practical tasks [50], recent research papers indicate that the FA implementation exhibits some drawbacks. The FA lacks raw exploration power and also exhibits poor intensification-diversification trade-off, as shown in [12, 49, 56]. During the early rounds of the execution of the algorithm its shortcomings can be perceived, where, during some runs, the basic FA is observed as incapable of finding and converging to the optimal areas of the search domain, resulting in inferior mean values. The FA search mechanism, given by Eq. (6), that drives the intensification procedure, is not able to focus the exploration to the promising areas in those situations. Contrariwise, if the random solutions have been created in the optimal and near-optimal areas by chance while the initialization of the algorithm is performed, FA is able to achieve excellent results.

Additionally, by observing the FA search mechanism given with Eq. (6), it is possible to conclude that it does not enclose an adequate exploration process. Several FA versions use the dynamic parameter α to solve this problem. This parameter decreases from the starting value α_0 to the defined threshold α_{min}, as presented in Eq. (8). This allows to emphasize the exploration during the initial rounds of execution, while in later stages, the balance is shifted towards the exploitation [55]. Nevertheless, extensive experiments have shown that even with the utilization of the dynamic α parameter, FA exploration capabilities are not improved enough.

$$\alpha^{t+1} = \alpha^t \cdot \left(1 - \frac{t}{T}\right), \tag{8}$$

where the current and following round are represented as t and $t+1$, and the T represents the maximal amount of iterations in one algorithm's run.

Finally, recent work suggests the efficiency of FA exploitation power in solving a large variety of types of tasks, and FA is renowned for its robustness and powerful exploitation as one of the best [12, 49, 56].

3.3 Proposed Chaotic-Based FA Metaheuristics

The chaos is the property of non-linear and deterministic systems that is very reactive with respect to its initial condition [5]. Mathematically speaking, chaotic-based local search (CLS) exhibits more power than the ergodic search, as observed by [47], resulting in a large number of possible sequences that could be generated by simply modifying the initial conditions.

Since a large number of chaotic maps exist in the available literature, by performing extensive empirical simulations, it was judged that for the suggested improved FA, the logistic map was able to achieve the most encouraging results. It should also be noted that the logistic map approach was already recognized and used in a number of swarm intelligence methods, such as [22, 36, 37] to mention a few. The logistic map used by the suggested FA algorithm is executed in K steps as follows:

$$\sigma_{i,j}^{k+1} = \mu\sigma_{i,j}^{k}(1 - \sigma_{i,j}), \; k = 1, 2, ...K, \tag{9}$$

in which $\sigma_{i,j}^{k}$ and $\sigma_{i,j}^{k+1}$ denote the chaotic variables for the j-th component from the i-th solution during iterations k and $k+1$, while the control parameter is denoted by μ.

The $\sigma_{i,j} \neq 0.25, 0.5$ and 0.75, $\sigma_{i,j} \in (0, 1)$, the μ is calibrated to 4, a value previously resolved in an empirical way [37].

The suggested modified FA encompass the global best (gBest) CLS mechanism since the chaos-based search is executed in the proximity of the solution x^*. During every step k, novel x^*, marked with x'^*, is being created by using Eqs. (10) and (11), executed for the j component of x^*:

$$x_j'^* = (1 - \lambda)x_j^* + \lambda S_j \tag{10}$$

$$S_j = l_j + \sigma_j^k(u_j - l_j) \tag{11}$$

where σ_j^k is resolved by using the Eq. (9) and λ represents the dynamic shrinking parameter dependent of the ongoing fitness function evaluation (FFE) and of the maximal amount of fitness function evaluations ($maxFFE$) during the run:

$$\lambda = \frac{maxFFE - FFE + 1}{maxFFE} \tag{12}$$

The usage of the dynamic λ parameter allows establishing better exploitation-exploration balance in the proximity of the x^*. In early rounds of the chaotic-based FA algorithm execution, a wide encircling around the x^* will be examined, while in the later rounds, exploitation is narrowed down. Finally, it is possible to replace FFE and $maxFFE$ by t and T if the termination condition is chosen to be the maximal number of iterations.

The proposed CLS strategy allows the improvement of x^* in K steps, and in case that the better fitness is achieved by the x'^* compared to the x^*, the CLS mechanism stops, while the replacement of x^* with x'^* is performed. On the other hand, if the x^* does not improve over K steps, it stays within the population.

The proposed FA algorithm uses the dynamic α parameter, according to Eq. (8). With all the above-mentioned concepts in place, the proposed Chaotic Best Firefly Algorithm—CBFA pseudo-code is shown in Algorithm 1.

Algorithm 1 The CBFA pseudo-code

Initialize main metaheuristics control parameters N and T
Initialize search space parameters D, u_j and l_j
Initialize CBFA control parameters γ, β_0, α_0, α_{min}, K and ϕ
Generate initial random population in the search space
while $t < T$ **do**
  for $i = 1$ to N **do**
    for $z = 1$ to i **do**
      if $I_z < I_i$ **then**
        Move solution z in the direction of individual i in D dimensions (Eq. (6))
        Attractiveness changes with distance r as $\exp[-\gamma r]$ (Eq. (4))
        Evaluate new solution, replace the worse individual with better one and update intensity of light (fitness)
      end if
    end for
  end for
  for $k = 1$ to K **do**
    Perform gBest CLS around the x^* using Eqs. (9)–(11) and generate $x'*$
    Retain better solution between x^* and x'^*
  end for
  Update α and λ according to Eqs. (8) and (12), respectively
end while
Return the best individual x_* from the population

4 Experimental Setup

The aim of this part of the paper is to explain the process of fake news detection. It is divided into five phases which are the collection of data, its preprocessing and feature extraction, FS, model development, and finally the evaluation and the assessment. The authors have used the unchanged setup as in the paper [3] out of the reasons for having the foundation for comparative analysis.

4.1 Dataset Collection

The development of a detection mechanism for fake news as with other artificial intelligence solutions requires training on a labeled set of data. In the case of the problem that is tackled in this research, a dataset that consists of a true and false set of news on the topics regarding the COVID-19 and the pandemic. The name of the used dataset is the COVID-19 Fake News Dataset [1], made by Koirala [35] and published on Mendeley. Furthermore, the dataset consists of more than 6,000 news published in the period of December 2019 ending with July 2020. The news have been published on various platforms and the following keywords were used to identify the relations of news to the topic: COVID-19, coronavirus, and corona. The

Table 1 Datasets description

Datasets	Tokenization	Stemming	Number of features
Data 1	Binary	No	1231
Data 2	TF-IDF	No	1231
Data 3	TF	No	1231
Data 4	TF-IDF	Snowball stemmer	1240
Data 5	TF-IDF	Lovins stemmer	1223
Data 6	Bag-of-words	Dawson stemmer	611

news were then put into three categories. Firstly the clearly false news. Secondly, the partially false news (both with label 0). Finally, the true news (with label 1).

4.2 Data Preprocessing and Feature Extraction

The first step is data cleaning, likewise as in [3]. For this goal firstly, the feature extraction is performed by the tokenization of extracted tweets. The transformations on the text are performed to make it into the rich text because in this format the classification output is better. The organization of the data is divided into groups for words, phrases, and sentences. Three different methods are employed for the extraction of features and they are the Bag of Words (BoW), Term Frequency (TF), and Term Frequency-Inverse Document Frequency (TF-IDF). After the tokenization process, the next in the code cleaning process is to remove the stop words like "and", "but", "or", and "the". Due to the high frequency of these types of words the dimensionality of the data is reduced. Additionally, common and exclusive words are also removed. To further decrease the dimensionality stemming is performed which also detects various variants of similar words. The archetypes of words are obtained by Dawson, Lovins, and the snowball stemmers. Furthermore, the figures that do not hold any information are removed. These figures can be punctuation marks, symbols, extra spaces, etc. As a result, the evolutionary classifier is improved. The next step is feature extraction with the three methods mentioned before. BoW allows for operation on groups of text without regarding the syntax and semantics. The reason for doing so is so the classifiers can operate with them. TF method is self-explanatory, providing the density of a term in the selected part of the text over all documents. TF-IDF additionally considers the weight of the words but only in the current file. The word weight is based on the importance of it, since some words that are rare can hold more information. This totals up to the six variations of the datasets that are shown in Table 1.

4.3 Feature Selection

FA is used as the search algorithm in the FS, and the selected features are evaluated by the k-nearest neighbors (kNN) classifier. The previous research suggests in the paper [3] that the produced subset of features is considered as a vector with the length n (the number of solutions of the produced selection) and values are 1 and 0 depict if the feature was selected or not, respectively. The retrieved accuracy of the classifier while the selected features number was used to determine the subset quality.

4.4 Model Development

The Eq. (13) describes the fitness function of the projected model:

$$\text{Fitness} = \alpha \times (1 - \text{accuracy}) + (1 - \alpha * \frac{|S|}{|W|}) \tag{13}$$

the accuracy is the correctly predicted news ratio of either true or fake, and the total classifications number including the wrong and right predictions, the $|S|$ shows selected features number, while the total number of features in the dataset is denoted with $|W|$, and the $\alpha \in [0, 1]$. The split of the dataset into training and testing was achieved through five-fold cross-validation.

4.5 Evaluation and Assesment

The evaluation parameters of the fake news identification are precision, accuracy, g-mean, and recall. For the reason of performing calculations that return these values a confusion matrix is created. The news that are correctly predicted to be fake are the true positives (TP). Following the same principle are the true negatives (TN) identified as false with the difference of true news now being the subject. On the other hand, the false positives (FP) and false negatives (FN), depict incorrect predictions of true news and false news, respectively.

The ratio of correctly predicted fake news and the total news identified as fake (both correct and incorrect predictions) is the precision.

$$\text{Precision} = \frac{TP}{FP + TP} \tag{14}$$

The correct predictions of both true and fake news against the total number of all predictions (TP, TN, FP, and FN) measures the accuracy.

Table 2 Parameter settings

Method	Parameter	Value
BSSA	c1; c2;	[0-1]; [0-1];
BPSO	Acceleration constants; Inertia w;	[2.1, 2.1]; [0.9, 0.6];
BGA	Single-point crossover; Mutation;	0.9; 0.01;
FA and CBFA	α; μ;	5;0.5

$$\text{Accuracy} = \frac{TP + TN}{TP + TN + FP + FN} \tag{15}$$

The rated average of precision and recall is the F-measure is calculated as:

$$F - \text{Measure} = \frac{(2 \times \text{Precision} \times \text{Recall})}{(\text{Precision} + \text{Recall})} \tag{16}$$

The recall or otherwise identified as sensitivity is the ratio of true positives and the total number of falsely identified true news and correctly identified fake news.

$$\text{Recall} = \frac{TP}{FN + TP} \tag{17}$$

5 Results and Discussion

The [3] lays the foundation for the simulations which was applied in this research as well. We have implemented FA and conducted experiments for the k-NN-FA method, as well as the proposed chaotic-based firefly k-NN-CBFA method. The results were also compared to additional algorithms. All of the used algorithms are, k-NN-BSSA (binary salp swarm algorithm), k-NN-BPSO (binary particle swarm algorithm), k-NN-BGA (genetic algorithm), and k-NN by itself, obtained likewise from [3]. The parameters of these methods are shown in Table 2. The experiments described in [3] have been recreated, where the algorithms were executed with 100 iterations and 15 solutions. The termination condition for the proposed CBFA method was set to 1500 FFE (100×15) to provide fair grounds for the comparisons. Algorithms have been executed in 30 independent runs.

For the dataset using BoW representation (data 1), the proposed k-NN-CBFA performed the best, with 0.75 accuracy and 642 features that were selected from the original dataset.

The suggested k-NN-CBFA has shown outstanding results, outperforming other metaheuristics methods in terms of accuracy and precision, and it was second-best (behind k-NN-BPSO) for recall and the F-measure, as shown in Table 3, where the best results for each indicator are marked in bold.

Table 3 Standard k-NN (without FS) and observed methods with FS classification results comparison over BoW representation (data 1)

Method	Accuracy	Precision	Recall	F-measure	No. of Selected Features
k-NN-BSSA	0.7264	0.6045	0.5605	0.5817	790.4
k-NN-BPSO	0.7258	0.6194	**0.6030**	**0.6111**	619.8
k-NN-BGA	0.7348	0.6468	0.5142	0.5729	631.2
k-NN	0.7053	0.5628	0.5888	0.5755	All
k-NN-FA	0.7135	0.5944	0.5659	0.5723	685.3
k-NN-CBFA	**0.7485**	**0.6453**	0.5937	0.6124	643.5

Table 4 Standard k-NN (without FS) and observed methods with FS classification results comparison over the TF-IDF representation dataset (data 2)

Method	Accuracy	Precision	Recall	F-measure	No. of Selected Features
k-NN-BSSA	0.6161	0.4587	0.7297	0.5633	753.6
k-NN-BPSO	0.6639	0.6194	0.6030	0.6111	613.8
k-NN-BGA	0.6764	0.6468	0.5142	0.5729	619.2
k-NN	**0.7053**	0.5628	0.5888	0.5755	All
k-NN-FA	0.6537	0.6125	0.5579	0.5844	652.4
k-NN-CBFA	0.6823	**0.6574**	**0.7311**	**0.6252**	631.2

The results with the dataset that uses TF-IDF representation (data 2) are given in Table 4. In this experimental instance, the pure k-NN method obtained the best accuracy of 0.70, closely followed by the proposed k-NN-CBFA technique, which achieved 0.68 and outperformed all other metaheuristic approaches. On the other hand, the proposed k-NN-CBFA returned the best results for other metrics (precision, recall, and F-measure).

The results of the dataset with TF representation are given in Table 5 for (data 3). The proposed k-NN-CBFA approach obtained the best accuracy of 0.741, closely outperforming the k-NN-BGA and k-NN-BPSO in second place with 0.734. The proposed k-NN-CBFA achieved the best result for precision and recall metrics as well, slightly outperforming the k-NN-BGA and k-NN-BSSA methods, respectively. Concerning the F-measure, k-NN-CBFA obtained the second-best result, closely behind the k-NN-BSSA method.

The results on the dataset with TF-IDF with snowball stemming (data 4) are shown in Table 6. The proposed k-NN-CBFA approach obtained the best results for

Table 5 Standard k-NN (without FS) and observed methods with FS classification results comparison over the TF representation dataset (data 3)

Method	Accuracy	Precision	Recall	F-measure	No. of Selected Features
k-NN-BSSA	0.7332	0.6095	0.5945	**0.6019**	794
k-NN-BPSO	0.7348	0.6039	0.5577	0.5799	613.4
k-NN-BGA	0.7348	0.6442	0.4877	0.5551	606.4
k-NN	0.7053	0.5628	0.5888	0.5755	All
k-NN-FA	0.7285	0.6024	0.5149	0.5796	648.5
k-NN-CBFA	**0.7414**	**0.6497**	**0.6085**	0.5839	617.3

Table 6 Standard k-NN (without FS) and observed methods with FS classification results comparison over TF-IDF and snowball stemming representation dataset (data 4)

Method	Accuracy	Precision	Recall	F-measure	No. of Selected Features
k-NN-BSSA	0.7220	0.5782	0.6673	**0.6196**	808.6
k-NN-BPSO	0.7187	0.5810	0.6134	0.5968	625.8
k-NN-BGA	0.7251	0.5949	0.5955	0.5952	602.4
k-NN	0.6963	0.5408	**0.6957**	0.6085	All
k-NN-FA	0.7114	0.5793	0.6026	0.5897	652.3
k-NN-CBFA	**0.7339**	**0.6012**	0.6358	0.6039	611.2

the accuracy and precision metrics. The basic k-NN scored best for the recall, while the k-NN-BSSA obtained the best F-measure score.

For the dataset with TF-IDF and Lovins stemming (data 5), results are observable from Table 7. The proposed k-NN-CBFA method obtained the best results for accuracy and precision metrics, outperforming the k-NN-BGA method that was a close second. For the recall and F-measure metrics, k-NN-BSSA and k-NN-BGA achieved the best results, respectively.

Table 8 shows results over the dataset with BoW and Dawson stemming (data 6). The proposed k-NN-CBFA approach again outperformed other approaches for accuracy and precision metric, while the k-NN-BGA method was on second and k-NN-BSSA on third. Moreover, for the recall and F-measures, the k-NN and k-NN-BSSA the best scores were obtained, while the proposed method was on the third and second positions, respectively.

Similarly, as in [3], the obtained results for the k-NN-CBFA method were compared to the results achieved by the most popular traditional approaches from the recent literature, namely J48, RF, and SVM. The results for these traditional

Table 7 Standard k-NN (without FS) and observed methods with FS classification results comparison over the TF-IDF and Lovins stemming representation dataset (data 5)

Method	Accuracy	Precision	Recall	F-measure	No. of Selected Features
k-NN-BSSA	0.7261	0.5962	**0.5974**	0.5968	789.4
k-NN-BPSO	0.7312	0.6198	0.5378	0.5759	616.2
k-NN-BGA	0.7543	0.6622	0.5633	**0.6088**	605.4
k-NN	0.7065	0.5642	0.5936	0.5785	All
k-NN-FA	0.7226	0.5973	0.5692	0.5717	645.4
k-NN-CBFA	**0.7768**	**0.6849**	0.5835	0.5912	616.8

Table 8 Standard k-NN (without FS) and observed methods with FS classification results comparison over the BoW and Dawson stemming representation dataset (data 6)

Method	Accuracy	Precision	Recall	F-measure	No. of Selected Features
k-NN-BSSA	0.7338	0.6111	0.5926	**0.6017**	400
k-NN-BPSO	0.7393	0.6319	0.5548	0.5908	301.6
k-NN-BGA	0.7402	0.6351	0.5510	0.5901	300
k-NN	0.6985	0.5499	**0.6144**	0.5804	All
k-NN-FA	0.7325	0.6024	0.5513	0.5910	325.6
k-NN-CBFA	**0.7558**	**0.6396**	0.5643	0.5996	308

approaches on the dataset with TF-IDF and Lovins (data 5) stemming were extracted from [3] and combined with results from Table 7. The proposed k-NN-CFBA method significantly outperformed the traditional approaches on this particular dataset (Table 9).

Comparison of results among various approaches is not enough to prove superiority of one method over others, therefore results comparison in terms of statistics need to be conducted. The Friedman test and two-way variance analysis by ranks [26] have been employed to determine if there is a significant results' difference among the proposed k-NN-CBFA and opponent methods. The Friedman test is employed for six compared algorithms over six datasets (data1 - data6). According to test results, proposed k-NN-CBFA established superior performance when compared to other algorithms with an average rank value of 1.16. Conversely, basic k-NN-FA obtained an average rank of 5. Moreover, the Friedman statistics ($\chi_r^2 = 21.07$) is larger than the χ^2 critical value with 10 °C of freedom (18.3), at significance level $\alpha = 0.05$ and as conclusion, the null hypothesis (H_0) can be rejected and it can be stated that

Table 9 Best classification results comparison (TF-IDF and Lovins stemming representation used—data 5) for the algorithms with FS and the standard k-NN (no FS) on the dataset and traditional classifiers

Method	Accuracy	Precision	Recall	F-measure	No. of Selected Features
k-NN-BSSA	0.7261	0.5962	**0.5974**	0.5968	789.4
k-NN-BPSO	0.7312	0.6198	0.5378	0.5759	616.2
k-NN-BGA	0.7543	0.6622	0.5633	**0.6088**	605.4
k-NN	0.7065	0.5642	0.5936	0.5785	All
J48	0.7229	0.5960	0.5690	0.5822	All
RF	0.7041	0.6342	0.3015	0.4087	All
SVM	0.7075	0.5665	0.5879	0.5770	All
k-NN-FA	0.7226	0.5973	0.5692	0.5717	645.4
k-NN-CBFA	**0.7768**	**0.6849**	0.5835	0.5912	616.8

the suggested k-NN-CBFA achieved results that are significantly different than the other six algorithms.

Finally, to better visualize the search ability of proposed method, the convergence graphs of the CFBA algorithm in comparison to the basic FA are shown in Fig. 1, while graphical representation of k-NN-CFBA performance indicators (accuracy, precision, recall, and F-measure) for six datasets is shown in the Fig. 2.

6 Conclusion

The proportions of the COVID-19 pandemic took humanity by surprise and the consequences were unpredictable, of which some are still yet to be observed. The nature of the situation resulted in massive amounts of false information and news on the topic, including the transmission, cures, and vaccination. The confusion that was created resulted in panic among people.

When regarding health topics, especially the pandemics, wrongful information on the subject can negatively influence the global situation and even result in an increase of number of deaths. To battle this problem, various fake news detection models have been developed by different scientists, with the goal of identifying and isolating incorrect information. The presented work suggests the CBFA-powered k-NN system that detects fake news on the COVID-19. The algorithm CBFA was employed as a wrapper method for performing the feature selection which increased the accuracy of the k-NN. The proposed solution is identified as k-NN-CBFA and the implementation and the testing were performed on the novel Koirala dataset. The evaluation of methods performance was taken out by comparing it to previous well-

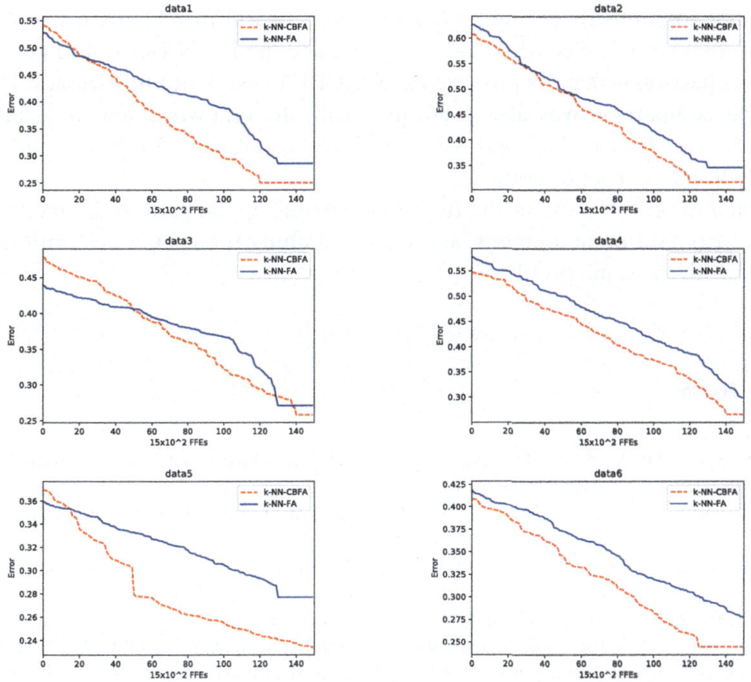

Fig. 1 Comparison of the convergence graphs of the CBFA and FA methods on all six datasets

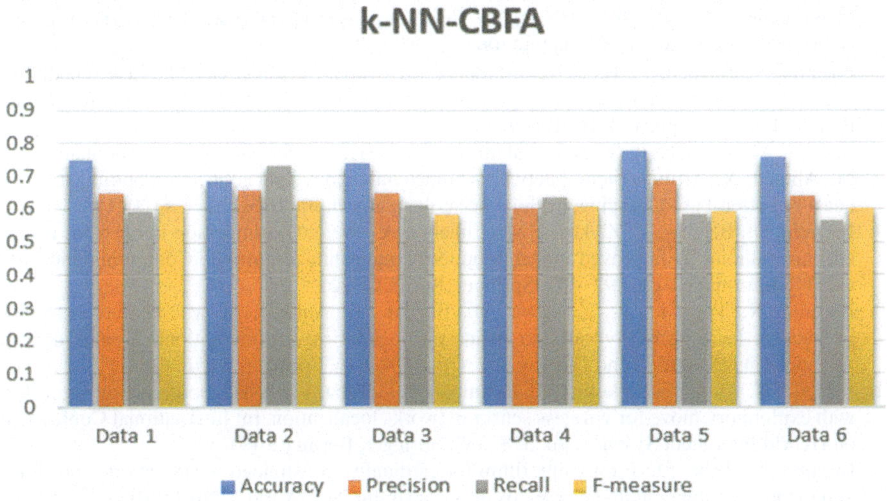

Fig. 2 Accuracy, Precision, Recall and F-measure achieved by k-NN-CBFA approach on 6 observed datasets

performing solutions including k-NN, and other swarm metaheuristics approaches to the problem like k-NN-BSA, k-NN-BPSO, and the k-NN-BGA. The conclusion of the comparison is that the proposed k-NN-CBFA is superior in accuracy. Furthermore, the comparison was also performed with the renowned standard classifiers like the random forest, J48, and SVM. The proposed method repeated its success similarly to the first set of comparison.

Planned improvements for the future include the application of the method with scenarios from different domains, as well as searching for a better performing metaheuristic with the same problem.

References

1. Mendeley data—covid-19 fake news dataset. https://data.mendeley.com/datasets/zwfdmp5syg/1. Accessed 27 Aug 2021
2. Abualigah, L., Diabat, A., Mirjalili, S., Abd Elaziz, M., Gandomi, A.H.: The arithmetic optimization algorithm. Comput. Methods Appl. Mech. Eng. **376**, 113609 (2021)
3. Al-Ahmad, B., Al-Zoubi, A., Abu Khurma, R., Aljarah, I.: An evolutionary fake news detection method for covid-19 pandemic information. Symmetry **13**(6), 1091 (2021)
4. Al-Tashi, Q., Kadir, S.J.A., Rais, H.M., Mirjalili, S., Alhussian, H.: Binary optimization using hybrid grey wolf optimization for feature selection. IEEE Access **7**, 39496–39508 (2019)
5. Alatas, B.: Chaotic bee colony algorithms for global numerical optimization. Expert. Syst. Appl. **37**(8), 5682–5687 (2010)
6. Bacanin, N., Alhazmi, K., Zivkovic, M., Venkatachalam, K., Bezdan, T., Nebhen, J.: Training multi-layer perceptron with enhanced brain storm optimization metaheuristics. Comput., Mater. Contin. **70**(2), 4199–4215 (2022). https://doi.org/10.32604/cmc.2022.020449, http://www.techscience.com/cmc/v70n2/44706
7. Bacanin, N., Bezdan, T., Tuba, E., Strumberger, I., Tuba, M., Zivkovic, M.: Task scheduling in cloud computing environment by grey wolf optimizer. In: 2019 27th Telecommunications Forum (TELFOR), pp. 1–4. IEEE (2019)
8. Bacanin, N., Bezdan, T., Venkatachalam, K., Zivkovic, M., Strumberger, I., Abouhawwash, M., Ahmed, A.: Artificial neural networks hidden unit and weight connection optimization by quasi-refection-based learning artificial bee colony algorithm. IEEE Access (2021)
9. Bacanin, N., Bezdan, T., Zivkovic, M., Chhabra, A.: Weight optimization in artificial neural network training by improved monarch butterfly algorithm. In: Mobile Computing and Sustainable Informatics, pp. 397–409. Springer, Berlin (2022)
10. Bacanin, N., Petrovic, A., Zivkovic, M., Bezdan, T., Antonijevic, M.: Feature selection in machine learning by hybrid sine cosine metaheuristics. In: International Conference on Advances in Computing and Data Sciences, pp. 604–616. Springer, Berlin (2021)
11. Bacanin, N., Tuba, E., Zivkovic, M., Strumberger, I., Tuba, M.: Whale optimization algorithm with exploratory move for wireless sensor networks localization. In: International Conference on Hybrid Intelligent Systems, pp. 328–338. Springer, Berlin (2019)
12. Bacanin, N., Tuba, M.: Firefly algorithm for cardinality constrained mean-variance portfolio optimization problem with entropy diversity constraint. Sci. World J. **2014** (2014)
13. Bacanin, N., Zivkovic, M., Bezdan, T., Cvetnic, D., Gajic, L.: Dimensionality reduction using hybrid brainstorm optimization algorithm. In: Proceedings of International Conference on Data Science and Applications, pp. 679–692. Springer, Berlin (2022)
14. Basha, J., Bacanin, N., Vukobrat, N., Zivkovic, M., Venkatachalam, K., Hubálovský, S., Trojovský, P.: Chaotic harris hawks optimization with quasi-reflection-based learning: An application to enhance cnn design. Sensors **21**(19), 6654 (2021)

15. Bezdan, T., Cvetnic, D., Gajic, L., Zivkovic, M., Strumberger, I., Bacanin, N.: Feature selection by firefly algorithm with improved initialization strategy. In: 7th Conference on the Engineering of Computer Based Systems, pp. 1–8 (2021)
16. Bezdan, T., Milosevic, S., Venkatachalam, K., Zivkovic, M., Bacanin, N., Strumberger, I.: Optimizing convolutional neural network by hybridized elephant herding optimization algorithm for magnetic resonance image classification of glioma brain tumor grade. In: 2021 Zooming Innovation in Consumer Technologies Conference (ZINC), pp. 171–176. IEEE (2021)
17. Bezdan, T., Petrovic, A., Zivkovic, M., Strumberger, I., Devi, V.K., Bacanin, N.: Current best opposition-based learning salp swarm algorithm for global numerical optimization. In: 2021 Zooming Innovation in Consumer Technologies Conference (ZINC), pp. 5–10. IEEE (2021)
18. Bezdan, T., Stoean, C., Naamany, A.A., Bacanin, N., Rashid, T.A., Zivkovic, M., Venkatachalam, K.: Hybrid fruit-fly optimization algorithm with k-means for text document clustering. Mathematics **9**(16), 1929 (2021)
19. Bezdan, T., Zivkovic, M., Tuba, E., Strumberger, I., Bacanin, N., Tuba, M.: Glioma brain tumor grade classification from mri using convolutional neural networks designed by modified fa. In: International Conference on Intelligent and Fuzzy Systems, pp. 955–963. Springer, Berlin (2020)
20. Bezdan, T., Zivkovic, M., Tuba, E., Strumberger, I., Bacanin, N., Tuba, M.: Multi-objective task scheduling in cloud computing environment by hybridized bat algorithm. In: International Conference on Intelligent and Fuzzy Systems, pp. 718–725. Springer, Berlin (2020)
21. Brezočnik, L., Fister, I., Podgorelec, V.: Swarm intelligence algorithms for feature selection: a review. Appl. Sci. **8**(9), 1521 (2018)
22. Chen, H., Xu, Y., Wang, M., Zhao, X.: A balanced whale optimization algorithm for constrained engineering design problems. Appl. Math. Model. **71**, 45–59 (2019)
23. Chou, J.S., Truong, D.N.: A novel metaheuristic optimizer inspired by behavior of jellyfish in ocean. Appl. Math. Comput. **389**, 125535 (2021)
24. Cuk, A., Bezdan, T., Bacanin, N., Zivkovic, M., Venkatachalam, K., Rashid, T.A., Devi, V.K.: Feedforward multi-layer perceptron training by hybridized method between genetic algorithm and artificial bee colony. In: Data Science and Data Analytics: Opportunities and Challenges, p. 279 (2021)
25. Dorigo, M., Birattari, M., Stutzle, T.: Ant colony optimization. IEEE Comput. Intell. Mag. **1**(4), 28–39 (2006)
26. Friedman, M.: A comparison of alternative tests of significance for the problem of m rankings. Ann. Math. Stat. **11**(1), 86–92 (1940)
27. Gajic, L., Cvetnic, D., Zivkovic, M., Bezdan, T., Bacanin, N., Milosevic, S.: Multi-layer perceptron training using hybridized bat algorithm. In: Computational Vision and Bio-Inspired Computing, pp. 689–705. Springer, Berlin (2021)
28. Hande, A., Puranik, K., Priyadharshini, R., Thavareesan, S., Chakravarthi, B.R.: Evaluating pretrained transformer-based models for covid-19 fake news detection. In: 2021 5th International Conference on Computing Methodologies and Communication (ICCMC), pp. 766–772. IEEE (2021)
29. Heidari, A.A., Mirjalili, S., Faris, H., Aljarah, I., Mafarja, M., Chen, H.: Harris hawks optimization: algorithm and applications. Futur. Gener. Comput. Syst. **97**, 849–872 (2019)
30. Igel, C., Hansen, N., Roth, S.: Covariance matrix adaptation for multi-objective optimization. Evol. Comput. **15**, 1–28 (2007). https://doi.org/10.1162/evco.2007.15.1.1
31. Jnr, E.O.N., Ziggah, Y.Y., Relvas, S.: Hybrid ensemble intelligent model based on wavelet transform, swarm intelligence and artificial neural network for electricity demand forecasting. Sustain. Cities Soc. **66**, 102679 (2021)
32. Karaboğa, D., Okdem, S.: A simple and global optimization algorithm for engineering problems: differential evolution algorithm. Turk. J. Electr. Eng. Comput. Sci. **12**, 53–60 (2004)
33. Karaboga, D., Basturk, B.: A powerful and efficient algorithm for numerical function optimization: artificial bee colony (abc) algorithm. J. Glob. Optim. **39**(3), 459–471 (2007)
34. Kennedy, J., Eberhart, R.: Particle swarm optimization. In: Proceedings of ICNN'95-International Conference on Neural Networks, vol. 4, pp. 1942–1948. IEEE (1995)

35. Koirala, A.: Covid-19 Fake News Classification with Deep Learning. Preprint (2020)
36. Li, C., Zhou, J., Xiao, J., Xiao, H.: Parameters identification of chaotic system by chaotic gravitational search algorithm. Chaos, Solitons Fractals **45**(4), 539–547 (2012)
37. Liang, X., Cai, Z., Wang, M., Zhao, X., Chen, H., Li, C.: Chaotic oppositional sine–cosine method for solving global optimization problems. Eng. Comput. pp. 1–17 (2020)
38. Lichtblau, D., Stoean, C.: Cancer diagnosis through a tandem of classifiers for digitized histopathological slides. PLOS ONE **14**(1), 1–20 (2019). https://doi.org/10.1371/journal.pone.0209274
39. Milosevic, S., Bezdan, T., Zivkovic, M., Bacanin, N., Strumberger, I., Tuba, M.: Feed-forward neural network training by hybrid bat algorithm. In: Modelling and Development of Intelligent Systems: 7th International Conference, MDIS 2020. Sibiu, Romania. Revised Selected Papers 7, pp. 52–66. Springer International Publishing (2021)
40. Mirjalili, S.: Sca: a sine cosine algorithm for solving optimization problems. Knowl. Based Syst. **96**, 120–133 (2016)
41. Mirjalili, S., Gandomi, A.H., Mirjalili, S.Z., Saremi, S., Faris, H., Mirjalili, S.M.: Salp swarm algorithm: a bio-inspired optimizer for engineering design problems. Adv. Eng. Softw. **114**, 163–191 (2017)
42. Mirjalili, S., Lewis, A.: The whale optimization algorithm. Adv. Eng. Softw. **95**, 51–67 (2016)
43. Mirjalili, S., Mirjalili, S.M., Lewis, A.: Grey wolf optimizer. Adv. Eng. Softw. **69**, 46–61 (2014)
44. Nguyen, B.H., Xue, B., Zhang, M.: A survey on swarm intelligence approaches to feature selection in data mining. Swarm Evol. Comput. **54**, 100663 (2020)
45. Paka, W.S., Bansal, R., Kaushik, A., Sengupta, S., Chakraborty, T.: Cross-sean: a cross-stitch semi-supervised neural attention model for covid-19 fake news detection. Appl. Soft Comput. **107**, 107393 (2021)
46. Postavaru, S., Stoean, R., Stoean, C., Joya Caparros, G.: Adaptation of deep convolutional neural networks for cancer grading from histopathological images. In: Rojas, I., Joya, G., Catala, A. (eds.) Advances In Computational Intelligence, Iwann 2017, PT II. Lecture Notes in Computer Science, vol. 10306, pp. 38–49. Univ Granada; Univ Malaga; Polytechn Univ Catalonia (2017). https://doi.org/10.1007/978-3-319-59147-6_4, 14th International Work-Conference on Artificial Neural Networks (IWANN), Cadiz, SPAIN, JUN 14-16, 2017
47. dos Santos Coelho, L., Mariani, V.C.: Use of chaotic sequences in a biologically inspired algorithm for engineering design optimization. Exp. Syst. Appl. **34**(3), 1905–1913 (2008)
48. Stoean, R.: Analysis on the potential of an ea-surrogate modelling tandem for deep learning parametrization: an example for cancer classification from medical images. Neural Comput. Appl. **32**, 313–322 (2018)
49. Strumberger, I., Bacanin, N., Tuba, M.: Enhanced firefly algorithm for constrained numerical optimization. In: 2017 IEEE Congress on Evolutionary Computation (CEC), pp. 2120–2127. IEEE (2017)
50. Strumberger, I., Tuba, E., Bacanin, N., Zivkovic, M., Beko, M., Tuba, M.: Designing convolutional neural network architecture by the firefly algorithm. In: 2019 International Young Engineers Forum (YEF-ECE), pp. 59–65. IEEE (2019)
51. Unler, A., Murat, A., Chinnam, R.B.: mr2pso: a maximum relevance minimum redundancy feature selection method based on swarm intelligence for support vector machine classification. Inf. Sci. **181**(20), 4625–4641 (2011)
52. Varma, R., Verma, Y., Vijayvargiya, P., Churi, P.P.: A systematic survey on deep learning and machine learning approaches of fake news detection in the pre-and post-covid-19 pandemic. Int. J. Intell. Comput. Cybern. (2021)
53. Wang, G.G.: Moth search algorithm: a bio-inspired metaheuristic algorithm for global optimization problems. Memetic Comput. **10**(2), 151–164 (2018)
54. Wang, G.G., Deb, S., Cui, Z.: Monarch butterfly optimization. Neural Comput. Appl. **31**(7), 1995–2014 (2019)
55. Wang, H., Zhou, X., Sun, H., Yu, X., Zhao, J., Zhang, H., Cui, L.: Firefly algorithm with adaptive control parameters. Soft Comput. **21**(17), 5091–5102 (2017)

56. Xu, G.H., Zhang, T.W., Lai, Q.: A new firefly algorithm with mean condition partial attraction. Appl. Intell. 1–14 (2021)
57. Yang, X.S.: Firefly algorithms for multimodal optimization. In: International Symposium on Stochastic Algorithms, pp. 169–178. Springer, Berlin (2009)
58. Yang, X.S.: Firefly algorithms for multimodal optimization. In: Watanabe, O., Zeugmann, T. (eds.) Stochastic Algorithms: Foundations and Applications, pp. 169–178. Springer, Berlin (2009)
59. Yang, X.S., He, X.: Firefly algorithm: recent advances and applications. Int. J. Swarm Intell. **1**(1), 36–50 (2013)
60. Zivkovic, M., Bacanin, N., Tuba, E., Strumberger, I., Bezdan, T., Tuba, M.: Wireless sensor networks life time optimization based on the improved firefly algorithm. In: 2020 International Wireless Communications and Mobile Computing (IWCMC), pp. 1176–1181. IEEE (2020)
61. Zivkovic, M., Bacanin, N., Venkatachalam, K., Nayyar, A., Djordjevic, A., Strumberger, I., Al-Turjman, F.: Covid-19 cases prediction by using hybrid machine learning and beetle antennae search approach. Sustain. Cities Soc. **66**, 102669 (2021)
62. Zivkovic, M., Bacanin, N., Zivkovic, T., Strumberger, I., Tuba, E., Tuba, M.: Enhanced grey wolf algorithm for energy efficient wireless sensor networks. In: 2020 Zooming Innovation in Consumer Technologies Conference (ZINC), pp. 87–92. IEEE (2020)
63. Zivkovic, M., Venkatachalam, K., Bacanin, N., Djordjevic, A., Antonijevic, M., Strumberger, I., Rashid, T.A.: Hybrid genetic algorithm and machine learning method for covid-19 cases prediction. In: Proceedings of International Conference on Sustainable Expert Systems: ICSES 2020, vol. 176, p. 169. Springer Nature (2021)

Other Applications

Artificial Bee Colony and Genetic Algorithms for Parameters Estimation of Weibull Distribution

Muhammet Burak Kılıç

Abstract Weibull distribution has been widely applied in many scientific disciplines such as modeling wind speed or failure rates. Various estimation methods such as maximum likelihood (ML), moment (MOM) inferences are proposed for the Weibull parameters estimation, but there has not been a comparison of performances based on swarm optimization algorithms. Therefore, this chapter investigates the performances of the artificial bee colony (ABC) and the genetic algorithm (GA) for the Weibull parameters estimation. The estimation problem of the Weibull shape parameter is solved by the proposed functions using the ABC and GA. The contribution of this chapter is to provide a comparison of the ABC and the GA for estimating the Weibull parameters using the proposed functions. To identify the performances of the ABC and the GA methods for Weibull parameters estimation, a comprehensive simulation study, and a failure data example are performed.

Keywords Weibull distribution · Parameters estimation · Artificial Bee Colony · Genetic algorithm

1 Introduction

Weibull distribution proposed by [1] has a wide field of problems and plays an important practical role in modeling lifetime and wind speed datasets [2, 3]. Various methods can be employed for the Weibull parameters estimation such as the maximum likelihood (ML) estimation, moment (MOM) estimation, and the modified moment estimation [4, 5]. Among the parameters estimation methods, the ML is an attractive method and is frequently used for estimating the Weibull parameters. However, the ML with standard optimization techniques may present difficult problems such as inequality constraints and gradient problems when it has no explicit solutions of the estimators of the parameters. A good alternative to conventional

M. B. Kılıç (✉)
Department of Business Administration, Quantitative Methods Unit, Mehmet Akif Ersoy University, Burdur, Turkey
e-mail: mburak@mehmetakif.edu.tr

A. Biswas et al. (eds.), *Advances in Swarm Intelligence*, Studies in Computational Intelligence 1054, https://doi.org/10.1007/978-3-031-09835-2_17

numerical optimization algorithms is swarm optimization algorithms. These algorithms have been also seen as a popular tool for the Weibull parameters estimation because of their practical implementation and simplicity.

Swarm intelligence methods are inspired by self-organized system, the migration of birds, the colony of inspects such as ants or bees [6]. These methods have been commonly used for different optimization problems. A class of swarm algorithms is known as population-based approaches and works with a set of solutions [7]. The main advantage of swarm intelligence methods uses a search space rather initial points over the classical iterative procedures. Therefore, this chapter is considered the popular swarm intelligence methods as the artificial bee colony (ABC) and the genetic algorithm (GA) to estimate the Weibull shape parameter using the ML and the MOM inferences.

One of the most population-based swarm algorithms is the ABC algorithm proposed by [8] and this algorithm is widely used in many disciplines for different optimization problems such as neural network [9], function optimization [10], scheduling of multi-skilled manpower [11] and signal processing [12]. The ABC algorithm is also employed for the Weibull parameters estimation [13] and the results of [13] show that the parameters estimation incorporating the ABC algorithm outperforms the classical iterative methods for the Weibull parameters estimation using the ML and the MOM inferences with wind speed data. Recently, the variants of the ABC algorithm is commonly used for parameter estimation of statistical distributions. For instance, the ABC and adaptive hybrid ABC algorithms are proposed for estimating the q-Weibull parameters based on the ML inference in [14] and the ABC with Levy flights for the ML inference of a three-parameter Weibull distribution is used in [15]. One of the popular swarm optimization algorithms among other algorithms is the GA proposed by Holland [16] and this algorithm is a class of swarm intelligence methods motivated by biological systems and consists of the natural phenomena included crossover, selection and mutation operators. The GA is widely used in many disciplines for solving different optimization problems such as image processing [17], pipeline control systems [18], and machine learning systems [19]. The ML inference using the GA was applied in [20, 21] and performed to maximize the log-likelihood function for the Weibull parameters estimation. The results of [20] show that the GA approach is a promising alternative estimation to the Newton-Raphson technique. The GA was also performed for estimating the parameters of various statistical distributions such as a mixture of normal distributions [22], the negative binomial gamma [23], the skew-normal [24], and a mixture of two Weibull distributions [25]. It was concluded from the aforementioned studies that the GA approach generally outperforms the other classical parameters estimation methods. The other swarm algorithms are also used for the estimation method of a three-parameter Weibull distribution (see also [26–28]).

The current focus of this chapter is to investigate the performances of the ABC and the GA swarm methods for the Weibull parameters estimation using proposed functions. To show the efficiency of the ML and MOM inferences using the ABC and the GA with proposed functions, a simulation study is performed with different scenarios. The remainder of this chapter is structured as follows. Section 2 provides the probability density function and the ML and the MOM inferences for the Weibull distribution and presents the proposed functions to estimate the Weibull shape parameter. Section 3 gives the steps of the ABC and the GA. Section 4 illustrates a comprehensive simulation study and failure data to compare the performances of the ABC and the GA. Section 5 ends with possible extensions.

2 Weibull Distribution

The probability density and the corresponding cumulative distribution functions for a two-parameter Weibull distribution are presented as follows:

$$f(x) = \frac{k}{c}\left(\frac{x}{c}\right)^{k-1} \exp\left\{-\left(\frac{x}{c}\right)^{k}\right\}, \qquad x > 0,\ k > 0,\ c > 0 \tag{1}$$

and

$$F(x) = 1 - \exp\left\{-\left(\frac{x}{c}\right)^{k}\right\}, \tag{2}$$

respectively. k and c denote the shape and the scale parameters, respectively. x denotes an observed dataset.

2.1 Maximum Likelihood Inference

The ML inference for the Weibull parameters estimation is performed by maximizing of the log-likelihood function. This function, for this reason, is as follows:

$$\ln L = n \ln k - n\, k \ln c + (k-1)\sum_{i=1}^{n} \ln x_i - c^{-k}\sum_{i=1}^{n} x_i^k. \tag{3}$$

Differentiating of the ln$L$ function with shape parameter k and scale parameter c given in Eq. (3), and equating them to zero, then the ML estimators for the Weibull distribution are obtained by

$$\widehat{k}_{ML} = \left[\frac{\sum_{i=1}^{n} x_i^{\widehat{k}} \ln x_i}{\sum_{i=1}^{n} x_i^{\widehat{k}}} - \frac{\sum_{i=1}^{n} \ln x_i}{n}\right]^{-1}, \tag{4}$$

and

$$\widehat{c}_{ML} = \left[\frac{\sum_{i=1}^{n} x_i^{\widehat{k}}}{n} \right]^{\frac{1}{\widehat{k}}}, \tag{5}$$

respectively. From Eq. (4), it can be observed that there is no explicit solution due to the nonlinear equations of the ML inference for the Weibull shape parameter k. Therefore, various iterative techniques can be applied for estimating the shape parameter, but classical iterative techniques lead to computational difficulties such as the initial value problem and these numerical methods have been discussed by numerous authors. This chapter presents the parameters estimation of the Weibull distribution using swarm intelligence approaches of the ABC algorithm motivated by the movement of a honeybee swarm [10], and the GA inspired by biological principles [16]. After the ML inference for the Weibull shape parameter k using the ABC and the GA, the ML inference for the Weibull scale parameter c is obtained from Eq. (5).

2.2 *Moment Inference*

One of the classical parameters estimation methods is the MOM inference about statistical distributions using their moments. The MOM inference is performed by equating the theoretical moments to their sample moments, then solving the corresponding equations to obtain the Weibull parameters estimation [5]. The first and second non-central moments for the Weibull distribution is given by

$$M_1 = c\,\Gamma\left(1 + \frac{1}{k}\right) \text{ and } M_2 = c^2\,\Gamma\left(1 + \frac{2}{k}\right), \tag{6}$$

where Γ shows the gamma function. To estimate Weibull shape parameter, the quantity is considered by:

$$\frac{M_2 - M_1^2}{M_1^2} = \frac{\Gamma(1 + \frac{2}{k})}{\left[\Gamma(1 + \frac{1}{k})\right]^2} - 1 = \frac{n \sum_i^n x_i^2}{\left(\sum_i^n x_i\right)^2} - 1, \tag{7}$$

after Eq. (7) is solved by numerical methods, the Weibull scale parameter estimate is computed by:

$$\widehat{c}_{MOM} = \left[\frac{\frac{1}{n}\sum_{i=1}^{n} x_i}{\Gamma(1 + \frac{1}{\widehat{k}})} \right]. \tag{8}$$

The MOM inference is performed to estimate the Weibull shape parameter k using the ABC and the GA. Equation (8) is employed to estimate the Weibull scale parameter c.

2.3 Proposed Functions

The ABC algorithm for the Weibull parameters estimation is used by [13]. Here, the difference of these functions presented in [13] is that the square function is preferred as the proposed functions. The GA for the Weibull parameters estimation is performed using the log-likelihood function given by Eq. (3) (see also [20, 21]). This chapter reports the ABC and the GA for estimating the Weibull parameters with ML and MOM inferences using the proposed functions as follows:

$$f_{ML} = \left(\widehat{k} - \left[\frac{\sum_{i=1}^{n} x_i^{\widehat{k}} \ln x_i}{\sum_{i=1}^{n} x_i^{\widehat{k}}} - \frac{\sum_{i=1}^{n} \ln x_i}{n}\right]^{-1}\right)^2, \tag{9}$$

and

$$f_{MOM} = \left(\frac{n\sum_i^n x_i^2}{\left(\sum_i^n x_i\right)^2} - \frac{\Gamma(1+\frac{2}{k})}{\left[\Gamma(1+\frac{1}{k})\right]^2}\right)^2. \tag{10}$$

The proposed functions in Eqs. (9) and (10) are minimized by using the ABC and the GA and then compared the performances of these algorithms. In the next section, these swarm algorithms are summarized for the Weibull parameters estimation.

3 Swarm Intelligence Methods

The popular population-based algorithms are the ABC and the GA swarm intelligence methods. This section presents the steps of the ABC and the GA methods as follows.

3.1 Artificial Bee Colony

The ABC algorithm motivated by the nature of honey bees is one of the popular swarm intelligent methods. It has many advantages including local search, memory, and improvement solutions [11]. To solve optimization problems using the ABC algorithm, a candidate food position corresponds to a possible solution and the nectar amount of a food source determines the quality of this corresponding solution. The colony of artificial bees for the ABC algorithm includes three groups of bees as employed bees, onlookers, and scouts and the number of solutions in the population corresponds to the number of employed or onlooker bees [7]. The number of food sources is set to the number of employed bees and every food source has only one employed bee and the search established by the artificial bees can be outlined as shown below [29]:

- Employed bees identify a food source in the neighbourhood of the food source in their mind.
- The information with respect to the nectar amounts of a food source is obtained by the employed bees and it is shared by onlookers within the hive and then one of the food sources is chosen by the onlookers.
- Onlooker bees is selected a food source in the neighbourhood of the food source.
- An employed bee for food search has been abandoned and it becomes a scout for discovering a new food source.

The ABC is a swarm algorithm inspired by the movement of a honeybee and the steps of the algorithm can be summarized as shown below [10]:

Step 1. Deliver the employed bees to the food source sites and measure their nectar amounts.
Step 2. Compute the probability value of the food source sites by onlookers.
Step 3. Deliver the onlookers to the food source sites and identify their nectar amounts.
Step 4. Stop the procedure of exploitation and the food source sites run out by the bees.
Step 5. Choose the scouts and deliver the search field for exploring new food source sites, randomly.
Step 6. Remind the best food source found until convergence criteria are satisfied.

The phases of the ABC algorithm presented in [30] can be outlined, respectively. The initialization phase is that initial food sources are generated randomly based on the search space of the parameters and this algorithm starts with randomly producing the food source as follows:

$$x_{ij} = x_j^{\min} + \text{rand}(0, 1)(x_j^{\max} - x_j^{\min}), \quad i = 1, 2, ..., SN \quad j = 1, 2, ..., D \tag{11}$$

where SN presents the number of food sources and D is the number of optimization parameters. The search process of food sources is conducted by the employed bees, the onlooker bees and the scout bees after the initialization phase of this algorithm. Employed bee phase of this algorithm is to find the neighbourhood of food sources via the employed bees, then the simulation of the process is defined as follows:

$$v_{ij} = x_{ij} + u_{ij}(x_{ij} - x_{kj}) \tag{12}$$

where j is a random integer in $[1, D]$, $k = 1, 2, ..., SN$ is randomly determined index and it is different from i index. u_{ij} is randomly simulated in [-1,1]. After simulating v_i within the search space of the optimization parameters, a fitness value of a minimization problem is defined as follows:

$$\text{fitness}_i = \begin{cases} 1/(1 + f_i) & \text{if } f_i \geq 0 \\ 1 + \text{abs}(f_i) & \text{if } f_i < 0 \end{cases} \tag{13}$$

where f_i is known as the value of the objective function for the solution v_i. Onlooker bee phase of this algorithm is conducted by the roulette-wheel scheme as follows:

$$p_i = \frac{\text{fitness}_i}{\sum_{i=1}^{SN} \text{fitness}_i}. \tag{14}$$

The main advantage of the probabilistic scheme is to increase the nectar amount of food sources, then the number of onlooker bees is increased too. Scout bee phase of this algorithm is related to any exhausted source to be abandoned after the employed bees and the onlooker bees are finished their searches. Finally, the stopping criterion of this algorithm is a maximum cycle number or acceptable error rate. A pseudo-code of the ABC algorithm [31] is given in Algorithm 1.

Algorithm 1 Pseudo-code of the artificial bee colony (ABC)

```
Initialize the population solutions
cycle=1
while (cycle < MaximumCycleNumber) or (StoppingCriterion) do
    Produce new solutions for the employed bees using Eq. (12)
    Use greedy procedure for the employed bees
    Compute the probability values p_i for the solutions using Eq. (14)
    Produce new solutions for the onlooker bees based on the probability values p_i
    Use greedy procedure for the onlooker bees
    Identify the abandoned source and if it exists, replace new ones with randomly produced
  solution for the scout using Eq. (11)
    Remind the best food source found until now
    cycle=cycle+1
end while
Processing results
```

3.2 *Genetic Algorithms*

The GA proposed by [16] has been widely used by researchers for different optimization problems. The GA is well-known among other swarm intelligence algorithms and based on nature of biological systems. Initialization phase for this algorithm begins with the generation of the initial population with search space of the optimization parameters. The initial population for each generation is conducted by the natural selection, crossover and mutation operators. In every generation, a set of solutions represents as the population. Each solution has a fitness value and the quality of this solution is related to the corresponding chromosome. The possible solutions are called also individuals.

The main steps for estimating the Weibull parameters using the GA are given below:

Step 1. Determine the search space, fitness function, convergence criteria and the GA's parameters consisting of population size, crossover probability, selection rate, and mutation probability. The fitness functions are considered as Eqs. (9) and (10).

Step 2. Produce an initial population of N chromosomes in given search space using initialization strategy. The initial population is represented by $k_1^{(0)}, k_2^{(0)}, \ldots, k_N^{(0)}$.

Step 3. The values of the fitness function, $f^{(t)}$ at any iteration t for each chromosome, are computed.

Step 4. The individuals with the worst fitness value were replaced by new individuals and applied the roulette wheel based on a selection proportion.

Step 5. Identify best potential individuals using crossover and mutation operators. Crossover parents are performed to determine new offspring individuals and apply random mutation for identifying new individuals.

Step 6. Continue steps 2–5 with the evaluation of the fitness functions until the evaluation requirements are satisfied.

A pseudo-code for the GA is presented in Algorithm 2.

Algorithm 2 Pseudo-code of the genetic algorithm (GA)

```
Identify the fitness function, convergence criterion, and GA's parameters
Select a initial random population of individuals
while (t < MaxGeneration) or (StoppingCriterion) do
    Identify the individuals using genetic operators
    Produce new individuals using the crossover and mutation operators
    Compute the fitness function by the new individuals;
    Substitute the worst individuals with the best individuals in the population;
end while
Processing results
```

4 Simulation Study

The aim of the simulation study is to evaluate the efficiencies of the ML and the MOM inferences for the Weibull parameters estimation using the ABC and the GA. The simulation study consists of the different sample sizes as 50, 100, 200 and 500 and the true parameters $(k, c) = (0.5, 1), (1, 1), (5, 1)$ and $(2, 5)$. The number of replications is set as $n_{rep} = 1000$ and these simulation scenarios are presented in Fig. 1.

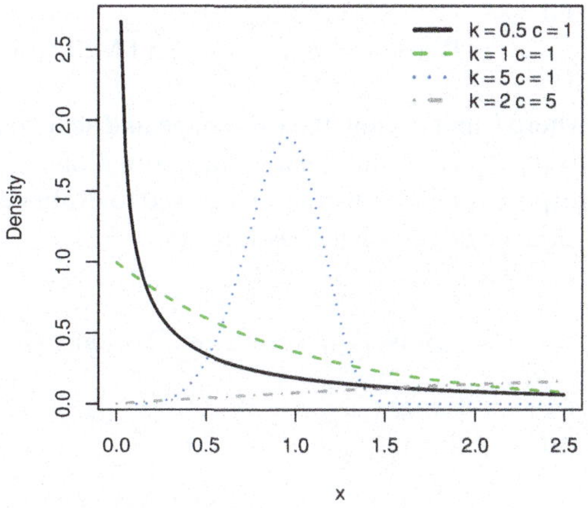

Fig. 1 Probability density functions for different Weibull parameters

4.1 Swarm Parameter Settings

The swarm parameter settings of the ABC algorithm consist of the number of food sources and a maximum number of cycles. The number of food sources is also known as the population size and is set to 125. The maximum number of cycles is also known as the maximum number of generations and is equal to 1000. These parameter values are also consistently used in [7, 32]. Initialization search for the proposed function presented in Eq. (9) using the ML inference is [0, 100] and initialization search for the proposed function presented in Eq. (10) using the MOM inference is [0.05, 100]. This search can be explained that the function $\Gamma(\frac{1}{x})$ is complex infinity when the lower limit of search space is to zero.

The swarm parameter settings of the GA algorithm consist of the population size, crossover probability and mutation probability. The following GA scheme is preferred as: single point uniform crossover with the probability of 0.95, roulette-wheel selection, and random mutation with the probability of 0.1. The population size is to set 125. These values are the same as the numerical optimization of different benchmark functions presented in [7]. Initialization searches for the proposed functions are the same as the ABC algorithm.

4.2 Evaluation Criteria

To determine the performances of the Weibull parameters estimation methods using the ABC and the GA swarm optimization algorithms, the bias and mean square error (MSE) are defined as follows:

$$\text{Bias}(\hat{k}) = E(\hat{k} - k), \quad \text{and } \text{MSE}(\hat{k}) = \text{Variance}(\hat{k}) + (\text{Bias}(\hat{k}))^2,$$

and

$$\text{Bias}(\hat{c}) = E(\hat{c} - c), \quad \text{and MSE}(\hat{c}) = \text{Variance}(\hat{c}) + (\text{Bias}(\hat{c}))^2$$

where k and c show the true shape and scale parameters for the Weibull distribution, respectively. $\hat{k}$ and $\hat{c}$ show the estimations of the true shape and scale parameters, respectively. These criteria are used to evaluate the efficiency of estimators. Bias and variance are computed as follows:

$$\text{Bias}(\hat{k}) = \bar{\hat{k}} - k, \text{ and Variance}(\hat{k}) = \frac{1}{n_{rep} - 1} \sum_{i=1}^{n_{rep}} (\hat{k}_i - \bar{\hat{k}})^2,$$

and

$$\text{Bias}(\hat{c}) = \bar{\hat{c}} - c, \text{ and Variance}(\hat{c}) = \frac{1}{n_{rep} - 1} \sum_{i=1}^{n_{rep}} (\hat{c}_i - \bar{\hat{c}})^2,$$

where $\bar{\hat{k}}$ and $\bar{\hat{c}}$ are the average estimates of $\hat{k}$ and $\hat{c}$, respectively. To evaluate the joint efficiencies of $\hat{k}$ and $\hat{c}$, the deficiency criterion (Def) [33] is used as follows:

$$\text{Def}(\hat{k}, \hat{c}) = \text{MSE}(\hat{k}) + \text{MSE}(\hat{c}). \tag{15}$$

The lowest Def values identify the best performance of the algorithms.

4.3 Computational Implementation

Optimization is an important task in a variety of scientific fields, including economics, environmental sciences, and engineering problems. One of the popular methodologies used in optimization is the use of population-based swarm intelligence algorithms. In the R programming language, the `ABCoptim` [34] and the `GA` [35] packages of R software are employed for the Weibull parameters estimation using the ABC and the GA approaches.

4.4 Simulation Results

The simulation results for the Weibull parameters estimation methods using the ABC and the GA are presented in Table 1. The biases, the MSEs and the Def of the estimators $\hat{k}$ and $\hat{c}$ obtained from the ML and the MOM inferences are compared. Comparisons are performed by using different sample sizes as moderate ($n = 50$, $n = 100$) and large ($n = 200$ and $n = 500$). The results from simulation study about

Table 1 Simulated Mean, MSE and Def values of the parameter estimates for the Weibull distribution

f	(k,c)	Alg.	ABC					GA				
		n	$\hat{k}$		$\hat{c}$			$\hat{k}$		$\hat{c}$		
			Mean	MSE	Mean	MSE	Def	Mean	MSE	Mean	MSE	Def
f_{ML}	(0.5,1)	50	**0.5172**	**0.0038**	**1.0449**	**0.1095**	**0.1133**	0.5178	0.0041	1.0464	0.1112	0.1153
		100	0.5086	**0.0017**	1.0182	**0.0469**	**0.0486**	**0.5080**	0.0020	**1.0179**	0.0484	0.0504
		200	**0.5039**	**0.0008**	**1.0099**	**0.0216**	**0.0224**	0.5042	0.0011	1.0105	0.0224	0.0235
		500	**0.5015**	**0.0003**	**1.0029**	**0.0090**	**0.0093**	0.5019	0.0006	1.0037	0.0097	0.0103
f_{MOM}		50	0.5869	0.0164	1.2267	0.2232	0.2396	0.5894	0.0169	1.2349	0.2276	0.2445
		100	0.5548	0.0087	1.1423	0.0999	0.1086	0.5583	0.0095	1.1531	0.1094	0.1189
		200	0.5322	0.0043	1.0909	0.0533	0.0576	0.5361	0.0048	1.1033	0.0586	0.0634
		500	0.5166	0.0020	1.0477	0.0270	0.0290	0.5193	0.0024	1.0558	0.0313	0.0337
f_{ML}	(1,1)	50	1.0345	0.0153	**1.0101**	0.0247	0.0400	**1.0344**	0.0153	1.0102	0.0247	0.0400
		100	1.0172	**0.0069**	1.0034	0.0113	**0.0182**	**1.0168**	0.0071	**1.0033**	0.0113	0.0184
		200	1.0078	**0.0032**	**1.0023**	**0.0053**	**0.0085**	1.0078	0.0035	1.0024	0.0055	0.0090
		500	1.0031	**0.0012**	1.0003	**0.0022**	**0.0034**	**1.0026**	0.0016	**1.0001**	0.0023	0.0039
f_{MOM}		50	1.0594	0.0220	1.0170	0.0259	0.0479	1.0612	0.0223	1.0176	0.0259	0.0482
		100	1.0304	0.0106	1.0067	0.0118	0.0224	1.0313	0.0109	1.0069	0.0119	0.0228
		200	1.0152	0.0051	1.0042	0.0056	0.0107	1.0168	0.0053	1.0047	0.0057	0.0110
		500	1.0064	0.0020	1.0013	0.0024	0.0044	1.0065	0.0024	1.0011	0.0025	0.0049
f_{ML}	(5,1)	50	5.1724	**0.3829**	1.0001	0.0010	**0.3839**	5.1712	0.3834	1.0001	0.0010	0.3844
		100	5.0860	**0.1720**	0.9998	0.0005	**0.1725**	5.0856	0.1729	0.9998	0.0005	0.1734
		200	5.0392	**0.0806**	1.0000	0.0002	**0.0808**	**5.0389**	0.0816	1.0000	0.0002	0.0818
		500	5.0153	**0.0310**	0.9999	0.0001	**0.0311**	5.0156	0.0312	0.9999	0.0001	0.0313

(continued)

Table 1 (continued)

f	(k,c)	Alg.	ABC					GA				
		n	$\hat{k}$		$\hat{c}$			$\hat{k}$		$\hat{c}$		
			Mean	MSE	Mean	MSE	Def	Mean	MSE	Mean	MSE	Def
f_{MOM}		50	**5.1631**	0.4367	0.9999	0.0010	0.4377	5.1640	0.4372	0.9999	0.0010	0.4382
		100	**5.0847**	0.1991	0.9997	0.0005	0.1996	5.0849	0.1993	0.9997	0.0005	0.1998
		200	5.0391	0.0949	1.0000	0.0002	0.0951	5.0391	0.0955	1.0000	0.0002	0.0957
		500	**5.0152**	0.0357	0.9999	0.0001	0.0358	5.0153	0.0360	0.9999	0.0001	0.0361
f_{ML}	(2,5)	50	2.0690	0.0613	5.0101	0.1519	0.2132	2.0696	0.0612	5.0105	**0.1516**	0.2128
		100	2.0344	**0.0275**	5.0015	0.0708	**0.0983**	2.0341	0.0281	5.0013	0.0710	0.0991
		200	2.0157	**0.0129**	5.0024	0.0331	**0.0460**	2.0161	0.0132	5.0026	0.0332	0.0464
		500	2.0061	**0.0050**	4.9995	0.0139	**0.0189**	2.0071	0.0055	**4.9999**	0.0139	0.0194
f_{MOM}		50	**2.0680**	**0.0610**	5.0093	0.1517	**0.2127**	2.0682	0.0616	**5.0092**	0.1517	0.2133
		100	**2.0331**	0.0279	5.0006	0.0708	0.0987	2.0345	0.0282	5.0006	0.0708	0.0990
		200	2.0152	0.0132	5.0020	0.0331	0.0463	**2.0149**	0.0133	5.0020	0.0331	0.0465
		500	**2.0058**	0.0051	4.9993	0.0139	0.0190	2.0065	0.0055	4.9992	0.0139	0.0194

the efficiencies of $\hat{k}$ and $\hat{c}$ in respect of the bias, the MSE and the Def criteria are summarized as follows.

For ($k = 0.5, c = 1$): The MOM estimators have larger the biases, the MSEs and the Def values than the ML estimators for all sample sizes and the ML estimators using the ABC algorithm generally outperform the ML estimators using the GA.

For ($k = 1, c = 1$): The MOM estimators have larger the biases, the MSEs and the Def values than the ML estimators for all sample sizes. The ML estimators using the GA algorithm in terms of the biases generally outperform the other estimators. The ML estimators using the ABC in terms of the MSEs and the Def values outperform the ML estimators using the GA.

For ($k = 5, c = 1$): The MOM estimators using the ABC in terms of the biases generally outperform the other estimators. The ML estimators using the ABC in terms of the MSEs and the Def values generally outperform the other estimators.

For ($k = 2, c = 5$): The MOM estimators using the ABC have generally smaller the biases than the ML estimators using the ABC and have generally larger the MSEs and the Def values than the ML estimators using the ABC.

An overall comparison of the results obtained from the ML and the MOM inferences using the ABC and the GA is shown in Table 2. Accordingly, the ML inference using the GA yields the best results in 9 out of 32 cases (28%), the ML inference using the ABC yields the best results in 8 out of 32 cases (25%), the MOM inference using the ABC yields the best results in 6 out of 32 cases (19%) and the MOM inference using the GA yields the best results in 2 out of 32 cases (6%) with respect to the bias criterion. The ML inference using the ABC provides the best results in 20 out of 32 cases (63%) with respect to the MSE criterion and provides the best results in 14 out of 16 cases (88%) with respect to the Def criterion. These results are also shown in Fig. 2.

4.5 *Real Data Example*

A real dataset is used to illustrate the Weibull parameters estimation methods using the ABC and the GA. The dataset includes 101 data points and represents the stress-rupture life of kevlar 49/epoxy strands failure at 90 % stress level. Previously, this dataset was analyzed by [36–39].

The parameter estimates, Kolmogorov-Smirnov (K-S) test statistic, and the p-values are presented in Table 3. It can be seen that the results of the four methods are very similar, but the K-S test statistic and the p-values identify the ML inference using the GA as the best fitting parameters estimation method. Figure 3 denotes the corresponding probability density functions of Weibull distribution obtained from the four methods. Accordingly, all fitted densities by using different parameters estimation methods with the ABC and the GA are superimposed as a histogram of the failure data.

Table 2 An overall comparison of the results obtained from the ML and MOM inferences

(*k*,*c*)	Method	ML(ABC)		MOM(ABC)		ML(GA)		MOM(GA)	
		Best	Equal	Best	Equal	Best	Equal	Best	Equal
(0.5,1)	Mean	6	0	0	0	2	0	0	0
	MSE	8	0	0	0	0	0	0	0
	Def	4	0	0	0	0	0	0	0
(1,1)	Mean	2	1	0	0	5	1	0	0
	MSE	5	3	0	0	0	3	0	0
	Def	3	1	0	0	0	1	0	0
(5,1)	Mean	0	4	3	3	1	4	0	3
	MSE	4	4	0	4	0	4	0	4
	Def	4	0	0	0	0	0	0	0
(2,5)	Mean	0	0	3	2	1	0	2	2
	MSE	3	3	1	3	1	1	0	3
	Def	3	0	1	0	0	0	0	0
Overall	Mean	8	5	6	5	9	5	2	5
	MSE	20	10	1	7	1	8	0	7
	Def	14	1	1	0	0	1	0	0

Fig. 2 An overall comparison of the results of the Bias, the MSE's and the Def for the ML and MOM inferences using the ABC and the GA

Table 3 The parameter estimates and Kolmogorov-Smirnov test statistics and p-values

f	Alg. ABC				GA			
	$\hat{k}$	$\hat{c}$	K-S	p-value	$\hat{k}$	$\hat{c}$	K-S	p-value
f_{ML}	0.9259	0.9900	0.0906	0.3781	0.9473	1.0500	**0.0789**	**0.5562**
f_{MOM}	0.9211	0.9861	0.0926	0.3512	0.9208	0.9859	0.0928	0.3498

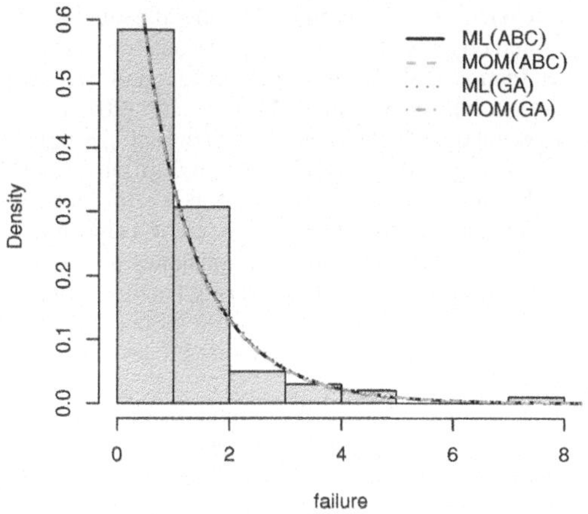

Fig. 3 The fitted probability density functions for failure data using the ML and the MOM inferences with the ABC and the GA

5 Conclusion

This chapter has been considered a comparative study of the ABC and the GA methods for the Weibull parameters estimation. These methods are also known as swarm intelligence approaches and popular population-based algorithms. To show the performances of the ABC and GA methods, the simulation study was conducted for the Weibull parameters estimation. From the simulation study, it was concluded that the ABC and the GA methods for the ML inferences have been successfully used for the Weibull parameters estimation. The MOM inferences using the ABC algorithm generally outperform the MOM inferences using the GA. From the failure data result, the ML inference using the GA algorithm outperforms the other variations of the Weibull parameters estimation methods. The swarm parameter settings of the ABC and the GA methods are also useful for the Weibull parameters estimation. One possible extension of this chapter could be considered by the parameters estimation of a mixture of the Weibull distributions in expectation and maximization algorithm.

References

1. Weibull, W.: A statistical distribution function of wide applicability. J. Appl. Mech. **18**, 293–297 (1951)
2. Seguro, J.V., Lambert, T.W.: Modern estimation of the parameters of the Weibull wind speed distribution for wind energy analysis. J. Wind. Eng. Ind. Aerodyn. **85**(1), 75–84 (2000)
3. Carrasco, J.M., Ortega, E.M., Cordeiro, G.M.: A generalized modified Weibull distribution for lifetime modeling. Comput. Stat. Data Anal. **53**(2), 450–462 (2008)
4. Cohen, A.C.: Maximum likelihood estimation in the Weibull distribution based on complete and on censored samples. Technometrics **7**(4), 579–588 (1965)
5. Teimouri, M., Hoseini, S.M., Nadarajah, S.: Comparison of estimation methods for the Weibull distribution. Statistics **47**(1), 93–109 (2013)
6. Karaboga, D., Gorkemli, B., Ozturk, C., Karaboga, N.: A comprehensive survey: artificial bee colony (ABC) algorithm and applications. Artif. Intell. Rev. **42**, 21–57 (2014)
7. Karaboga, D., Basturk, B.: A powerful and efficient algorithm for numerical function optimization: artificial bee colony (ABC) algorithm. J. Glob. Optim. **39**, 459–471 (2007)
8. Karaboga, D.: An idea based on honey bee swarm for numerical optimization. Technical report, Erciyes University, Engineering Faculty, Computer Engineering Department (2005)
9. Karaboga, D., Akay, B.: Artificial bee colony (ABC) algorithm on training artificial neural networks. In: 15th IEEE Signal Processing and Communications Applications, pp. 1–4. IEEE (2007)
10. Karaboga, D., Akay, B.: A comparative study of artificial bee colony algorithm. Appl. Math. Comput. **214**(1), 108–132 (2009)
11. Seidgar, H., Kiani, M., Fazlollahtabar, H.: Genetic and artificial bee colony algorithms for scheduling of multi-skilled manpower in combined manpower-vehicle routing problem. Prod. Manuf. Res. **4**(1), 133–151 (2016)
12. Zhang, Z., Lin, J., Shi, Y.: Application of artificial bee colony algorithm to maximum likelihood DOA estimation. J. Bionic Eng. **10**, 100–109 (2013)
13. Ravindra, K., Rao, R.S., Narasimham, S.V.L.: Wind distribution analysis incorporating artificial bee colony algorithm. In: 2012 International Conference on Advances in Power Conversion and Energy Technologies (APCET), pp. 1–6. IEEE (2012)
14. Xu, M., Droguett, E.L., Lins, I.D., das Chagas Moura, M.: On the q-Weibull distribution for reliability applications: an adaptive hybrid artificial bee colony algorithm for parameter estimation. Reliab. Eng. Syst. Saf. **158**, 93–105 (2017)
15. Yonar, A., Yapici Pehlivan, N.: Artificial bee colony with levy flights for parameter estimation of 3-p Weibull distribution. Iran. J. Sci. Technol., Trans. A: Sci. **44**(3), 851–864 (2020)
16. Holland, J.H.: Adaptation in Natural and Artificial Systems. University of Mich. Press, Ann Arbor (1975)
17. Booker, L.B.: Intelligent Behavior as an Adaptation to the Task Environment. University of Mich. Press, Ann Arbor (1982)
18. Goldberg, D.E., Kuo, C.H.: Genetic algorithms in pipeline optimization. J. Comput. Civ. Eng. **1**(2), 128–141 (1987)
19. Goldberg, D.E., Holland, J.H.: Genetic algorithms and machine learning. Mach. Learn. **3**, 95–99 (1988)
20. Thomas, G.M., Gerth, R., Velasco, T., Rabelo, L.C.: Using real-coded genetic algorithms for Weibull parameter estimation. Comput. Ind. Eng. **29**(1–4), 377–381 (1995)
21. Koca, M.B., Kılıç, M.B., Şahin, Y.: Using genetic algorithms for estimating Weibull parameters with application to wind speed. Int. J. Optim. Control.: Theor. Appl. (IJOCTA) **10**(1), 137–146 (2020)
22. Shin, J.Y., Heo, J.H., Jeong, C., Lee, T.: Meta-heuristic maximum likelihood parameter estimation of the mixture normal distribution for hydro-meteorological variables. Stoch. Environ. Res. Risk Assess. **28**(2), 347–358 (2014)
23. Gençtürk, Y., Yiğiter, A.: Modelling claim number using a new mixture model: negative binomial gamma distribution. J. Stat. Comput. Simul. **86**(10), 1829–1839 (2016)

24. Yalçinkaya, A., Şenoğlu, B., Yolcu, U.: Maximum likelihood estimation for the parameters of skew normal distribution using genetic algorithm. Swarm Evol. Comput. **38**, 127–138 (2018)
25. Kılıç, M.B., Şahin, Y., Koca, M.B.: Genetic algorithm approach with an adaptive search space based on EM algorithm in two-component mixture Weibull parameter estimation. Comput. Stat. **36**, 1219–1242 (2021)
26. Örkcü, H.H., Özsoy, V.S., Aksoy, E., Dogan, M.I.: Estimating the parameters of 3-p Weibull distribution using particle swarm optimization: a comprehensive experimental comparison. Appl. Math. Comput. **268**, 201–226 (2015)
27. Acitas, S., Aladag, C.H., Senoglu, B.: A new approach for estimating the parameters of Weibull distribution via particle swarm optimization: an application to the strengths of glass fibre data. Reliab. Eng. Syst. Saf. **183**, 116–127 (2019)
28. Zhou, D., Zhuang, X., Zuo, H.: A novel three-parameter Weibull distribution parameter estimation using chaos simulated annealing particle swarm optimization in civil aircraft risk assessment. Arab. J. Sci. Eng. **46**(9), 8311–8328 (2021)
29. Karaboga, D., Basturk, B.: On the performance of artificial bee colony (ABC) algorithm. Appl. Soft Comput. **8**, 687–697 (2008)
30. Akay, B., Karaboga, D.: A modified artificial bee colony algorithm for real-parameter optimization. Inf. Sci. **192**, 120–142 (2012)
31. Karaboga, D., Basturk, B.: Artificial bee colony (ABC) optimization algorithm for solving constrained optimization problems. In: International Fuzzy Systems Association World Congress, pp. 789-798. Springer, Berlin (2007)
32. Srinivasan, D., Seow, T.H.: Evolutionary computation. CEC **3**, 2292–2297 (2003)
33. Gebizlioglu, O.L., Şenoğlu, B., Kantar, Y.M.: Comparison of certain value-at-risk estimation methods for the two-parameter Weibull loss distribution. J. Comput. Appl. Math. **235**(11), 3304–3314 (2011)
34. Vega Y.G., Muñoz, E.: ABCoptim: an implementation of the artificial bee colony (ABC) algorithm. R package version (2018)
35. Scrucca, L.: GA: a package for genetic algorithms in R. J. Stat. Softw. **53**(4), 1–37 (2013)
36. Andrews, D.F., Herzberg, A.M.: Data: A Collection of Problems from Many Fields for the Student and Research Worker. Springer, New York (1985)
37. Cooray, K., Ananda, M.M: A generalization of the half-normal distribution with applications to lifetime data. Commun. Stat.-Theory Methods **37**(9), 1323–1337 (2008)
38. Paranaiba, P.F., Ortega, E.M., Cordeiro, G.M., Pascoa, M.A.D.: The Kumaraswamy Burr XII distribution: theory and practice. J. Stat. Comput. Simul. **83**(11), 2117–2143 (2013)
39. Babacan, E.K., Kaya, S: comparison of parameter estimation methods in Weibull distribution. Sigma J. Eng. Nat. Sci. **38**(3), 1609–1621 (2020)

Graph Structure Optimization for Agent Control Problems Using ACO

Mohamad Roshanzamir, Mahdi Roshanzamir, Navid Hoseini Izadi, and Maziar Palhang

Abstract Ant Colony Optimization (ACO) is one of the powerful swarm intelligence algorithms capable of solving various problems. In this research, ACO is used to optimize individuals with graph structure. This structure is exactly like the approach taken by genetic network programming (GNP) for individual representation for solving agent control problems. However, in some types of environments such as stochastic environments, calculated fitness of an individual is not the same in each evaluation. Therefore, to estimate the true fitness of an individual, several times of evaluation are needed leading to increased process time of the evolution. In this research, a method is proposed to avoid slowing down the progress speed of the algorithm using ACO. This method can be well adapted on graph structures and in each iteration, it enhances the fitness of individuals using a constructive mechanism. In constructive mechanism, an individual is produced according to the experience of previous generations. In this research, the experience is the achieved fitness and it is distributed on the corresponding paths in the graph structure. This new method was used to solve an agent control problem called Pursuit-Domain while the environment is deterministic or stochastic. The experimental results showed high capabilities of this algorithm in generation of efficient strategies for agents in an agent control problem.

M. Roshanzamir (✉)
Department of Computer Engineering, Faculty of Engineering, Fasa University, 74617-81189 Fasa, Iran
e-mail: roshanzamir@fasau.ac.ir

M. Roshanzamir
Department of Electrical and Computer Engineering, University of Tabriz, Tabriz, Iran
e-mail: roshanzamir@tabrizu.ac.ir

N. Hoseini Izadi · M. Palhang
Department of Electrical and Computer Engineering, Isfahan University of Technology, Isfahan 84156-83111, Iran
e-mail: navid.hoseini1@ec.iut.ac.ir

M. Palhang
e-mail: palhang@cc.iut.ac.ir

A. Biswas et al. (eds.), *Advances in Swarm Intelligence*, Studies in Computational Intelligence 1054, https://doi.org/10.1007/978-3-031-09835-2_18

Keywords Ant colony optimization · Agent control problems · Stochastic environments · Genetic programming · Genetic network programming

1 Introduction

Intelligent agents can be used in a variety of environments with different features. The problem of controlling the behavior of these agents is one of the most challenging issues in the field of artificial intelligence. Due to the capabilities of evolutionary algorithms and swarm intelligence, these algorithms are one of the main options for solving such problems. These types of algorithms have very good capabilities in solving optimization problems. So, they can be used to optimize the behavior of intelligent agents. These algorithms are divided into two categories: evolutionary algorithms and swarm intelligence algorithms. Evolutionary algorithms are a set of algorithms in which next generation individuals are created by combining current generation individuals. This mechanism of successive generation production is called the productive structure. In swarm intelligence algorithms, the production of the next generation individuals is based on the experiences of the current generation. This mechanism is called the constructive structure.

The most famous evolutionary algorithms are genetic algorithms, genetic programming, evolutionary strategy, and evolutionary programming. These algorithms have differences in the structure of individuals as well as in the individuals evolution process. For example, the individuals in genetic programming have tree structures. So, this algorithm can be used to automatically generate computer programs or decision trees. Decision trees can be used to control the behavior of the agents in the environments. Therefore, genetic programming can be considered as a proper tool for optimizing the behavior of agents in the environment [1, 2]. It has been shown that the graph structure can be used instead of tree structure. This modification led to a new version for genetic programming named as genetic network programming (GNP) [3–10]. The purpose of GNP is to create a structure that has a better expression ability to model the decision-making process. Indeed, a feature that distinguishes the structure of GNP from many other evolutionary structures is its high expressiveness, which is due to its graph structure. Using graph structure, it is possible to model the conditions in different environments better than tree structure, especially in complex environments, and determine what should be done in each situation.

The structure of individuals in GNP has more capabilities. For example, the ability to reuse nodes is one of them. This leads to the creation of some kind of memory in this structure. Also, using directional graph structure, repetitive processes can be managed very well.

Despite the strengths of GNP, this algorithm has some weaknesses. For example, to perform the evolution process, recombination and mutation operators are used. Using these operators on graph structures breaks the individuals' structure. Although this often leads to better individuals structures, it can also destroy the useful ones that have been created so far. This reduces the efficiency of the algorithm, because in this type of structure, the nodes in relation with each other can solve a problem and breaking them can significantly reduce the efficiency of the algorithm. This problem is an important weakness of the GNP, which has been the subject of a lot of researches [10–14].

Another major problem of the GNP algorithm appears when used in stochastic environments. In this type of environments, a single evaluation of an individual cannot accurately calculate the fitness of that individual and it is necessary to repeat the evaluation process several times. The higher the number of the repetitions, the more accurate the calculated fitness will be. Because these types of algorithms are population based, repeating the evaluation of individuals, which commonly is the most time consuming section of these algorithms, increases the execution time of these algorithms significantly. On the other hand, if the fitness of individuals is not calculated correctly, parent and survivor selections are not performed properly. Therefore, the whole process of evolution will be disrupted.

In this chapter, to overcome these weaknesses of the GNP, a solution based on the ACO algorithm is proposed in which, instead of using recombination and mutation operators, the accumulated experience of the best individuals of each generation is used to create the next generation. This solution will solve both the problems of breaking down the useful structures and the need for repeating evaluation of individuals in stochastic environments. This is the main novelty of this research. In addition, it is for the first time that ACO is used to improve the efficiency of GNP in stochastic environments. ACO is selected because of the graph structure used in it.

In this chapter, first in Sect. 2, the studies that have been performed so far in this field are reviewed. In Sect. 3, first, the basic concepts needed to study this chapter are described (Sect. 3.1). Next, the proposed algorithms are explained (Sect. 3.2). In Sect. 4, the proposed algorithms are applied on one of the most common problems in the field of controlling the behavior of agents and the results are analyzed. These studies have been performed in both stochastic and deterministic environments. Finally, in Sect. 5, the conclusion will be presented.

2 Previous Works

Various modifications have been proposed on GNP and it was used in a variety of applications [15–18]. Modifications on its recombination and mutation and combining it with other machine learning algorithms are some of them. Mabu et al. [19] added online learning to this algorithm. In [20], Q learning [21] was used in GNP for faster adaptation to dynamic environments. In other researches such as [22], reinforcement learning has been used again. The difference is that instead of Q learning, another method of reinforcement learning was used, namely SARSA [21].

In [23], Mabu and Hirasawa tried to extract several programs simultaneously from a GNP structure using several start nodes. There have been fundamental improvements in how the next generation is created in GNP. For this purpose, in [14], an estimation of distribution algorithm (EDA) was used for the first time. In this algorithm, the structure of the superior individuals was used to create a probabilistic model. Then, this probabilistic model was used to produce next generation. In [10], in addition to EDA, reinforcement learning was also used to generate a probabilistic model made by better individuals. Li et al. [24] proposed a method in which the algorithm looked for suitable substructures in undesirable individuals. More experiments and analysis of this research can be found in [25] in which more details such as testing the effect of parameters in improving the efficiency of the algorithm and also testing the algorithm in different conditions have been done.

An example of combining GNP and ACO was presented in [26, 27] to increase the exploitation ability of GNP. According to these studies, the GNP algorithm is inherently biased to exploration, while ACO is biased to exploitation. Therefore, in order to better utilize these two basic capabilities, one out of every ten iterations of the GNP algorithm has been dedicated to ACO. Another example of this type of research [7] is the combination of GNP algorithm and artificial bee colony (ABC) algorithm [28]. In this research, ABC algorithm has been used to perform the evolution process. Accordingly, the recombination and mutation operators have been completely eliminated and the evolution method in the ABC algorithm has been used.

3 Proposed Method

In this section, at first, GNP is completely described including a complete description of the structure of individuals and how the evolution process is performed. Then, the proposed method will be explained.

3.1 GNP Algorithm

In GNP, each individual is represented as a directed graph with three different types of nodes in it. The phenotype and genotype of this structure is shown in Fig. 1. This structure can be interpreted as a flowchart. Each node in this structure has an identification number i. NT_i specifies the node type. If the value of NT_i is equal to zero, one, or two, it identifies the nodes as the start, judgment, or processing nodes, respectively. NF_i shows the function number that indicates which judgment or process must be executed. d_i indicates the time delay when executing this function. C_{ik} determines the identification number of the node connected from the kth branch of node i. d_{ik} specifies the transition time to the node C_{ik}.The start node specifies that the work should be started from which node; so its Node Gene section contains no information.

Accordingly, each individual consists of three types of node. A start node, several judgment nodes and several processing nodes form the structure of an individual in GNP. Each judgment node has several conditional branches and based on the results of the environmental conditions, it is determined which branch should be

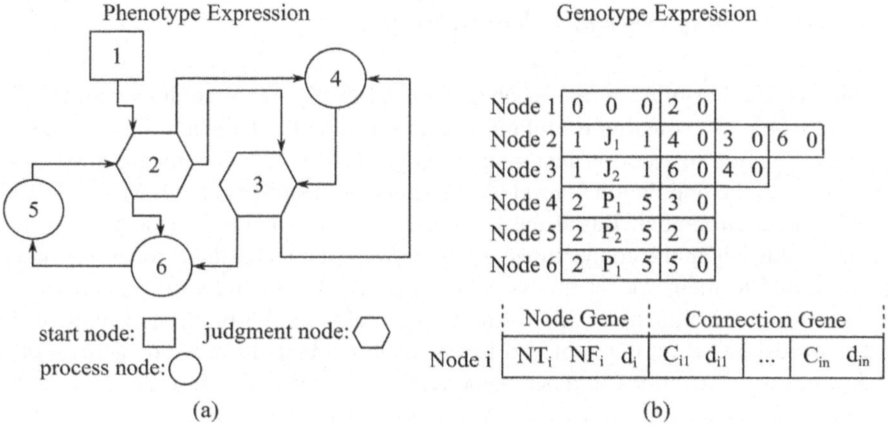

Fig. 1 Phenotype **a** and genotype **b** expressions of an individual in GNP

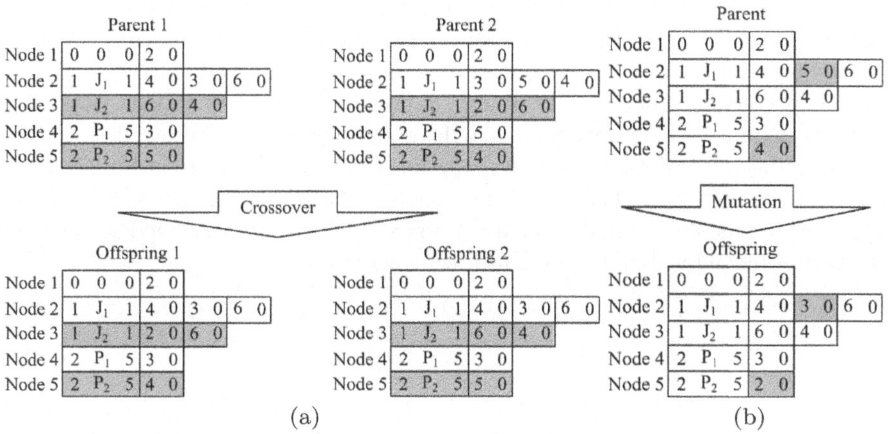

Fig. 2 An example of **a** crossover and **b** mutation in GNP

followed. Each processing node has one output and specifies what the agent should do. A flowchart is generated by connecting these three types of nodes. This flowchart specifies the behavior of agents in the environment.

To perform crossover, two individuals were selected using one of the selection methods used in evolutionary algorithms [29]. Then, as it is shown in Fig. 2a, the connections of corresponding nodes in these two individuals are swapped with the probability of p_c. For mutation, as it is shown in Fig. 2b, the selected individual's connections are randomly changed with a probability p_m. Since the number of nodes is constant, the bloat problem [29] in the genetic programming algorithm does not happen here.

3.2 Ant Colony Network Programming

In this section, inspired by the mechanism used in the ACO, an algorithm is proposed capable of extracting useful information from individuals in each generation. This information is used for next generation production [5]. For this purpose, each individual in the population is used by the agents as a strategy that they should follow in the environment. Based on this strategy, agents make their decision in the environment. Then, the connections of that individual are assigned values which are proportional to their role in fitness achievement. Consequently, higher values are assigned to the connections that form the more efficient parts of an individual. In other words, based on the achieved fitness of individuals in previous generations, the values of connections in a new generation are determined. In order to save the fitness of individuals in successive generations, a structure like Fig. 3 is used, which is called the experience table.

Experience table showed in Fig. 3 is generated based on the individual's structure shown in Fig. 1. In this table, one row is created for each possible connection from each node and one column is created for each node except the start node. There is no column for start node because the start node has no input. Each entry $v_{i(j),n}$ of the table represents the value of connecting jth branch of node i to node n. According to this table, the value of connections in the population can be estimated.

The initial value of each cell in the experience table can be set to identical values. These values are then updated in each iteration of the algorithm in order to estimate the values of the connections. For this purpose, $v_{i(j),n}$ values are updated in each iteration of the algorithm using the following equations:

		Nodes				
		2	3	4	5	6
Branches	*1(1)*	$v_{1(1),2}$	$v_{1(1),3}$	$v_{1(1),4}$	$v_{1(1),5}$	$v_{1(1),6}$
	2(1)	$v_{2(1),2}$	$v_{2(1),3}$	$v_{2(1),4}$	$v_{2(1),5}$	$v_{2(1),6}$
	2(2)	$v_{2(2),2}$	$v_{2(2),3}$	$v_{2(2),4}$	$v_{2(2),5}$	$v_{2(2),6}$
	2(3)	$v_{2(3),2}$	$v_{2(3),3}$	$v_{2(3),4}$	$v_{2(3),5}$	$v_{2(3),6}$
	3(1)	$v_{3(1),2}$	$v_{3(1),3}$	$v_{3(1),4}$	$v_{3(1),5}$	$v_{3(1),6}$
	3(2)	$v_{3(2),2}$	$v_{3(2),3}$	$v_{3(2),4}$	$v_{3(2),5}$	$v_{3(2),6}$
	4(1)	$v_{4(1),2}$	$v_{4(1),3}$	$v_{4(1),4}$	$v_{4(1),5}$	$v_{4(1),6}$
	5(1)	$v_{5(1),2}$	$v_{5(1),3}$	$v_{5(1),4}$	$v_{5(1),5}$	$v_{5(1),6}$
	6(1)	$v_{6(1),2}$	$v_{6(1),3}$	$v_{6(1),4}$	$v_{6(1),5}$	$v_{6(1),6}$

Fig. 3 The structure of experience table which is created according to number of nodes and their connections in the individual shown in Fig. 1

$$v_{i(j).n}|_{g+1} = v_{i(j).n}|_g + \left(q_1 \times \Delta v_{i(j).n}^{BestSoFar} + q_2 \times \sum_{k=1}^{M} \Delta v_{i(j).n}^k\right) \tag{1}$$

$$\Delta v_{i(j).n}^k = \begin{cases} \log\left(\sigma_{i(j).n}^k \times f(k) + 1\right), & \text{if } i \neq n \\ 0, & \text{if } i = n \end{cases} \tag{2}$$

$$\sigma_{i(j).n}^k = \begin{cases} 1, & \text{if the branch is used during runtime} \\ 0, & \text{Otherwise} \end{cases} \tag{3}$$

where $i(j)$ represents the jth branch of node i, n is the node's identity, g is the generation index. $v_{i(j).n}|_g$ represents the value of jth branch of node i if it is connected to node n in gth generation. $\Delta v_{i(j).n}^{BestSoFar}$ indicates the rate of change of $v_{i(j).n}$ by the best individual that has ever been found. $\Delta v_{i(j).n}^k$ shows the rate of change of $v_{i(j).n}$ by kth best individual found in this generation of algorithm. M is the number of top individuals selected in each generation to update the value of $v_{i(j).n}$ based on equation (1). q_1 and q_2 show the impact rate of the best individual ever created and M best individuals created in this generation respectively. $f(k)$ shows the fitness of kth individual. $\sigma_{i(j).n}^k$ indicates whether the transition from jth branch of node i to node n occurs when agents make decisions based on the kth individual.

To calculate $v_{i(j).n}$, the experience of the best individual ever generated i.e. $\Delta v_{i(j).n}^{BestSoFar}$ and the experiences of the M best individuals created in the current generation i.e. $\sum_{k=1}^{M} \Delta v_{i(j).n}^k$ will be used with coefficients q_1 and q_2. The sum of these two factors is added to the current value of $v_{i(j).n}$. In order to calculate $\Delta v_{i(j).n}^k$, it is checked whether the jth branch of node i connected to node n is used (i.e. $\sigma_{i(j).n}^k = 1$) or not (i.e. $\sigma_{i(j).n}^k = 0$). If this connection is used, the value of that connection increases proportional to the fitness of that individual. This method helps for finding useful sections of high performance individuals and using them to create individuals of the next generations.

In Eq. (2), when the value of i is equal to n, the value of the variable $\Delta v_{i(j).n}^k$ is set to zero. This prevents the creation of direct infinite loops. The log function in this equation prevents the rapid growth of $v_{i(j).n}$ and consequently prevents from premature convergence. If this function is not used, the algorithm quickly converges to local optimums because some connections quickly gain more value than others and the difference between their values and other connections will increase. These types of connections subsequently lead the evolution process towards themselves and will take the opportunity from other connections to be used and do not give them the opportunity to interfere in the process of evolution. This reduces the population diversity and causes the premature convergence problem for evolution process.

Finally, in order to prevent convergence to connections that have been assigned high values by chance, Eq. (4) was used.

$$v_{i(j).n}|_{g+1} = (1 - \rho) \times v_{i(j).n}|_g \tag{4}$$

As it is clear in this equation, in each iteration, all connections lose a percentage of their values. This assures us that the connections accidentally valued will lose their values over time, because it is unlikely for an improper individual to achieve high fitness in many iterations. Consequently, after a while, these types of connections lose their values that have been gained by chance. So, in general, in each iteration of the algorithm, some values are added to the connections that were useful. Then, the values of all connections are reduced by $0 \leq (1 - \rho) < 1$.

New individuals are created using Eq. (5). $p_{i(j).n}$ shows the probability of connecting j^{th} branch of node i to node n. This way, all connections in an individual are determined.

$$p_{i(j).n} = \frac{v_{i(j).n}}{\sum_{m=2}^{N} v_{i(j).n}} \tag{5}$$

There are N nodes in each individual. In the first iteration of the algorithm, the values in the experience table are the same. Consequently, the connections between the nodes in the initial population are made completely randomly and with equal probability. The variable m starts from 2 because the node number one is the start node that has no input. Therefore, there is no column for it in the experience table. After each individual is generated, the proposed strategy suggested by that individual is used by the agents to determine their behavior. This is done for all individuals to determine their fitness.

Overall, the main goal of the proposed method is to distribute the fitness of an individual on the branches that using them, the agents could achieve this level of fitness. In each iteration of the algorithm, the fitness values of the best individuals created in this generation are added to the previous values of experience table according to Eq. (1). In the last step of each iteration of the algorithm, a percentage of the values in experience table are decreased according to Eq. (4).

It should be noted that in Eq. (1) if only the existence of a connection between two nodes is considered, it will neglect the usefulness of that connection in problem solving. In order to solve this problem, the value of the connections in experience table is updated based on the fitness of individuals.

This algorithm is named as Ant Colony Network Programming (ACNP). The ACNP algorithm is proposed in two regular and online versions. The difference between the regular and online versions is in experience table updating. In the regular version of the ACNP algorithm, the experience table is updated after creating all the individuals in a generation and selecting the best individual in that generation. But in the online version, updating the experience table is done by each individual after its creation if it is better than all the individuals which have been created in that generation so far. The online version of this algorithm has more exploration ability while in the regular version, the exploration ability has been reduced a bit, but its exploitation ability has been emphasized more than the online version. The pseudo codes of these two versions are shown in Algorithms 1 and 2.

Algorithm 1: ACNP Pseudo code

```
Initializing the parameters and variables
while termination condition is not met do
    According to the number of individuals:
        * An individual is generated based on the values in experience table
    using Eq. (5).
        * Its fitness is calculated by the agents.
    The best individual ever created and M best individual in this iteration
    update the experience table according to Eq. (1).
    Update the experience table according to Eq. (4).
```

Algorithm 2: ACNP_Online Pseudo code

```
Initializing the parameters and variables
while termination condition is not met do
    According to the number of individuals:
        * An individual is generated based on the values in experience table
    using Eq. (5).
        * Its fitness is calculated by the agents
        * The best individual ever created update the experience table according
    to Eq. (1).
    Update the experience table according to Eq. (4).
```

4 Experimental Results

In order to evaluate the efficiency of the ACNP and its variantes i.e. ACNP_V02, ACNP_V03 and ACNP_Online, they were compared with GNP and its three different versions in which swarm intelligence has been used to enhance the evolution process. They were tested on the pursuit domain [30] while the environment was deterministic or stochastic. ACNP_V02 and ACNP_V03 are different from ACNP only in their parameters. Three different versions of GNP used in this research were GNPwithACO [26], GNPcleanupACO [27] and GNPwithABC [7]. In GNPwithACO and GNPcleanupACO, in some iterations GNP and in some others ACO were used. For better analyses, we also tested ACO without GNP on our benchmark. The parameter values of all of these algorithms are listed in Table 1. All tests were implemented using MATLAB 2016a.

4.1 Pursuit Domain

The environment used in this chapter is called the pursuit domain [30], which is also known as the prey and predator world. The pursuit domain is one of the popular

Table 1 Parameters setting of different algorithms

Parameter Name	GNP	GNP withACO	GNP cleanupACO	ACO	GNP withABC	ACNP	ACNP_V02	ACNP_V03	ACNP_ Online
Population Size	50	50	50	50	50	50	50	50	50
# of Elite transfer directly	1	1	1	–	–	–	–	–	–
Crossover generated individuals	20	20	20	–	–	–	–	–	–
Mutation generated individuals	29	29	29	–	–	–	–	–	–
Tournament size	2	2	2	–	2	–	–	-	–
Crossover rate	0.90	0.90	0.90	–	–	–	–	–	–
Mutation rate	0.01	0.01	0.01	–	–	–	–	–	–
S_{UB}	60	60	60	60	60	60	60	60	60
ρ	–	0.1	0.1	0.1	–	0.1	0.1	0.1	0.1
q_1	–	–	–	–	–	1.0	0.0	0.0	1.0
q_2	–	–	–	–	–	1.0	1.0	1.0	1.0
M	–	–	–	–	–	10	10	10	–
modif_rate	–	–	–	–	0.02	–	–	–	–
limit	–	–	–	–	200	–	–	–	–

benchmarks for evaluating the efficiency of algorithms in multi-agent systems [31–34]. This environment includes a set of adjacent cells. These cells are empty or they are occupied by a predators, a pray or an obstacle. The pursuit domain used in this report is a 20×20 two-dimensional environment. A 10×10 example of this environment is shown in Fig. 4a. In this environment, a predator is able to sense its surrounding environment and decide what to do according to its perceptions. The list of its sensors and actuators are shown in Table 2. If the predator takes the action of moving forward, it will be transferred to the next cell in front of itself only if the cell is empty. But if there is another predator, prey or obstacle in that cell and the predator moves forward, there will be no change in its location. If predators could place the prey in a position similar to Fig. 4b, they would have succeeded in hunting the prey. If a predator is in the cell adjacent to a pray, it is in the capture position. Therefore, if four predators are in the capture position or prey is trapped between predators and obstacles and unable to move, it is hunted.

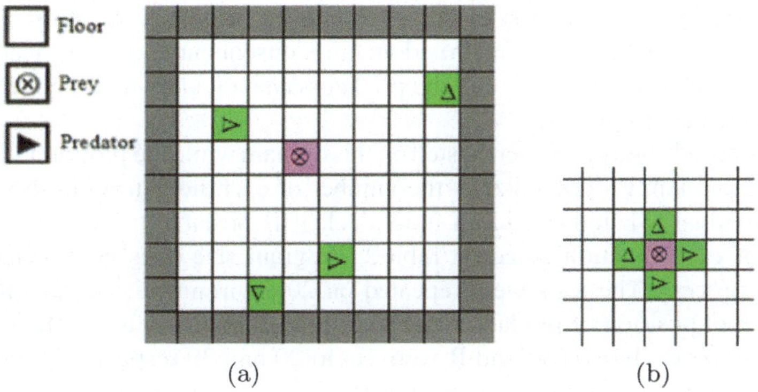

Fig. 4 **a** an example of 10×10 two-dimensional pursuit domain **b** four predators in capture position

Table 2 List of sensors and actuators of a predator

Node type	Description	Outputs
Judgment nodes	Judge forward	Floor, Obstacle, Prey, Predator
	Judge left	
	Judge backward	
	Judge right	
	Direction of nearest prey form a predator	Forward, Backward, redundant line
		Left, Right, Unknown
Processing nodes	Stay	
	Turn left	
	Move forward	
	Turn right	

In this research, the prey speed is half that of predators. The first four judgment functions shown in Table 2 enable the predator to perceive adjacent cells in four directions. The fifth judgment function is used to perceive the environment at farther distances. This function indicates the direction of the nearest prey. Predators behave according to a strategy specified by an individual. This individual specifies what to be sensed, and what to do next. The fitness of each individual is evaluated as mentioned in Ref. [5]. Since the prey moves in the environment randomly, in order to more accurately evaluate the fitness of each individual, W randomly generated environments are produced. Then, each individual is used R times on each environment. Finally, the sum of fitness values is considered as the final fitness of an individual.

4.2 *Experimental Results and Analysis*

Since pursuit domain is a dynamic environment, it is considered as a complex problem. The pray moves randomly in this environment. So, the paths of its movement are not the same in each run of this domain. Consequently, quickly learning the behavior of the pray is not easy and the predators cannot adopt their behaviors in an ever-changing environment.

The proposed algorithms were tested on this domain while the program size is one, three, five and ten. Program size is the number of each node used in the structure of an individual [3]. Indeed, in an individual, it is possible to be more than one instance of each function listed in Table 2. Program size does not change during evolution process. The tests were repeated on 20 environments of pursuit domain with different position of predators and prey. For each environment, the tests were repeated 30 times. Indeed W and R were set to 20 and 30 respectively. Predefined number of fitness evaluation was used as the termination condition in these tests. It was set to 20000. In Fig. 5, in each iteration the fitness of best individuals was summed up and shown.

According to Fig. 5a, when program size is 1, ACNP_Online is by far the best while ACO and GNPwithABC are the worst. Other algorithms almost have the same rank. Figure 5b shows that by increasing the program size to 3, ACNP variants (ACNP, ACNP_V02, ACNP_Online, ACNP_V03) and GNP have almost the same performance. They are better than other algorithms. GNP_with_ACO and GNP_cleanup_ACO are in the next ranks respectively and GNP_with_ABC and ACO were in the last ranks.

As it is clear in Fig. 5c, by increasing the program size to 5, the performance of all algorithms but ACNP_V02 and ACNP_V03 was decreased because of search space growth in a dynamic space. The same phenomenon occurs when the program size was increased to 10. In this condition, performance reduction happened in all algorithms. However, this reduction is less for two versions of ACNP algorithms i.e. ACNP_V02 and ACNP_V03. They were the bests in this condition.

For easier and more accurate comparison of the algorithms, the final results of them were reported in Tables 3, 4, 5 and 6. According to the results in Table 3,

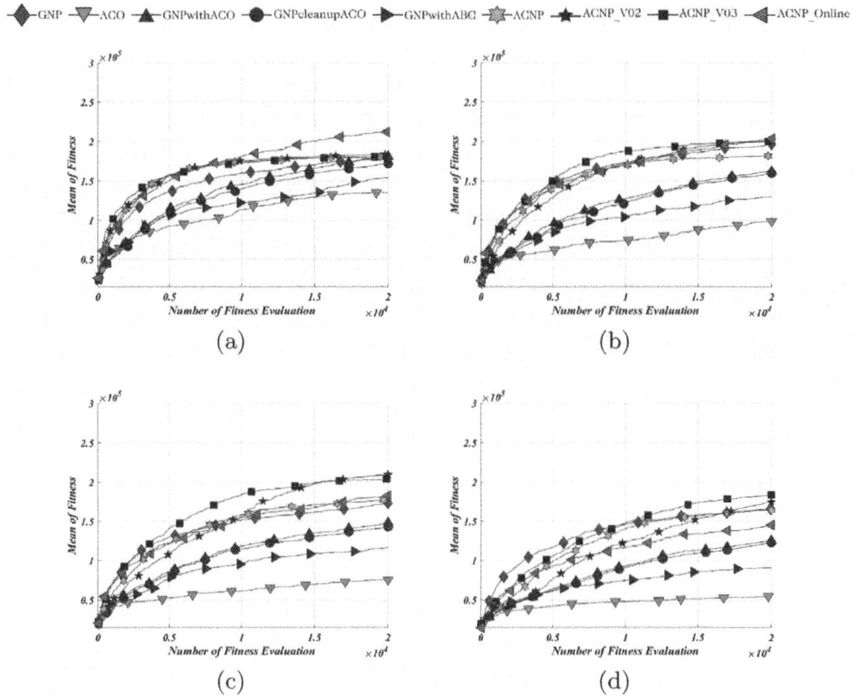

Fig. 5 Mean of fitness in different algorithms while program size is **a** 1, **b** 3, **c** 5 and **d** 10

ACNP_Online showed the best efficiency in all reported factors mean, minimum and maximum of fitness while its standard deviation was the lowest. Table 4 shows that this algorithm remained as the best algorithm by increasing the program size to 5. However, the p-value of Wilcoxon rank-sum test [35] shows that there was no significant difference between this algorithm and the algorithms in the second and third ranks. According to Tables 5 and 6, increasing the program size to 5 or 10 leads to some changes in the rank of algorithms. ACNP_V02 and ACNP_V03 achieved the first and second ranks respectively. According to these results, almost all algorithms have high standard deviation because of dynamic nature of the investigated problem.

In stochastic environments [36], GNP faces with a big challenge. The fitness of an individual cannot be calculated precisely when it is evaluated only once in the environment. Because of the stochastic nature of the environment, different fitness values will be obtained in each evaluation. For better evaluation of each individual, its average fitness must be computed over multiple times of evaluation. In population based algorithms, it is a time consuming process. If it is not done, the fitness of individuals is not precise. Consequently, parent selection and survivor selection cannot be performed correctly and evolution process cannot be applied efficiently. This challenge was solved by the mechanism used in ACNP and its various versions using the experience table. The proposed algorithms use these experiences to produce the

Table 3 Final results of different algorithms in which the program size was set to 1

Algorithm name	Rank	Mean of fitness	STD of fitness	Max of fitness	Min of fitness	p-value
ACNP_Online	1	159696.8	28081.84	203594.5	96990.8	–
GNPwithACO	2	126642.2	32360.6	186614	67634.5	2.39E–04
ACNP_V02	3	121067.2	37125.54	189921.5	14219.19	7.66E–05
GNPclearupACO	4	118267.2	37119.55	199075.6	58941.68	4.94E–05
ACNP	5	117527.4	37280.85	192914.2	34694.09	2.77E–05
GNP	6	115786.7	37079.35	187103.8	63243.97	1.25E–05
ACNP_V03	7	109586.8	31867.07	160258.6	37877.47	4.44E–07
GNPwithABC	8	102767.6	28911.3	151645.4	46062.8	1.85E-08
ACO	9	99670.37	19924.66	139308.1	72537.37	2.44E–09
Mean:		119001.36		178937.30	54689.1	

Table 4 Final results of different algorithms in which the program size was set to 3

Algorithm name	Rank	Mean of fitness	STD of fitness	Max of fitness	Min of fitness	p-value
ACNP_Online	1	158295.9	32978.41	220624.5	107670.2	–
ACNP_V02	2	144346.1	33232.13	216216.2	87705.23	0.17145
ACNP_V03	3	137902.6	31840.50	204075.5	85050.34	0.015014
GNP	4	136031.3	32106.34	178665.4	70435.32	2.71E–02
ACNP	5	120992.2	35365.14	194127	47619.92	1.89E–04
GNPwithACO	6	108121.2	28861.52	171632.8	54118.66	6.53E–07
GNPclearupACO	7	104283.1	31843.26	169503	61646.12	7.04E–07
ACO	8	76094.37	16122.84	122611.4	53457.63	6.07E–11
GNPwithABC	9	75192.06	20616.59	121574.8	36851.98	4.98E–11
Mean:		117917.65		177670.04	67172.82	

next generation. Consequently, it is not necessary to know the fitness of individuals precisely.

To test the performance of different algorithms in a stochastic environment, the algorithms were also tested on the pursuit domain when deterministic parameter [3] was set to 0.75 and 0.50 and the program size was set to 5 and 10. Deterministic parameter shows how likely an action leads to its desired result. Other parameters were not different from parameters in deterministic environment. The results were shown in Tables 7, 8, 9 and 10. In Table 7, ACNP_V03, ACNP and ACNP_V02 were in the first, second and third ranks, respectively. ACNP_Online is in the fourth rank. It is clear that different versions of ACNP had the best performance in stochastic environment as the individuals in these algorithms can save their experience during each iteration. This accumulated experience can efficiently be used for next generation production. The p-values in these tables reveal that the first rank algorithms are

Table 5 Final results of different algorithms in which the program size was set to 5

Algorithm name	Rank	Mean of fitness	STD of fitness	Max of fitness	Min of fitness	p-value
ACNP_V02	1	150778.4	33768.92	208698.7	90872.04	–
ACNP_V03	2	140912.4	34883.71	206105.1	71780.79	0.283778
ACNP_Online	3	135088	36075.17	191638.9	60092.08	0.133454
ACNP	4	114423.4	26767.18	162442.7	57407.91	1.89E–04
GNP	5	106197.7	26394.38	180614	59729.87	3.09E–06
GNPwithACO	6	98726.25	25624.84	154876.9	57707.35	4.44E–07
GNPclearupACO	7	90285.28	22660.31	158519.5	54325.18	4.57E–09
GNPwithABC	8	63476.75	21384.49	106389.8	30409.02	6.07E–11
ACO	9	56977.75	11363.06	79612.97	32666.44	3.02E–11
Mean:		106318.44		160988.72	57221.19	

Table 6 Final results of different algorithms in which the program size was set to 10

Algorithm name	Rank	Mean of fitness	STD of fitness	Max of fitness	Min of fitness	p-value
ACNP_V02	1	121865.1	34579.30	176071	55737.37	–
ACNP_V03	2	121835.7	33627.34	194175.7	63312.16	0.888303
GNP	3	105364.1	28546.91	172822.8	58511.66	7.24E–02
ACNP	4	101487.2	30353.70	188676	64345.47	0.033874
ACNP_Online	5	94415.76	25766.40	177180.3	55899.19	0.001518
GNPclearupACO	6	75883.81	21235.61	118906.5	42036	2.32E–06
GNPwithACO	7	74377.48	17416.43	111432	41627.33	3.26E–07
GNPwithABC	8	45267.4	14169.04	80161.19	26553.54	8.15E–11
ACO	9	45099.47	11520.53	81213.74	27118.61	5.49E–11
Mean:		87288.4477		144515.47	48349.04	

significantly better than those which do not use accumulated fitness in offspring generation. In almost all tests ACNP and ACNP_Online are ranked after ACNP_V02 and ACNP_V03 because in each iteration they only use the fitness of the best individual ever found for updating the experience table. Therefore, if an improper individual achieves randomly to good fitness, the evolution process will be misled because of its destructive effect on the experience table. Hence, if the experience table is updated by a group of better individuals, it will reduce this risk.

The performance of GNP and its various versions i.e. GNPcleanupACO and GNPwithACO are not good because of inaccurate fitness of individuals. Inaccurate fitness disrupts the crossover and mutation operators because of unsuitable parent and survival selection. As ACO suffers from biasing toward exploitation [26, 27], it could not find any suitable solution for the tested benchmark. The main weakness of GNPwithABC is its slow evolution speed [7]. Consequently, it takes more times to

Table 7 Final results of using different algorithms in which deterministic parameter and program size were set to 0.75 and 5, respectively

Algorithm	Rank	Fitness mean	Fitness STD	Max fitness	Min fitness	p-value
ACNP_V03	1	71150.92	11609.36	87357.17	50318.82	–
ACNP	2	71091.51	13409.34	100386.1	50104.43	0.641424
ACNP_V02	3	70904.32	11957.36	92324.8	47202.2	0.899995
ACNP_Online	4	62976.02	12141.02	85755.57	43690.89	0.007617
GNP	5	62853.31	9484.59	78549.55	44912.25	0.003034
GNPclearupACO	6	60131.72	10644.79	80326.74	40036.35	0.00062
GNPwithACO	7	55957.5	10018.32	74644.08	40208.64	9.51E–06
ACO	8	36114.38	11676.83	78237.64	14110.48	3.47E–10
GNPwithABC	9	34564.03	14862.3	80550.8	19897.03	3.20E–09
Mean:		58415.97		84236.94	38942.34	

Table 8 Final results of using different algorithms in which deterministic parameter and program size were set to 0.5 and 5, respectively

Algorithm	Rank	Fitness mean	Fitness STD	Max fitness	Min fitness	p-value
ACNP_V02	1	31692.55	6304.52	43843.52	14239.87	–
ACNP_V03	2	30628.53	4523.51	39603.4	20641.87	0.379036
ACNP_Online	3	29462.2	5893.55	41704.65	18635.76	0.195791
ACNP	4	27355.89	7312.78	37502.71	8968.16	0.040595
GNPwithACO	5	27139.44	5439.86	37649.44	10466.69	0.002157
GNPwithABC	6	26655.68	14565.47	76360.72	8686.55	0.006669
GNP	7	24171.91	10184.97	44234.17	8888.8	0.004427
GNPclearupACO	8	23510.46	5843.39	33033.34	11073.48	4.42E–06
ACO	9	15995.33	4781.24	25214.21	8944.92	3.47E—10
Mean:		26290.22		42127.35	12282.9	

find suitable solutions. In a stochastic dynamic environment, this will decrease the performance of algorithm significantly because more time is needed when learning is performed in these types of environments.

Table 9 Final results of using different algorithms in which deterministic parameter and program size were set to 0.75 and 10, respectively

Algorithm	Rank	Fitness mean	Fitness STD	Max fitness	Min fitness	p-value
ACNP_V02	1	66526.99	9829.71	94167.7	52812.51	–
ACNP_V03	2	62848.87	14039.79	98204.85	35828.44	0.222573
ACNP	3	61823.17	10545.84	82057.14	43249.63	0.195791
ACNP_Online	4	60246.01	12488.82	88310.93	41628.91	0.029205
GNP	5	56426.06	13914.34	78644.06	9295.2	0.004033
GNPwithACO	6	53206.36	10311.96	71327.15	34031.34	1.75E–05
GNPclearupACO	7	47156.73	11100.7	72536.56	30937.3	1.16E–07
ACO	8	28448.4	6618.97	41671.21	16228.03	3.02E–11
GNPwithABC	9	24712.64	9552.38	58583.94	14515.39	6.70E–11
Mean:		51266.14		76167.06	30947.42	

Table 10 Final results of using different algorithms in which deterministic parameter and program size were set to 0.5 and 10, respectively

Algorithm	Rank	Fitness mean	Fitness STD	Max fitness	Min fitness	p-value
ACNP_V03	1	30891.29	7967.13	57687.58	13481.91	–
ACNP_V02	2	19834.02	7548.08	38621.69	8757.16	1.73E–06
ACNP	3	19101.61	8295.17	35061.48	8548.66	3.09E–06
GNPclearupACO	4	17856.74	7897.15	32011.55	7162.28	5.19E–07
GNPwithACO	5	17180.79	6100.48	28938.84	6811.19	8.48E–09
GNP	6	17099.7	8286.49	37709.2	8816.61	1.87E–07
ACNP_Online	7	15746.62	7120.23	34738.81	8387	3.08E–08
ACO	8	11873.93	3850.79	25579.23	6844.01	1.09E–10
GNPwithABC	9	11038.3	3518.27	18966.8	3275.97	6.70E–11
Mean:		17847		34368.35	8009.42	

According to Tables 8 and 10, it is clear that decreasing the deterministic parameter reduces the performance of the algorithms because learning in more stochastic environment is harder. However, their ranks do not change significantly. In addition, just like in deterministic environment, when the program size was increased to 10, learning process takes more time because of the search space growth.

5 Conclusions

In summary, in the proposed algorithms, the entire process by which the agent behaves is valued in proportion to the reward it receives for that behavior. This procedure is repeated many times. With the suitable use of the learned knowledge, it is possible to create individuals with better efficiency for the next generation. One of the most important advantages of using this method is that it could be used to identify the appropriate sub-structures, even if the whole flowchart is not suitable. These appropriate sub-structures can be used to create better flowcharts in next generations. Another feature that makes proposed algorithm better than the standard GNP in improving the structure of individuals is that ant-based methods are inherently constructive. This means that individuals in each generation are constructed based on the experience gained in the previous generation while algorithms such as GNP are inherently productive. It means that individuals in each generation are created from a combination of the previous generation's individuals. This feature of the proposed algorithm makes it superior compared to others.

References

1. Miller, J.F.: Cartesian genetic programming: its status and future. Genet. Program. Evolvable Mach. **21**(1), 129–168 (2020)
2. La Cava, W., Silva, S., Danai, K., Spector, L., Vanneschi, L., Moore, J.H.: Multidimensional genetic programming for multiclass classification. Swarm Evol. Comput. **44**, 260–272 (2019)
3. Roshanzamir, M., Palhang, M., Mirzaei, A.: Efficiency improvement of genetic network programming by tasks decomposition in different types of environments. Genet. Program. Evolvable Mach. 1–38 (2021)
4. Roshanzamir, M., Palhang, M., Mirzaei, A.: Tasks decomposition for improvement of genetic network programming. In: 2019 9th International Conference on Computer and Knowledge Engineering (ICCKE), pp. 201–206. IEEE (2019)
5. Roshanzamir, M., Palhang, M., Mirzaei, A.: Graph structure optimization of genetic network programming with ant colony mechanism in deterministic and stochastic environments. Swarm Evol. Comput. **51**, 100581 (2019)
6. Li, X., Yang, H., Yang, M.: Revisiting genetic network programming (gnp): towards the simplified genetic operators. IEEE Access **6**, 43274–43289 (2018)
7. Li, X., Yang, G., Hirasawa, K.: Evolving directed graphs with artificial bee colony algorithm. In: 2014 14th International Conference on Intelligent Systems Design and Applications, pp. 89–94. IEEE (2014)
8. Li, X., He, W., Hirasawa, K.: Learning and evolution of genetic network programming with knowledge transfer. In: 2014 IEEE Congress on Evolutionary Computation (CEC), pp. 798–805. IEEE (2014)
9. Li, X., He, W., Hirasawa, K.: Adaptive genetic network programming. In: 2014 IEEE Congress on Evolutionary Computation (CEC), pp. 1808–1815. IEEE (2014)
10. Li, X., Mabu, S., Hirasawa, K.: A novel graph-based estimation of the distribution algorithm and its extension using reinforcement learning. IEEE Trans. Evol. Comput. **18**(1), 98–113 (2013)
11. Murata, H., Koshino, M., Kimura, H.: K-cut crossover using graph theory in genetic network programming. Int. J. Innov. Comput. Inf. Control **9**(2), 641–650 (2013)

12. Li, X., He, W., Hirasawa, K.: Genetic network programming with simplified genetic operators. In: International Conference on Neural Information Processing, pp. 51–58. Springer, Berlin (2013)
13. Li, X., Li, B., Mabu, S., Hirasawa, K.: A novel estimation of distribution algorithm using graph-based chromosome representation and reinforcement learning. In: 2011 IEEE Congress of Evolutionary Computation (CEC), pp. 37–44. IEEE (2011)
14. Li, X., Mabu, S., Zhou, H., Shimada, K., Hirasawa, K.: Genetic network programming with estimation of distribution algorithms for class association rule mining in traffic prediction. In: IEEE Congress on Evolutionary Computation, pp. 1–8. IEEE (2010)
15. Ramezanian, R., Peymanfar, A., Ebrahimi, S.B.: An integrated framework of genetic network programming and multi-layer perceptron neural network for prediction of daily stock return: an application in tehran stock exchange market. Appl. Soft Comput. **82**, 105551 (2019)
16. Mabu, S., Higuchi, T., Kuremoto, T.: Semisupervised learning for class association rule mining using genetic network programming. IEEJ Trans. Electr. Electron. Eng. **15**(5), 733–740 (2020)
17. Sendari, S., Nur Afandi, A., Ari Elbaith Zaeni, I., Dwi Mahandi, Y., Hirasawa, K., Lin, H.-I.: Exploration of genetic network programming with two-stage reinforcement learning for mobile robot. Telkomnika **17**(3), 1447–1454 (2019)
18. Madokoro, H., Nix, S., Sato, K.: Automatic calibration of piezoelectric bed-leaving sensor signals using genetic network programming algorithms. Algorithms **14**(4), 117 (2021)
19. Mabu, S., Hirasawa, K., Jinglu, H., Murata, J.: Online learning of genetic network programming. IEEJ Trans. Electron., Inf. Syst. **122**(3), 355–362 (2002)
20. Mabu, S., Hirasawa, K., Hu, J.: Genetic network programming with reinforcement learning and its performance evaluation. In: Genetic and Evolutionary Computation Conference, pp. 710–711. Springer, Berlin (2004)
21. Sutton, R.S., Barto, A.G.: Reinforcement Learning: An Introduction. MIT Press (2018)
22. Park, S.G., Mabu, S., Hirasawa, K.: Robust genetic network programming using sarsa learning for autonomous robots. In: 2009 ICCAS-SICE, pp. 523–527. IEEE (2009)
23. Mabu, S., Hirasawa, K.: Evolving plural programs by genetic network programming with multi-start nodes. In: 2009 IEEE International Conference on Systems, Man and Cybernetics, pp. 1382–1387. IEEE (2009)
24. Li, X., Mabu, S., Hirasawa, K.: Use of infeasible individuals in probabilistic model building genetic network programming. In: Proceedings of the 13th Annual Conference on Genetic and Evolutionary Computation, pp. 601–608 (2011)
25. Li, X., Mabu, S., Hirasawa, K.: An extended probabilistic model building genetic network programming using both of good and bad individuals. IEEJ Trans. Electr. Electron. Eng. **8**(4), 339–347 (2013)
26. Yu, L., Zhou, J., Mabu, S., Hirasawa, K., Hu, J., Markon, S.: Elevator group control system using genetic network programming with aco considering transitions. In: SICE Annual Conference 2007, pp. 1330–1336. IEEE (2007)
27. Yu, L., Zhou, J., Mabu, S., Hirasawa, K., Hu, J., Markon, S.: Double-deck elevator group supervisory control system using genetic network programming with ant colony optimization. In: 2007 IEEE Congress on Evolutionary Computation, pp. 1015–1022. IEEE (2007)
28. Karaboga, D., Basturk, B.: On the performance of artificial bee colony (abc) algorithm. Appl. Soft Comput. **8**(1), 687–697 (2008)
29. Eiben, A.E., Smith, J.E., et al.: Introduction to Evolutionary Computing, vol. 53. Springer, Berlin (2003)
30. Benda, M.: On Optimal Cooperation of Knowledge Sources: An Empirical Investigation. Technical Report, Boeing Advanced Technology Center (1986)
31. Naeini, A.T., Ghaziasgar, M.: Improving coordination via emergent communication in cooperative multiagent systems: a genetic network programming approach. In: 2009 IEEE International Conference on Systems, Man and Cybernetics, pp. 589–594. IEEE (2009)
32. Naeini, A.T., Palhang, M.: Evolving a multiagent coordination strategy using genetic network programming for pursuit domain. In: 2008 IEEE Congress on Evolutionary Computation (IEEE World Congress on Computational Intelligence), pp. 3102–3107. IEEE (2008)

33. Eguchi, T., Hirasawa, K., Hu, J., Ota, N.: A study of evolutionary multiagent models based on symbiosis. IEEE Trans. Syst., Man, Cybern., Part B (Cybernetics) **36**(1), 179–193 (2006)
34. Mabu, S., Hirasawa, K., Jinglu, H.: A graph-based evolutionary algorithm: Genetic network programming (gnp) and its extension using reinforcement learning. Evol. Comput. **15**(3), 369–398 (2007)
35. Wilcoxon, F.: Individual comparisons by ranking methods. Biom. Bull. **1**(6), 80–83 (1945)
36. Stuart, R., Peter, N.: Artificial Intelligence: A Modern Approach (2020)

A Bumble Bees Mating Optimization Algorithm for the Discrete and Dynamic Berth Allocation Problem

Eleftherios Tsakirakis, Magdalene Marinaki, and Yannis Marinakis

Abstract In maritime supply chain, there is a large number of significant optimization problems that have to be modeled and solved. One of them is the Berth Allocation Problem (BAP). In BAP, the assignment and scheduling of incoming ships to berth positions are studied. In literature, several variations and connections with other maritime problems can be found. In this paper, we present an efficient methodology for solving the discrete and dynamic version of BAP (DDBAP). We formulate the DDBAP as a heterogeneous vehicle routing problem with time windows (HVRPTW). In this formulation, berths correspond to vehicles, ships correspond to customers/nodes and the sequence of serviced ships at a particular berth represent a vehicle route. Each one of these sequences must start and end at a specific point such as the depot in VRP. The sequence of serviced ships starts at the origin and finishes at the destination point. Time windows can be imposed on each one of the ships. The time windows for each ship correspond to the availability time in order to be served properly in one of the available berths. In order to solve DDBAP, we apply a modified Bumble Bees Mating Optimization (BBMO) algorithm. BBMO belongs to the nature inspired optimization algorithms and it simulates the mating behavior of the bumble bees. It is an algorithm that its basis is the Honey Bees Mating Optimization (HBMO) algorithm but with a number of different steps that make the algorithm a completely different and competitive algorithm with other nature inspired algorithms. We applied the algorithm to benchmark instances from the literature and the obtained computational results proved the efficiency of BBMO with respect to the quality of solutions and the computational time needed to find these solutions.

Keywords Bumble Bees mating optimization · Berth allocation problem · Vehicle routing problem

E. Tsakirakis · M. Marinaki · Y. Marinakis (✉)
School of Production Engineering and Management, Technical University of Crete, Kounoupidiana, Greece
e-mail: marinakis@ergasya.tuc.gr

E. Tsakirakis
e-mail: etsakirakis@isc.tuc.gr

M. Marinaki
e-mail: magda@dssl.tuc.gr

A. Biswas et al. (eds.), *Advances in Swarm Intelligence*, Studies in Computational Intelligence 1054, https://doi.org/10.1007/978-3-031-09835-2_19

1 Introduction

The study in the field of swarm intelligence algorithms has been increased significantly, either by modifying an existing algorithm or by simulating the behavior of another insect or animal in the nature and producing a new, and in the most cases, a very efficient algorithm. Although there is a number of algorithms that simulate the swarming/movement behavior (Particle Swarm Optimization [14], Firefly Algorithm [41], etc.), the hunting behavior (i.e. Grey Wolf Optimizer [35], Whale Optimization Algorithm [34]) or the food searching behavior (i.e. Ant Colony Optimization [6], Artificial Bee Colony [13]) of some species, little work has been performed for algorithms that simulate the mating behavior of these species. The most important of these algorithms are the Honey Bees Mating Optimization [28] algorithm and the Bumble Bees Mating Optimization [25] algorithm which simulate the mating behavior of two different kind of bees, the Honey Bees and the Bumble Bees, respectively. Recently, a new algorithm that simulates the mating behavior of Fireflies has been proposed. This algorithm has been used in the literature for the solution of two different problems (for continuous optimization problems [38] and for the internet of underwater wireless sensor networks-based events monitoring application [7]).

In this paper, the Discrete and Dynamic Berth Allocation Problem (DDBAP), which is one of the most known, popular and well studied maritime logistics problem, is solved using the Bumble Bees Mating Optimization algorithm. BAP reflects the strategical planning in container terminals, which aims to serve accurately a predefined set of ships. Furthermore, the fleet of vessels must be moored under the restrictions of specific timetables, while checking the availability of the current berthing position. In literature, BAP is categorized variously [1, 2] according to different factors, such as the type of berth layout, the timetable attribute of service process, the handling time of ships and the performance measure of the problem. Characterized as a NP-hard problem due to its computational complexity, the applications of BAP draw the attention of researchers to propose innovative models and solution methodologies.

In the next sections, initially, the berth allocation problem is described by presenting the formulation used in this paper. In the third section, a complete presentation of the solution algorithm is given. Next, the computational results of the method are presented and compared with the best computational results from the literature and with the results of the HBMO algorithm [40]. The reason of comparing the computational results with the computational results of the HBMO is that HBMO is the only mating nature inspired algorithm that was ever used for solving this problem and, thus, the computational results of these two algorithms are directly comparable. Finally, in the last section some conclusions and future directions are given.

2 Berth Allocation Problem

In container ports, an increasing demand for efficient maritime logistics services is observed. For this reason, in port terminals the berth quays areas are frequently used to properly forward the transported freights from the incoming ships. Due to the real world applications, which occur in the ports, managers and researchers have shown significant interest in providing reliable methods for particularly servicing and scheduling the cargo ships by exploiting port infrastructures such as berths. One of the most well studied maritime problems is the berth allocation problem (BAP). The main objective of BAP is the reduction of total servicing time of the incoming ships in the container terminals by allocated them in the most proper berth areas.

This paper is focused on the most popular variant, the dynamic berth allocation problem (DBAP). In the literature, DABP was firstly introduced by Imai et al. [12]. They presented a discrete layout of berths for assigning and scheduling the incoming ships. As a result, numerous methodologies such as heuristic and metaheuristic approaches [5, 16, 17, 19, 20, 33, 37] and computational models [3, 4, 15, 36] have been proposed. The proposed formulations for DBAP in Buhrkal et al. [3] and Kramer et al. [15] works, summarize all the models including important details for further improvement and computational performance. To solve efficiently DBAP, we have selected the formulation of Heterogeneous Vehicle Routing Problem with Time Windows (HVRPTW).

The static Berth Allocation Problem was introduced by Imai et al. [11] and later Imai et al. [12] introduced its dynamic version. The main objective of BAP aims on the assignments of the incoming container ships to the available berths in order to minimize the time they spend waiting for berths. The main difference between the two approaches is that in the static approach, the set of ships is predefined while in the dynamic approach, the ships arrive anytime during the working period of the day.

A number of different formulations [3, 4, 11, 36] have been proposed for the solution of the DDBAP. There are two formulations of the DDBAP that are based on the Vehicle Routing Problem. In [4] the authors formulate the DDBAP as a MultiDepot Vehicle Routing Problem with Time Windows (MDVRPTW) while in [3] the authors formulate the DDBAP as an Heterogeneous Fleet Vehicle Routing Problem with Time Windows (HVRPTW). The two models differ between them in that in the first, a berth corresponds to a different depot in a multigrap while in the second, a berth corresponds to an heterogeneous vehicle in a single graph where all vehicles are assigned to the same one depot, which is the origin o and the destination d point. In both models, ships correspond to customers/nodes. In the second model, a sequence of serviced ships corresponds to a vehicle route that starts and ends in a specific point, the depot in VRP. Time windows for each ship can be imposed and correspond to the availability time in the available berth, meaning the arrival time of the ship to the port and the departure time of the ship from the port.

DBAP as a HVRPTW [3] is defined as a graph $G = (V, A)$, where in $V = N \cup \{o, d\}$ all the ships with the origin and the destination points of the graph are included. Each ship $i \in N$ is linked with a time window $[a_i, b_i]$, which indicates the arrival and the departure time of the current ship, a service priority v_i and a handling time h_i^k. The origin and destination points have, also, time windows $[s^k, e^k]$, which depend on the available berth k, as berth can be available on a different time. The decision variables are divided into binary variables ($x_{ij}^k \in M$, $(i, j) \in A$) and continuous variables ($T_i^k, i \in V, k \in M$). If ship j immediately follows up ship i, then x_{ij}^k is equal to 1, otherwise is equal to 0. T_i^k denotes the time for ship i to moor at berth k, or takes value a_i if berth k is unused. T_o^k and T_d^k show the start and the finish time of activities at berths $k \in M$. The HVRPTW as DDBAP mathematical model is described as follows :

$$\min \sum_{i \in N} \sum_{k \in M} v_i (T_i^k - a_i + h_i^k \sum_{j \in N \cup \{d\}} x_{ij}^k) \tag{1}$$

subject to

$$\sum_{k \in M} \sum_{j \in N \cup \{d\}} x_{ij}^k = 1, \qquad \forall i \in N \tag{2}$$

$$\sum_{j \in N \cup \{d\}} x_{o,j}^k = 1, \qquad \forall k \in M \tag{3}$$

$$\sum_{i \in N \cup \{o\}} x_{i,d}^k = 1, \qquad \forall k \in M \tag{4}$$

$$\sum_{j \in N \cup \{d\}} x_{ij}^k = \sum_{j \in N \cup \{o\}} x_{ji}^k, \qquad \forall k \in M, i \in N \tag{5}$$

$$T_i^k + h_i^k - T_j^k \leq (1 - x_{ij}^k) M_{ij}^k, \qquad \forall k \in M, (i, j) \in A \tag{6}$$

$$a_i \leq T_i^k \qquad \forall k \in M, i \in N \tag{7}$$

$$T_i^k + h_i^k \sum_{j \in N \cup \{d\}} x_{ij}^k \leq b_i \qquad \forall k \in M, i \in N \tag{8}$$

$$s^k \leq T_o^k \qquad \forall k \in M, \tag{9}$$

$$T_d^k \leq e^k \qquad \forall k \in M, \tag{10}$$

$$x_{ij}^k \in \{0, 1\}, \qquad \forall k \in M, (i, j) \in A \tag{11}$$

$$T_i^k \in \Re^+, \qquad \forall k \in M, (i, j) \in A \tag{12}$$

The objective function (1) is used for the minimization of the total cost including the waiting and the handling times of the berthed ships. Constraints (2) are used in order to ensure that each ship must be assigned to exactly one berth k. Constraints (3) and (4) state that every sequence of served ships in each berth k must begin at the origin o and finish at destination d. The flow conservation for all of the ships is guaranteed by constraints (5). Constraints (6) note the consistency for berthing time and mooring sequence on each berth k. Constraints (7) and (8) ensure that the vessels must be served within the time windows limits. The berth availability time is enforced by the constrains (9) and (10). Lastly, constraints (11) and (12) define the respective domains of the decision variables.

3 Bumble Bees Mating Optimization

The Bumble Bees Mating Optimization (BBMO) is an algorithm that was presented before thirteen years in 2009 in a hybridized form [29] for the solution of the clustering problem, a classic combinatorial optimization problem. In this first version of the algorithm, the Greedy Randomized Adaptive Search Procedure (GRASP) was used in order to create the initial solutions [8]. In the next publication, that it could be thought as the first complete publication of the algorithm without any hybridizations, the BBMO algorithm was used for solving global unconstrained optimization problems [30]. If we would like to describe the algorithm we could say that the solutions are divided in three parts, the queens, the workers and the drones (the existence of the drones gave us the first difference from the HBMO where the solutions are divided in two parts, the only one queen and the drones). Initially, a number of solutions are created without the use of any other algorithm (this is the difference with the first publication of the BBMO algorithm where the initial solutions are created using GRASP). The best of these solutions played the role of the initial queen and all the other solutions were the drones.

In nature, almost every species of bees, in order for the queen bee to begin the mating, it needs to fly out of the hive in order to begin the so-called mating flight. We use the same procedure in the algorithm but without using the speed and energy equations of the HBMO algorithm (second difference of the BBMO algorithm from the HBMO algorithm). This difference was chosen as we would like to have less equations in order to be easier for someone to use the algorithm in more applications. Thus, only the best drones will mate with the queen bee. In the next steps, the spermatheca of the queen will be filled, the queen bee will return to the hive and it will begin to lay eggs producing:

- new queens. The new queens will be created using a multiparent crossover operator where the one parent is always the queen bee and the other parents are selected between the drones using the following equation:

$$b_{ij}(t) = \begin{cases} q_j(t), & \text{if } rand_i(0,1) \le Cr_1 \\ d_{kj}(t), & \text{otherwise.} \end{cases} \tag{13}$$

where j is the dimension of the problem, t is the iteration number, Cr_1 is a parameter that controls which elements of the broods $b_{ij}(t)$ have been selected from the solution of the queen bee, $q_j(t)$, and which from the selected drone k, $d_{kj}(t)$. The value of Cr_1 is between 0 and 1.

- workers, using the same procedure as in the production of the new queens, i.e., the new bees that could not be the new queens using the previous procedure are the workers.
- drones, using a random mutation operator either in an old queen or in one of the fittest workers.

This procedure is very important for understanding the difference between the HBMO and the BBMO. The BBMO begins as in the HBMO with one queen but in the end of the first iteration a number of new queens have been produced, meaning that we take the most promising solutions and we give them the characterization of new queens. This procedure was used as the idea behind the algorithm was to propose an algorithm that would not stagnate quickly in a local optimum and the use of a number of different queens was one way to avoid it. In the HBMO algorithm, the mating process is performed always with the queen bee and this is the main reason that HBMO algorithm converges fast to a local optimum. Thus, we begin the algorithm with the same procedure as in the HBMO (one queen bee that performs mating flight) and, then, we produce a number of different queen bees in order to have the possibility to explore more places in the solution space.

The feeding procedure of the new queens is done using Eq. (14):

$$\begin{aligned} nq_{ij} = nq_{ij} + (b_{\max} - \frac{(b_{\max} - b_{\min}) * lsi}{lsi_{\max}}) * (nq_{ij} - q_j) + \\ \frac{1}{M} * \sum_{k=1}^{M} (b_{\min} - \frac{(b_{\min} - b_{\max}) * lsi}{lsi_{\max}}) * (nq_{ij} - w_{kj}) \end{aligned} \tag{14}$$

The meaning of the feeding procedure is that in the beginning, old queens (q_j) fed the new queens (nq_{ij}) but with the increase of the iterations, the role of the old queens in the feeding of the new queens is decreasing and, at the same time, the role of the workers (w_{kj}) is increasing as the solution of the workers is improving through the iterations. Each new queen can select from the set of workers, a number M of different workers that will be used for the feeding procedure. The $b_{\min}$, $b_{\max}$ (the values of the parameters are in the interval (0, 1)), the lsi (the current local search iteration) and the $lsi_{\max}$ (the maximum number of local search iterations) are used for controlling the feeding procedure. This is the classic equation that almost every swarm intelligence algorithm uses and it is inspired from Particle Swarm Optimization (PSO). However, this equation has some interesting parts which conclude that we have a different equation from the classic velocity equation of PSO. The part of the equation that concerns the solution of the old queen has, initially, the most important role and this role is decreased during the iterations, while the opposite holds for the part of the solution that concerns the workers (small role in the beginning which is increased during the iterations). Also, the workers are completely different solutions than the new queen. Finally, a random old queen and a random worker are used for the feed of the new queen. Thus, a new solution could easily choose a very different way to move in the solution space, avoiding to have one swarm around the best solution, but having small swarms in different places in the solution space. Another interesting feature of these swarms is that, as every solution could choose a solution from the feeding procedure from a different swarm, a continuous migration process is performed and information between different swarms is always exchanged.

The final procedure of the BBMO is the leaving of the drones from the hive in a swarm. This is performed as the drones are looking for the best places in the solution space and wait there the new queens, which they have left from the hive after the feeding procedure, in order to mate. The following equation is used (in the initial version of the BBMO algorithm, this equation is the last different procedure from the HBMO algorithm):

$$d_{ij} = d_{ij} + \alpha * (d_{kj} - d_{lj}) \tag{15}$$

where d_{ij}, d_{kj} and d_{lj} represent the solutions of drones i, k, l, respectively and the parameter α determines in which percentage the drone i is affected by the two other drones k and l.

As the main research interest of our group concerns the solution of routing problems with the use of metaheuristic and swarm intelligence algorithms, in the next publication of the BBMO, the Capacitated Vehicle Routing Problem (CVRP) [23] was solved using this method. The problem of all swarm intelligence algorithms, except the ones that are based on Ant Colony Optimization, is that they are not suitable for this kind of problems as most of them are proposed for the solution of continuous optimization problems. Thus, a number of modifications have to be made in order for these algorithms to be suitable for solving CVRP. Initially, path representation of the routes are used for mapping the solutions. In the feeding equation all elements of each solution are transformed into a floating point interval. Afterwards, the feeding equation of all bees is calculated and, then, each element of the solution is converted back into the integer domain using relative position indexing [18]. The local search algorithm that was used in the mutation equation replacing the random mutation operator of the previous version is the Expanding Neighborhood Search [32]. Finally, in each external generation, the number of the drones per colony was calculated by:

$$\text{number of drones per colony} = \frac{\text{number of drones}}{\text{number of queens}} \tag{16}$$

Afterwards, a number of improvements of the algorithm have been published solving, mainly, different variants of routing problems. Initially, in [25], the BBMO algorithm was used to solve the Open Vehicle Routing Problem (OVRP). The difference with the previous version of the algorithm concerned the movement of the drones outside the hive. The equation corresponding to this movement was replaced by the Iterated Local Search (ILS) algorithm [21] in order to avoid the transformation of each solution of the drones in continuous values and vice versa. Although the algorithm without this modification had proved its effectiveness, with the addition of the ILS algorithm instead of the equation of the movement of the drones, the algorithm became faster and had better performance in less computational time.

In [27], a discrete version of the BBMO algorithm, the Combinatorial Neighborhood Topology Bumble Bees Mating Optimization (CNTBBMO), was presented and used for solving the Vehicle Routing Problem with Stochastic Demands. In this algorithm, no transformation at all in continuous values is performed, thus, a local search procedure was used for all the main equations of the initially proposed BBMO algorithm. In the initially proposed algorithm, there were three equations, the mutation operator, the movement of the drones outside the hive and the feeding procedure. For applying the algorithm to the solution of routing problems, in the previous two publications, the first two equations were replaced by the Iterated Local Search (ILS) [21] algorithm and the Expanding Neighborhood Search, respectively. In this version of the algorithm, the Expanding Neighborhood Search algorithm is replaced by the Variable Neighborhood Search (VNS) [10] algorithm and the feeding procedure is replaced from a new, at that time, strategy denoted as Combinatorial Neighborhood Topology (CNT) [24] which includes a Path Relinking (PR) [9] procedure. The CNT procedure replaced the feeding procedure of each new queen where the new queen decided from which of the two, the old queen or the worker, it would be fed from. Following of what happens in real life, the new queen, initially, was fed from the old queen and as the local search iterations increased, it was fed more from the workers and, at the end, it was fed almost only from the workers.

This procedure is working as follows: a random number, different for each element of the new queen's solution, in the interval $(l, b) = (l_{\text{bound}}, u_{\text{bound}})$, is created [27]. We have the following possibilities:

- this value to be greater than a number L_2, where in this case, either no Path Relinking (NOPR) is performed or a path relinking with a second random worker is performed (PRRW2), (with equal probability),
- this value to be between L_1 and L_2, where in this case, a path relinking (PRPW) of the new queen bee with a random worker is performed, and,
- finally, the value to be less than L_1, where in this case, a path relinking with a randomly selected old queen is realized (PROQ).

Initially, L_2 and L_1 have large values which are decreased during the iterations in order to perform path relinking with the old queens in the beginning of the algorithm and, as the iterations are increasing, to perform the path relinking with larger probability with a worker. As we would like the algorithm not to converge very fast, a solution is selected if the percentage of difference from the best queen is greater than 15% except if this solution improves the best solution. The L_1 and L_2 are given from:

$$L_1 = (u_{\text{bound}} - l_{\text{bound}}) \times (w_1 - \frac{w_1}{iter_{\max}} \times t) + l_{\text{bound}} \tag{17}$$

and

$$L_2 = (u_{\text{bound}} - l_{\text{bound}}) \times (w_2 - \frac{w_2}{2 * iter_{\max}} \times t) + l_{\text{bound}} \tag{18}$$

where, l_{bound} and u_{bound} are the lower and upper bounds, $iter_{\max}$ is the maximum number of iterations and t is the current iteration. As we would like to have L_1 and L_2 with large values in the beginning of the algorithm (with L_2 always greater than L_1), the w_1 and w_2 have, also, large values (w_2 is selected to be equal to 0.9 and w_1 equal to 0.8).

Afterwards, the feeding procedure is realized as follows:

$$tnq_i(t) = \begin{cases} q_i(t), & \text{if } rand_i(l, b) \le L_1 \\ wr_i(t), & \text{if } L_1 \le rand_i(l, b) \le L_2 \\ nq_i(t) \ or \ wr_j(t), & \text{otherwise.} \end{cases} \tag{19}$$

where, tnq is the auxiliary solution that has been produced from the PR procedure (a possible new queen), q is the solution of the old queen bee, nq is the solution of the new queen bee and with wr_i and wr_j are denoted the solutions of two different workers. This version of the algorithm proved to be the most effective of all.

In [22], the initially proposed algorithm was used for the solution of the Feature Selection Problem. As in the solution vector of this problem, only two values exist, 0 if the feature was not selected and 1, otherwise, we used all the initially proposed equations and transformed suitably the solutions using the sigmoid function:

$$sig(nq_{ij}) = \frac{1}{1 + \exp(-nq_{ij})} \tag{20}$$

and, then, the activated features are given from:

$$y_{ij} = \begin{cases} 1, & \text{if } rand_j(0, 1) < sig(nq_{ij}) \\ 0, & \text{if } rand_j(0, 1) \ge sig(nq_{ij}) \end{cases} \tag{21}$$

where y_{ij} is the transformed solution.

Finally, for solving three other problems, the Permutation Flowshop Scheduling Problem [26, 31], the Multicast Routing Problem [31] and the Probabilistic Traveling Salesman Problem [31], an adaptive version of the CNTBBMO algorithm was proposed and used. In this version of the algorithm, an optimization of all the parameters during the iterations was performed. Initially, for each one of the parameters, random values are given, with the only restriction the solutions to be feasible. This was guaranteed if we have $w_1 < w_2$ and $u_{\text{bound}} > l_{\text{bound}} > 0$. The parameters that were used and optimized in this procedure are the w_1, w_2, u_{bound} and l_{bound}, the number of bees, the number of Variable Neighborhood Search (VNS) iterations and the number of iterations. The parameters were updated using three different procedures.

- **First procedure**. If the best solution had not been improved for a consecutive number of iterations (the value of this number was selected equal to 20 after thorough investigation), the following parameters were updated:

$$w_1 = w_{1opt} + \frac{w_1 - w_{1opt}}{w_{1opt}} \tag{22}$$

$$w_2 = w_{2opt} + \frac{w_2 - w_{2opt}}{w_{2opt}} \tag{23}$$

$$u_{\text{bound}} = UB + \frac{u_{\text{bound}} - UB}{UB} \tag{24}$$

$$l_{\text{bound}} = LB + \frac{l_{\text{bound}} - LB}{LB} \tag{25}$$

$$Local\ Search\ iter = LS + \frac{Local\ Search\ iter - LS}{LS} \tag{26}$$

  where LS, LB, UB, w_{2opt} and w_{1opt} are the best performed values so far for the local search iterations, the lower bound, the upper bound, the w_2 and the w_1, respectively. In the first iteration of the algorithm, for these parameters, random values are selected.
- **Second procedure**. If the average best solutions of all bumble bees was not improving more than 2% and, in addition, no improvement of the best solution as in the first procedure had performed, for a consecutive number of iterations, the number of the bumble bees was increasing by the use of the following equation:

$$Bees = NB + \frac{Bees - NB}{NB} \tag{27}$$

  where NB is the best value of the bumble bees.
- **Third procedure**. The decrease of the number of bumble bees was performed in the this procedure taking into account the fact that a bumble bee with a not very good solution, probably, would not give a solution better than the best queen in the next iterations. Thus, if a solution had a solution 5% worst than the solution of the best queen, for a consecutive number of iterations, it was deleted from the hive.

The final outcomes of the algorithm were the best solutions as well as the final best values of the parameters of the algorithm for the corresponding studied problem.

4 Bumble Bees Mating Optimization for the Discrete and Dynamic Berth Allocation Problem

The proposed Bumble Bees Mating Optimization algorithm is suitably modified inspired from genetic-based algorithms, which differs from the approach introduced by Marinakis et al. [30], to solve optimally the discrete and dynamic berth allocation problem. It is implemented a modified approach of the Greedy Randomized Adaptive Search Procedure (GRASP) to form the initial solutions. A multi-crossover operator is used for the creation of new solutions and a number of local search (LS) methods are applied for the improvement of these solutions. During this phase of the LS, a 1–0 relocate procedure is applied followed by a Large Scale Neighborhood Search (LSNS) technique and the heuristic 1–1 exchange. The algorithmic procedure concludes with the completion when a maximum number of iterations has been achieved.

BBMO algorithm starts with the creation of the initial solutions (s). Applying GRASP, the algorithm sorts in ascending order the vessels according to their arrival times and creates a temporary list, in which the candidate vessels for insertion are included. The size of the list depends on the amount of available berths, which is given. For instance, the list will contain up to 5 vessels when the total berths are equal to five. Each one of the candidate vessels can be selected with equal probability and the selected one is inserted in the suitable berth with the lowest horizon time. If the candidate vessel cannot be serviced from the chosen berth due to time limitations, the vessel will be inserted to the next available berth with the lower horizon time. A new solution is completed, when all vessels are serviced with respect to the time windows and the horizon time of each berth. The GRASP process continues until all the initial s are created.

To enter the phase of crossover, the most profitable solution from the s is found. The solution with the best quality (q) is used for the multi-parent crossover operator in order to create a new set of solutions (s'). The feasibility of new solutions (s') is checked by taking into account, that some vessels can be inserted in only specific berths and cannot be served twice. In order the new solution to be completed, all the vessels must be served. Thus, the algorithm inserts the remain vessels into the berths with the lower costs. Crossover phase ends when all vessels are properly inserted.

In order to expand even further the search in solution area, three local search methods are applied. The heuristics 1–0 relocate, 1–1 exchange and the technique Large Scale Neighborhood Search (LSNS) are implemented. In the LSNS procedure, two different berths are selected and a large part of these two solutions is exchanged. The length of these parts can be equal or not, depending on the constraints of the problem, while it ranges from 20 to 60% of the current berths. Before the exchange begins, the algorithm performs a permutation into the picked parts to achieve a further exploitation in the rest of the solution area. The exchange is feasible, if the total cost of the solution is reduced or if it remains unchanged.

5 Computational Results

In this section, the experimental results conducted by HBMO and BBMO algorithms in the benchmark instances provided by Lalla-Ruiz et al. [16] are presented. The given data are divided and illustrated into three tables according to their computational complexity. Through Tables 1, 2 and 3 the obtained results from the algorithmic executions are summarized. We compare the performance of both algorithms, the best known results from the literature (BKS), the best results of the executions, the solution gap from the BKS, the mean values and the execution times. The following Tables 1, 2 and 3 are organized in ten columns. The first two columns give the name ($Name$) for each one OF the instances followed by the best known solution from the literature denoted by BKS. Columns 3 up to 6 give the obtained results ($HBMO$), the efficiency gap (Dif(%)) in percentage, the average value (Avg) from five executions and the total time (T) from the experimental procedures of the Honey Bees Mating Optimization (HBMO) algorithm. Respectively, the results of the proposed BBMO algorithm are presented in the last four columns of the tables.

In Table 1, it can be observed, that in the set of instances with 30 and 40 vessels both algorithms perform quite satisfactorily, while they reach most of the best known solutions. HBMO obtains 18 out of 30 best known solutions, while BBMO reaches BKS in 26 out of 30. The average efficiency gap is calculated equal to 0,086% and 0,021% respectively. The largest gap of HBMO can be found in the instance f40x5-09 with 0,575%. Considering BBMO, its largest efficiency gap is observed in f40x5-02 with 0,247%. Additionally, both algorithms obtain most of the results in quite short time, while the average computational time is calculated 26,73 and 43,33 s for HBMO and BBMO, respectively.

Table 2 presents the computational results of the second set of instances with number of vessels between 40 and 55. It can be observed, that the proposed algorithm performs better in all cases than the HBMO algorithm. The effectiveness of HBMO is slightly decreased, thus, the gap between its obtained values and the best known solutions exceeds 1% in five instances. More importantly, HBMO reaches only one best known solution, while the proposed BBMO algorithm finds the best known solutions in 11 out of 40 instances. For average, the efficiency gaps are equal to 0,535% and 0,288% for HBMO and BBMO, respectively. Considering the computational results of the BBMO algorithm, the obtained values of the efficiency gap, in almost all instances, are below 0,5%.

In Table 3, the results obtained for the large size instances considering 60 vessels to be served are presented. The HBMO algorithm cannot reach any of the best known solutions. The average efficiency gap is calculated to be equal to 0,464% and in one only case in the instance f60x7-07 is larger than 1%. The performance of the proposed BBMO algorithm is satisfactorily compared to the performance of the HBMO, as the computational results are in the most cases near to the best known solutions while in some cases (instances f60x5-3, 5, 7, 8) it reaches the best known solutions. The average efficiency gap from the best known solutions of the BBMO algorithm is calculated to be equal to 0,260%.

Table 1 Set with Small size instances

Name	BKS	HBMO	Dif (%)	Avg	T (s)	BBMO	Dif′ (%)	Avg′	T′ (s)
f30x3-01	1763	1763	0,000	1763	9,27	1763	0,000	1763	12,46
f30x3-02	2090	2090	0,000	2090	12,44	2090	0,000	2090	29,12
f30x3-03	2186	2186	0,000	2186	16,03	2186	0,000	2186	38,48
f30x3-04	1538	1538	0,000	1538	22,47	1538	0,000	1538	59,06
f30x3-05	2114	2114	0,000	2114	15,58	2114	0,000	2114	40,12
f30x3-06	2185	2185	0,000	2185	17,12	2185	0,000	2185	57,35
f30x3-07	1845	1847	0,108	1849,6	20,23	1845	0,000	1845	22,29
f30x3-08	1271	1271	0,000	1271	16,33	1271	0,000	1271	25,07
f30x3-09	1595	1595	0,000	1595	22,52	1595	0,000	1595	45,58
f30x3-10	2195	2195	0,000	2195	18,48	2195	0,000	2195	67,14
f30x5-01	1149	1150	0,087	1151,4	28,31	1150	0,087	1152,8	39,09
f30x5-02	1475	1475	0,000	1475	14,36	1475	0,000	1475	53,07
f30x5-03	1542	1544	0,130	1545,7	19,44	1475	0,000	1542	44,51
f30x5-04	1075	1075	0,000	1075	11,09	1075	0,000	1075	51,36
f30x5-05	1463	1463	0,000	1463	14,27	1464	0,068	1468,4	78,58
f30x5-06	1580	1580	0,000	1580	22,37	1580	0,000	1580	50,39
f30x5-07	1276	1276	0,000	1276	18;45	1276	0,000	1276	18,03
f30x5-08	870	870	0,000	870	12,55	870	0,000	870	35,48
f30x5-09	1134	1134	0,000	1134	30,39	1134	0,000	1135,2	70,04
f30x5-10	1527	1528	0,065	1531,4	39,22	1527	0,000	1529	42,01
f40x5-01	2301	2301	0,000	2301	34,11	2301	0,000	2301	77,39
f40x5-02	2829	2838	0,318	2841,8	51,38	2836	0,247	2838,8	12,58
f40x5-03	2880	2881	0,035	2883,6	67,47	2880	0,000	2880	43,46
f40x5-04	2001	2004	0,150	2007,2	29,05	2001	0,000	2001	54,29
f40x5-05	2815	2825	0,355	2828,7	48,38	2815	0,000	2824,4	28,37
f40x5-06	2934	2943	0,307	2948,1	40,38	2934	0,000	2936	53,32
f40x5-07	2632	2640	0,304	2644,6	48,58	2632	0,000	2632,4	35,14
f40x5-08	1835	1836	0,054	1837,4	34,06	1835	0,000	1835	39,29
f40x5-09	2086	2098	0,575	2101,2	38,25	2091	0,240	2099,2	41,58
f40x5-10	2962	2962	0,000	2962	29,17	2962	0,000	2974	35.17

Through Tables 1, 2 3 the obtained computational results from the two algorithms, the HBMO algorithm and the proposed BBMO algorithm are presented. Both methodologies achieve quality results despite the high complexity of the discrete and dynamic Berth Allocation Problem. The impact of the HBMO and BBMO algorithms is shown in the obtained results, while the average efficiency gap is calculated 0,369% and 0,193% for HBMO and BNMO, respectively for the two algorithms. The average computational times are calculated 46,02 and 49,96 s for HBMO and BNMO, respectively.

Table 2 2nd set of instances

Name	BKS	HBMO	Dif (%)	Avg	T (s)	BBMO	Dif′ (%)	Avg′	T′ (s)
f40x7-01	1458	1471	0,892	1476,9	50,28	1465	0,480	1470,6	53,47
f40x7-02	1375	1378	0,218	1380,7	43,04	1380	0,364	1381,8	35,29
f40x7-03	2119	2133	0,661	2142,2	57,21	2133	0,661	2138,2	53,1
f40x7-04	1591	1602	0,691	1606	68,13	1596	0,314	1597	88,33
f40x7-05	1847	1863	0,866	1869,6	33,55	1860	0,704	1865,2	30,37
f40x7-06	2080	2088	0,385	29,33	2091,8	2080	0,000	2083,8	67,57
f40x7-07	1841	1841	0,000	1844,6	25,48	1841	0,000	1841,4	48,31
f40x7-08	2025	2026	0,049	2032	31,41	2027	0,099	2030,6	64,38
f40x7-09	1880	1890	0,532	1892,1	66,44	1880	0,000	1880	49,32
f40x7-10	1883	1905	1,168	1911,5	49,22	1890	0,372	1897,8	62,47
f55x5-01	4689	4705	0,341	4709,4	46,57	4689	0,000	4693	77,16
f55x5-02	5467	5476	0,165	5478,2	38,18	5479	0,219	5482,8	18,48
f55x5-03	5499	5501	0,036	5511,3	27,45	5499	0,000	5501	46,07
f55x5-04	4165	4192	0,648	4201	53,22	4165	0,000	4170,6	68,02
f55x5-05	5478	5496	0,329	5499,4	41,10	5478	0,000	5484,8	50,01
f55x5-06	5595	5603	0,143	5609,6	36,43	5595	0,000	5600,6	55,59
f55x5-07	4870	4886	0,329	4898,2	44,06	4884	0,287	4888,4	65,39
f55x5-08	3552	3567	0,422	3571,7	67,45	3556	0,113	3560	58,47
f55x5-09	4273	4289	0,374	4301	59,52	4279	0,140	4287,1	43,10
f55x5-10	5739	5753	0,244	5760,8	66,55	5739	0,000	5747,4	59,33
f55x7-01	2846	2871	0,878	2875,2	43,17	2861	0,527	2871,4	52,23
f55x7-02	2883	2899	0,555	2904,4	71.49	2893	0,347	2897,4	82,17
f55x7-03	3825	3855	0,784	3861,3	58,13	3842	0,444	3857,2	40,25
f55x7-04	2951	2978	0,915	2990,6	83,11	2973	0,136	2985,4	78,42
f55x7-05	3797	3825	1,110	3832,1	39,23	3807	0,263	3811,2	49,56
f55x7-06	3783	3795	0,317	3799,8	36,29	3783	0,000	3790,8	33,09
f55x7-07	3774	3783	0,238	3794	46,38	3779	0,132	3784,4	41,56
f55x7-08	3862	3876	0,363	3878,2	48,14	3876	0,363	3876,8	87,01
f55x7-09	3591	3609	0,501	3611,6	53,41	3601	0,278	3607,8	71,32
f55x7-10	3623	3665	1,159	3671,3	85,33	3642	0,524	3653,6	52,22
f55x10-01	2742	2754	0,438	2759,5	34,19	2751	0,328	2755,8	52,22
f55x10-02	2527	2546	0,752	2553,7	56,33	2543	0,633	2547,2	33,57
f55x10-03	2544	2556	0,472	2560,1	72,04	2552	0,314	2564,6	64,27
f55x10-04	3315	3325	0,302	3329	37,41	3324	0,271	3327,2	26,32
f55x10-05	3109	3142	1,061	3146,4	49,44	3131	0,708	3141	64,38
f55x10-06	2283	2289	0,263	2294,6	60,43	2283	0,000	2287,4	64,38
f55x10-07	2144	2146	0,093	2159,2	29,39	2152	0,373	2157,2	57,54
f55x10-08	2720	2749	1,066	2756,3	64,53	2739	0,699	2742,8	87,34
f55x10-09	2149	2170	0,977	2179,8	54,46	2157	0,372	2171,4	39,13
f55x10-10	2814	2833	0,675	2841,4	62,35	2826	0,426	2837,2	87,34

Table 3 3rd set of instances

Name	BKS	HBMO	Dif (%)	Avg	T(s)	BBMO	Dif′ (%)	Avg′	T′ (s)
f60x5-01	5753	5763	0,174	5768,9	34,56	5759	0,104	5760,6	22,4
f60x5-02	6884	6897	0,189	6902,2	57,16	6888	0,058	6899	35,41
f60x5-03	6780	6791	0,162	6798	68,39	6780	0,000	6786,8	37,04
f60x5-04	5092	5118	0,511	5125,1	77,41	5100	0,157	5111,2	46,11
f60x5-05	6715	6731	0,238	6744,3	55,29	6715	0,000	6727,4	24,57
f60x5-06	6616	6624	0,121	6631,9	42,31	6626	0,151	6634,4	19,03
f60x5-07	6011	6017	0,100	6021,2	49,25	6011	0,000	6024,6	71,43
f60x5-08	4385	4397	0,274	4401,7	67,57	4385	0,000	4389,6	55,27
f60x5-09	5235	5245	0,191	5248,6	58,15	5245	0,191	5249	59,48
f60x5-10	7255	7291	0,496	7301	57,07	7260	0,069	7278,2	43,54
f60x7-01	3707	3732	0,674	3738,4	80,48	3731	0,647	3734	55,27
f60x7-02	4146	4185	0,941	4188,7	100,50	4164	0,434	4181,2	95,00
f60x7-03	4273	4302	0,679	4307	58,34	4300	0,632	4304,8	40,40
f60x7-04	3910	3934	0,614	3939,2	71,33	3924	0,358	3926	44,32
f60x7-05	4251	4290	0,917	4303,3	54,38	4269	0,423	4282,2	65,37
f60x7-06	5727	5739	0,210	5746,4	47,46	5743	0,279	5756,6	85,03
f60x7-07	3719	3759	1,076	3772,7	88,24	3745	0,699	3763,6	62,18
f60x7-08	4582	4606	0,524	4609,8	96,45	4596	0,306	4600,8	54,04
f60x7-09	3979	4006	0,679	4012,1	73,29	3989	0,251	3995,8	28,58
f60x7-10	4107	4128	0,511	4133,6	82,38	4125	0,438	4127,8	43,38

5.1 *Overall Results*

To evaluate the impact of Bumble Bees Mating Optimization, the obtained results of the proposed method are compared with algorithms from the literature. We compared the computational results of our algorithm with the computational results of ten algorithms from the literature. In Table 4 all the algorithms used for the comparisons are presented. The second column indicates the author/authors that have proposed the algorithm. Third column illustrates the complete name of each of the algorithms. Fourth column contains the abbreviation names of the algorithms. In the last column the year of publication is depicted.

In Tables 5 and 6, all the obtained results from the eleven optimization algorithms are presented. As we can see from the two Tables the proposed algorithm is a competitive algorithm based on its computational results. The proposed algorithm performs better than the versions of Tabu Search algorithm and the Honey Bees Mating Optimization algorithm. In the comparisons with the other algorithms, the proposed algorithm gives in the majority of instances the same results. However, in some instances the proposed algorithm gives slightly inferior results. Thus, we can say that the proposed algorithm can be used effectively for the solution of the studied problem.

Table 4 Algorithms used in the comparisons

#	Authors	Algorithm	Abbreviation	Year
1	Lalla-Ruiz et al. [16]	Tabu Search	T^2S	2012
2	Lalla-Ruiz et al. [16]	Tabu Search	T^2S^*	2012
3	Lalla-Ruiz et al. [16]	Tabu Search with Path Relinking	$T^2S^* + PR$	2012
4	de Oliveira et al. [5]	Clustering Search	Cs	2012
5	Mauri et al. [33]	Adaptive Large Neighborhood Search	$ALNS$	2016
6	Lin et al. [19]	Simulated Annealing with restarting strategy	SA_{RS}	2014
7	Lin et al. [20]	Iterated Greedy Heuristic	IG	2014
8	Lalla-Ruiz et al. [17]	Partial Optimization Metaheuristic Under Special Intensification Conditions	$POPMUSIC$	2016
9	Nishi et al. [37]	Iterated Enhanced Dynasearch	IED	2016
10	Tsakirakis et al. [40]	Honey Bees Mating Optimization	$HBMO$	2020
11	Tsakirakis et al.	Bumble Bees Mating Optimization	$BBMO$	2022

6 Conclusions

The Bumble Bees Mating Optimization (BBMO) algorithm, initially, was applied for solving global optimization problems, and, afterwards, was used for solving a large number of combinatorial optimization problems, mainly, routing problems. Until now, there was not any application of the algorithm for solving maritime logistic problems. Thus, in this paper, we propose a new version of the BBMO algorithm for the solution of the Discrete and Dynamic Berth Allocation Problem, one of the classic maritime logistics problems. Instead of making a full comparison with all other algorithms that have been applied for the solution of this problem, we decided to perform a thorough comparison with the Honey Bees Mating Optimization algorithm, the only other well known nature inspired mating algorithm. These two algorithms share a number of common characteristics as both algorithms simulate the mating procedure of a species of bees, the Honey Bees and the Bumble Bees. The BBMO proved that it is a slightly more efficient algorithm than the HBMO algorithm in the studied problem. Our future research will focus in the application of this modified

Table 5 Comparison results in Set with small size instances

#	BKS	T^2S	T^2S^*	$T^2S^* + PR$	Cs	$ALNS$	SA_{RS}	IG	$POPMUSIC$	IED	$HBMO$	$BBMO$
f30x3-01	1763	1763	1763	1763	1763	1763	1763	1763	–	–	1763	1763
f30x3-02	2090	2090	2090	2090	2090	2090	2090	2090	–	–	2090	2090
f30x3-03	2186	2188	2186	2186	2186	2186	2186	2186	–	–	2186	2186
f30x3-04	1538	1544	1538	1538	1538	1538	1538	1538	–	–	1538	1538
f30x3-05	2114	2114	2114	2114	2114	2114	2114	2114	–	–	2114	2114
f30x3-06	2185	2187	2187	2185	2185	2185	2185	2185	–	–	2185	2185
f30x3-07	1845	1849	1847	1845	1845	1845	1845	1845	–	–	1847	1847
f30x3-08	1271	1278	1271	1271	1271	1271	1271	1271	–	–	1271	1271
f30x3-09	1595	1595	1595	1595	1595	1595	1595	1595	–	–	1595	1595
f30x3-10	2195	2197	2195	2195	2195	2195	2195	2195	–	–	2195	2195
f30x5-01	1149	1153	1149	1149	1149	1149	1149	1149	–	–	1150	1150
f30x5-02	1475	1480	1476	1475	1475	1475	1475	1475	–	–	1475	1475
f30x5-03	1542	1547	1542	1542	1542	1542	1542	1542	–	–	1544	1542
f30x5-04	1075	1077	1075	1075	1075	1075	1075	1075	–	–	1075	1075
f30x5-05	1463	1475	1463	1463	1463	1463	1463	1463	–	–	1463	1464
f30x5-06	1580	1587	1581	1580	1580	1580	1580	1580	–	–	1580	1580
f30x5-07	1276	1279	1276	1276	1276	1276	1276	1276	–	–	1276	1276
f30x5-08	870	877	870	870	870	870	870	870	–	–	870	870
f30x5-09	1134	1156	1153	1134	1134	1134	1134	1134	–	–	1134	1134
f30x5-10	1527	1536	1527	1527	1527	1527	1527	1527	–	–	1528	1527
f40x5-01	2301	2317	2307	2303	2301	2301	2301	2301	2303	–	2301	2301
f40x5-02	2829	2839	2835	2834	2829	2829	2829	2829	2829	–	2838	2836
f40x5-03	2880	2886	2880	2880	2880	2880	2880	2880	2880	–	2881	2880
f40x5-04	2001	2033	2001	2001	2001	2001	2001	2001	–	–	2004	2001
f40x5-05	2815	2834	2815	2815	2815	2815	2815	2815	–	–	2825	2815

(continued)

Table 5 (continued)

#	BKS	T^2S	T^2S^*	$T^2S^* + PR$	Cs	$ALNS$	SA_{RS}	IG	$POPMUSIC$	IED	$HBMO$	$BBMO$
f40x5-06	2934	2945	2934	2934	2934	2934	2934	2934	–	–	2943	2934
f40x5-07	2632	2656	2632	2632	2632	2632	2632	2632	–	–	2640	2632
f40x5-08	1835	1852	1836	1835	1835	1835	1835	1835	–	–	1836	1835
f40x5-09	2086	2119	2095	2089	2086	2086	2086	2086	–	–	2098	2091
f40x5-10	2962	2976	2964	2962	2962	2962	2962	2962	–	–	2962	2962
f40x7-01	1458	1489	1467	1460	1460	1458	1458	1458	–	–	1471	1465
f40x7-02	1375	1423	1381	1375	1399	1375	1375	1375	–	–	1378	1380
f40x7-03	2119	2149	2119	2119	2119	2119	2119	2119	2119	–	2133	2133
f40x7-04	1591	1618	1600	1597	1595	1591	1591	1591	–	–	1602	1596
f40x7-05	1847	1885	1849	1847	1869	1847	1847	1847	–	–	1863	1860
f40x7-06	2080	2104	2080	2080	2080	2080	2080	2080	–	–	2088	2080
f40x7-07	1841	1863	1845	1841	1841	1841	1841	1841	–	–	1841	1841
f40x7-08	2025	2040	2026	2026	2044	2025	2025	2025	–	–	2026	2027
f40x7-09	1880	1901	1888	1880	1880	1880	1880	1880	–	–	1890	1880
f40x7-10	1883	1922	1905	1892	1888	1883	1884	1883	–	–	1905	1890
f55x5-01	4689	4701	4693	4689	4690	4689	4689	4689	4689	4689	4705	4689
f55x5-02	5467	5496	5483	5467	5467	5467	5467	5467	5467	5467	5476	5479
f55x5-03	5499	5523	5499	5499	5517	5499	5499	5499	5499	5499	5501	5499
f55x5-04	4165	4249	4189	4179	4200	4165	4165	4165	4165	4165	4192	4165
f55x5-05	5478	5590	5484	5478	5478	5478	5478	5478	5478	5478	5496	5478

(continued)

Table 5 (continued)

#	BKS	T^2S	T^2S^*	T^2S^*+PR	Cs	$ALNS$	SA_{RS}	IG	$POPMUSIC$	IED	$HBMO$	$BBMO$
f55x5-06	5595	5609	5599	5595	5595	5595	5595	5595	5595	5595	5603	5595
f55x5-07	4870	4914	4902	4882	4878	4870	4870	4870	4870	4870	4886	4884
f55x5-08	3552	3585	3565	3552	3561	3552	3552	3552	3552	3552	3567	3556
f55x5-09	4273	4301	4277	4275	4285	4273	4273	4273	4273	4273	4289	4279
f55x5-10	5739	5831	5739	5739	5739	5739	5739	5739	5739	5739	5753	5739
f55x7-01	2846	2871	2846	2846	2855	2846	2846	2846	–	–	2871	2861
f55x7-02	2883	2941	2887	2883	2913	2883	2883	2883	–	–	2899	2893
f55x7-03	3825	3853	3840	3833	3830	3825	3831	3825	3825	–	3855	3842
f55x7-04	2951	3022	2977	2971	2963	2955	2953	2951	–	–	2978	2973
f55x7-05	3797	3845	3803	3801	3801	3799	3797	3797	3979	–	3825	3807
f55x7-06	3783	3833	3783	3783	3783	3783	3783	3783	3783	–	3795	3783
f55x7-07	3774	3844	3774	3774	3774	3774	3774	3774	–	–	3783	3779
f55x7-08	3862	3893	3864	3863	2865	3862	3862	3862	–	–	3876	3876
f55x7-09	3591	3627	3597	3591	3591	3591	3591	3591	–	–	3609	3601
f55x7-10	3623	3699	3658	3635	3635	3623	3630	3623	–	–	3665	3642
f55x10-01	2742	2777	2745	2745	2756	2745	2744	2742	–	–	2754	2751
f55x10-02	2527	2577	2549	2534	2534	2527	2527	2527	–	–	2546	2543
f55x10-03	2544	2570	2545	2545	2546	2544	2544	2544	–	–	2556	2552
f55x10-04	3315	3351	3315	3315	3315	3315	3315	3315	–	–	3325	3324
f55x10-05	3109	3157	3147	3123	3121	3109	3111	3109	–	–	3142	3131
f55x10-06	2283	2293	2283	2283	2283	2283	2283	2283	–	–	2289	2283
f55x10-07	2144	2172	2146	2146	2146	2144	2144	2144	–	–	2146	2152
f55x10-08	2720	2783	2743	2726	2730	2720	2720	2720	–	–	2749	2739
f55x10-09	2149	2212	2162	2162	2153	2149	2152	2149	–	–	2170	2157
f55x10-10	2814	2894	2815	2815	2814	2814	2814	2814	–	–	2833	2826

Table 6 Comparison Results in Set with Large size instances

#	BKS	T^2S	T^2S^*	$T^2S^* + PR$	Cs	$ALNS$	SA_{RS}	IG	$POPMUSIC$	IED	$HBMO$	$BBMO$
f60x5-01	5753	5763	5761	5753	5755	5753	5753	5753	5753	5753	5763	5759
f60x5-02	6884	6932	6884	6884	6892	6884	6884	6884	6884	6884	6897	6888
f60x5-03	6780	6795	6782	6780	6780	6780	6780	6780	6780	6780	6791	6780
f60x5-04	5092	5172	5105	5105	5105	5092	5092	5092	5092	5092	5118	5100
f60x5-05	6715	6747	6715	6715	6715	6715	6715	6715	6715	6715	6731	6715
f60x5-06	6616	6637	6618	6616	6616	6616	6616	6616	6616	6616	6624	6626
f60x5-07	6011	6073	6011	6011	6017	6011	6011	6011	6011	6011	6017	6011
f60x5-08	4385	4415	4406	4385	4385	4385	4385	4385	4385	4385	4397	4385
f60x5-09	5235	5263	5235	5235	5241	5235	5235	5235	5235	5235	5245	5245
f60x5-10	7255	7350	7281	7281	7279	7255	7255	7255	7255	7255	7291	7260
f60x7-01	3707	3735	3724	3715	3717	3707	3709	3707	3707	3707	3732	3731
f60x7-02	4146	4279	4191	4172	4166	4148	4150	4147	4146	4147	4185	4164
f60x7-03	4273	4291	4290	4281	4280	4275	4275	4273	4273	4274	4302	4300
f60x7-04	3910	3926	3916	3916	3917	3913	3911	3910	3910	3910	3934	3924
f60x7-05	4251	4294	4264	4261	4261	4252	4251	4251	4251	4251	4290	4269
f60x7-06	5727	5741	5731	5729	5736	5729	5727	5727	5727	5727	5739	5743
f60x7-07	3719	3825	3749	3743	3743	3733	3721	3719	3719	3719	3759	3745
f60x7-08	4582	4735	4600	4586	4586	4583	4582	4582	4582	4582	4606	4596
f60x7-09	3979	4049	4011	4004	4014	3979	3985	3979	3979	3979	4006	3989
f60x7-10	4107	4222	4125	4115	4121	4111	4107	4107	4107	4107	4128	4125
Total	–	4/90	34/90	55/90	48/90	79/90	78/90	89/90	22/22	13/15	20/90	38/90

version of the BBMO algorithm in other problems and domains in order to see its efficiency in different kind of problems.

Acknowledgements The research work was supported by the Hellenic Foundation for Research and Innovation (HFRI) under the HFRI PhD Fellowship grant (Fellowship Number: 1330).

References

1. Bierwirth, C., Meisel, F.: A survey of berth allocation and quay crane scheduling problems in container terminals. Eur. J. Oper. Res. **202**(3), 615–627 (2010)
2. Bierwirth, C., Meisel, F.: A follow-up survey of berth allocation and quay crane scheduling problems in container terminals. Eur. J. Oper. Res. **244**, 675–689 (2015)
3. Buhrkal, K., Zuglian, S., Ropke, S., Larsen, J., Lusby, R.: Models for the discrete berth allocation problem: a computational comparison. Transp. Res., Part E **47**, 461–473 (2011)
4. Cordeau, J.F., Laporte, G., Legato, P., Moccia, L.: Models and tabu search heuristics for the berth allocation problem. Transp. Sci. **39**, 526–538 (2005)
5. de Oliveira, R.M., Mauri, G.R., Lorena, L.A.N.: Clustering search for the berth allocation problem. Exp. Syst. Appl. **39**(5), 5499–5505 (2012)
6. Dorigo, M., Birattari, M., Stutzle, T.: Ant colony optimization. IEEE Comput. Intell. Mag. **1**(4), 28–39 (2006)
7. Faheem, M., Butt, R.A., Raza, B., Alquhayz, H., Ashraf, M.W., Raza, S., Ngadi, M.A.B.: FFRP: dynamic firefly mating optimization inspired energy efficient routing protocol for internet of underwater wireless sensor networks. IEEE Access **8**, 39587–39604 (2020)
8. Feo, T.A., Resende, M.G.: Greedy randomized adaptive search procedures. J. Glob. Optim. **6**(2), 109–133 (1995)
9. Glover, F., Laguna, M., Marti, R.: Scatter search and path relinking: advances and applications. In: Glover, F., Kochenberger, G.A. (eds.) Handbook of Metaheuristics, pp. 1–36. Kluwer Academic Publishers, Boston (2003)
10. Hansen, P., Mladenovic, N.: Variable neighborhood search: principles and applications. Eur. J. Oper. Res. **130**, 449–467 (2001)
11. Imai, A., Nagaiwa, K., Tat, C.W.: Efficient planning of berth allocation for container terminals in Asia. J. Adv. Transp. **31**(1), 75–94 (1997)
12. Imai, A., Nishimura, E., Papadimitriou, S.: The dynamic berth allocation problem for a container port. Transp. Res. Part B **35**, 401–417 (2001)
13. Karaboga, D., Basturk, B.: On the performance of Artificial Bee Colony (ABC) algorithm. Appl. Soft Comput. **8**(1), 687–697 (2008)
14. Kennedy, J., Eberhart, R.: Particle swarm optimization. In: Proceedings of ICNN'95-International Conference on Neural Networks, Vol. 4, pp. 1942–1948. IEEE (1995)
15. Kramer, A., Lalla-Ruiz, E., Iori, M., Voß, S.: Novel formulations and modeling enhancements for the dynamic berth allocation problem. Eur. J. Oper. Res. **278**(1), 170–185 (2019)
16. Lalla-Ruiz, E., Melin-Batista, B., Moreno-Vega, J.M.: Artificial intelligence hybrid heuristic based on tabu search for the dynamic berth allocation problem. Eng. Appl. Artif. Intell. **25**(6), 1132–1141 (2012)
17. Lalla-Ruiz, E., Voss, S.: POPMUSIC as a matheuristic for the berth allocation problem. Ann. Math. Artif. Intell. **76**(1), 173–189 (2016)
18. Lichtblau, D.: discrete optimization using mathematica. in world multi-conference on systemics, cybernetics and informatics (SCI 2002). Int. Inst. Inform. Syst. **16**, 169–174 (2002)
19. Lin, S.W., Ting, C.J.: Solving the dynamic berth allocation problem by simulated annealing. Eng. Optim. **46**(3), 308–327 (2014)

20. Lin, S.W., Ying, K.C., Wan, S.Y.: Minimizing the total service time of discrete dynamic berth allocation problem by an iterated greedy heuristic. Sci. World J. (2014)
21. Lourenco, H.R., Martin, O., Stützle, T.: Iterated local search. In: Handbook of Metaheuristics. Vol. 57 of Operations Research and Management Science, pp. 321–353. Kluwer Academic Publishers (2002)
22. Marinaki, M., Marinakis, Y.: A bumble bes mating optimization algorithm for the feature selection problem. Int. J. Mach. Learn. Cybern. **7**(4), 519–538 (2016)
23. Marinakis, Y., Marinaki, M.: Bumble bees mating optimization algorithm for the vehicle routing problem. In: Panigrahi, B.K., Shi, Y., Lim, M.H. (eds.) Handbook of Swarm Intelligence—Concepts, Principles and Applications, Series on Adaptation, Learning, and Optimization, vol. 8, pp. 347–369. Springer , Berlin (2011)
24. Marinakis, Y., Marinaki, M.: Combinatorial neighborhood topology particle swarm optimization algorithm for the vehicle routing problem. In: Middendorf, M., Blum, C. (Eds.) EvoCOP 2013, LNCS 7832, pp. 133–144 (2013)
25. Marinakis, Y., Marinaki, M.: a bumble bees mating optimization algorithm for the open vehicle routing problem. Swarm Evol. Comput. **15**, 80–94 (2014)
26. Marinakis, Y., Marinaki, M.: an adaptive bumble bees mating optimization algorithm for the hierarchical permutation flowshop scheduling problem. In: Dorigo, M., et al. (eds.) ANTS 2014, LNCS 8667, pp. 282–283 (2014)
27. Marinakis, Y., Marinaki, M.: Combinatorial neighborhood topology bumble bees mating optimization for the vehicle routing problem with stochastic demands. Soft Comput. **19**, 353–373 (2015)
28. Marinakis, Y., Marinaki, M., Dounias, G.: Honey bees mating optimization algorithm for the euclidean traveling salesman problem. Inf. Sci. **181**(20), 4684–4698 (2011)
29. Marinakis, Y., Marinaki, M., Matsatsinis, N.: A hybrid bumble bees mating optimization-GRASP algorithm for clustering. In: Corchado, E., Wu, X., Oja, E., Herrero, A., Baruque, B. (eds.) HAIS 2009, LNAI, 5572, pp. 549–556. Springer, Berlin (2009)
30. Marinakis, Y., Marinaki, M., Matsatsinis, N.: A bumble bees mating optimization algorithm for global unconstrained optimization problems. In: Gonzalez, J.R, Pelta, D.A., Cruz, C., Terrazas, G., Krasnogor, N. (eds.) Nature Inspired Cooperative Strategies for Optimization—NICSO 2010, Studies in Computational Intelligence, vol. 284, pp. 305–318, Springer, Berlin (2010)
31. Marinakis, Y., Marinaki, M., Migdalas, A.: An adaptive bumble bees mating optimization algorithm. Appl. Soft Comput. **55**, 13–30 (2017)
32. Marinakis, Y., Migdalas, A., Pardalos, P.M.: Expanding neighborhood GRASP for the traveling salesman problem. Comput. Optim. Appl. **32**(3), 231–257 (2005)
33. Mauri, G.R., Ribeiro, G.M., Lorena, L.A.N., Laporte, G.: An adaptive large neighborhood search for the discrete and continuous berth allocation problem. Comput. Oper. Res. **70**, 140–154 (2016)
34. Mirjalili, S., Lewis, A.: The whale optimization algorithm. Adv. Eng. Softw. **95**, 51–67 (2016)
35. Mirjalili, S., Mirjalili, S.M., Lewis, A.: Grey wolf optimizer. Adv. Eng. Softw. **69**, 46–61 (2014)
36. Monaco, M.F., Sammarra, M.: The berth allocation problem: a strong formulation solved by a lagrangean approach. Transp. Sci. **41**(2), 265–280 (2007)
37. Nishi, T., Okura, T., Lalla-Ruiz, E., Voß, S.: A dynamic programming-based matheuristic for the dynamic berth allocation problem. Ann. Oper. Res. **286**(1), 391–410 (2020)
38. Ritthipakdee, A., Thammano, A., Premasathian, N., Jitkongchuen, D.: Firefly mating algorithm for continuous optimization problems. Comput. Intell. Neurosci. (2017)
39. Ting, C.J., Wu, K.C., Chou, H.: Particle swarm optimization algorithm for the berth allocation problem. Expert. Syst. Appl. **41**(4), 1543–1550 (2014)
40. Tsakirakis, E., Marinaki, M., Matsatsinis, N., Marinakis, Y.:Honey bees mating optimization algorithm for the berth allocation problem. In: Proceedings of the XIV Balcan Conference on Operational Research, pp. 377–381 (2020) (Virtual Balcor - 30 September - 3 October 2020)
41. Yang, X.S.: Firefly algorithms for multimodal optimization. In: International Symposium on Stochastic Algorithms, pp. 169–178. Springer, Berlin (2009)

Applying the Population-Based Ant Colony Optimization to the Dynamic Vehicle Routing Problem

Michalis Mavrovouniotis, Georgios Ellinas, Iaê S. Bonilha, Felipe M. Müller, and Marios Polycarpou

Abstract The population-based ant colony optimization (P-ACO) algorithm is a variant of the ant colony optimization metaheuristic specifically designed to address dynamic optimization problems. Whenever a change in the environment occurs, P-ACO repairs the pheromone trails affected by the change using previous solutions maintained in a population-list. Typically, change-related information are utilized for repairing these solutions. The change-related information for this dynamic vehicle routing problem (DVRP) case are the nodes removed and inserted when a change in the environment occurs. In this chapter, the operators of the unstringing and stringing (US) heuristic are utilized for repairing the solutions. Experimental results demonstrate that P-ACO embedded with the US heuristic outperforms other peer methods in a series of DVRP test cases.

Keywords Ant colony optimization · Dynamic optimization · Vehicle routing

---

M. Mavrovouniotis (✉) · G. Ellinas · M. Polycarpou
KIOS Research and Innovation Center of Excellence, Department of Electrical and Computer Engineering, University of Cyprus, Nicosia, Cyprus
e-mail: mavrovouniotis.michalis@ucy.ac.cy

G. Ellinas
e-mail: gellinas@ucy.ac.cy

M. Polycarpou
e-mail: mpolycar@ucy.ac.cy

F. M. Müller
Department of Applied Computing, Federal University of Santa Maria, 97105-900 Santa Maria, Brazil
e-mail: felipe@inf.ufsm.br

I. S. Bonilha
PPGEP, Federal University of Santa Maria, 97105-900 Santa Maria, Brazil
e-mail: iaesb@hotmail.com

A. Biswas et al. (eds.), *Advances in Swarm Intelligence*, Studies in Computational Intelligence 1054, https://doi.org/10.1007/978-3-031-09835-2_20

1 Introduction

The vehicle routing problem (VRP) is the problem of finding the best possible routes for a fleet of vehicles to satisfy the customers' delivery demands. Ant colony optimization (ACO) algorithms have been shown to be successful in addressing stationary VRPs, including the capacitated VRP [1, 2], the VRP with time windows [3, 4], the VRP with backhuals [5], the electric VRP [6, 7], and many other VRP variations. However, in most real-world VRPs their environments change over time. In contrast to stationary VRPs, in which the objective of the optimizer is to discover the stationary optimum quickly, the aim of dynamic VRPs (DVRPs) is to quickly track the changing optimum.

The dynamic changes in DVRPs are typically small to medium, and thus, the environments before and after the dynamic change will have similarities [8]. For example, changes in the state of the traffic on some roads (e.g., due to congestion, accidents, etc.) or when new customer demands arrive, will usually cause small to medium changes to the current optimal solution. The intrinsic characteristics of ACO algorithms (i.e., adapting to changes) have shown promising performance in addressing dynamic VRPs, such as the VRP with varying orders [9] and the VRP with traffic factors [10, 11] (a comprehensive survey of DVRPs is available in [12]). In particular, ACO algorithms are capable of transferring knowledge from previously optimized environments using the information accumulated in the pheromone trails [13].

Guntsch and Middendorf [14, 15] proposed an ACO variation, namely the population-based ACO (P-ACO), which utilizes change-related information to repair solutions from the previous environment when a dynamic change occurs. This feature enables P-ACO to have an advantage against evaporation-based ACO variations when addressing dynamic environments. For the case of the dynamic version of the popular traveling salesman problem, the nodes affected by the dynamic change are removed and are inserted to the best possible positions to repair the previous solutions, with these operations having an immediate effect on the pheromone trails. On the contrary, in evaporation-based ACO variations, a constant amount of pheromone is continuously removed from pheromone trails, thus requiring more time for evaporation to express its effect, as demonstrated in [13]. For the DVRP case, the inserted nodes represent new customer orders and the removed nodes represent already-visited customers. In this work, the Unstringing and Stringing (US) [16, 17] insertion and removal moves are utilized to heuristically repair the solutions of P-ACO, rather than the aforementioned repair strategy.

The outline of this chapter is as follows. Section 2 presents the dynamic benchmark generator adopted to construct DVRP test cases for the experiments, while Sect. 3 describes P-ACO, which is an ACO variation especially designed for dynamic optimization problems [15]. Further, in Sect. 3, the insertion and removal operators of the US heuristic used to repair a solution when a dynamic change occurs are described. In Sect. 4, the experimental results are presented, with a comparative analysis regarding the dynamic behavior and performance of P-ACO algorithms. The conclusions are stated in Sect. 5.

2 Dynamic Vehicle Routing Problem

2.1 Problem Formulation

The VRP is an $\mathcal{NP}$-hard combinatorial optimization problem appearing in modern logistics systems [18]. The capacitated variant of the VRP consists of a vehicle fleet that has limited cargo load capacity and a customers set with specific delivery demands. The objective is to optimally route the vehicles comprising the fleet in order to satisfy the delivery demands of all customers. The origin and final destination of each vehicle's route is the central depot.

A VRP instance can be modeled by a weighted graph $G = (N, A)$, where $N = \{1, \ldots, n\} \cup \{0\}$ is a set of $n + 1$ nodes and $A = \{(i, j) \mid i, j \in N, i \neq j\}$ is a set of arcs fully connecting the graph nodes. A weight $w_{ij} \in \mathbb{R}^+$ is associated with each arc, that defines the cost for traveling from node i to node j. Node $\{0\}$ represents the central depot, while the rest of the nodes represent the customers. Each node (i.e., customer) $j \in N$ is associated with a delivery demand defined by δ_j. For all customers the delivery demand is greater than zero whereas for the central depot is zero, i.e., $\delta_0 = 0$. Each vehicle h of the fleet of size $|H|$ has a maximal cargo capacity C^h. The VRP instances used for the experiments in Sect. 4 consider a homogeneous fleet (i.e., C^h value is the same for all vehicles).

VRP is formulated as an integer linear program utilizing a binary decision variable x_{ij}^h defined as follows [19, 20]:

$$x_{ij}^h = \begin{cases} 1, & \text{if arc } (i, j) \text{ traversed by vehicle } h \text{ belongs to the solution;} \\ 0, & \text{otherwise.} \end{cases} \tag{1}$$

The objective function [Eq. (2)] minimizes the routes' total cost as follows:

$$\min \sum_{h \in H} \sum_{i,j \in N, i \neq j} w_{ij} x_{ij}^h, \tag{2}$$

s.t.

$$\sum_{i,j \in N, i \neq j} \delta_j x_{ij}^h \leq C^h, \forall h \in H, \tag{3}$$

$$\sum_{i \in N, i \neq j} x_{ij}^h - \sum_{i \in N, i \neq j} x_{ji}^h = 0, \forall j \in N, \forall h \in H, \tag{4}$$

$$\sum_{h\in H}\sum_{i\in N, i\neq j} x_{ij}^h = \begin{cases} 1, & \forall j \in N\backslash\{0\}, \\ |H|, & j = 0, \end{cases} \quad (5)$$

$$\sum_{i,j\in S, i\neq j} x_{ij}^h \leq |S| - 1, \forall S \subseteq \{1, \ldots, n\}, \forall h \in H, \quad (6)$$

$$x_{ij}^h \in \{0, 1\}, \ \forall h \in H, \ \forall i \in N, \ \forall j \in N, i \neq j, \quad (7)$$

where Eq. (3) guarantees that the vehicles' cargo capacities are not exceeded, Eq. (4) makes sure that a vehicle departs from the visited customer, Eq. (5) ensures that all customers are visited exactly once, and that the depot is left $|H|$ times, Eq. (6) eliminates subtours that do not originate at the central depot, and Eq. (7) is the binary decision variable defined in Eq. (1).

2.2 Generating Dynamic Test Cases

The dynamic benchmark generator proposed in [13] is adopted to convert a stationary VRP instance into a dynamic VRP. Initially, a spare set of random nodes is constructed in the range of the problem instance node set $N(0)$. At each environmental period $T = \lceil t/f \rceil$, the node set $N(T)$ of the problem instance is modified by replacing nodes with new nodes selected from the spare node set, where t is the algorithmic evaluation count and f is a constant parameter of the benchmark generator that defines the frequency of the dynamic changes. Another important parameter of the benchmark generator is the degree of change $m \in (0, 1]$, which is expressed by the number of nodes to be replaced. In every dynamic change exactly $\lceil mn \rceil$ nodes are randomly selected from the spare node set for replacing exactly $\lceil mn \rceil$ nodes that are randomly selected from node set $N(T)$ of the problem instance.

When node set $N(T)$ is changing to $N(T+1)$ the weights on the arcs reconnecting the newly generated node set will affect the quality of the output (e.g., the current best solution may not be the best when a dynamic change occurs). When a DVRP test case is generated with dynamic changes defined by a low m value the quality of the output will be slightly affected, whereas with a high m value the quality of the output will be severely affected.

The described dynamic changes can be found in real-world logistics systems. Consider the following scenario, where a logistics company has already generated the routes of its fleet of vehicles to deliver goods to its customers. However, while the vehicles are in-route, new orders may arrive, and hence, the currently optimal routes are now outdated. Therefore, in such a case, the optimal routes must be updated with these new orders (i.e., inserting new nodes) while ignoring customers already visited (i.e., removing current nodes).

Algorithm 1 Population-Based ACO in Dynamic Environments

```
I ← 1 and t ← 0
P(I) ← ∅
for all (i, j) ∈ A do
    τ_ij ← τ_0
end for
while termination condition is not satisfied do
    for each ant k ∈ {1, ..., ω} do
        π^k ← solution constructed by ant k
        φ(π^k, t) ← solution quality of π^k
        t ← t + 1
    end for
    π^ib ← find iteration-best solution
    if |P(I)| = K then
        π^old ← find the oldest solution in P(I)
        P(I) ← P(I)\{π^old}
    end if
    π^new ← π^ib
    P(I) ← P(I) ∪ {π^new}
    for all (i, j) ∈ A do
        τ_ij ← τ_0
    end for
    for each ant r ∈ {1, ..., |P(I)|} do
        if dynamic change occurs then
            π^r' ← repair solution π^r
            π^r ← π^r'
        end if
        for all (i, j) ∈ π^r do
            τ_ij ← τ_ij + (τ_max − τ_0)/K
        end for
    end for
    if φ(π^ib, t) is better than φ(π^bs, t) then
        π^bs ← π^ib
    end if
    I ← I + 1
end while
output: the best-so-far solution π^bs
```

3 Population-based Ant Colony Optimization

The basic principle of P-ACO is to maintain a list of solutions, called the population-list $P(I)$ that is updated at every iteration I [14]. The solutions stored in the population-list are utilized to update the pheromone trails of P-ACO. Initially, the population-list is empty and all pheromone trails' values associated with the solution components of the problem (i.e., nodes) are set with an equal amount $\tau_0 \leftarrow 1/(n-1)$. At each iteration, a colony of ω ants will construct solutions, update the population-list, and update the pheromone trails as described in the follows sub-sections. An algorithmic outline of P-ACO is given in Algorithm 1.

3.1 Constructing Solutions

At the beginning of the iteration, each ant k will consist of a partial solution including only the depot component. Then, the next component (either customer or depot) will be selected based on the existing pheromone trail and heuristic information values until all customers are selected. At the end of the iteration, each ant k will represent a feasible VRP solution (i.e., π^k). Ant k chooses the next solution component j, when being at solution component i, based on the following decision rule [21]:

$$j = \begin{cases} \arg\max_{l \in \mathcal{N}_i^k} [\tau_{il}]^\alpha [\eta_{il}]^\beta, & q \leq q_0; \\ J, & \text{otherwise}; \end{cases} \tag{8}$$

where τ_{il} is the existing pheromone trail value associated with arc (i, l) and $\eta_{il} = 1/w_{il}(T)$ is the heuristic information value associated with arc (i, l), $\mathcal{N}_i^k$ is the set of feasible solution components for ant k when being on solution component i, $q \in [0, 1]$ is a random number obtained from uniform distribution, q_0 $(0 \leq q_0 \leq 1)$ is a decision rule parameter, and J is a random variable selected from the probability distribution defined as follows:

$$p_{ij}^k = \begin{cases} \frac{[\tau_{ij}]^\alpha [\eta_{ij}]^\beta}{\sum_{l \in \mathcal{N}_i^k} [\tau_{il}]^\alpha [\eta_{il}]^\beta}, & \text{if } j \in \mathcal{N}_i^k; \\ 0, & \text{otherwise}. \end{cases} \tag{9}$$

In particular, $\mathcal{N}_i^k$ contains solution components (i.e., customers) that satisfy all the following criteria: (a) customers that are not yet visited, (b) customers that are within the nearest neighbor set of size nn, and (c) customers that do not violate the capacity constraint of vehicles as defined in Eq. (3). If $\mathcal{N}_i^k$ is empty, then the depot component will be added to the solution constructed by ant k. This corresponds to the closure of one vehicle route and the beginning of a new vehicle route. In case all customers are visited then the solution construction is completed.

3.2 Updating Population-List

At each iteration the solutions built by the ants are evaluated and the solution of the iteration-best ant is inserted into the population-list of size K. As a result, at iteration K the population-list will be full. Therefore, at iteration $K + 1$ the solution which was inserted in the population-list first (i.e., π^{old}), will be replaced by the solution of the iteration-best ant (i.e., π^{new}) found at iteration $K + 1$ (see lines 12–18 in Algorithm 1). At the following iterations of the P-ACO algorithm, the same update procedure occurs, that is, solution π^{old} will be replaced by solution π^{new} in a first-in-first-out fashion [14].

3.3 Updating Pheromone Trails

The pheromone trails of P-ACO are directly associated with the population-list (see lines 19–30 in Algorithm 1). Whenever an update occurs in the population-list the corresponding update will be made in the pheromone trails as follows:

$$\tau_{ij} \leftarrow \tau_0 + \sum_{r=1}^{|P(I)|} \Delta\tau_{ij}^r, \quad \forall (i, j) \in A, \tag{10}$$

where $P(I)$ is the population-list in the I-th iteration and $\Delta\tau_{ij}^r$ is the amount of pheromone with which ant r will update the trails associated with the arcs it has visited. $\Delta\tau_{ij}^r$ is defined as follows:

$$\Delta\tau_{ij}^r = \begin{cases} \frac{\tau_{max}-\tau_0}{K}, & \text{if arc}(i, j) \in \pi^r; \\ 0, & \text{otherwise;} \end{cases} \tag{11}$$

where K is the maximum number of solutions in the population-list, π^r is the r-th solution of the population-list and τ_{max} is a constant value. In other words, whenever an ant enters the population-list the pheromone trails associated with the arcs belonging to its solution will be increased by a constant value, whereas whenever an ant leaves the population-list the pheromone trails associated with the arcs belonging to its solution will be decreased by a constant value.

3.4 Responding to Dynamic Changes

A straightforward method to address dynamic changes for the DVRP with P-ACO is to reinitialize the pheromone trails to τ_0 and remove all the solutions from the population-list. This will correspond to a complete restart of P-ACO whenever an environmental change occurs. However, typically the dynamic changes are small to medium, and hence, the environment before the change will be similar to the environment after the change. Therefore, the solutions maintained in the population-list can be used to speed up the re-optimization process.

Guntsch and Middendorf [15] repaired the solutions utilizing change-related information. In particular, all customers removed from the problem instance are also deleted from the solution reconnecting the arcs of their successor and predecessor nodes. Then, the new customers added to the problem instances are individually added to the solution at the best possible position. Recall, that the pheromone trails are updated using the population-list. Therefore, the changes occurring by the repairing procedure of the solutions (see lines 23–26 in Algorithm 1), will have an effect on the existing pheromone trails of P-ACO.

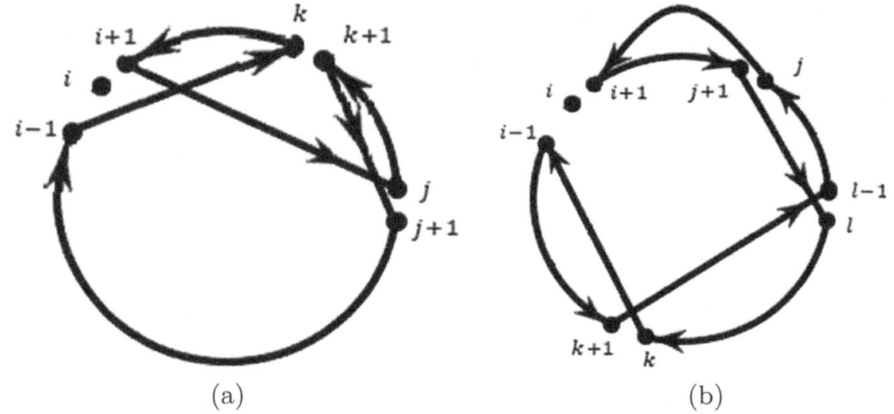

Fig. 1 Illustration of **a** Type I and **b** Type II removals on single VRP route

In this work, the removal (or unstringing) and insertion (or stringing) operators of the US heuristic [16] are utilized to repair the solutions of the population-list by better reconnecting the arcs of the removed and inserted nodes. Specifically, the best unstringing moves will be utilized to delete the nodes affected by the dynamic changes, and the best stringing moves will be utilized to replace these nodes with new ones. The key idea of the US heuristic is that when nodes are removed a local search is performed and when nodes are inserted a local search is also performed [16, 22]. Suppose that node i is the node to be removed and that nodes $i+1$ and $i-1$ are the successor and predecessor nodes of node i. The re-connection of the successor and predecessor nodes to repair the solution can be done with different types of removals. Also, the insertion of node x between two nodes i and j can be done with different types of insertions. Note that each route of a VRP solution is repaired individually as described in [17]. The different types of removals and insertions are described in the following.

3.4.1 Removal of Nodes

The unstringing phase consists of two types of removals [16, 23]: Type I shown in Fig. 1a and Type II shown in Fig. 1b. The removals are described below:

- Type I removal: Suppose that node i is the one to be removed, nodes j and k belong to the neighborhood of $i+1$ and $i-1$, respectively, and also, node k belongs to the subtour $(i+1, \ldots, j-1)$. The removal of node i will also remove arcs $(j, j+1)$, $(k, k+1)$, $(i, i+1)$, and $(i-1, i)$; insert arcs $(k+1, j+1)$, $(i+1, j)$, and $(i-1, k)$; and reverse subtours $(k+1, \ldots, j)$ and $(i+1, \ldots, k)$.
- Type II removal: Suppose that i is the node to be removed. Let nodes j, k, and l belong to the neighborhood of $i+1$, $i-1$, and $k+1$, respectively, and also, let nodes k and l belong to the subtours $(j+1, \ldots, i-2)$ and $(j, \ldots, k-1)$,

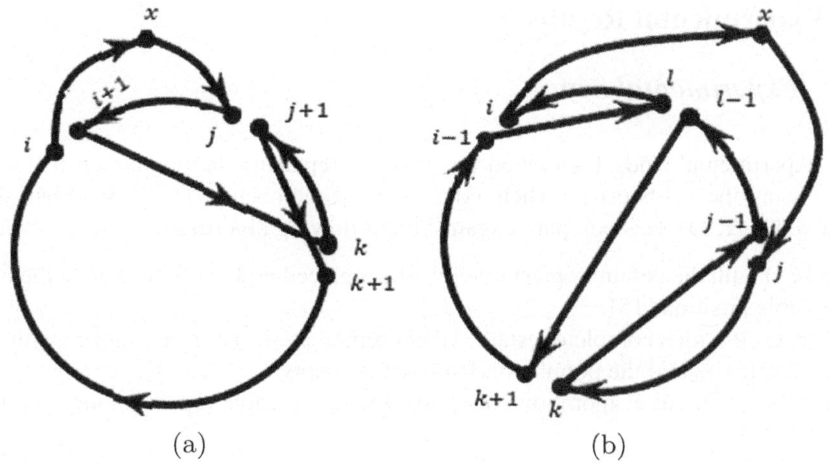

Fig. 2 Illustration of **a** Type I and **b** Type II insertions on single VRP route

respectively. The removal of node i will also remove arcs $(l, l+1)$, $(k, k+1)$, $(j-1, j)$, $(i, i+1)$, and $(i-1, i)$; insert arcs $(l, k+1)$, $(i+1, j)$, $(l+1, j-1)$, and $(i-1, k)$; and reverse subtours $(l+1, \ldots, k)$ and $(i+1, \ldots, j-1)$.

3.4.2 Insertion of Nodes

The stringing phase consists of two types of insertions [16, 23]: Type I shown in Fig. 2a and Type II shown in Fig. 2b. The insertions are described below:

- Type I insertion: Suppose that x is the node to be inserted, $k \neq i$ and $k \neq j$. The insertion of node x will also insert arcs $(j+1, k+1)$, $(i+1, k)$, (x, j), and (i, x); remove arcs $(k, k+1)$, $(j, j-1)$, and $(i, i+1)$; and reverse subtours $(j+1, \ldots, k)$ and $(i+1, \ldots, j)$.
- Type II insertion: Suppose that x is the node to be inserted, $k \neq j, k \neq j+1, l \neq i$, and $l \neq i+1$. The insertion of node x will also insert arcs $(i+1, k)$, $(k-1, l-1)$, $(l, j+1)$, (x, j), and (i, x); remove arcs $(k-1, k)$, $(j, j+1)$, $(l-1, l)$, and $(i, i+1)$; and reverse subtours $(l, \ldots, j)$ and $(i+1, \ldots, l-1)$.

4 Experimental Results

4.1 Experimental Setup

The experimental study focuses on the effect of repairing the population-list of P-ACO using the US heuristic when a dynamic change occurs. The P-ACO with US (denoted P-ACO+H) is compared against the following algorithms:

- P-ACO with the default repair strategy. The replaced nodes are inserted in the best possible position [15].
- P-ACO+R with a complete restart. All pheromone trails are re-initialized with the same value τ_0 and the population-list is set to empty.
- ACO+H [17]. An evaporation-based ACO with the same repair strategy used in P-ACO+H.

The algorithmic parameters for all P-ACO algorithms are set as follows: $\alpha = 1$, $\beta = 2$, $q_0 = 0.5$, $K = 3$, $\tau_{max} = 1$, $nn = 20$ and $\omega = 50$. The parameters of ACO+H are set as in [17]. Six stationary benchmark problem instances of different sizes are used as the base to generate DVRP test cases [24]. The following problem instances are obtained from CVRPLIB[1]: `X-n101-k25`, `X-n143-k7`, `X-n219-k73`, `X-n313-k71`, `X-n429-k61`, and `X-n561-k42`. The parameters of the dynamic benchmark generator are set as follows: $m \in \{0.1, 0.25, 0.5, 0.75\}$ and f equals to $10e4$ algorithmic evaluations. As a result, for each problem instance 4 DVRP test cases are systematically generated to analyze the performance of the ACO algorithms. For each ACO, 30 independent executions are performed on each DVRP test case. Each independent execution uses a different random seed number and consists of 50 environmental changes. The solution quality of the best-so-far ant after each dynamic change is taken as an observation. For each dynamic change the same number of evaluations (including the partial evaluations required to repair the solutions of the population-list) are allowed for all P-ACO algorithms to ensure fair comparison.

The *offline performance* [25] is considered as evaluation metric defined as follows:

$$\bar{P}_{\text{offline}} = \frac{1}{R}\sum_{t=1}^{R}\phi(\pi^{bs}, t), \tag{12}$$

where R is the number of algorithmic evaluations and $\phi(\pi^{bs}, t)$ is the quality of the solution of the best-so-far ant after a change.

[1] Available from http://vrp.galgos.inf.puc-rio.br/index.php/en/.

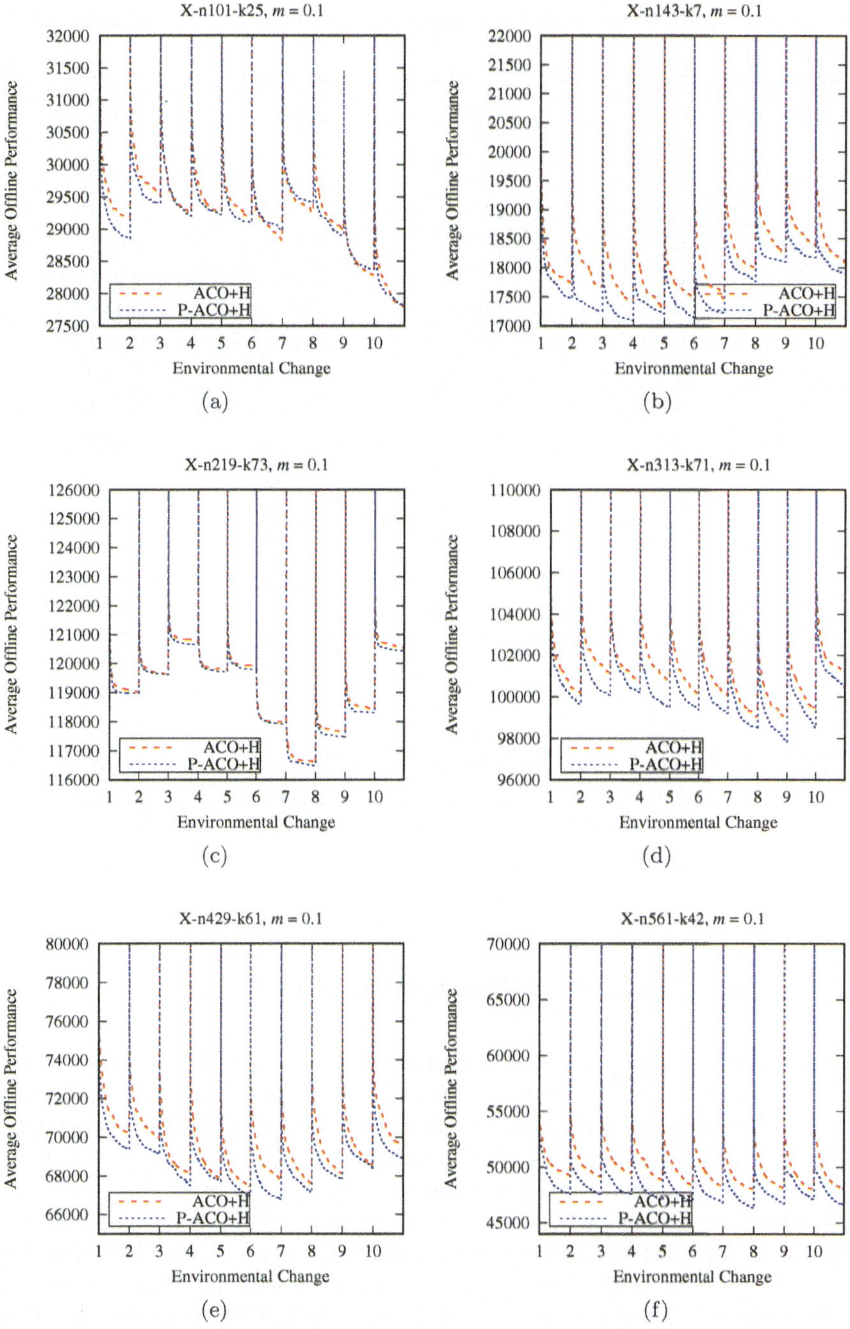

Fig. 3 Dynamic offline performance (mean values from 30 runs) of P-ACO+H and ACO+H on the **a** X-n101-k25, **b** X-n143-k7, **c** X-n219-k73, **d** X-n313-k71, **e** X-n429-k61, and **f** X-n561-k42 problem instances for the first ten dynamic changes with $m = 0.1$

Table 1 Offline performance results (mean values from 30 runs) of P-ACO algorithms on DVRP test cases

Problem Instance	m	P-ACO+H	P-ACO+R	P-ACO
X-n101-k25	0.1	29089	29148	29078
	0.25	29370	29398	29387
	0.5	29466	29498	29507
	0.75	29310	29348	29330
X-n143-k7	0.1	17896	18107	18032
	0.25	17874	18060	18069
	0.5	17915	18079	18118
	0.75	17989	18140	18190
X-n219-k73	0.1	121328	121326	121320
	0.25	121506	121480	121491
	0.5	121954	121924	121957
	0.75	121559	121509	121530
X-n313-k71	0.1	101104	101298	101183
	0.25	100613	100748	100723
	0.5	99761	99817	99886
	0.75	99462	99534	99554
X-n429-k61	0.1	69627	69938	69873
	0.25	69771	70051	70169
	0.5	70218	70491	70605
	0.75	70147	70469	70546
X-n561-k42	0.1	47931	48527	48502
	0.25	48299	48996	49116
	0.5	48450	49096	49303
	0.75	48537	49208	49362

4.2 Comparison of P-ACO+H Against ACO+H

Figure 3 presents the dynamic offline performance of P-ACO+H and ACO+H on the first environmental changes (with $m = 0.1$) for all problem instances. The dynamic changes of the selected DVRP test cases are small to ensure that the changing environments will have similarities, so as to better compare the behavior of the two algorithms when heuristic repair is applied.

From the comparisons in Fig. 3 it can be observed that P-ACO+H maintains a better output than the ACO+H in most environmental changes for all problem instances. Although the two algorithms have a common repair strategy (i.e., the US) that affects their pheromone trails when a change occurs, they have different pheromone update policies and solution construction decision rules. Therefore, their difference in the performance can be caused by several reasons.

First, the decision rule of P-ACO+H is more exploitative (i.e., the decision rule parameter in Eq. (8) is set to $q_0 = 0.5$) whereas the decision rule of ACO+H is more explorative (i.e., the decision rule parameter in Eq. (8) is set to $q_0 = 0.0$). Therefore, ACO+H may have slower recovery than P-ACO+H. Second, the recovery time is also affected by the pheromone update policy of ACO+H which uses evaporation to constantly remove a small amount of the pheromone trails. On the contrary, the pheromone trails of P-ACO+H are directly removed based on the limited lifetime defined by the size of the population-list (e.g., the pheromone trails associated with a solution will be removed after K iterations).

Nevertheless, we can conclude that P-ACO+H has better dynamic behavior than ACO+H because it can recover faster due to any of the reasons defined above or due to a combinations of these reasons. In addition, the repair strategy is beneficial for both algorithms (for ACO+H is demonstrated in [17] and for P-ACO+H in the following subsection).

4.3 Comparison of P-ACO+H Against Other P-ACO Algorithms

Table 1 presents the offline performance results (mean values from 30 runs) of P-ACO+H, P-ACO+R, and P-ACO for all DVRPs. Table 2 presents the statistical comparisons of the algorithms in which "$s+$" denotes that the first algorithm of the pair is significantly better than the second one, "$s-$" denotes that the second algorithm of the pair is significantly better than the first one, and "$\sim$" denotes that the difference between the algorithms of the pair is not statistically significant. The multiple comparisons are performed using *Kruskal-Wallis* statistical tests. The pairwise comparisons are performed using *Wilcoxon rank-sum* statistical tests. Since multiple comparisons take place, the p-values are corrected by Bonferroni correction. Figure 4 presents the dynamic offline performance of P-ACO+H, P-ACO+R, and P-ACO on the first ten environmental changes (with $m = 0.25$) on the `X-n429-k61` and `X-n516-k42` problem instances. From the comparisons of the algorithms the following remarks can be highlighted.

The first observation is that P-ACO+H performs significantly better than P-ACO for most of the DVRP test cases. These results were expected because the repair procedure performed by the US moves does not only reconnect the arcs associated with the affected nodes as is done in the default repair procedure performed by P-ACO. Instead, as it can be observed from Figs. 1 and 2, local search exchanges of type 3-*opt* and 4-*opt* are performed in the arcs of the neighbor nodes of the affected nodes, that further improve the quality of the repaired solutions.

The second observation is that P-ACO+H performs significantly better than P-ACO+R for most of the DVRP test cases. These results demonstrate that adapting to dynamic changes is more efficient than restarting the optimization process from

Table 2 Statistical comparisons of P-ACO algorithms on DVRP test cases

Problem Instance	m	P-ACO+H versus P-ACO+R	P-ACO+H versus P-ACO	P-ACO+R versus P-ACO
X-n101-k25	0.1	$s+$	$s-$	$s-$
	0.25	$s+$	$\sim$	$\sim$
	0.5	$s+$	$s+$	$\sim$
	0.75	$s+$	$\sim$	$\sim$
X-n143-k7	0.1	$s+$	$s+$	$s-$
	0.25	$s+$	$s+$	$\sim$
	0.5	$s+$	$s+$	$s+$
	0.75	$s+$	$s+$	$s+$
X-n219-k73	0.1	$\sim$	$\sim$	$\sim$
	0.25	$s-$	$\sim$	$s+$
	0.5	$s-$	$\sim$	$s+$
	0.75	$s-$	$s-$	$s+$
X-n313-k71	0.1	$s+$	$s+$	$s-$
	0.25	$s+$	$s+$	$\sim$
	0.5	$\sim$	$s+$	$s+$
	0.75	$s+$	$s+$	$\sim$
X-n429-k61	0.1	$s+$	$s+$	$s-$
	0.25	$s+$	$s+$	$s+$
	0.5	$s+$	$s+$	$s+$
	0.75	$s+$	$s+$	$s+$
X-n561-k42	0.1	$s+$	$s+$	$\sim$
	0.25	$s+$	$s+$	$s+$
	0.5	$s+$	$s+$	$s+$
	0.75	$s+$	$s+$	$s+$

the beginning. This is evident from Fig. 4, which shows that P-ACO+R has a slower recovery than P-ACO+H in most environmental changes.

The final observation is that P-ACO significantly outperforms P-ACO+R is most DVRP test cases with $m = 0.1$, while it is outperformed in the remaining DVRP test cases with $m > 0.1$. This observation confirms again that restarting the optimization process from the beginning makes sense only when the dynamic changes are severe. However, this observation is only valid when P-ACO+R is compared with P-ACO. As discussed previously, P-ACO+R is outperformed by P-ACO+H in DVRPs test cases for any value of m, since, although both P-ACO and P-ACO+H transfer knowledge from previous environments, the latter algorithm performs additional local search moves which enhance the adaptation process.

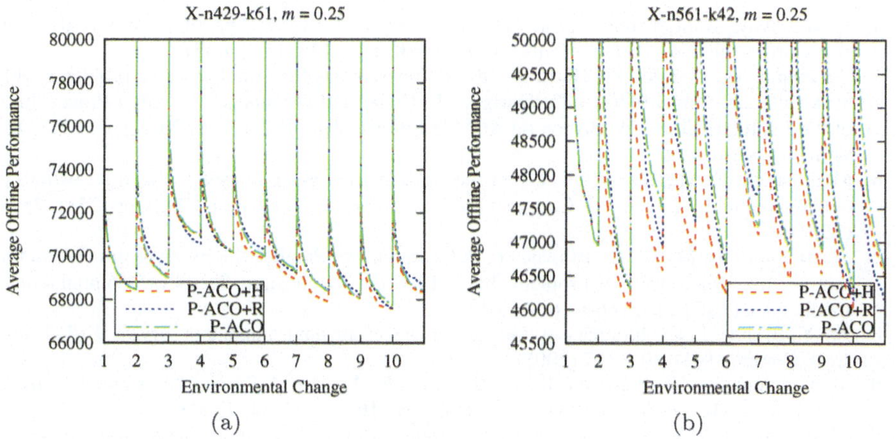

Fig. 4 Dynamic offline performance (mean values from 30 runs) of P-ACO algorithms on the **a** `X-n429-k61` **b** and `X-n561-k42` problem instances for the first ten dynamic changes

5 Conclusions

In this chapter, P-ACO is used to address the DVRP in which the set of customers is subject to changes. This particular ACO variation repairs previous solutions when a dynamic change occurs utilizing change-related information (e.g., for the particular DVRP variation this information are determined by the inserted and removed customers). To improve the performance of P-ACO the removal and insertion operators of the US heuristic [16] are utilized to repair the solutions stored in the population-list. The experimental results on several DVRP test cases demonstrate the improvement on the solution quality of P-ACO when heuristic repair is used.

Acknowledgements This work was supported by the European Union's Horizon 2020 research and innovation programme under grant agreement No 739551 (KIOS CoE—TEAMING) and from the Republic of Cyprus through the Deputy Ministry of Research, Innovation and Digital Policy.

References

1. Bullnheimer, B., Hartl, R., Strauss, C.: Applying the ant system to the vehicle routing problem. In: Voß, S., Martello, S., Osman, I., Roucairol, C., (eds.) Meta-Heuristics: Advances and Trends in Local Search Paradigms for Optimization, pp. 285–296. Kluwer Academic (1997)
2. Mavrovouniotis, M., Yang, S.: An ant system with direct communication for the capacitated vehicle routing problem. In: 2010 UK Workshop on Computational Intelligence (UKCI), pp. 14–19 (2011)
3. Gambardella, L.M., Taillard, E.D., Agazzi, C.: MACS-VRPTW: A multicolony ant colony system for vehicle routing problems with time windows. In: New Ideas in Optimization, pp. 63–76 (1999)

4. Rizzoli, A.E., Montemanni, R., Lucibello, E., Gambardella, L.M.: Ant colony optimization for real-world vehicle routing problems. Swarm Intell. **1**(2), 135–151 (2007)
5. Reinmann, M., Doerner, K., Hartl, R.: Insertion based ants for vehicle routing problems with backhauls and time windows. In: Dorigo, M., Caro, G.D., Sampels, M., (eds.) Proceedings of the 3rd International Workshop on Ant Algorithms. Volume 2463 of LNCS., pp. 135–148. Springer, Berlin (2002)
6. Mavrovouniotis, M., Ellinas, G., Polycarpou, M.: Electric vehicle charging scheduling using ant colony system. In: 2019 IEEE Congress on Evolutionary Computation (CEC), pp. 2581–2588 (2019)
7. Mavrovouniotis, M., Li, C., Ellinas, G., Polycarpou, M.: Parallel ant colony optimization for the electric vehicle routing problem. In: 2019 IEEE Symposium Series on Computational Intelligence (SSCI), pp. 1660–1667 (2019)
8. Jin, Y., Branke, J.: Evolutionary optimization in uncertain environments-a survey. IEEE Trans. Evol. Comput. **9**(3), 303–317 (2005)
9. Montemanni, R., Gambardella, L.M., Rizzoli, A.E., Donati, A.V.: Ant colony system for a dynamic vehicle routing problem. J. Comb. Optim. **10**(4), 327–343 (2005)
10. Mavrovouniotis, M., Yang, S.: Ant colony optimization with memory-based immigrants for the dynamic vehicle routing problem. In: 2012 IEEE Congress on Evolutionary Computation (CEC), pp. 2645–2652 (2012)
11. Mavrovouniotis, M., Yang, S.: Ant algorithms with immigrants schemes for the dynamic vehicle routing problem. Inf. Sci. **294**, 456–477 (2015)
12. Psaraftis, H.N., Wen, M., Kontovas, C.A.: Dynamic vehicle routing problems: three decades and counting. Networks **67**(1), 3–31 (2016)
13. Mavrovouniotis, M., Yang, S., Van, M., Li, C., Polycarpou, M.: Ant colony optimization algorithms for dynamic optimization: A case study of the dynamic travelling salesperson problem [research frontier]. IEEE Comput. Intell. Mag. **15**(1), 52–63 (2020)
14. Guntsch, M., Middendorf, M.: A population based approach for ACO. In: EvoWorkshops 2002: Applications of Evolutionary Computing. Volume 2279 of LNCS., pp. 72–81. Springer, Berlin (2002)
15. Guntsch, M., Middendorf, M.: Applying population based ACO to dynamic optimization problems. In: Dorigo, M., Di Caro, G., Sampels, M. (eds.) Ant Algorithms. Volume 2463 of LNCS., pp. 111–122. Springer, Berlin (2002)
16. Gendreau, M., Hertz, A., Laporte, G.: New insertion and postoptimization procedures for the traveling salesman problem. Oper. Res. **40**(6), 1086–1094 (1992)
17. Bonilha, I.S., Mavrovouniotis, M., Müller, F.M., Ellinas, G., Polycarpou, M.: Ant colony optimization with heuristic repair for the dynamic vehicle routing problem. In: 2020 IEEE Symposium Series on Computational Intelligence (SSCI), pp. 313–320 (2020)
18. Garey, M., Johnson, D.: Computer and intractability: A guide to the theory of $\mathcal{NP}$ -completeness. Freeman, San Francisco (1979)
19. Dantzig, G.B., Ramser, J.H.: The truck dispatching problem. Manag. Sci. **6**(1), 80–91 (1959)
20. Clarke, G., Wright, J.W.: Scheduling of vehicles from a central depot to a number of delivery points. Oper. Res. **12**(4), 568–581 (1964)
21. Dorigo, M., Gambardella, L.M.: Ant colony system: A cooperative learning approach to the traveling salesman problem. IEEE Trans. Evol. Comput. **1**(1), 53–66 (1997)
22. Pesant, G., Gendreaul, M., Rousseau, J.M.: GENIUS-CP: A generic single-vehicle routing algorithm. In: Smolka, G. (ed.) Principles and Practice of Constraint Programming-CP97, pp. 420–434. Springer, Berlin (1997)
23. Mavrovouniotis, M., Müller, F.M., Yang, S.: Ant colony optimization with local search for the dynamic travelling salesman problems. IEEE Trans. Cybern. **47**(7), 1743–1756 (2017)
24. Uchoa, E., Pecin, D., Pessoa, A., Poggi, M., Vidal, T., Subramanian, A.: New benchmark instances for the capacitated vehicle routing problem. Eur. J. Oper. Res. **257**(3), 845–858 (2017)
25. Branke, J., Schmeck, H.: Designing evolutionary algorithms for dynamic optimization problems. In: Ghosh, A., Tsutsui, S. (eds.) Advances in Evolutionary Computing, Natural Computing Series, pp. 239–262. Springer, Berlin (2003)

An Improved Cuckoo Search Algorithm for the Capacitated Green Vehicle Routing Problem

Kenan Karagul and Yusuf Sahin

Abstract Vehicle routing is an optimization problem that determines the shortest path to take while delivering products to consumers. Recently, increasing environmental awareness has led scientists and practitioners to find solutions to reduce the emission gases that cause environmental pollution. These environmental concerns are taken into account in the solution of the green vehicle routing problem, a variant of the vehicle routing problem. For this reason, it has become a frequently studied subject in the literature in recent years. The Modified Cuckoo Search Algorithm, a swarm-based optimization approach, was used in this study to provide a solution to the green vehicle routing problem. When the solutions obtained as a result of the experiments with two different solution models were examined, it was concluded that the method could not show the performance it showed in continuous optimization problems despite the changes made on it.

Keywords Modified cuckoo search algorithm · Green vehicle routing problem

1 Introduction

As a concept, logistics refers to the planned management of the flow that occurs during the transportation of raw materials, materials, or finished products needed by individual customers or companies from the first point of departure to the final consumption point. To carry out production and deliver the finished products to the customer, raw materials, semi-finished and finished products must be physically transported from one place to another. Supply and materials management refers to the flow of these flows towards the production process, while distribution refers

K. Karagul
Honaz Vocational High School, Logistics Department, Pamukkale University, Honaz, Turkey
e-mail: kkaragul@pau.edu.tr

Y. Sahin (✉)
Faculty of Economics and Administrative Sciences, Department of Business Administration, Burdur Mehmet Akif Ersoy University, Burdur, Turkey
e-mail: ysahin@mehmetakif.edu.tr

A. Biswas et al. (eds.), *Advances in Swarm Intelligence*, Studies in Computational Intelligence 1054, https://doi.org/10.1007/978-3-031-09835-2_21

to the movement of the finished product to the customer [1, 2]. Today, companies focus on processes in their main fields of activity and try to reduce the depth of production processes. At this point, efforts to reduce non-value-added activities such as transportation while increasing the quality and variety of products have gained importance.

Logistics management is the part of functional areas that are relevant to the company's performance. The success of each functional area provides increased quality in terms of customer service while reducing the costs of the company [3]. Logistics companies play a very important role in the realization of fast, economical, and perfect transportation systems. To ensure cost-effectiveness, activities related to the transportation and storage of these products must be strategically managed. Logistics activities are divided into several functional categories since they have a substantial influence on the performance of a company's production and distribution operations. Improved service quality and lower operational costs are guaranteed through performance in these functional areas. Effective logistics management also entails providing high-quality service at a low or acceptable cost [4, 5]. As a result of this, intense relations are established with logistics companies both during the bringing of the raw material to the production unit and the supply of the finished product to the market.

After the orders are prepared, shipment processes are carried out. Distribution plans are prepared to determine the route to be followed by vehicles with homogeneous or heterogeneous capacity from one or more warehouses. Within the scope of these plans, issues such as how many vehicles are needed and which vehicle will be transported to which customers are planned. This problem, which is linked to determining the most suitable routes that reduce the overall distance traveled by distribution trucks under certain constraints, is defined in the literature as the Vehicle Routing Problem (VRP) [6]. VRP is a very popular problem in the NP-Hard class and is extensively studied in the field of optimization. VRP, which is a combination of the traveling salesman and bin packing problems, can be defined as the problem of finding the least-cost routes for deliveries from a central warehouse to customers in geographically different locations [7].

VRP consists of key components such as warehouses, road networks, drivers, and customers, and is often defined by capacity or route constraints. VRP, in which only the capacity constraint is taken into account, is called the Capacitated VRP (CVRP). Distance and capacity constrained (DCVRP), time windows (VRPTW), pickup and delivery (VRPPD) vehicle routing are other types of VRP studied in the literature. Recently, with the addition of new and current constraints, types such as open (OVRP), multiple warehouses (MDVRP), split delivery (SDVRP), periodic (PVRP), heterogeneous fleet (HFVRP), green (GVRP), and fuzzy vehicle routing problem (FVRP) have emerged [8].

The research aims to evaluate the performance of the Cuckoo Search, a new generation algorithm inspired by nature, in solving GVRP. However, this study also attempts to answer the following research questions:

1 Is there any difference between the emission values calculated after the traditional VRP solution and the model solution proposed by Adiba et al. [9]?
2 Do the modifications made to the cuckoo algorithm affect the solution quality?

The remainder of this study is structured as follows: A literature review was carried out in Sect. 2. The details of the GVRP model and the solution approach are provided in the third section. Section 4 covers the application of the solution approaches used to benchmark problems, and the last section includes conclusions and recommendations.

2 Literature Review

Traditional VRP's major goal is to reduce the overall distance traveled by the vehicle, which will return to the starting point after only visiting each customer on the list once. In GVRP, it is aimed to determine the routes where the amount of CO_2 released into the atmosphere is minimized as well as the minimum distance. GVRP is an NP-hard class problem that is an extension of the standard VRP and is one of the current green supply chain applications. Since it is a popular optimization problem, a very large literature can be mentioned on this subject.

In recent years, many literature review studies on the subject have been conducted to determine the direction of research on GVRP and to shed light on future studies. Lin et al. [10] provided a literature review to assess the state of the art in GVRP research, explore how classical VRP variations may connect with GVRP, and provide insights for future GVRP research domains. The distribution of the studies by years and the methods used in the studies are included. Koç and Özceylan [11] conducted a literature review in which Electrical VRP studies were also examined together with GVRP. Asgari et al. [12] have created different analytical summary tables for each variant (internal combustion, alternative energy source, and hybrid vehicles) to highlight some of the main features driving the development of research. Belbağ [13] reviewed 57 research published between 2007 and 2016 to assess the elements influencing environmental considerations such as energy usage and emissions associated with GVRP, as well as the methodology used. Moghdani et al. [14] reviewed 309 studies published between 2006 and 2019 and introduced several new research fields based on objective function approaches for problem classification, uncertainties, solution methodologies, and finally further research directions. The study's findings suggest that research on GVRPs is still in its early stages, and there are still opportunities for major improvements in a variety of fields.

Since GVRP is difficult to solve with analytical methods, especially for medium and large-sized problems, there are few studies in the literature using analytical methods. Andelmin and Bartolini [15] presented a GVRP model as a cluster division problem in a multi-graph corresponding to simple circuits in which the appropriate routes are shown in the columns. They strengthened their proposed approach for the cluster division problem by adding valid inequalities, including the k-path cutting approach, and produced solutions for problem examples of up to 110 customers. Koç and Karaoglan [16] suggested a simulated annealing heuristic-based exact solution approach for G-VRP, which is formed by adding a limited refueling infrastructure and a limited driving range to the classical vehicle routing problem. In the proposed exact solution method, a few valid inequalities derived from the literature are combined to improve the lower bounds, and a branch and cut algorithm, which presents a heuristic algorithm based on simulated annealing, is used to obtain the upper bounds. Leggieri and Haouari [17] developed a solution approach that can solve medium-sized problems, using a mixed-integer linear formulation and reduction procedure for solving GVRP. He concluded that the proposed exact solution approach is more consistent and performs better than the branch and cut algorithm, creating an attractive and practical alternative to solve GVRPs in the best way. Bruglieri et al. [18] proposed a new mixed-integer programming formulation for GVRP in which visits to alternative fuel stations (AFSs) were only considered indirectly, and performed experiments for CPLEX benchmark samples. Afshar-Bakeshloo et al. [19] suggested a mixed-integer linear programming (MILP) model to improve management by balancing customer satisfaction, total cost, and emission levels. They demonstrated the applicability of the model through a numerical example. Zhou and Lee [20] defined the GVRP as a nonlinear integer programming problem in which gravity, three-dimensional customer coordinates, vehicle speed, and vehicle operating time were all considered to minimize emission values, and the Lagrangian relaxation approach was used as a solution method.

It is very difficult to determine the exact solution of vehicle routing and derivative problems in a mathematically reasonable time [21]. Since this problem is in the NP-hard class, which is frequently encountered in real life, heuristic and meta-heuristic approaches are widely utilized in the solution of VRP to provide high-quality solutions in a short time [22]. The block recombination algorithm [23], modified savings algorithm and density-based clustering algorithm [24], route-then-cluster-based method and the cluster decomposition algorithm [25], and two-stage heuristics and CCR algorithm [26] are some of the heuristics in solving GVRP.

In the solution of GVRP, the biology, physics, and swarm-inspired computational intelligence methods are used more frequently. Among the biology-based methods used, evolutionary methods stand out. Jemai et al. [27] used the NSGA-II evolutionary algorithm to solve the multi-purpose GVRP for the minimization of distance and CO_2 emission values. Sruthi et al. [28] produced solutions that provide a remarkable reduction in emission values compared to conventional models by using the genetic algorithm heuristic for GVRP, which takes into account the energy consumption and emission values simultaneously for the capacity vehicle routing problem. Peng et al. [29] proposed a method for solving GVRP consisting of the local search procedure, crossover operator, and population update strategy.

In physics-inspired solution methods, the simulated annealing method comes to the fore. Kucukoglu et al. [30] used the memory structure adapted simulated annealing meta-heuristic algorithm for the GVRPTW when formulated as a mixed-integer linear programming model. In their proposed model, they have integrated a fuel consumption and CO_2 emission calculation algorithm that takes into account vehicle technical characteristics, vehicle load, and transport distance. Normasari et al. [31] used the simulated annealing method to solve the capacitated GVRP, which they expressed as a mixed-integer programming model, and obtained good solutions in terms of solution quality and time. Karagul et al. [7] suggested a simulated annealing-based solution approach for homogeneous capacity GVRP. As a result of the statistical analysis, they showed that three different models produced statistically significant solutions. Sahin et al. [2], considered the distribution activities of a company operating as a regional distributor as a GVRP with a heterogeneous fleet and used the simulated annealing algorithm to obtain environmentally friendly solutions that provide lower emission values.

Methods such as ant colony optimizaton (ACO) and particle swarm optimization (PSO), which are very suitable for solving the vehicle routing problem due to their structure, are frequently preferred in the solution of GVRP. Li et al. [32] used the ant colony algorithm in the solution of GVRP with multiple warehouses to maximize revenue, minimize costs, time and emissions. Jabir et al. [33, 34] proposed a hybrid method for multi-store GVRP, consisting of the ant colony and variable neighborhood search methods. Small-sized examples of the problem, which they modeled as integer programming, were solved with the LINGO program. While solving small and medium-sized problems with the ACO algorithm, the variable neighborhood search method is integrated with the ant colony method to improve the solution quality of large-sized problems. Zhang et al. [35] used a two-stage ant colony system to minimize total carbon emissions in a transport network with alternative fuel vehicles and multiple warehouses. The distinctive feature of the proposed method is the use of two different ant species for two different purposes. Prakash and Pushkar [36] used a multiple ACO algorithm to solve GVRP. The algorithm used in this study to solve GVRP shows remarkable flexibility and efficiency. Furthermore, multiple ant colony optimization yielded superior solution quality than standard ant colony optimization.

Poonthalir and Nadarajan [37] modeled GVRP as a goal programming model, which includes variables such as variable speed restriction to minimize fuel consumption along with route distance. The PSO algorithm, which includes the greedy mutation operator and the time-varying acceleration coefficient, is used to solve the proposed model. Li et al. [38] considered GVRP in cold chain logistics and used standard PSO and modified PSO methods for the solution. Srijaroon et al. [39] proposed an enhanced MILP model for GVRP that accounts for mixed and simultaneous pickup and delivery, time window, and route types. They applied a self-adaptive learning PSO algorithm to increase the quality of large problem solutions (SAL-PSO). Cai et al. [40] used vehicle speed as a decision variable in the GVRP, taking into account the variation in speed limitations for each time period and road type to minimize carbon emissions of connected and automated vehicles (CAVs). To solve the problem expressed as a nonlinear MILP model, a hybrid PSO algorithm is designed. Ng et al. [41] developed an objective approach based on the CS algorithm for generating green solutions to the vehicle routing problem in a firm that offers dishwashing services to small restaurants and fast-food chain stores.

In addition to the studies mentioned above, studies using traditional metaheuristic methods and swarm-based methods have also been carried out. Utama et al. [42] used the hybrid butterfly optimization algorithm (BOA) to minimize distribution costs in a single tank GVRP. Whale optimization algorithm (WOA), which is another approach inspired by nature, was used by Dewi and Utana [43] in solving GVRP. In the proposed hybrid solution method, tabu search (TS) and local search (LS) methods are adapted to the problem together with the WOA. In addition to the studies mentioned here, there are many studies in the literature in which heuristic and meta-heuristic methods are applied to different versions of GVRP. Affi et al. [44] used the variable neighborhood search (VNS) method for GVRP, in which the refueling stations and the fuel tank capacity limit are taken into account for the construction of a tour. Sadati and Çatay [45] developed a MILP model for multi-warehouse GVRP and solved it with a hybrid global VNS and TS algorithm using novel neighborhood structures. Yavuz [46] used Iterated Beam Search (IBS) algorithm for homogeneous capacity GVRP. A dominance strategy has been proposed to increase the efficiency of the algorithm, which can use different lower and upper bound strategies and operate as an exact or heuristic algorithm. Erdogdu and Karabulut [47] hybridized an adaptive wide neighborhood search method with two new local search heuristics for bi-objective GVRP. The solutions obtained for the benchmarking problems in the literature were compared with the results of the NSGA-II method. Amiri et al. [48] combined the weighted sum method and Adaptive LNS to solve the multi-objective GVRP. The proposed method was used to solve the short-haul delivery problem using their real-world location in Ontario, Canada.

When the studies in the literature are examined, it is clearly seen that heuristic and meta-heuristic methods are frequently used especially for the solution of medium and large sized problems. Within the scope of this study, GVRP was tried to be solved by making some changes in the modified CS Algorithm proposed by Liu et al. [49]. To show the effectiveness of the solution methodology used, the same data set was solved as a CVRP and the emission values were calculated and compared with the results of the proposed method. In this way, how the change in the objective function affects the solutions is also shown with the calculations made. The details of the mathematical model and the method used are given in the following section.

3 Material and Method

VRP is a popular integrated optimization problem for which exact solution methods and many general heuristics are used. This problem, which is a generalization of the traveling salesman problem, is characterized as the problem of defining the minimum cost routes for shipments to customers located at geographically different locations from a warehouse, and it belongs to the NP-Hard class. Vehicle capacity or route distance constraints are commonly used to define VRP. The problem is known as a capacity-constrained VRP if only capacity constraints are considered. Although many exact solution methods have been developed considering capacity constraints, some of them can be applied to distance-constrained problems with minor modifications. On the other hand, heuristic methods developed mostly take both constraints into account [50].

The GVRP, which has started to be studied more today, it is aimed to determine the routes where the amount of CO_2 released into the atmosphere is minimized as well as the distance. In the following section, the mathematical model of this problem, which is among the current green supply chain applications, as well as the methods used for its solution are presented.

3.1 A Classical Green VRP Model

The capacitated VRP is defined on the set of the undirected graph ($G = (V, A)$). Here, the corner points defined as $V = \{0, 1, 2, \ldots, n\}$, and $E = \{(i, j) : i, j \in V, i < j\}\}$ are the set of arcs that connect the corner points to each other [44]. In traditional VRP, it is tried to calculate the minimum travel distance where each customer is visited once and the vehicle capacity constraint is met. In real life, the amount of fuel consumed is more important than the distance traveled to save fuel costs and pollute the environment [45].When this condition is taken into consideration in the GVRP, a series of delivery routes are established to satisfy demand with the minimum amount of distance traveled and the lowest amount of CO_2 released. The GVRP, one of the green supply chain applications today, is an NP-hard problem like traditional VRP. The following variables are introduced to present the model for the GVRP [7, 9]:

q_i	: non-negative weight (demand and supply) of location i
s_i	: service time for location i
Q_k	: the vehicle capacity of vehicle k (truck)
e_{ij}	: emission value between locations i and j
d_{ij}	: distance between locations i and j
t_{ij}	: driving time between locations i and j
T_k	: the maximum allowed driving time for vehicle k.
m	: number of vehicles
n	: number of customers
$E_{ij}(q, d)$	: the CO_2 emissions from a vehicle in kg/km
e_{fl}	: emission (CO_2) value for the fully-loaded vehicle
e_{el}	: emission (CO_2) value for the empty vehiclee

The integer programming model model is defined as follows, with x_{ij}^k $(i \neq j)$ and y_i^k being the binary decision variable and $E_{ij}(q, d)$ emission amount calculation (Eq. 1) [9]

$$x_{ij}^k = \begin{cases} 1 & \text{if vehicle } k \text{ serves customer } j \text{after customer } i \\ 0 & \text{otherwise} \end{cases}$$

$$y_i^k = \begin{cases} 1 & \text{if vehicle } k \text{ serves customer } i \\ 0 & \text{otherwise} \end{cases}$$

$$E_{\text{ij}}(q, d) = d_{\text{ij}} \times \left[\left(\frac{e_f - e_e}{Q} \right) q_{\text{ij}} + e_e \right] \tag{1}$$

$$\min \text{CO}_2 Emission\ (g) = \sum_{i=0}^{n} \sum_{j=0}^{n} \sum_{k=1}^{K} e_{\text{ij}}.x_{\text{ij}}^k \tag{2}$$

Subject to:

$$\sum_{i=0}^{n} x_{0i}^k \leq 1 \quad \forall k \in \{1, \ldots, m\} \tag{3}$$

$$\sum_{i=0}^{n} x_{i0}^k \leq 1 \quad \forall k \in \{1, \ldots, m\} \tag{4}$$

$$\sum_{i=1}^{n} y_i^k \leq M. \sum_{j=1}^{n} x_{0j}^k \qquad \forall k \in \{1, \ldots, m\} \tag{5}$$

$$\sum_{i=1}^{n} y_i^k \leq M. \sum_{j=1}^{n} x_{j0}^k \qquad \forall k \in \{1, \ldots, m\} \tag{6}$$

$$\sum_{k=1}^{n} y_i^k = 1 \qquad \forall i \in \{1, \ldots, n\} \tag{7}$$

$$\sum_{i=1}^{n} x_{ij}^k = y_j^k \qquad \forall j \in \{1, \ldots, n\}\ i \neq j, \forall k \in \{1, \ldots, m\} \tag{8}$$

$$\sum_{i=1}^{n} x_{\text{ij}}^k = y_i^k \qquad \forall i \in \{1, \ldots, n\}\ i \neq j, \forall k \in \{1, \ldots, m\} \tag{9}$$

$$\sum_{i=1}^{n} q_i y_i^k \leq Q_k \qquad \forall k \in \{1, \ldots, m\} \tag{10}$$

$$\sum_{i=0}^{n} \sum_{j=0}^{n} t_{\text{ij}} x_{\text{ij}}^k + \sum_{i=1}^{n} s_i y_i^k \leq T_k i \neq j, \qquad \forall k \in \{1, \ldots, m\} \tag{11}$$

$$\sum_{i,\ j \in S} x_{ij}^{k} \leq |S| - 1 i \neq j, \forall k \in \{1, 2, \ldots, m\}, S \subset N, 2 \leq |S| \leq n - 1 \quad (12)$$

The total emission value is tried to be minimized With the objective function (g) shown in Eq. (2). The model's constraints (3), (4), (5), and (6) guarantee that all routes start and end in the depot. According to constraint (7), nodes outside the depot must be serviced by a single-vehicle. According to constraints (8) and (9), all nodes other than the depot must be connected to a pair of nodes, one before the other and the other following it. Furthermore, constraint (10) guarantees that no vehicle is overloaded, and constraint (11) prevents any vehicle from exceeding the maximum daily allowed driving duration T_k. Finally, constraint (12) is the sub-tour elimination constraint.

3.2 Solution Method

Metaheuristic methods are approaches that are inspired by natural phenomena to achieve a certain goal or to reach the determined goal, but their convergence to the optimum solution in the solution space cannot be fully proven. These methods are the most preferred today because they are easy to understand and apply, and can be easily applied to different problems with small changes. They are divided into different classes according to criteria such as the source of inspiration, the initial solution generation method, the objective function used, and the preferred neighborhood structure and memory status [22, 53]. Methods such as PSO, ACO, TS, GA, CS, and SA, can be given as examples of the methods frequently used in the literature.

The cuckoos are interesting animals, not only because of their amazing voices but also because of their aggressive breeding habits. One of the most surprising behaviors of cuckoos is host-parasite evolution, also known as brood parasitism, in which they deposit their eggs in the nests of other species and rely on them to grow their offspring [54]. Some cuckoo species, such as Ani and Guira, lay their eggs in suitable nests and discard any eggs that are likely to hatch. Quite a large number of host species lay their eggs in the nests of other bird species, forcing them to brood parasitism. The Cuckoo Search (CS) algorithm is based on the brood parasitism of some cuckoo species. Furthermore, rather than simple isotropic random walks, this framework was developed utilizing Lévy flights. [55].

The CS, like other swarm-based approaches, starts with a population of n nest. The first host nests are chosen at random by egg-bearing cuckoos and random Lévy flights. The nest's quality is then assessed and compared to the quality of another random host nest. If the new host nest outperforms the old one, it will take its position. The egg laid by the cuckoo is included in this new solution. If the host bird discovers the egg with a probability of $P(0, 1)$, the eggs are either discarded or the nest is abandoned and a new one is constructed. This is accomplished by substituting the new random solution for the numerous solutions [56]. The pseudo code of the CS algorithm is presented below as Algorithm 1 [57].

Algorithm 1 The CS algorithm

```
Fitness function f(x)
Produce initial population of n host nests
while (t < Max Generation) or (Stopping Criteria) do
    Get a random cuckoo via Lèvy flights and assess its fitness value
    Select a nest among n randomly
    if F_i < F_j then
        replace j by the new solution;
    end if
    A fraction p_a of worst nests are abandoned and new ones are built;
    Place the best solutions;
    Rank the solutions and obtain the best solution
end while
Results of post-processing and visualization
```

3.3 The VLCS Algorithm

The modified cuckoo search algorithm with variational parameters and logistic sequences (VLCS) [49], which is one of the new methods, was preferred as the solution method within the scope of the study. The details of the method are explained in this section.

Regarding the CS, there are some disadvantages in "initialization", "parameters (α and p_α) and "boundary issue". The first disadvantage of the method is that it uses random numbers for the initialization of nests. Random start causes the slot locations to be the same in some cases, and sometimes the nests are not evenly distributed in a particular area. This is a factor that increases the chance of repetitive calculations and getting stuck in the local optimal solution. Liu et al. [57] used a concept called "Logistics map" to cope with this situation. The logistics map is expressed by Eq. (13) and the pseudo-code of the initialization process is shown in Algorithm 2 [57].

$$x_{n+1} = \mu x_n \left(1 - x_n\right) \qquad \begin{array}{l} n = 0, 1, 2, 3 \ldots \\ \mu \in [0,\ 4] : controlparameter \\ n : n - th\ iteration \\ x_n \in (0, 1) \end{array} \tag{13}$$

The second disadvantage of the traditional CS is that the values of the parameters "α" (L/10 or L/100) and p_α are fixed. These parameter values should change later in the iteration as the method search local and global optimum solutions. To cope with this second disadvantage in the VLCS method, the following coefficient function

Algorithm 2 Initialization process

Input: dimension (d);
L_b is the lower bound of $f(x)$,
$L_b = (L_{b1}, L_{b2}, ..., L_{bd})^T$
U_b is the upper bound of $f(x)$,
$U_b = (U_{b1}, U_{b2}, ..., U_{bd})^T$
Output: nest
Generate global random number r,
the number of nests is equal to n;
for $k \leftarrow 1$ to n **do**
  $r \leftarrow 4 * r. * (1 - r)$ ▷ if $\mu = 4$ Logistic map creates chaos
  $nest(k, :) \leftarrow L_b + (U_b - L_b). * r$ ▷ Initial population
end for
return *nest*

is used. Equation (14) sets the parameter values α and p_α, where "cur$_{iteration}$" is the current iteration number and "total$_{iteration}$" is the total number of iterations. The coefficient function is used to arrange the "α" and "p_α" parameters in Eq. (16) [55].

$$\varepsilon_\alpha = 10^{(10*\tan(\arctan(0.2))*(2*r_{\text{evolving}}-1)))} \tag{14}$$

$$r_{\text{evolving}} = cur\_iteration/total\_iteration \tag{15}$$

$$p_{anew} = \begin{cases} p_{anewLowerValue} & if \quad \frac{p_\alpha}{\varepsilon} < p_{anewLower} \\ p_{anewUpperValue} & elseif\ p_\alpha \varepsilon > p_{anewUpper} \\ \frac{p_\alpha}{\varepsilon} & else \end{cases} \tag{16}$$

The third and final disadvantage of the traditional CS algorithm, which uses Lévy flights and random walks, is the nest's locations found. The use of boundary points for nests out of bounds can result in multiple nests in the same location. The VLCS method also uses the global random number r, expressed in the logistic map concept, for the boundary problem. The r values in the boundary problem and the algorithms proposed for the start are the same. With this value, the position of each nest is calculated only once, thus reducing the repetitive calculations and speeding up the convergence process. The pseudo-code of boundary processing is shown in Algorithm 3 [49].

In this study, the standard VLCS algorithm was applied with some modifications. The first change made is related to the "α" parameter. The "α" parameter was initially defined as "α=e" (e: Euler constant), unlike the standard VLCS. The second change is related to the coefficient function expressed in Eq. (15). Equations (17) and (18) shown below are used for coefficient function calculation. The calculation of α_{new} value in Eq. (16) has also been changed as shown in Eq. (19). The third change covers the values of the p_{new} parameter. In the stan-

Algorithm 3 The boundary processing

```
Input: dimension (d);
  Global random number r,
  nest(i, :) = [x1, ..., xd], i ∈ (1, n)
Output: nest
  outLbMatrix is 1xd matrix
      outLbMatrix = (OL1, OL2, ..., OLd), OLd = 1;
  outUbMatrix is 1xd matrix
      outUbMatrix = (OU1, OU2, ..., OUd), OUd = 1;
for j ← 1 to d do
    if the xd of nest_i out of Lb then
        OLj = 1;
    else
        OLj = 0;
    end if
end for
for j ← 1 to d do
    if the xd of nest_i out of Ub then
        OUj = 1;
    else
        OUj = 0;
    end if
end for
for i ← 1 to n do
    for j ← 1 to d do
        if the outLbMatrix(j) > 0 then
            r(j) = 4 * r(j) * (1 − r(i));
            nest(i, j) = Lb(j) + (Ub(j) − Lb(j)) * r(j);
        end if
        if the outUbMatrix(j) > 0 then
            r(j) = 4 * r(j) * (1 − r(i));
            nest(i, j) = Lb(j) + (Ub(j) − Lb(j)) * r(j);
        end if
    end for
end for
return nest
```

dard VLCS method, these values were taken as 0.25 and 0.75, respectively, while in this study, they were randomly generated in each iteration with the "*rand*" function ($p_{anew}LowerValue = rand(); \ p_{anew}UpperValue = rand()$).

$$r_{evolving} = \cosh(\frac{cur_{iteration}}{total_{iteration}}) \tag{17}$$

$$\varepsilon_{\alpha} = e^{(0.1*\tan(\arctan(0.2)\ )\ *(2*r_{evolving}-1)))}) \tag{18}$$

$$\alpha_{new} = \sqrt{\left|e^{\cos(e^{\alpha_{new}})}\right|)} \tag{19}$$

Table 1 Parameter values of the VLCS algorithm

Parameters	Values
Number of nests (n)	20
Discovery rate of aliena eggs/solutions (p_α)	0.25
Number of iterations	100
The lower bound (Lb)	*Lb*=-1**ones(nd)*
The upper bound (Ub)	*Ub*=1-*ones(nd)*
The control parameter $\mu \in [0, 4]$	4
$p_{\alpha new}$	0
$p_{\alpha new} Lower\ Value$	0.20
$p_{\alpha_{new}} Upper\ Value$	0.80

Details of the method can be found in Liu et al. [49]. The following section includes experimental studies in which modifications to the standard method were applied.

4 Experimental Study

This section details the computational studies performed to investigate the VLCS's performance. The method was coded in Julia 1.5.3, and it was tested on a computer with an Intel Xeon 2.60 GHz CPU and 16 GB of RAM. Different CVRP test problems [58] were used to evaluate the effectiveness of the VLCS algorithm. The parameter values used for the VLCS algorithm are shown in Table 1.

Two different calculations were made with the data sets used in the study. In the first calculation, the emission values were calculated after the problem was solved as conventional VRP. In the second calculation, the problem was tried to be solved as a GVRP, based on the direct emission value according to the mathematical model in the third section. While making this calculation, the objective function (Z) shown in Eq. (20) was used. G in the equation represents the amount of CO_2 emissions and F represent the total distance. If the parameter r_x is *1*, the objective function creates CVRP solutions, and when this parameter takes the value 10, the GVRP solutions are obtained.

$$Min\ Z = \left(\frac{1}{r_x}\right) * F + \left(\frac{r_x - 1}{r_x}\right) * G, r_x = \{1,\ 10\} \tag{20}$$

For the experimental study, 10 repetitive experiments were carried out using 64 data sets and the parameter values in Table 1. The emission values, travel distances, and solution times obtained for the solutions are summarized in Table 2.

Table 2 Solution results

			CO_2		Distance		Solution time	
Abbr.	Data Set	Optimal	GVRP	CVRP	GVRP	CVRP	GVRP	CVRP
A1	A_n32_k5	784	855,9	855,9	932,8	932,8	12,6	12,6
A2	A_n37_k5	669	755,9	**746,4**	817,3	**808,5**	**13,5**	15,3
A3	A_n37_k6	822	1153,1	**1148,7**	**1242,8**	1259,8	**18,5**	21,3
A4	A_n38_k5	730	**789,8**	817,9	**840,4**	874,3	**15,1**	17,3
A5	A_n39_k5	822	**959,0**	961,6	1044,0	**1039,2**	**15,0**	17,3
A6	A_n39_k6	831	910,3	**889,0**	999,1	**964,3**	**17,1**	19,6
A7	A_n44_k6	937	**1100,9**	1105,4	1197,1	**1183,1**	**17,9**	20,6
A8	A_n45_k6	944	1073,9	**1061,6**	1172,6	**1160,5**	**18,3**	21,1
A9	A_n45_k7	1146	1362,4	**1355,8**	1480,8	**1455,5**	**19,6**	22,8
A10	A_n46_k7	914	**1055,2**	1068,0	**1165,7**	1175,3	**19,2**	22,0
A11	A_n48_k7	1073	**1271,6**	1281,9	1390,0	**1386,7**	19,6	**19,2**
A12	A_n53_k7	1010	**1204,6**	1216,6	**1315,6**	1327,7	21,9	**21,8**
A13	A_n54_k7	1167	**1377,1**	1416,8	**1502,9**	1530,0	22,4	**22,3**
A14	A_n55_k9	1073	**1297,1**	1358,6	**1407,9**	1476,6	27,2	27,2
A15	A_n60_k9	1354	**1731,3**	1763,4	**1884,3**	1928,1	29,5	**29,4**
A16	A_n61_k9	1034	**1245,1**	1273,6	**1357,1**	1388,0	29,2	**29,0**
A17	A_n62_k8	1288	**1513,2**	1524,7	**1638,3**	1656,6	25,7	25,7
A18	A_n63_k9	1616	**1966,0**	1978,6	**2134,7**	2147,5	29,2	29,2
A19	A_n63_k10	1314	1666,2	**1622,1**	1810,3	**1760,2**	**30,0**	31,7
A20	A_n64_k9	1401	**1679,2**	1708,6	**1810,7**	1837,5	**29,3**	30,2
A21	A_n65_k9	1174	1425,4	**1412,0**	1553,7	**1527,0**	**27,6**	28,9
A22	A_n69_k9	1159	**1359,9**	1391,2	**1486,9**	1512,0	**28,1**	29,4
A23	A_n80_k10	1763	2134,8	**2124,2**	2325,3	**2288,0**	**35,8**	36,7
B1	B_n38_k6	805	875,6	**851,0**	954,5	**942,3**	**15,0**	15,7
B2	B_n39_k5	549	**634,5**	637,4	**701,6**	703,6	**13,4**	13,9
B3	B_n41_k6	829	**957,2**	963,9	**1029,4**	1036,9	**16,5**	17,1
B4	B_n43_k6	742	**827,8**	833,6	907,3	**906,0**	**16,3**	17,0
B5	B_n44_k7	909	1139,1	1139,1	1238,5	1238,5	**20,1**	21,1
B6	B_n45_k5	751	**784,0**	794,0	**848,8**	859,6	**15,2**	15,8
B7	B_n45_k6	678	**778,5**	782,3	861,9	**853,4**	**17,2**	18,2
B8	B_n50_k7	741	877,8	877,8	957,1	957,1	**20,0**	20,8
B9	B_n50_k8	1312	1584,1	**1582,9**	1733,7	**1732,4**	**23,7**	25,4
B10	B_n51_k7	1032	**1173,8**	1173,9	**1285,1**	1732,4	**20,2**	21,0
B11	B_n52_k7	747	841,6	**826,4**	911,1	**898,3**	**18,7**	19,3
B12	B_n56_k7	707	**792,7**	796,9	862,1	**854,8**	**19,9**	20,3
B13	B_n57_k7	1153	1510,9	**1509,6**	1650,6	**1638,9**	**23,9**	24,6
B14	B_n57_k9	1598	**1888,3**	1907,0	**2064,0**	2067,9	**25,2**	26,1

(continued)

Table 2 (continued)

			CO_2		Distance		Solution time	
Abbr.	Data Set	Optimal	GVRP	CVRP	GVRP	CVRP	GVRP	CVRP
B15	B_n63_k10	1496	**1870,3**	1883,5	**2037,7**	2053,2	**30,3**	31,7
B16	B_n64_k9	861	**1030,6**	1078,6	**1114,4**	1174,1	**28,4**	29,7
B17	B_n66_k9	1316	**1443,7**	1444,0	1571,1	**1569,1**	**27,5**	28,8
B18	B_n67_k10	1032	1216,1	**1214,4**	**1312,6**	1309,6	**29,0**	30,4
B19	B_n68_k9	1272	1497,9	**1495,5**	1620,6	**1614,9**	**29,3**	30,4
B20	B_n78_k10	1221	1460,0	**1451,8**	1582,1	**1577,0**	**34,0**	35,1
P1	P_n19_k2	212	**217,8**	220,4	**240,8**	244,2	6,4	**6,3**
P2	P_n20_k2	216	**225,8**	233,2	249,5	**246,3**	6,4	**6,3**
P3	P_n21_k2	211	205,0	205,0	223,0	223,0	6,4	6,4
P4	P_n22_k2	216	**230,2**	234,6	**245,0**	248,9	6,6	6,6
P5	P_n40_k5	458	**501,1**	505,6	**537,7**	547,6	**13,0**	13,1
P6	P_n45_k5	510	**559,1**	562,4	**608,7**	609,8	13,7	13,7
P7	P_n50_k7	554	**628,1**	638,0	**680,8**	692,4	18,3	18,3
P8	P_n50_k8	631	751,4	**728,5**	821,3	**800,8**	22,7	22,7
P9	P_n50_k10	696	**874,1**	877,9	**950,9**	955,3	28,0	**27,9**
P10	P_n51_k10	741	977,2	**976,6**	1068,0	**1065,5**	**30,0**	30,1
P11	P_n55_k7	568	**623,2**	634,0	**682,2**	689,3	18,3	18,3
P12	P_n55_k8	588	649,6	**645,5**	703,8	**699,8**	**19,2**	19,3
P13	P_n55_k10	694	833,8	**831,6**	902,5	**899,7**	26,0	26,0
P14	P_n55_k15	989	1467,7	**1464,9**	1608,5	**1596,0**	52,9	**52,3**
P15	P_n60_k10	744	**884,3**	891,0	972,2	**969,4**	**27,6**	27,8
P16	P_n60_k15	968	1385,9	**1381,9**	1520,3	**1510,1**	46,5	46,5
P17	P_n65_k10	792	**921,3**	932,4	**997,2**	1005,4	**27,7**	28,1
P18	P_n70_k10	827	984,9	**980,8**	1066,4	**1065,3**	**29,6**	29,9
P19	P_n76_k4	593	640,0	**639,2**	697,3	**692,7**	**17,7**	18,4
P20	P_n76_k5	627	672,8	**668,3**	735,7	**723,3**	**19,0**	19,9
P21	P_n101_k4	681	**696,2**	707,7	**749,0**	758,3	**26,0**	26,1
Average Deviation (%)			–	–	27,37	27,42	–	–
Average Distance			1053,76	1057,65	1147,04	1155,19	22,34	23,08
Better Solution			36	24	29	31	43	10
Equal Solution			4	4	4	4	11	11

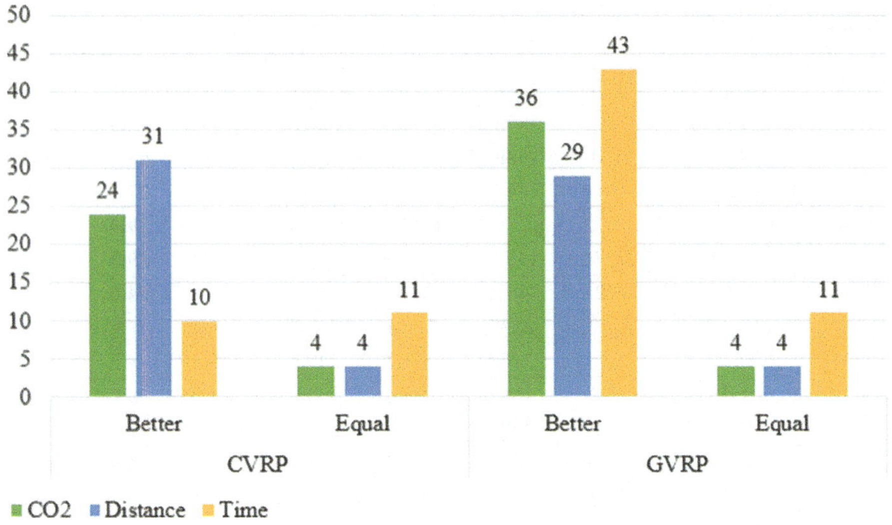

Fig. 1 Comparison of solutions (CVRP/GVRP)

The values in bold in Table 2 indicate better solutions for that data set. At the bottom of the table, data on the comparison of methods in terms of CO2, distance, and solution time are presented. For example, in terms of the amount of CO2, better results were obtained in 36 data sets with the GVRP model, while better results were found in 24 data sets as a result of the analysis with the traditional VRP model. Descriptive statistics about the solution results are shown in Fig. 1.

When the results in Table 2 are examined, GVRP solutions gave better results in 36 of the 64 experiments made in terms of emission values, while CVRP solutions provided better solutions in 24 of them. In terms of emission value, the obvious superiority of GVRP solutions stands out. When the travel distances obtained are analyzed, 29 of the GVRP solutions and 31 of the CVRP solutions have shorter distance solutions. In terms of solution time, GVRP solutions for 43 problems and CVRP solutions for 10 problems are faster. However, the solution times of both of these approaches are quite close to each other. The distance, emission, and solution time values obtained are shown in Figs. 2, 3, and 4, respectively.

When the distances obtained are analyzed how much they deviate from the optimal solution, it is seen that there is an average of 27% deviation. In some data sets, it has been observed that the deviation value is up to 62%. Although this is a very high value, it is also a clear indication that good solutions have not been achieved despite the modifications made to the CS algorithm. The graph of optimal deviation values for the data sets is shown in Fig. 5.

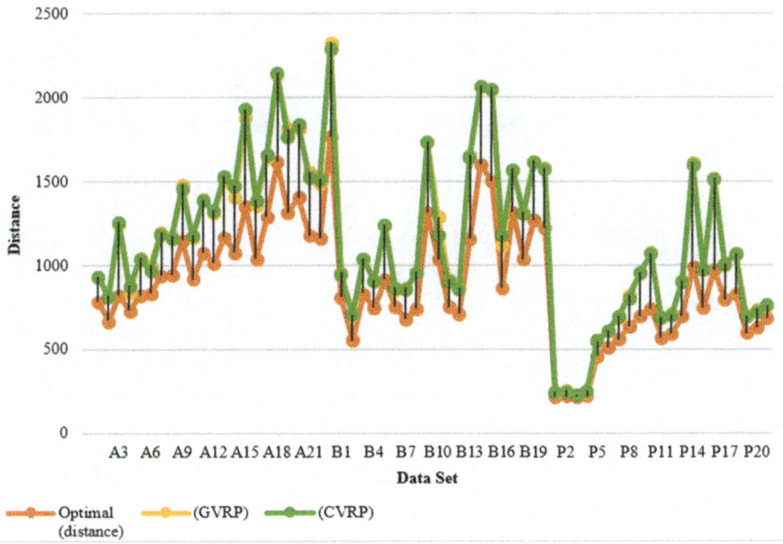

Fig. 2 Solution distances (optimal/GVRP/CVRP).

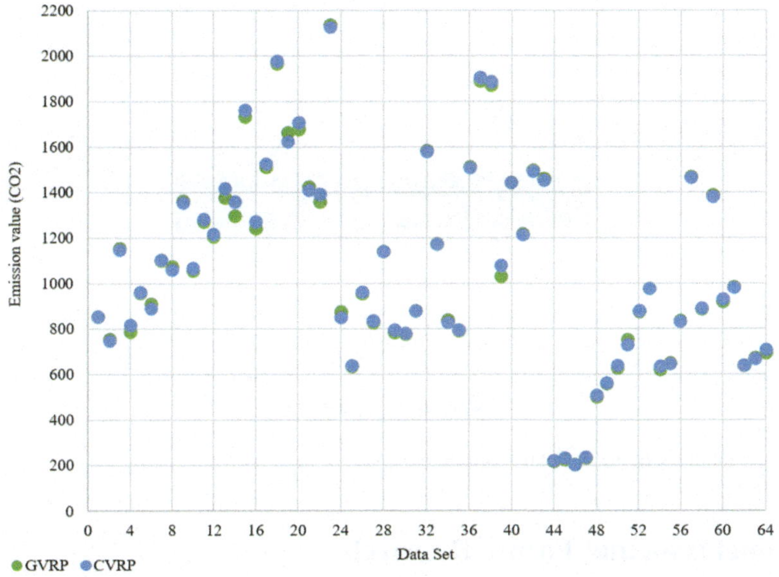

Fig. 3 Emission values of the solutions (GVRP-CVRP)

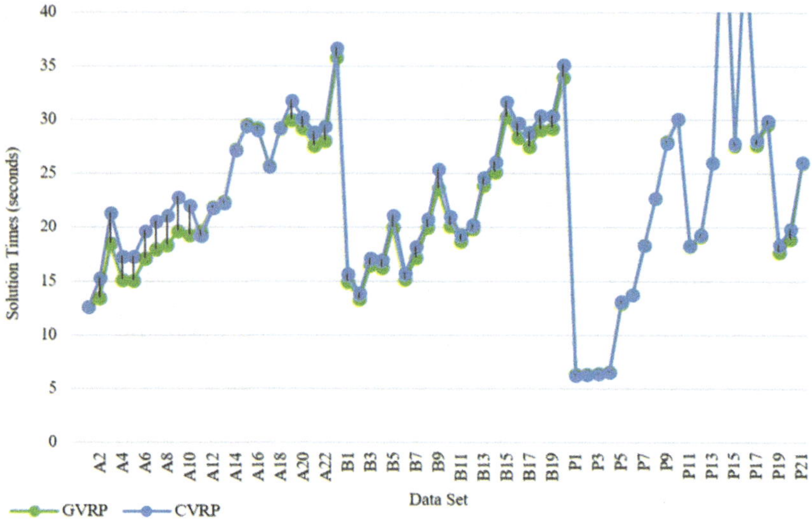

Fig. 4 Solution times (seconds)

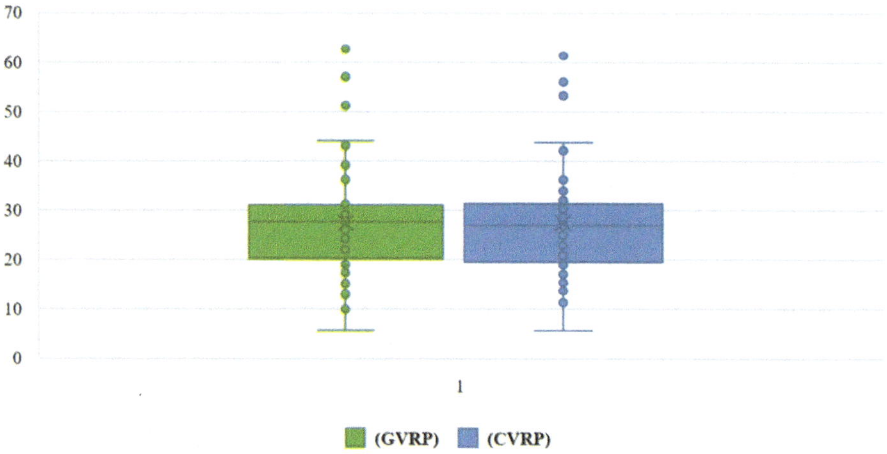

Fig. 5 Deviation from the optimal solution (%)

5 Conclusion and Future Research

Transportation operations, which are the most important components of logistics management, constitute an important part of the activities carried out during the transportation of the products that customers need. For these activities, which have high economic and environmental impacts, road transport is still the most widely used transportation mode. The high amount of fossil fuels consumed by vehicles is an important cost item for companies and an important threat to the environment that brings pollution. At this point, the way for companies to be sustainable companies is to plan their distribution activities with the least carbon emissions and the lowest cost.

In this study, the GVRP has been tried to be solved to eliminate environmental concerns caused by transportation-related pollution and to harmonize environmental and economic costs. For this purpose, two different objective functions and two different CS-based solution methods were used to solve the problem. Since the standard CS has some weaknesses in the initial solution, parameter values, and boundary processes, the modified cuckoo search algorithm (VLCS) was adapted to solve the problem. To speed up the solution of the problem and move the solution to better areas, some revisions were made in the parameter values and coefficient function in the original method. After these changes, the problem was solved in two different ways. In the first mathematical model (CVRP) used, the emission values were calculated after obtaining the solutions with the distance-based objective function. In the second model (GVRP), emission values were tried to be minimized directly with the objective function. The solutions obtained are not very close to the known optimal solutions in terms of distance value. However, when these two solutions are compared with each other, it has been shown that solutions with the same distance but lower emission values can be obtained with the GVRP formulation. The most important reason for this is to seek a solution to the problem by considering the relationship between vehicle load and total distance in the objective function of the model used for GVRP. When the results obtained for this problem and the results of previous studies in the literature are evaluated together, it is seen that CS is more effective in solving continuous optimization problems in terms of obtaining the optimal solution. In the future, studies can be conducted on solution approaches where different heuristics are hybridized so that CS can find optimal or very close optimal solutions.

References

1. Rushton, A., Croucher, P., Baker, P.: The Hand Book of Logistics and Distribution Management, 3rd edn. Kogan Page Limited, London (2006)
2. Sahin, Y., Karagül, K., Aydemir, E.: Heterogeneous fleet green vehicle routing problem with simulated annealing method. Düzce Univ. J. Sci. Technol. **9**(6), 65–82 (2021)
3. Sahin, Y.: Depo Operasyonları ve Sipariş Dağıtım Faaliyetlerinin Planlanması. Ekin Yayınevi, Bursa (2020)
4. Mason-Jones, R., Towill, D.R.: Total cycle time compression and the agile supply chain. Int. J. Prod. Econ. **62**(1–2), 61–73 (1999)
5. Dayıoğlu, E.G., Karagül, K., Şahin, Y., Kay, M.G.: Route planning methods for a modular warehouse system. An Int. J. Optim. Control.: Theor. Appl. **10**(1), 17–25 (2020)
6. Şahin, Y., Eroğlu, A.: Hierarchical Solution of Order Picking and Capacitated Vehicle Routing Problems. Suleyman Demirel Univ. J. Eng. Sci. Des. **3**(1), 15–28 (2015)
7. Karagul, K., Sahin, Y., Aydemir, E., Oral A.: A simulated annealing algorithm based solution method for a green vehicle routing problem with fuel consumption. In: Lean and Green Supply Chain Management, 1st ed., pp. 161–187. Cham, Switzerland (2019)

8. Daneshzand, F.: The vehicle-routing problem. In: Farahani, R.Z., Rezapour, S., Kardar, L. (eds.) Logistics Operations and Management Concepts and Models, pp. 127–145. Elsevier Insights, Waltham (2011)
9. Adiba, E.E., Aahmed, E.A., Youssef, B.: The green capacitated vehicle routing problem: optimizing of emissions of greenhouse gas. In: 2014 International Conference on Logistics Operations Management, pp. 161–167. IEEE Press, New York (2014)
10. Lin, C., Choy, K.L., Ho, G.T., Chung, S.H., Lam, H.Y.: Survey of green vehicle routing problem: past and future trends. Exp. Syst. Appl. **41**(4), 1118–1138 (2014)
11. Koç, Ç., Özceylan, E.: A literature review on green and electric vehicle routing problems and research perspectives. Gaziantep Univ. J. Soc. Sci. **17**(3), 1041–1053 (2018)
12. Asghari, M., Al-e-hashem, S.M.J.M.: Green vehicle routing problem: a state-of-the-art review. Int. J. Prod. Econ. **231**, 107899 (2021)
13. Belbağ, S.: Green capacitated vehicle routing problem: a literature review. Gazi Univ. J. Fac. Econ. Adm. Sci. **19**(1), 345–366 (2017)
14. Moghdani, R., Salimifard, K., Demir, E., Benyettou, A.: The green vehicle routing problem: a systematic literature review. J. Clean. Prod. **279**, 123691 (2021)
15. Andelmin, J., Bartolini, E.: An exact algorithm for the green vehicle routing problem. Transp. Sci. **51**(4), 1288–1303 (2017)
16. Koç, Ç., Karaoglan, I.: The green vehicle routing problem: a heuristic based exact solution approach. Appl. Soft Comput. **39**, 154–164 (2016)
17. Leggieri, V., Haouari, M.: A practical solution approach for the green vehicle routing problem. Transp. Res. Part E: Logist. Transp. Rev. **104**, 97–112 (2017)
18. Bruglieri, M., Mancini, S., Pezzella, F., Pisacane, O.: A new mathematical programming model for the green vehicle routing problem. Electron. Notes Discret. Math. **55**, 89–92 (2016)
19. Afshar-Bakeshloo, M., Mehrabi, A., Safari, H., Maleki, M., Jolai, F.: A green vehicle routing problem with customer satisfaction criteria. J. Ind. Eng. Int. **12**(4), 529–544 (2016)
20. Zhou, Y., Lee, G.M.: A Lagrangian relaxation-based solution method for a green vehicle routing problem to minimize greenhouse gas emissions. Sustainability **9**(5), 776 (2017)
21. Aydemir, E., Karagul, K.: Solving a periodic capacitated vehicle routing problem using simulated annealing algorithm for a manufacturing company. Braz. J. Oper. Prod. Manag. **17**(1), 1–13 (2020)
22. Şahin, Y., Eroğlu, E.: Metaheuristic methods for capacitated vehicle routing problem: literature review. Suleyman Demirel Univ. J. Fac. Econ. Adm. Sci. **9**(4), 337–355 (2014)
23. Tiwari, A., Chang, P.C.: A block recombination approach to solve green vehicle routing problem. Int. J. Prod. Econ. **164**, 379–387 (2015)
24. Erdoğan, S., Miller-Hooks, E.: A green vehicle routing problem. Transp. Res. Part E: Logist. Transp. Rev. **48**(1), 100–114 (2012)
25. Montoya, A., Guéret, C., Mendoza, J.E., Villegas, J.G.: A multi-space sampling heuristic for the green vehicle routing problem. Transp. Res. Part C: Emerg. Technol. **70**, 113–128 (2016)
26. Zhang, S., Gajpal, Y., Appadoo, S.S.: A meta-heuristic for capacitated green vehicle routing problem. Ann. Oper. Res. **269**, 753–771 (2018)
27. Jemai, J., Zekri, M., Mellouli, K.: An NSGA-II algorithm for the green vehicle routing problem. In: European Conference on Evolutionary Computation in Combinatorial Optimization, pp. 37–48. Springer, Berlin (2012)
28. Sruthi, A., Anbuudayasankar, S.P., Jeyakumar, G.: Energy-efficient green vehicle routing problem. Int. J. Inf. Syst. Supply Chain. Manag. (IJISSCM) **12**(4), 27–41 (2019)
29. Peng, B., Zhang, Y., Gajpal, Y., Chen, X.: A memetic algorithm for the green vehicle routing problem. Sustainability **11**(21), 6055 (2019)
30. Küçükoğlu, İ, Ene, S., Aksoy, A., Öztürk, N.: A memory structure adapted simulated annealing algorithm for a green vehicle routing problem. Environ. Sci. Pollut. Res. **22**(5), 3279–3297 (2015)
31. Normasari, N.M.E., Yu, V.F., Bachtiyar, C.: A simulated annealing heuristic for the capacitated green vehicle routing problem. Math. Probl. Eng. **2019**, 1–18 (2019)

32. Li, Y., Soleimani, H., Zohal, M.: An improved ant colony optimization algorithm for the multi-depot green vehicle routing problem with multiple objectives. J. Clean. Prod. **227**, 1161–1172 (2019)
33. Jabir, E., Panicker, V.V., Sridharan, R.: Multi-objective optimization model for a green vehicle routing problem. Procedia Soc. Behav. Sci. **189**, 33–39 (2015)
34. Jabir, E., Panicker, V.V., Sridharan, R.: Design and development of a hybrid ant colony-variable neighbourhood search algorithm for a multi-depot green vehicle routing problem. Transp. Res. Part D: Transp. Environ. **57**, 422–457 (2017)
35. Zhang, W., Gajpal, Y., Appadoo, S., Wei, Q.: Multi-depot green vehicle routing problem to minimize carbon emissions. Sustainability **12**(8), 350 (2020)
36. Prakash, R., Pushkar, S.: Implementation of GVRP in reducing environmental impacts and GHG emission. In: 5th International Conference on Electronics, Communication and Aerospace Technology (ICECA), pp. 172–178. IEEE (2021)
37. Poonthalir, G., Nadarajan, R.: A fuel efficient green vehicle routing problem with varying speed constraint (F-GVRP). Expert. Syst. Appl. **100**, 131–144 (2018)
38. Li, Y., Lim, M.K., Tseng, M.L.: A green vehicle routing model based on modified particle swarm optimization for cold chain logistics. Ind. Manag. Data Syst. **119**(3), 473–494 (2019)
39. Srijaroon, N., Sethanan, K., Jamrus, T., Chien, C.F.: Green vehicle routing problem with mixed and simultaneous pickup and delivery, time windows and road types using self-adaptive learning particle swarm optimization. Eng. Appl. Sci. Res. **48**(5), 657–669 (2021)
40. Cai, L., Lv, W., Xiao, L., Xu, Z.: Total carbon emissions minimization in connected and automated vehicle routing problem with speed variables. Expert. Syst. Appl. **165**, 113910 (2021)
41. Ng, C.Y., Lam, S.S., Liu, K.P.: Environmental consideration in heterogeneous vehicle fleet assignment using cuckoo search. Int. J. Sustain. Eng. **14**(6), 1907–1915 (2021)
42. Utama, D., Widodo, D., Ibrahim, M., Dewi, S.: A new hybrid butterfly optimization algorithm for green vehicle routing problem. J. Adv. Transp. **2020**, 1–14 (2020)
43. Dewi, S.K., Utama, D.M.: A new hybrid whale optimization algorithm for green vehicle routing problem. Syst. Sci. Control. Eng. **9**(1), 61–72 (2021)
44. Affi, M., Derbel, H., Jarboui, B.: Variable neighborhood search algorithm for the green vehicle routing problem. Int. J. Ind. Eng. Comput. **9**(2), 195–204 (2018)
45. Sadati, M.E.H., Çatay, B.: A hybrid variable neighborhood search approach for the multi-depot green vehicle routing problem. Transp. Res. Part E: Logist. Transp. Rev. **149**, 102293 (2021)
46. Yavuz, M.: An iterated beam search algorithm for the green vehicle routing problem. Networks **69**(3), 317–328 (2017)
47. Erdoğdu, K., Karabulut, K.: Bi-objective green vehicle routing problem. Int. Trans. Oper. Res. **29**(3), 1602–1626 (2022)
48. Amiri, A., Amin, S.H., Zolfagharinia, H.: A bi-objective green vehicle routing problem with a mixed-fleet of heavy-duty electric and conventional vehicles. In: IIE Annual Conference, pp. 181–186. Institute of Industrial and Systems Engineers (IISE) (2021)
49. Liu, L., Liu, X., Wang, N., Zou, P.: Modified cuckoo search algorithm with variational parameters and logistic map. Algorithms **11**(3), 30 (2018)
50. Cordeau, J.F., Laporte, G., Savelsbergh, M.W., Vigo, D.: Vehicle routing. Handb. Oper. Res. Manag. Sci. **14**, 367–428 (2007)
51. Toth, P., Vigo, D.: An overview of vehicle routing problems. In: The Vehicle Routing Problem, pp. 1–26. SIAM, Philadelphia (2002)
52. Xiao, Y., Zhao, Q., Kaku, I., Xu, Y.: Development of a fuel consumption optimization model for the capacitated vehicle routing problem. Comput. Oper. Res. **39**(7), 1419–1431 (2012)
53. Blum, C., Roli, A.: Metaheuristics in combinatorial optimization: overview and conceptual comparison. ACM Comput. Surv. (CSUR) **35**(3), 268–308 (2003)
54. Xing, B., Gao, W.J.: Innovative Computational Intelligence: A Rough Guide to 134 Clever Algorithms. Springer international publishing, New York, Cham (2014)
55. Karagül, K.: Cuckoo search algorithm: a plastic waste collection example. In: 15th International Symposium on Econometrics, Operations Research and Statistics, pp. 775–784. Suleyman Demirel University, Isparta (2014)

56. Shehab, M., Khader, A.T., Laouchedi, M.: A hybrid method based on cuckoo search algorithm for global optimization problems. J. Inf. Commun. Technol. **17**(3), 469–491 (2018)
57. Yang, X.S., Deb, S.: Engineering optimisation by cuckoo search. Int. J. Math. Model. Numer. Optim. **1**(4), 330–343 (2010)
58. Augerat, P., Belenguer, J.M., Benavent, E., Corberán, A., Naddef, D., Rinaldi, G.: Computational results with a branch and cut code for the capacitated vehicle routing problem. In: Rapport de recherche 1 RR949-M, ARTEMIS-IMAG, Grenoble, France (1995)

Multi-Objective Artificial Hummingbird Algorithm

Nima Khodadadi, Seyed Mohammad Mirjalili, Weiguo Zhao, Zhenxing Zhang, Liying Wang, and Seyedali Mirjalili

Abstract This chapter introduces Multi-Objective Artificial Hummingbird Algorithm (MOAHA), a multi-objective variation of the newly established Artificial Hummingbird Algorithm (AHA). The AHA algorithm simulates the specific flight skills and intelligent search strategies of hummingbirds in the wild. Three types of flight skills are used in food search strategies, including axial, oblique, and all-round flights. Multi-objective AHA is tested through 5 real-world engineering case studies. Various performance indicators, such as Spacing (S), Inverted Generational Distance (IGD), and Maximum Spread (MS), are used to compare the MOAHA to the MOPSO, MOWOA, and MOHHO. The suggested algorithm may produce quality Pareto fronts with appropriate precision, uniformity, and very competitive outcomes, according to the qualitative and quantitative.

Keywords Multi-objective artificial hummingbird algorithm · Artificial hummingbird algorithm · Real-world engineering

---

N. Khodadadi
Department of Civil and Environmental Engineering, Florida International University, Florida, USA

S. M. Mirjalili
Department of Engineering Physics, Polytechnique Montreal, Montreal, Canada

W. Zhao · L. Wang
School of Water Conservancy and Hydropower, Hebei University of Engineering, Handan 056038, Hebei, China

Z. Zhang
Illinois State Water Survey, Prairie Research Institute, University of Illinois at Urbana-Champaign, Champaign, IL 61820, USA

S. Mirjalili (✉)
Centre for Artificial Intelligence Research and Optimisation, Torrens University Australia, Fortitude Valley, Brisbane, QLD 4006, Australia
e-mail: ali.mirjalili@gmail.com

Yonsei Frontier Lab, Yonsei University, Seoul, South Korea

A. Biswas et al. (eds.), *Advances in Swarm Intelligence*, Studies in Computational Intelligence 1054, https://doi.org/10.1007/978-3-031-09835-2_22

1 Introduction

Meta-heuristics have become a very popular tools for scientists and practitioners to solve optimization problems in the last decade. Such algorithms are stochastic and derivative free. The former refers to random mechanism and operator used in meta-heuristics that provide stochastic behaviors. These algorithms tend to start with a set of random solutions. This set is then improved using different mechanisms, often inspired from nature, until a certain accuracy is achieved. The stochastic nature makes meta-heuristics unpredictable and imperfect. Such compromises are taken to find reasonably good solutions in reasonable time for problems with large search space, in which the exploration of all possible solutions is impossible. Various meta-heuristic algorithms based on natural processes, group behavior, or scientific rules have been proposed over the previous few decades such as: Genetic Algorithm (GA) [1], Differential Evolution (DE) [2], Grey Wolf Optimizer (GWO) [3], Cuckoo Search (CS) [4], Bat Algorithm (BA) [5], Flow Direction Algorithm (FDA), and Stochastic Paint Optimizer (SPO) [6].

Derivative free nature of meta-heuristics is due to the nature of their search behavior. To update a solution, meta-heuristics do not need to calculate the changes in the objective function using derivative to decide which step leads to improve the solutions. Meta-heuristics only evaluate solutions based on their objective values and employ certain mechanisms to intelligently combine solutions to increase the change finding better solutions. This is why they are often called black-box optimizers. A conceptual model of an optimization problem and algorithm is given in Fig. 1.

Figure 1 shows a constrained problem with one objective (f), once inequality constraint (g), and one equality constraint (h). The algorithm (robot in this figure)

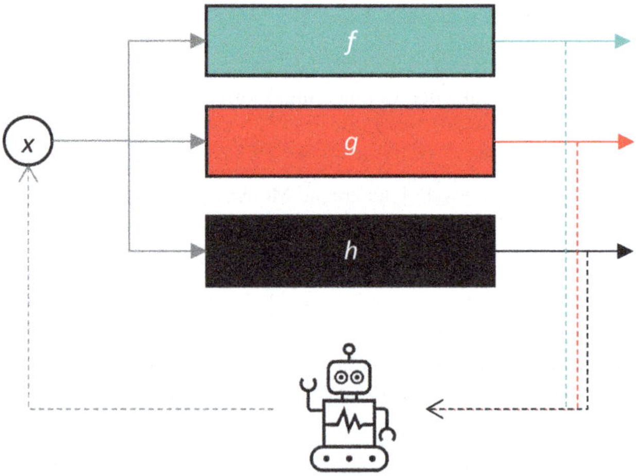

Fig. 1 A conceptual model of an optimization problem and algorithm

changes the value of the input (x) to minimize (or maximize in maximization problems) the objective function. If for any given input, the function g or h shows violation of constraints, the solution is considered infeasible. Without the loss of generality, the formulation of a minimization, constraint optimization problem is as follows:

$$\text{Minimize} : f(x) \tag{1}$$

$$\text{Subject to} : g_i(x) \leq 0 \quad \text{for}\, i = 1\, \text{to}\, n \tag{2}$$

$$h_j(x) = 0 \quad \text{for}\, j = 1\, \text{to}\, m \tag{3}$$

$$\text{Where} : x \in R^n \tag{4}$$

$$lb \leq x \leq ub \tag{5}$$

where $f : R^n \rightarrow R$, x is a real vector with n values where $n \geq 1$ is the number of variables, g_i shows ith inequality constraint, and h_j indicates jth equality constraint.

The set of all possible solutions for a given optimization problem is considered as a search space. Since the objective is a function (f), each solution is related to exactly one objective value. The set of all corresponding values is called objective space. The set of solutions and objective values are called search landscape. The search landscape of a unimodal and multimodal optimization problem with two variables can be seen in Fig. 2.

The first optimization problem in Fig. 2 is called unimodal due to the existence of one global optimum and lack of any local optimal. The second function in this figure, however, is considered to be multi-modal. In such optimization problems, there is a

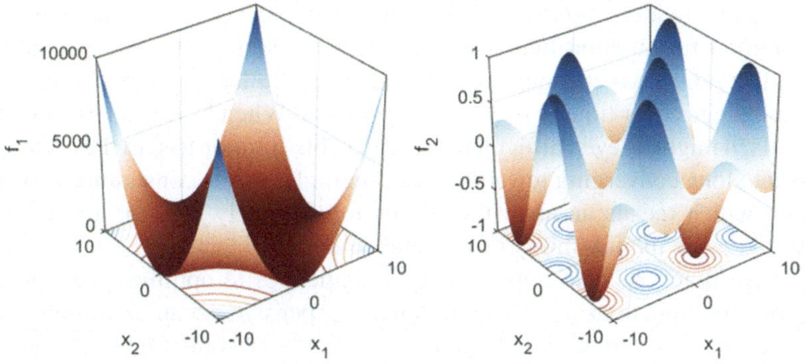

Fig. 2 Search landscapes of a unimodal ($f_1(x_1, x_2) = x_1^2 x_2^2$) and a multimodal ($f_2(x_1, x_2) = sinsin(0.5x_1)coscos(0.5x_2)$) optimization function

lot of local optima, which are good to find but not as good as the global optimum. Meta-heuristics are good for both types of optimization problems.

Based on the type of search, meta-heuristics are divided into two classes of local and global search algorithms. In the first class, the meta-heuristics have operators and mechanism to converge towards the nearest optimum. The search is local as there is no mechanism to explore other regions of the search space. One of the popular meta-heuristics in this class are hill climbing. This algorithm generates a set of "approachable" solutions in the neighborhood of an existing solution and choose the best one. This is done iteratively until no further improvement is done to the solution.

Meta-heuristics that are considered to be global search leverage of a wide range of stochastic mechanism to extensively search the search space. The divide the search process into two phases of exploration and exploitation. Exploration is the ability of a meta-heuristic in searching the search space extensively. Meta-heuristics are not complete search algorithm, so the goal of the exploration phase is to discover promising regions of the search space and increase the chance of finding the "valley" towards the global optimum. After performing the exploration, the algorithm moves into the phase of exploitation. The goal is to improve the accuracy of the most promising solutions found in the exploration. In other words, we want to ensure that the algorithm converges to the best possible solutions given the rough solutions found at the end of the exploration of the search space.

Despite the ease of solving optimization using meta-heuristics due to the above-mentioned advantages, we need to address several challenges related to the optimization problems. Such difficult are shown in Fig. 3. It can be seen that when it comes to variables, some of the challenges are dependency between variables, redundancy of variables, different types of variables (continuous, discrete, mixed), and exitance of perturbation. An optimization problem may have one or more than these difficulties.

In term of constraints, some of them can be quite critical, which is often called hard constraints and must be satisfied. Others can be considered as soft constraints and some violations are acceptable subject to applying penalty factors. As shown above, they are divided into two classes of inequality and equality constraints. There is also unified formulation that covert all equality constraints into inequality one as well. The other two changes are the existence of a large number of constraints and their changes over time in real-world problems. When having a large number of constraints, it might be even difficult to find a feasible solution to start the optimization process let alone converging towards an optimum. Dynamic constraints also change overtime, which means an algorithm needs to constantly monitor and track such changes to ensure the feasibility of the final solutions obtained.

Objective functions also impose several challenges to optimization algorithms. Some objective functions are computationally expensive, so an algorithm needs to be very cautious in minimizing the number of function evaluation to maximize the optimization speed. Another challenge is the dynamic nature of some objective function. This means that such functions change over time, so a global optimum can be changed to a local optimum over time. Some objective functions also inherently return noises, which should be considered to accurately optimization them. Finally, the main

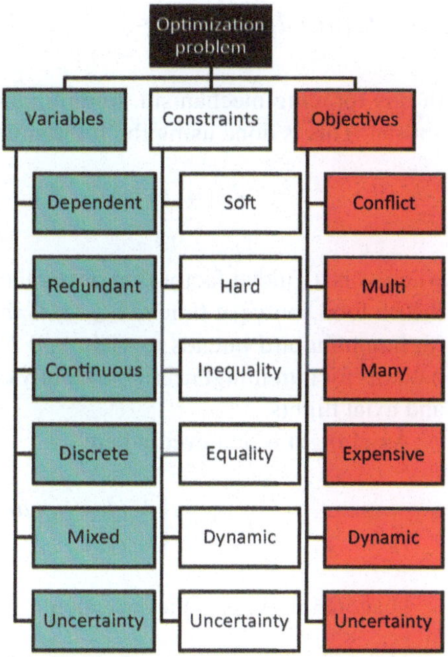

Fig. 3 Challenges related to different optimization problems that should be addressed by an optimization algorithm when solving them

challenge in the objective function is the existence of several of them in real-world problems. When dealing with multiple objectives, which are often in conflict, there is no longer a single solution. In fact, an optimization algorithm needs to find a set of solutions representing the best tradeoffs between the objectives. Some of the well-known multi-objective optimization algorithm are: Multi-objective Particle Swarm Optimization (MOPSO) [7], Multi-Objective Harris Hawks Optimizer (MOHHO) [8], Multi-objective Bonobo Optimizer (MOBO) [9], Multi-Objective Crystal Structure Algorithm (MOCryStAl) [10], Multi-Objective Whale Optimization Algorithm (MOWOA) [11], and Multi-objective Sine–Cosine Algorithm (MOSCA) [12]. In this chapter, the last above-mentioned difficulty, multiple objectives, is tracked by proposing a multi-objective meta-heuristic. Several engineering case studies are also solved.

2 Artificial Hummingbird Algorithm (AHA)

Artificial hummingbird algorithm (AHA) [13] is a recent meta-heuristic that mimics foraging behavior of hummingbirds in nature. The mechanisms modeled in this algorithm are guided foraging, territorial foraging, and migration foraging, which are discussed as follows:

2.1 Guided Foraging

In this foraging mechanism, hummingbirds, the best food source in the vicinity is chosen. This is done using the following equation:

$$v_i(\mathrm{t}+1) = x_{i,tar}(t) + D \cdot a \cdot (x_i(t) - x_{i,tar}(t)) \tag{6}$$

where a is a guided factor that obeys the normal distribution, $x_i(t)$ is the position of the ith food source at time t, $x_{i,tar}(t)$ is the position of the target food source that the ith hummingbird intends to visit, and D is the flight pattern. There are three flight patterns for hummingbirds to randomly choose, including omnidirectional, diagonal, and axial flights.

Axial flight is achieved when:

$$D^{(i)} = \begin{cases} 1 \; if \; i = randi([1, d]) \\ 0 \; else \end{cases} \quad i = 1, \cdots, d \tag{7}$$

Diagonal flight is achieved when:

$$D^{(i)} = \begin{cases} 1 \; if \; i = P(j), \, j \in [1, k], \, P = randperm(k), \\ \quad k \in [2, \lceil r_1 \cdot (d-2) \rceil + 1] \\ 0 \; else \end{cases} \quad i = 1, \cdots, d \tag{8}$$

Omnidirectional flight is achieved when:

$$D^{(i)} = 1 \quad i = 1, \cdots, d \tag{9}$$

where $randi([1, d])$ generates a random integer from 1 to d, $randperm(k)$ creates a random permutation of integers from 1 to k, and r_1 is a random number in [0, 1]. The diagonal flight in a d-D space is inside a hyperrectangle, which is bounded by any 2 to d-1 coordinate axes.

2.2 Territorial Foraging

In this type of foraging, the search is done in the vicinity of a hummingbird. This is modelled as follows:

$$v_i(\mathrm{t}+1) = x_i(t) + D \cdot b \cdot x_i(t) \tag{10}$$

where b is a territorial factor that obeys the normal distribution.

2.3 Migration Foraging

This type of foraging is where hummingbird search broadly and is expressed as follows:

$$x_{wor}(t+1) = Low + r \cdot (Up - Low) \tag{11}$$

where x_{wor} is the food source with the worst nectar-refilling rate in the population, and, *Low* and *Up* are the upper and lower boundaries, respectively.

The pseudocode of AHA is given in Fig. 4.

```
Input: n, d, f, Max_Iteration, Low, Up
Output: Globalminimum, Globalminimizer
Initialization:
  For ith hummingbird from 1 to n,
     Do x_i=Low+r(Up-Low),
     For jth food source from 1 to n, Do
        If i≠j
        Then Visit_table_{i,j}=1,
        Else Visit_table_{i,j}=null,
        End If
     End For
  End For
While t≤Max_Iteration Do
  For ith hummingbird from 1 to n, Do
    If rand≤0.5 Then
      If r<1/3 Then perform equation (7), //axial
       Else If r>2/3
        Then perform equation (8), //diagonal
        Else perform equation (9), //omnidirectional
        End If
      End If
      Perform equation (6), //guided foraging
      If f(v_i(t+1))<f(x_i(t))
      Then x_i(t+1)=v_i(t+1),
        For jth food source from 1 to n (j≠tar,i), Do
        Visit_table(i,j) =Visit_table(i,j) +1,
        End For
         Visit_table(i,tar)=0,
         For jth food source from 1 to n, Do
Visit_table(j,i)= max_{l∈n and l≠j} (Visit_table(j,l)) +1,
       End For
     Else
      For jth food source from 1 to n (j≠tar,i), Do
          Visit_table(i,j)=Visit_table(i,j)+1,
          End For
            Visit_table(i,tar)=0,
         End
       Else
         Perform equation (10), //territorial foraging
         If f(v_i(t+1))<f(x_i(t))
         Then x_i(t+1)= v_i(t+1),
            For jth food source from 1 to n (j≠i) Do
            Visit_table(i,j)=Visit_table(i,j)+1,
            End For
            For jth food source from 1 to n, Do
            Visit_table(j,i) = max_{l∈n and l≠j} (Visit_table(j,l)) +1,
            End For
          Else
            For jth food source from 1 to n (j≠i), Do
            Visit_table(i,j)=Visit_table(i,j)+1,
            End For
          End If
        End If
      End For
      If mod(t,2n)==0,
      Then perform equation (11), //migration foraging
        For jth food source from 1 to n (j≠wor), Do
        Visit_table(wor,j)=Visit_table(wor,j)+1,
        End For
        For jth food source from 1 to n, Do
        Visit_table(j,wor) = max_{l∈n and l≠j} (Visit_table(j,l)) +1,
        End For
      End If
End While
```

Fig. 4 Pseudocode of AHA algorithm

3 Multi-objective Artificial Hummingbird Algorithm

We have used three components to AHA in order to create a multi-objective version of mentioned algorithm in pervious section. MOPSO's components are quite similar to these. The No Free Lunch (NFL) theorem states that none of these methods will be able to handle all optimization problems. To put it another way, a particular algorithm may do well in some situations while failing miserably in others. In light of this, comparative research on a wide range of issues becomes even more critical. New in-sights into the performance of meta-heuristics can also be gained by conducting this type of comparative study. MOAHA algorithm is introduced and applied to five engineering problems in this chapter. Three main multi-objective optimization mechanisms are:

I. MOAHA has an archive component for keeping all of the non-dominated Pareto optimum solutions that have been found so far.
II. There is a grid mechanism in MOAHA that eliminates the most congested portions and enhances the non-dominant solutions in the archive.
III. The third component of MOAHA is the leader selection function, which uses the best solutions thus far to choose the optimal position.

In all non-dominated Pareto optimum solutions found so far, the archive serves as a data storage component. It has the ability to spread uniformly over the Pareto front if it is concave, convex, and unconnected. The archive controller, for example, is the sole component of this method. The primary role of this controller is to determine whether or not a solution should be added to the archive.

The distributed Pareto fronts are generated using a grid technique in this chapter. Objective function space is separated into several distinct areas. If a new member of the population is added to the population, the grid must be recalculated and each member of the population must be relocated. Using hypercubes, the grid creates a space for distributing solutions in a uniform manner.

Using leader function, the other search agents are directed to new and interesting areas of the search space where they can find a solution that is close to the global optimum. Although it is impossible to compare results in a multi-objective search space quickly due to the Pareto optimality, it is achievable in a single-objective search space. For this reason, the function for choosing a leader was developed. There is a collection of the best non-dominant solutions that have been compiled thus far. The least populated areas of the search space are selected by the leader selection component, and non-dominated solutions are provided from there.

Using a roulette-wheel approach, the hypercube with the following probability is found for each hypercube:

$$P_i = \frac{C}{N_i} \tag{12}$$

According to Eq. (12), hypercubes with fewer individuals are likely to nominate new leaders. In proportion to the decreasing variety of solutions that can be achieved by solving a hypercube, the likelihood of selecting a hypercube from which to select leaders grows.

The MOAHA algorithm is clearly derived from the AHA method. If we select one of the solutions from the archive, the MOAHA algorithm will almost likely be able to improve on its already great consistency. With a wide range of options, finding the Pareto optimal responses might be difficult. In order to solve this problem, we employed the leader function collection and archive maintenance.

4 Experimental Results and Discussions for Engineering Problems

Five benchmark test functions are used to validate MOAHA's performance in comparison to MOPSO, MOWOA and MOHHO. Table 1 lists the initial parameters for each algorithm. Each instance is limited to 100 populations and 1000 iterations (Max function evaluation 100,000). On an Intel Core i9 computer operating at 2.3 GHz with 16 GB of RAM and running MacOS Big Sur, the datasets were tested in 30 iterations with a sample size of 50 using MATLAB R2021a code. In this section, five real-world engineering design challenges are used to evaluate the technique's effectiveness.

Tables 2 and 3 show the findings of the IGD and MS performance metrics for five engineering design challenges. In four of the five examples when IGD measures were used, the MOAHA algorithm came out on top. The MOAHA outperforms most other algorithms except WELDED BEAM in terms of the IGD measure. This method's ability to produce consistent results over a large number of trials was demonstrated by the SD values. MOAHA got three out of five best results according to SD values. It is clear that the proposed MOAHA outperforms the MOPSO, MOWOA, and MOHHO in terms of MS performance. One exception to this was when a Welded Beam design was used as a test case. Thus, the results demonstrated the method's ability to solve complex multi-objective optimization problems in a mathematical

Table 1 The values for the controlling parameters of the algorithms

Parameters	MOHHO	MOWOA	MOPSO	MOAHA
Mutation Probability ($P_w, orpro$)	0.5	N/A	0.5	N/A
Population Size (N_{pop})	100	100	100	100
Archive Size ($N_{rep}, orTM$)	100	100	100	100
Number of Adaptive Grid (N_{grid})	30	30	30	30
Personal Learning Coefficient (C_1)	1	N/A	1	N/A
Global Learning Coefficient (C_2)	2	N/A	2	N/A
Inertia weight (w)	0.4	N/A	0.4	N/A
Beta	4	4	4	4
Gamma	2	2	2	2

context. Algorithm's capacity to provide better solutions that are closer to Pareto's front is shown in Fig. 5 for the engineering design issues considered. The proposed MOAHA, as depicted in this figure, exhibits perfect convergence toward all actual Pareto optimal fronts.

According to Table 4, when it comes to solving engineering design problems, the MOAHA outperforms MOPSO, MOHHO, and MOWOA in three of the five cases in terms of average results for S metric, while MOWOA and MOPSO outperforms the other algorithms in the DISK BRAKE and WELDED BEAM case, respectively. The MOAHA came out on top in terms of standard deviation. Using the IGD, S, and MS indices, this chapter demonstrates that the proposed MOAHA can outperform

Table 2 The IGD performance metric results for algorithms

Case Study		Algorithm			
		MOHHO	MOWOA	MOPSO	MOAHA
BNH	Ave	7.8786E-03	1.0003E−03	9.0098E−04	**8.1663E−04**
	SD	4.6734E−04	1.8976E−04	1.9047E−04	**1.5725E−04**
CONSTR	Ave	1.4563E−03	1.4343E−03	3.2338E−04	**2.3570E-04**
	SD	2.4536E−04	9.4536E−04	6.6718E−05	**4.5704E−05**
DISK BREAKE	Ave	1.4536E−03	5.7689E−04	4.4343E−04	**1.8426E−04**
	SD	2.4539E−04	8.3435E−05	**6.6595E−05**	1.8704E−04
WELDED BEAM	Ave	1.5690E−03	1.0033E−03	**7.9785E−04**	2.6492E−03
	SD	4.4890E−04	1.3465E−04	**6.3441E−05**	5.8224E−04
SRN	Ave	1.4565E−03	5.4563E−04	3.9846E−04	**3.0361E−04**
	SD	2.5690E−04	1.3457E−04	1.0986E−04	**3.8131E−05**

Table 3 The MS performance metric results for algorithms

Case Study		Algorithm			
		MOHHO	MOWOA	MOPSO	MOAHA
BNH	Ave	9.4587E−01	9.8330E−01	9.9997E−01	**1.0000E + 00**
	SD	1.1435E−02	3.3460E−03	9.5468E−09	**0.0000E + 00**
CONSTR	Ave	9.5435E−01	8.4563E−01	9.9283E−01	**9.9602E−01**
	SD	2.3245E−02	7.4432E−02	7.0602E−03	**5.2221E−03**
DISK BREAKE	Ave	9.8799E−01	9.9876E−01	9.9824E−01	**9.9993E−01**
	SD	2.5432E−02	1.3446E−03	9.3567E−04	**5.3592E−05**
WELDED BEAM	Ave	1.3457E + 00	1.0032E + 00	**1.4275E + 00**	9.9502E−01
	SD	2.5467E−01	5.5469E−02	7.3253E−02	**3.8019E−02**
SRN	Ave	8.9987E−01	9.0029E−01	9.2909E−01	**9.9064E−01**
	SD	3.4565E−02	3.3401E−02	5.5983E−02	**1.2896E−02**

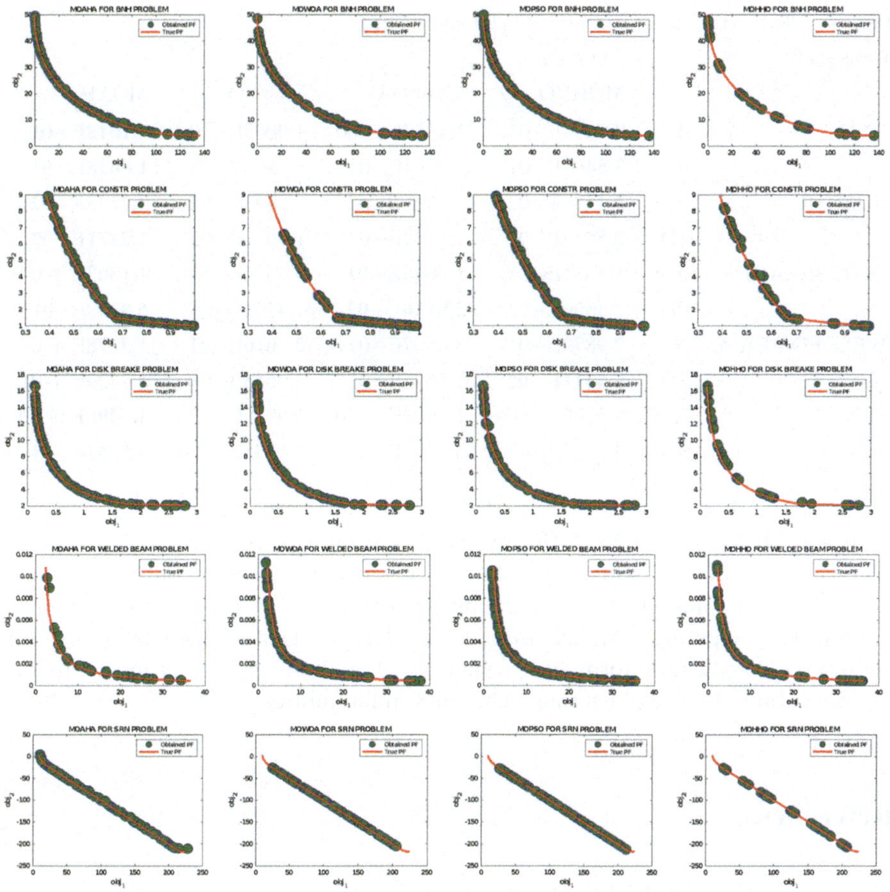

Fig. 5 MOHHO, MOWOA, MOPSO and MOAHA-based estimation of Pareto optimum solutions

existing approaches, with results that are potentially very competitive. After considering the true and achieved pareto fronts, the proposed MOAHA technique deliver better solutions closer to the Pareto front for the engineering issues.

5 Conclusion

To optimise for multiple objectives at once, the Multi-objective Artificial Hummingbird Algorithm (MOAHA), was presented in this chapter. A multi-objective problem necessitates the use of three components such as: leader selection, grid mechanism and archive mechanism. Five well-known benchmark test functions have been used to evaluate MOAHA. The results show that the MOAHA algorithm outperforms

Table 4 The S performance metric results for algorithms

Case Study		Algorithm			
		MOHHO	MOWOA	MOPSO	MOAHA
BNH	Ave	8.0084E−01	1.4359E + 00	1.0007E + 00	**6.1618E−01**
	SD	3.5467E−01	1.9094E−01	5.3263E−01	**1.8638E−01**
CONSTR	Ave	4.4536E−02	5.3005E−02	4.8794E−02	**3.2135E−02**
	SD	2.5557E−02	1.8342E−02	1.7997E−02	**1.2871E−02**
DISK BREAKE	Ave	1.3425E−01	**1.0048E−01**	3.3320E−01	1.1303E + 00
	SD	2.5696E−02	**2.0126E−02**	1.4502E−01	6.8359E−01
WELDED BEAM	Ave	2.9879E−01	3.4567E−01	**2.0010E−01**	1.3178E + 01
	SD	**6.4311E−02**	1.5675E−01	1.0613E−01	1.7729E + 01
SRN	Ave	1.5789E + 00	1.9902E + 00	1.9918E + 00	**1.3496E + 00**
	SD	4.8796E−01	4.4359E−01	1.5435E + 00	**1.0364E−01**

MOPSO, MOWOA and MOHHO in terms of overall performance. Many restrictions can be dealt with by the suggested approach, and statistical results show that it provides better solutions than existing optimizers. MOAHA outperforms the others in terms of computing cost, according to the data. It has been intended to use this approach to handle bioinformatics problems with multiple objectives and to broaden its applicability to data clustering techniques in the future.

References

1. Goldberg, D.E., Holland, J.H.: Genetic Algorithms and Machine Learning (1988)
2. Price, K.V.: Differential evolution. In: Handbook of Optimization, pp. 187–214. Springer, Berlin (2013)
3. Mirjalili, S., Mirjalili, S. M., Lewis, A.: Grey wolf optimizer. Adv. Eng. softw. **69**, 46–61 (2014)
4. Gandomi, A.H., Yang, X.-S., Alavi, A.H.: Cuckoo search algorithm: a metaheuristic approach to solve structural optimization problems. Eng. Comput. **29**(1), 17–35 (2013)
5. Yang, X.-S.: A new metaheuristic bat-inspired algorithm. In: Nature Inspired Cooperative Strategies for Optimization (NICSO 2010), pp. 65–74. Springer, Berlin (2010)
6. Kaveh, A., Talatahari, S., Khodadadi, N.: Stochastic paint optimizer: theory and application in civil engineering. Eng. Comput. 1–32 (2020)
7. Coello, C.A.C., Lechuga, M.S.: MOPSO: a proposal for multiple objective particle swarm optimization. In: Proceedings of the 2002 Congress on Evolutionary Computation. CEC'02 (Cat. No.02TH8600), vol. 2, pp. 1051–1056 (2002). https://doi.org/10.1109/CEC.2002.1004388
8. Yüzgeç, U., Kusoglu, M.: Multi-objective harris hawks optimizer for multiobjective optimization problems. BSEU J. Eng. Res. Technol. **1**(1), 31–41 (2020)
9. Das, A.K., Nikum, A.K., Krishnan, S.V., Pratihar, D.K.: Multi-objective Bonobo Optimizer (MOBO): an intelligent heuristic for multi-criteria optimization. Knowl. Inf. Syst. **62**(11), 4407–4444 (2020)
10. Khodadadi, N., Azizi, M., Talatahari, S., Sareh, P.: Multi-Objective Crystal Structure Algorithm (MOCryStAl): introduction and performance evaluation. IEEE Access (2021)

11. Aziz, M. A. E., Ewees, A. A., Hassanien, A. E.: Multi-objective whale optimization algorithm for content-based image retrieval. Multimedia tools and applications, **77**(19), 26135–26172 (2018)
12. Tawhid, M.A., Savsani, V.: Multi-objective sine-cosine algorithm (MO-SCA) for multi-objective engineering design problems. Neural Comput. Appl. **31**(2), 915–929 (2019)
13. Zhao, W., Wang, L., Mirjalili, S.: Artificial hummingbird algorithm: a new bio-inspired optimizer with its engineering applications. Comput. Methods Appl. Mech. Engrg. **388**, 114194 (2022)

The manufacturer's authorised representative in the EU is Springer Nature Customer Service Centre GmbH, Europaplatz 3, 69115 Heidelberg, Germany. If you have any concerns regarding our products, please contact ProductSafety@springernature.com

Printed and bound by CPI Group (UK) Ltd, Croydon, CR0 4YY
30/09/2026
02225242-0017